Lecture Notes in Computer Science 10913

Commenced Publication in 1973
Founding and Former Series Editors:
Gerhard Goos, Juris Hartmanis, and Jan van Leeuwen

Editorial Board

David Hutchison
Lancaster University, Lancaster, UK
Takeo Kanade
Carnegie Mellon University, Pittsburgh, PA, USA
Josef Kittler
University of Surrey, Guildford, UK
Jon M. Kleinberg
Cornell University, Ithaca, NY, USA
Friedemann Mattern
ETH Zurich, Zurich, Switzerland
John C. Mitchell
Stanford University, Stanford, CA, USA
Moni Naor
Weizmann Institute of Science, Rehovot, Israel
C. Pandu Rangan
Indian Institute of Technology Madras, Chennai, India
Bernhard Steffen
TU Dortmund University, Dortmund, Germany
Demetri Terzopoulos
University of California, Los Angeles, CA, USA
Doug Tygar
University of California, Berkeley, CA, USA
Gerhard Weikum
Max Planck Institute for Informatics, Saarbrücken, Germany

More information about this series at http://www.springer.com/series/7409

Gabriele Meiselwitz (Ed.)

Social Computing and Social Media

User Experience and Behavior

10th International Conference, SCSM 2018
Held as Part of HCI International 2018
Las Vegas, NV, USA, July 15–20, 2018
Proceedings, Part I

Editor
Gabriele Meiselwitz
Department of Computer and Information
Sciences
Towson University
Towson, MD
USA

ISSN 0302-9743 ISSN 1611-3349 (electronic)
Lecture Notes in Computer Science
ISBN 978-3-319-91520-3 ISBN 978-3-319-91521-0 (eBook)
https://doi.org/10.1007/978-3-319-91521-0

Library of Congress Control Number: 2018944281

LNCS Sublibrary: SL3 – Information Systems and Applications, incl. Internet/Web, and HCI

© Springer International Publishing AG, part of Springer Nature 2018
This work is subject to copyright. All rights are reserved by the Publisher, whether the whole or part of the material is concerned, specifically the rights of translation, reprinting, reuse of illustrations, recitation, broadcasting, reproduction on microfilms or in any other physical way, and transmission or information storage and retrieval, electronic adaptation, computer software, or by similar or dissimilar methodology now known or hereafter developed.
The use of general descriptive names, registered names, trademarks, service marks, etc. in this publication does not imply, even in the absence of a specific statement, that such names are exempt from the relevant protective laws and regulations and therefore free for general use.
The publisher, the authors and the editors are safe to assume that the advice and information in this book are believed to be true and accurate at the date of publication. Neither the publisher nor the authors or the editors give a warranty, express or implied, with respect to the material contained herein or for any errors or omissions that may have been made. The publisher remains neutral with regard to jurisdictional claims in published maps and institutional affiliations.

Printed on acid-free paper

This Springer imprint is published by the registered company Springer International Publishing AG part of Springer Nature
The registered company address is: Gewerbestrasse 11, 6330 Cham, Switzerland

Foreword

The 20th International Conference on Human-Computer Interaction, HCI International 2018, was held in Las Vegas, NV, USA, during July 15–20, 2018. The event incorporated the 14 conferences/thematic areas listed on the following page.

A total of 4,373 individuals from academia, research institutes, industry, and governmental agencies from 76 countries submitted contributions, and 1,170 papers and 195 posters have been included in the proceedings. These contributions address the latest research and development efforts and highlight the human aspects of design and use of computing systems. The contributions thoroughly cover the entire field of human-computer interaction, addressing major advances in knowledge and effective use of computers in a variety of application areas. The volumes constituting the full set of the conference proceedings are listed in the following pages.

I would like to thank the program board chairs and the members of the program boards of all thematic areas and affiliated conferences for their contribution to the highest scientific quality and the overall success of the HCI International 2018 conference.

This conference would not have been possible without the continuous and unwavering support and advice of the founder, Conference General Chair Emeritus and Conference Scientific Advisor Prof. Gavriel Salvendy. For his outstanding efforts, I would like to express my appreciation to the communications chair and editor of *HCI International News*, Dr. Abbas Moallem.

July 2018

Constantine Stephanidis

HCI International 2018 Thematic Areas and Affiliated Conferences

Thematic areas:

- Human-Computer Interaction (HCI 2018)
- Human Interface and the Management of Information (HIMI 2018)

Affiliated conferences:

- 15th International Conference on Engineering Psychology and Cognitive Ergonomics (EPCE 2018)
- 12th International Conference on Universal Access in Human-Computer Interaction (UAHCI 2018)
- 10th International Conference on Virtual, Augmented, and Mixed Reality (VAMR 2018)
- 10th International Conference on Cross-Cultural Design (CCD 2018)
- 10th International Conference on Social Computing and Social Media (SCSM 2018)
- 12th International Conference on Augmented Cognition (AC 2018)
- 9th International Conference on Digital Human Modeling and Applications in Health, Safety, Ergonomics, and Risk Management (DHM 2018)
- 7th International Conference on Design, User Experience, and Usability (DUXU 2018)
- 6th International Conference on Distributed, Ambient, and Pervasive Interactions (DAPI 2018)
- 5th International Conference on HCI in Business, Government, and Organizations (HCIBGO)
- 5th International Conference on Learning and Collaboration Technologies (LCT 2018)
- 4th International Conference on Human Aspects of IT for the Aged Population (ITAP 2018)

Conference Proceedings Volumes Full List

1. LNCS 10901, Human-Computer Interaction: Theories, Methods, and Human Issues (Part I), edited by Masaaki Kurosu
2. LNCS 10902, Human-Computer Interaction: Interaction in Context (Part II), edited by Masaaki Kurosu
3. LNCS 10903, Human-Computer Interaction: Interaction Technologies (Part III), edited by Masaaki Kurosu
4. LNCS 10904, Human Interface and the Management of Information: Interaction, Visualization, and Analytics (Part I), edited by Sakae Yamamoto and Hirohiko Mori
5. LNCS 10905, Human Interface and the Management of Information: Information in Applications and Services (Part II), edited by Sakae Yamamoto and Hirohiko Mori
6. LNAI 10906, Engineering Psychology and Cognitive Ergonomics, edited by Don Harris
7. LNCS 10907, Universal Access in Human-Computer Interaction: Methods, Technologies, and Users (Part I), edited by Margherita Antona and Constantine Stephanidis
8. LNCS 10908, Universal Access in Human-Computer Interaction: Virtual, Augmented, and Intelligent Environments (Part II), edited by Margherita Antona and Constantine Stephanidis
9. LNCS 10909, Virtual, Augmented and Mixed Reality: Interaction, Navigation, Visualization, Embodiment, and Simulation (Part I), edited by Jessie Y. C. Chen and Gino Fragomeni
10. LNCS 10910, Virtual, Augmented and Mixed Reality: Applications in Health, Cultural Heritage, and Industry (Part II), edited by Jessie Y. C. Chen and Gino Fragomeni
11. LNCS 10911, Cross-Cultural Design: Methods, Tools, and Users (Part I), edited by Pei-Luen Patrick Rau
12. LNCS 10912, Cross-Cultural Design: Applications in Cultural Heritage, Creativity, and Social Development (Part II), edited by Pei-Luen Patrick Rau
13. LNCS 10913, Social Computing and Social Media: User Experience and Behavior (Part I), edited by Gabriele Meiselwitz
14. LNCS 10914, Social Computing and Social Media: Technologies and Analytics (Part II), edited by Gabriele Meiselwitz
15. LNAI 10915, Augmented Cognition: Intelligent Technologies (Part I), edited by Dylan D. Schmorrow and Cali M. Fidopiastis
16. LNAI 10916, Augmented Cognition: Users and Contexts (Part II), edited by Dylan D. Schmorrow and Cali M. Fidopiastis
17. LNCS 10917, Digital Human Modeling and Applications in Health, Safety, Ergonomics, and Risk Management, edited by Vincent G. Duffy
18. LNCS 10918, Design, User Experience, and Usability: Theory and Practice (Part I), edited by Aaron Marcus and Wentao Wang

19. LNCS 10919, Design, User Experience, and Usability: Designing Interactions (Part II), edited by Aaron Marcus and Wentao Wang
20. LNCS 10920, Design, User Experience, and Usability: Users, Contexts, and Case Studies (Part III), edited by Aaron Marcus and Wentao Wang
21. LNCS 10921, Distributed, Ambient, and Pervasive Interactions: Understanding Humans (Part I), edited by Norbert Streitz and Shin'ichi Konomi
22. LNCS 10922, Distributed, Ambient, and Pervasive Interactions: Technologies and Contexts (Part II), edited by Norbert Streitz and Shin'ichi Konomi
23. LNCS 10923, HCI in Business, Government, and Organizations, edited by Fiona Fui-Hoon Nah and Bo Sophia Xiao
24. LNCS 10924, Learning and Collaboration Technologies: Design, Development and Technological Innovation (Part I), edited by Panayiotis Zaphiris and Andri Ioannou
25. LNCS 10925, Learning and Collaboration Technologies: Learning and Teaching (Part II), edited by Panayiotis Zaphiris and Andri Ioannou
26. LNCS 10926, Human Aspects of IT for the Aged Population: Acceptance, Communication, and Participation (Part I), edited by Jia Zhou and Gavriel Salvendy
27. LNCS 10927, Human Aspects of IT for the Aged Population: Applications in Health, Assistance, and Entertainment (Part II), edited by Jia Zhou and Gavriel Salvendy
28. CCIS 850, HCI International 2018 Posters Extended Abstracts (Part I), edited by Constantine Stephanidis
29. CCIS 851, HCI International 2018 Posters Extended Abstracts (Part II), edited by Constantine Stephanidis
30. CCIS 852, HCI International 2018 Posters Extended Abstracts (Part III), edited by Constantine Stephanidis

http://2018.hci.international/proceedings

10th International Conference on Social Computing and Social Media

Program Board Chair(s): **Gabriele Meiselwitz, *USA***

- James Braman, USA
- Cristóbal Fernández Robin, Chile
- Nick V. Flor, USA
- Panagiotis Germanakos, Germany
- Sara Anne Hook, USA
- Rushed Kanawati, France
- Carsten Kleiner, Germany
- Niki Lambropoulos, UK
- Marilia Mendes, Brazil
- Hoang Nguyen, Singapore
- Anthony Norcio, USA
- Michiko Ohkura, Japan
- Cristian Rusu, Chile
- Christian Scheiner, Germany
- Shubhi Shrivastava, USA
- Abraham Van der Vyver, South Africa
- Giovanni Vincenti, USA
- Jose Viterbo, Brazil
- Yuanqiong (Kathy) Wang, USA
- June Wei, USA
- Brian Wentz, USA

The full list with the Program Board Chairs and the members of the Program Boards of all thematic areas and affiliated conferences is available online at:

http://www.hci.international/board-members-2018.php

HCI International 2019

The 21st International Conference on Human-Computer Interaction, HCI International 2019, will be held jointly with the affiliated conferences in Orlando, FL, USA, at Walt Disney World Swan and Dolphin Resort, July 26–31, 2019. It will cover a broad spectrum of themes related to Human-Computer Interaction, including theoretical issues, methods, tools, processes, and case studies in HCI design, as well as novel interaction techniques, interfaces, and applications. The proceedings will be published by Springer. More information will be available on the conference website: http://2019.hci.international/.

General Chair
Prof. Constantine Stephanidis
University of Crete and ICS-FORTH
Heraklion, Crete, Greece
E-mail: general_chair@hcii2019.org

http://2019.hci.international/

Contents – Part I

Privacy and Ethical Issues in Social Media

Contents – Part II

Social Network Analysis

Agents, Models and Algorithms in Social Media

Social Media User Experience

A Study of the Influence of Images on Design Creative Stimulation

Mengjiao Chen[1], Tianjiao Zhao[1(✉)], Hechen Zhang[1], and Shijian Luo[2]

[1] Tianjin University, Tianjin, China
zhaotianjiao@tju.edu.cn
[2] Zhejiang University, Zhejiang, China

Abstract. Creative idea generation is the core of the design process. Although existing literatures show that image stimulation plays a significant role in the innovation design process, it is generally unclear how to evaluate the effect of image stimuli in the design process and where the applicable point image stimulus should be set. This paper carries out a study to investigate the effect of image stimuli under different phase of product design with 28 industrial design students. We get three findings. Firstly, we find that sketching is a phase where a lot of ideas are generated by tracking designer's total design flow. Secondly, by analyzing the sketching patterns of design students, we find that at the beginning of sketching their thinking model can be divided into two types, i.e., inspirational and rational. Under different thinking model, the participants, however, report positive benefits from the availability of image stimulus during creative idea generated in different levels, and express higher satisfaction to the image-stimulating work as well. The final finding is that applying the stimulus in the middle stage of the sketch design, which is the convergent stage of thinking, can stimulate creativity more effectively and avoid the limitations of thinking from insufficient stimulus to some extent. Based on the above findings, the research on creative stimulation theory and method driven by image materials can be established, which provides a reference for the development of intelligent design mechanism.

Keywords: Industrial design · Sketching · Image · Creative idea stimulation
Computer aided innovative design

1 Introduction

Creative idea generation is both primary and most important step in creative design. Stimulating creative inspiration timely and effectively can have a positive effect on the design process. The current methods of creative stimulation mainly include artificial-based creative stimulation method and computer-assisted creative stimulation method.

1.1 Method of Creative Idea Stimulation

Artificial-Based Creative Stimulation Method. The development of creativity first appeared in the United States in 1906 when Prindle's idea of training engineers in

© Springer International Publishing AG, part of Springer Nature 2018
G. Meiselwitz (Ed.): SCSM 2018, LNCS 10913, pp. 3–18, 2018.
https://doi.org/10.1007/978-3-319-91521-0_1

creativity became the beginning of creative theory. In 1938, known as "the father of creating engineering" Osbern developed a "brainstorm", on the basis of this method, he raised a "going-stopping method" which made the divergent thinking and convergence thinking cross over and over again [1]. In the 19th century forty or fifty's, "synecdoche" and "equivalent transformation theory" emerged in succession, on behalf of analogy techniques; the 5W2H method, innovation from the perspective of raising questions; "six caps of thinking", which stimulate human creativity potential, and so on.

As people-centered design extends from design-in-kind (industrial design) to design experience (including interactive design and service design), the researchers' understanding on the meaning of innovation design is more concrete and in-depth. After analyzing the importance of FFE (Fuzzy Front End), Moon and others proposed a user experience-driven creative idea generation method [2]. After exploring the connection between curiosity and innovation, Hardy analyzed the potential of curiosity which can be a source of creative inspiration and problem-solving [3]. While, Goncalves explored the inspirational effects of images, objects and words from the perspective of creative sources [4]. IDEO, a famous creative company, looks for ways to innovate design by using the way of drawing design situation story version and role-playing. Delft University of Technology proposed "Context mapping" to stimulate designers' creativity by looking for participatory points where users and stakeholders actively participate in the design [5]. According to statistics, the number of artificial-based creative stimulation methods is as many as 100 kinds.

The advantage of artificial-based creative stimulation method is giving full play to the importance of the human brain in the process of innovation, not emphasizing efficiency but imagination, having a high degree of dependence on people, so it requires a higher quality of creative personality. While, the disadvantage is that most of these methods are mind-driven and do not have many rules to follow. It is generally required that creative creators have a wide range of knowledge reserves themselves, so this method has some limitations for inexperienced designers.

Computer-Assisted Creative Stimulation Method. With the widespread popularization and application of computers, the stimulus for creativity has also from depending artificial-based method only transformed into computer-aided method gradually. From the initial CAD (Computer aided design) that computer is visualization of the designer's creative results, to the CACD (Computer aided concept design) that computer participate requirements analysis to solution concept design process, then the algorithmic-driven innovative design of the AI era is coming, which is automated intelligent design tools backed by rules and algorithms. The computers moving from the end of the design process to the heart of the design process.

Although computer-aided technology offers a wide range of opportunities for innovative design and creativity, it still has its drawbacks. Firstly, computer-aided innovation design emphasizes the solution to the evolutionary contradictions of technological systems, focusing on innovative solutions and results, but in terms of promoting problems' discovery and driving the stage of inspiration generation is not enough. Secondly, many methods place too much emphasis on design automation. It leads to the simplification of innovative thinking and lack of a collaborative model of human-computer interaction, mutual assistance and mutual encouragement. Finally, when computer-aided design combined with big data, most of the researches focus on the information of patented technologies, such as knowledge map [6], trend analysis [7–9], patent evasion [10], etc., however, it lacks research on large-scale design image materials and creative knowledge mining.

1.2 Images and Design Creative Idea Stimulation

There are two types of stimuli that produce creative inspiration, one is internal stimulus and the other is external stimulus. Internal stimuli exist in human's memory, the stimulation of metacognitive knowledge. External stimuli are stimuli from the surrounding environment, including audible, perceptible information such as graphics, language, and objects [1].

A comparative study of the stimulating effects on text, graphics, and objects, found that images are the best element to stimulate design inspiration, both for professional and non-professional designers [11]. In the design process, design beginners watch a large number of creative and inspiring image design material, which is an effective way to promoting creative thinking and inspiring design inspiration. Cognitive psychology shows that after inputting morphological problems, people will generate similar mappings under the stimulus of design resources, so as to output behaviors [12]. Therefore, finding out a method of selecting motivating design material and establishing the connection between image content and creative knowledge is the core of the image material-driven creative stimulus method.

At present, many design resource and generative design materials exist. However, in the design process, the difficulties still exist in finding the design material, and the found design materials' creative excitation is always weak or even invalid. A lot of image materials with potential creative stimulation abilities are still asleep. They do not provide powers in promoting the idea generation process for designers. Therefore, in big data era, we need to promote an image-driven creative idea generation method by adopting the deep learning technique. In this study, we focused on the first step of promoting an image driven creative method, which is exploring the influence of Images on Design Creative Stimulation. Only by defining the relationship between image, creative Stimulation and design process, can we solve the problem that "how image influence the idea generation process, how to select high quality images of stimuli creative idea, and how to apply the selected images into design process". The Image driven creative idea generation method based on deep learning technology can be developed in the next step.

In this study, through class experiment, we focus on the following two points: Firstly, during the design process, where is the applicable point that image stimulus should be set? Secondly, after the stimulation, how to evaluate the effect of image stimuli? Based on these two points, the research can help us to establish the image material-driven creative design theory and method. Then lay the foundation for the iterative evolution and innovation of intelligent design that leads from creative thinking.

2 Research Method

2.1 Image Stimulate Experiment

Our research takes product design as an example. Product design process, can be divided into five stages including design research, design elements mining, design and development, program revision and program output. New ideas are constantly being generated in the process and illustrated through sketch exploration. This project intends to apply stimulation of creative elements to all stages of the design process, and then, gradually reduce the scope of the experiment. Choose the applicable point that image stimulus should be set, by effective evaluating method of image stimuli. The whole research plan is shown in Fig. 1.

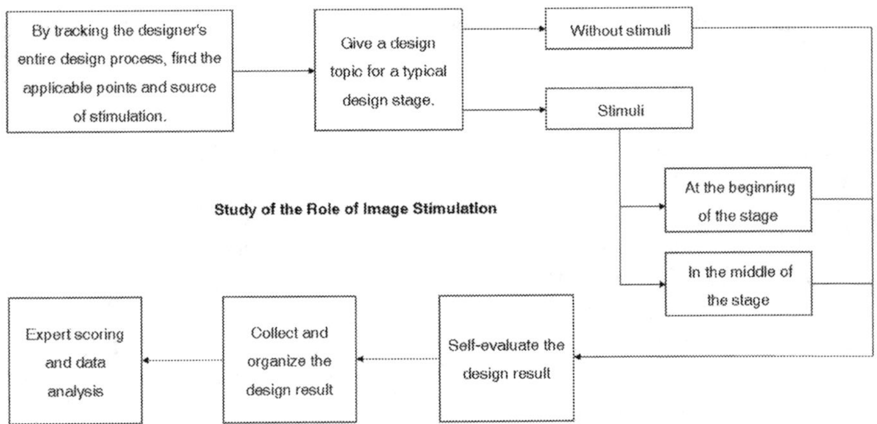

Fig. 1. Study process of the role of image stimulation

2.2 Demonstration Method of Design Results

Nowadays, designers have a variety of ways to express their ideas, ranging from the use of traditional words, numbers and sketches to the use of ever-expanding computer graphics and modeling software. The computer software may be only used to demonstrate the program, words and numbers can only express ideas, however, the sketches can express the functional characteristics of the idea and emotional content simultaneously. Gather the sketches in the design process, so that demonstrate the designer's thought process completely.

Sketch is accurate enough to express ideas that not be solved in fine language and in precise calculations accurately; sketch is ambiguous so that readers can get more information in the sketch even beyond itself; sketch is questionable, It solves a problem, at the same time, it can trigger new thinking; sketch is exploratory, it gives a problem a variety of possible solutions. Sketches are produced from current knowledge, and then, read or interpreted to produce new knowledge [13]. In such a repetitive process, ideas are continuously generated and iterated.

Many existing design studies show a great deal of importance in sketches. Chen Shi's research on design knowledge analytical method during sketching, analyzed the differences in the composition of knowledge in the exploration stage and knowledge expression stage of designers [14]. Simon Laing and Masood Masoodian, in studying the role of visual images in graphic design, also used sketches as research priorities [15].

Based on the characteristics of the sketch above, we will collect, analyze and evaluate the sketches which can use as the display of creative design during the experiment. In addition, in the process of experiment, we will ask the participant to make full use of sketches in all stages of design so that we can grasp the participant's thinking process and creative design results better.

2.3 Evaluation Method of Stimulate Effect of Image Material

In order to determine whether these material elements are effective in stimulating creativity or not, this paper intends to adopt a combination of quantitative and qualitative evaluation. Psychological studies show that when people look for a certain goal, human cognition and perception will be judged in different dimensions, including good and bad, positive and negative and so on. The result of this judgment will be expressed through the physiological and psychological information of the person. This study intends to use subjective evaluation scoring, vocal thinking, observation and depth of interview method to complete the evaluation of stimulate effects of material. Evaluation process as shown below (Fig. 2).

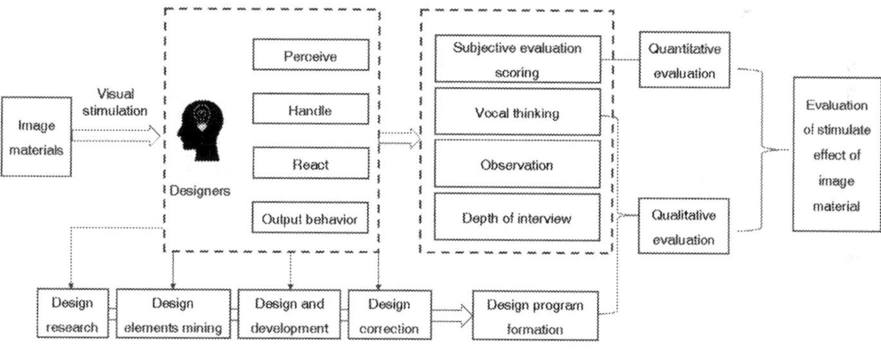

Fig. 2. Evaluation scheme of image creative stimulation effect

Subjective Evaluation Scoring. Let the set of test pictures be $C = \{c_1, c_2, c_3, \ldots, c_m\}$, the set of tested persons is $B = \{b_1, b_2, b_3, \ldots, b_n\}$, where m is the number of the material groups and n is the number of participation. The material group m' is used as the stimulus material, and the test object person n' evaluates the creative effect of m' as $P_{n', m'}$, then the subjects $(b_1, b_2, b_3, \ldots, b_n)$ to $(c_1, c_2, c_3, \ldots, c_m)$'s evaluation set as:

$$P = \begin{bmatrix} p_{1,1} & p_{1,2} & \cdots & p_{1,m} \\ p_{2,1} & p_{2,2} & \cdots & p_{2,m} \\ \vdots & \vdots & \vdots & \vdots \\ p_{n,1} & p_{n,2} & \cdots & p_{n,m} \end{bmatrix}$$

The comprehensive evaluation of the material group *i* is:

$$P_i = \frac{p_{1,i} + p_{2,i} + \cdots + p_{n,i}}{n}$$

The subjective evaluation score of each material group is:

$$P^T = \left[P_{c1}, P_{c2}, \ldots P_{cm}\right]$$

Qualitative Evaluation. The qualitative evaluation method can supplement the quantitative evaluation results and increase the reliability of the evaluation results. In design creative research, vocal thinking is an effective way to reflect on the designer's thinking. At the same time, in-depth interviewing of subjects at the end of the design process leads to the most direct and real message.

Firstly, analyze the content of the vocal thinking of the participants, and explore the tacit knowledge in the linguistic content and potential implications. Not only analyze the material evaluation stimulating effect, but also find out the potential demand in the design process. Secondly, according to the method of design investigation, we conducted in-depth interviews to find out the potential problems in the design process and the true effect of creative stimulation. Combined with the quantitative and qualitative evaluation results, the evaluation of the stimulating effect of the design material is completed, the evaluation method is established at the same time.

3 Images Influence on Creative Idea Generation in Design Process

As shown in Fig. 1, this paper divides our experiment into two parts:

- By tracking the designer's entire design process, find the applicable points and source of stimulation.
- Under three different cases-no image stimulation, stimulation at the beginning of the process and stimulation in the middle of the process, study a certain stage in the whole design process. Then collect the sketches and make an artificial evaluation.

3.1 Stage 1: Study of the Whole Process of Design

Use in-depth interviews, vocal thinking, observation to study the stage that needs for image stimulation and the type of image stimulation in the whole product design process.

3.1.1 Process of Experiments

We choose an industrial design graduate student, with five-year industrial design learning experience and six months of product design work experience, as the experimenters. Track and observe the process from him received the mission of the enterprise - high-end lamp design to his works won the bidding finally, the design process lasted two months. In most cases, his place of work is in the studio, while, in a small part of the time, he did lamp design in the dormitory. Our scope of observation is his studio. After the pre-training, we asked him to describe his thought by talking as much as possible in the lamp design, and record the sound with his permission. Due to the limit of recording conditions, observers recorded the design process by taking notes and photographing. After he finished the design, the interviewers conducted in-depth interviews, focusing on the psychological activities as received the image stimulus in different stages. Finally, organize and analyze the collected material according to the design process (Fig. 3).

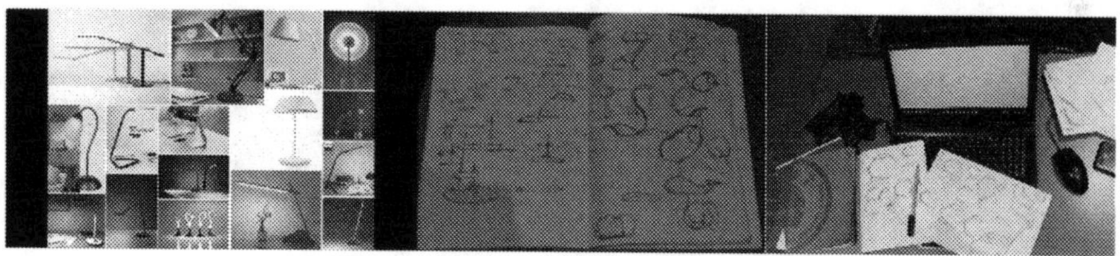

Fig. 3. Photographing of experiments progress

3.1.2 Sketch Data Analysis

Classify the sketches as the design stage, then combine it with the results of observation, vocal thinking and in-depth interviews to analyze the characteristic in each stage. The results as shown below.

According to the discussion, we can find the following three things. Firstly, the image materials that can stimulate the experimenter to produce creative lamp works, mostly are the existing lamp design pictures, while, others are some pictures of well-designed decoration. Secondly, during the stage 2 and the stage between stage 2 and stage 3 in the Fig. 4, the experimenter had the highest demand for the image material and the absorption effect (it can be seen from the sketch). Thirdly, in other stages, most of the image materials needed by the experimenter are in-depth analysis of the collected materials, and the collection of one certain materials. In the meantime, the absorption effect on other materials is decreasing obviously.

To sum up, creativity stimulated by image stimulus is consistent throughout the design process, and most clearly in the design research phase and the element mining phase. We will summarize this two phase as the initial stage of design production, and then, we will carry out in-depth study of this stage.

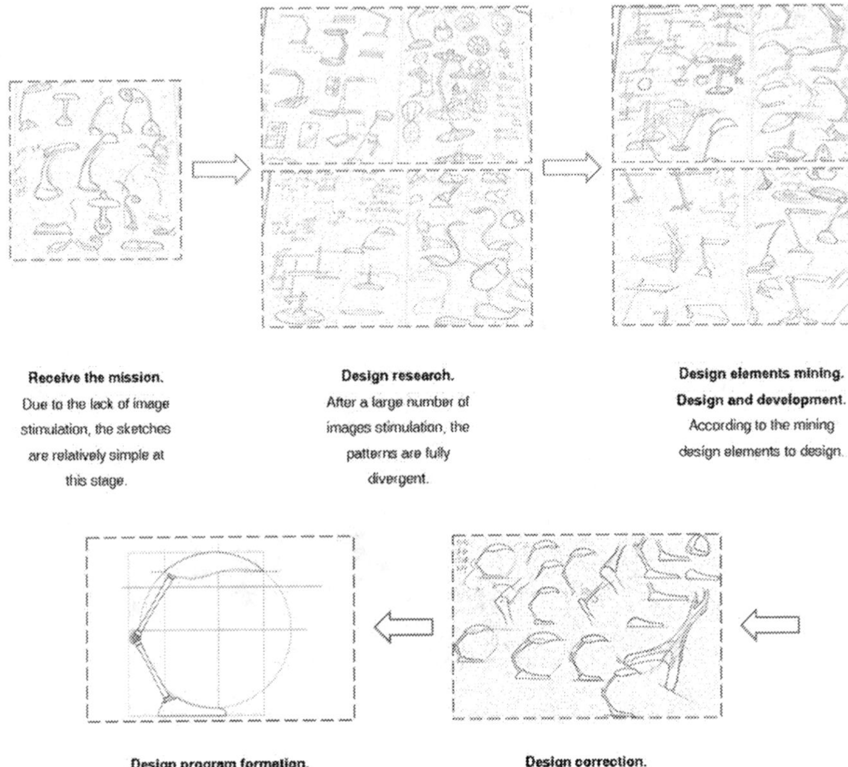

Receive the mission.
Due to the lack of image stimulation, the sketches are relatively simple at this stage.
Design research.
After a large number of images stimulation, the patterns are fully divergent.
Design elements mining.
Design and development.
According to the mining design elements to design.
Design program formation.
Design correction.

Fig. 4. Classify the sketches as the design stage

3.2 Stage 2: Study of Initial Idea Stage of Design

By setting up a couple of control experiment groups, that is, with or without stimulation, different stages of stimulation, study the creativity stimulation at initial idea stage of design. Evaluate the stimulation effect through self-evaluation and experts' evaluation.

3.2.1 Participants

There are 28 industrial design students voluntarily participated in this study. They were all undergraduates from the same class, with 1–3 years' industrial design learning experience. There were 16 female and 12 male participants, with most being in the 18–22 year age range.

3.2.2 Experiment Process

All sessions were conducted within design classrooms that featured a video camera, bare walls, and the projector to provide visual stimuli (Fig. 5). Participants worked on a bare desk using A4 drawing paper and a variety of pencils to sketch with.

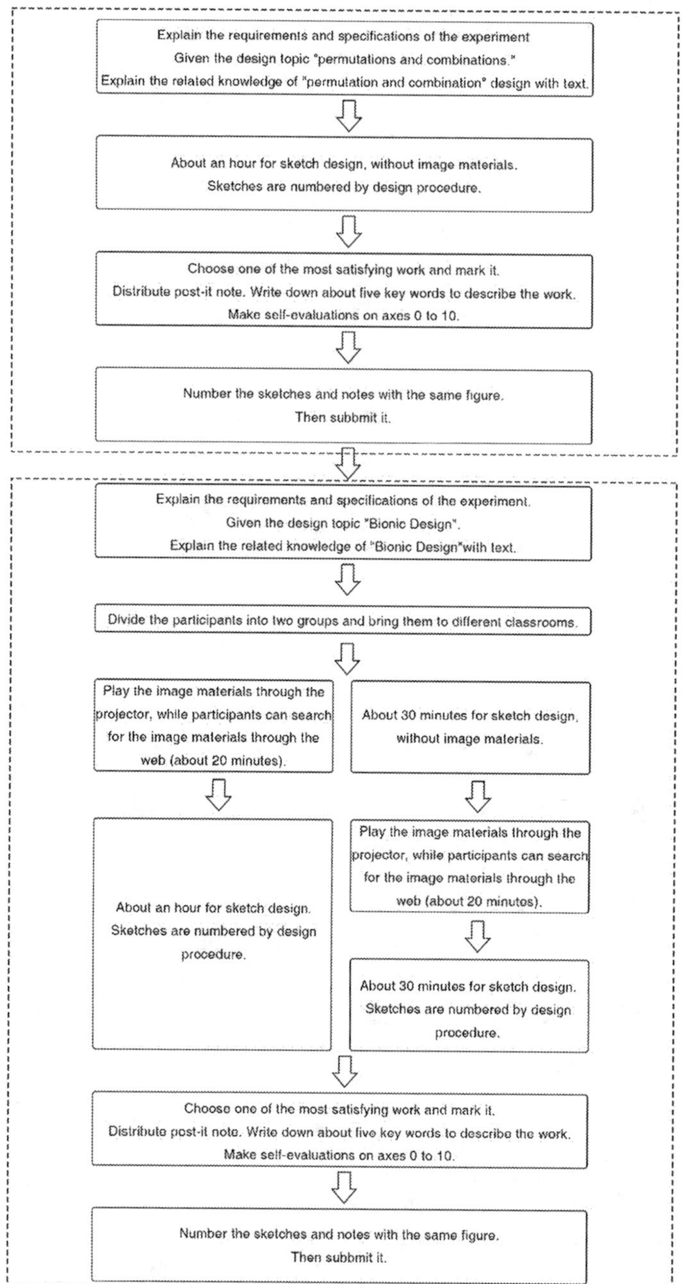

Fig. 5. Study session

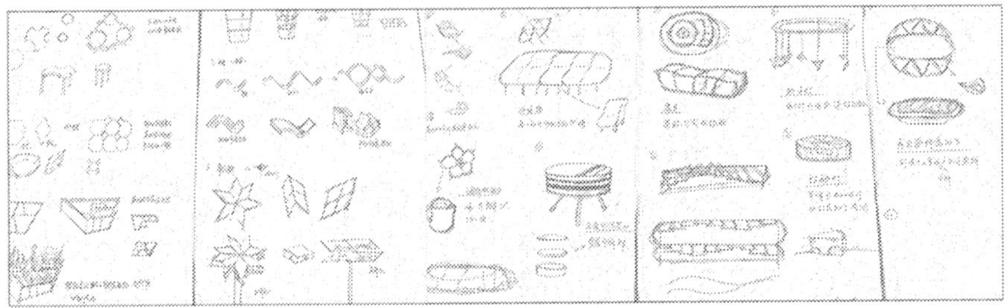

1. starting from the shape

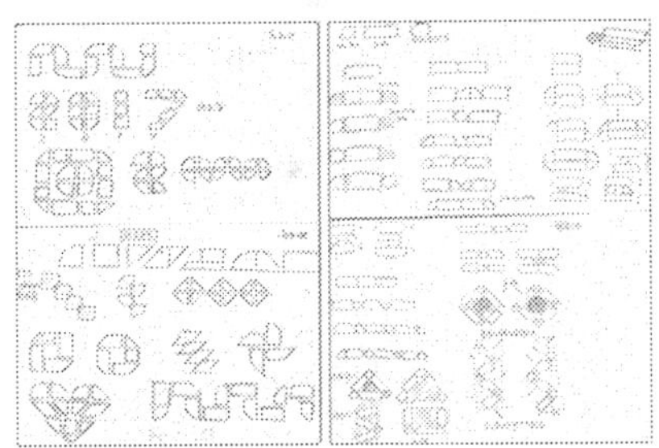

2. starting from the product

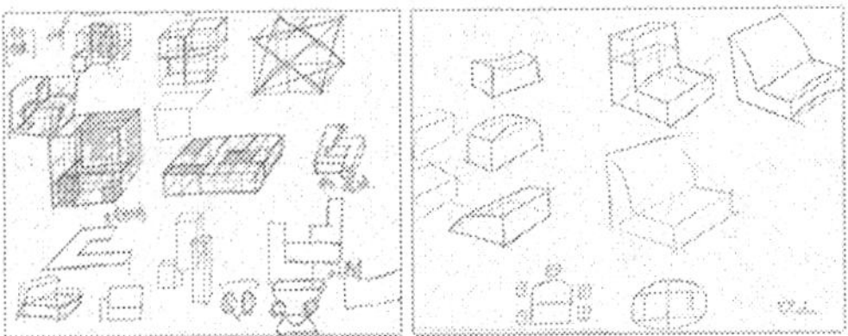

3. starting from the theory

Fig. 6. Three kinds of sketch thinking

3.2.3 Sketch Data Analysis

As can be seen from the sketch of the first set of experiments, the participants' design thinking for a given topic can be broadly divided into three types. First, associating the possible shapes from the perceptual words, in the analysis of the shape, confirming the possible combination methods and applicable design products; the second is, starting from the product, determine the product to be designed, and then divergent possible both shapes and combinations; the third is, starting from the theory, analyze the concept of permutation and combination, then diverge scheme from the cutting of the basic form.

As can be seen from the sketches of the second experiment group, the group that received the image stimulus at the beginning has the accurate shape and sooth line. However, due to the limitation caused by the image stimulus, the design concept is rather single and the divergence is not strong. The group receiving image stimulus in the middle was slightly deficient in styling and sketching, but the diverging effect was better, the shape was more varied and paying more attention to the expression of function and emotion at the same time (Fig. 7).

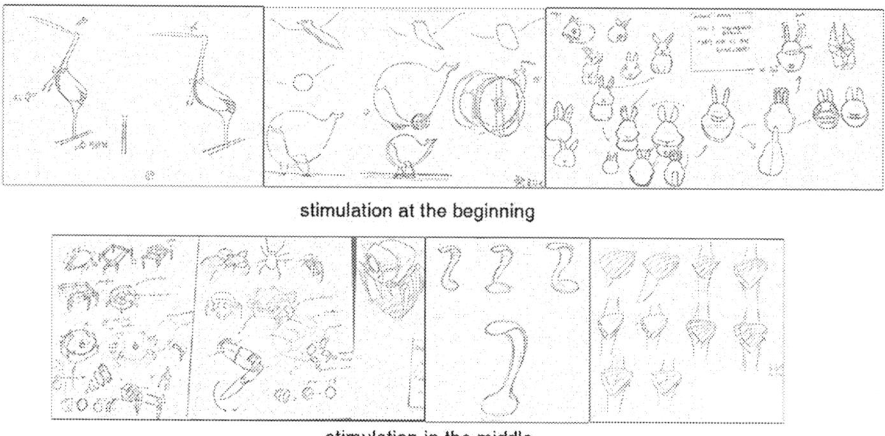

Fig. 7. Stimulation in different points

4 Evaluation Result of Stimulate Effect of Image Material

Our data collected from the study took a variety of forms: sound recordings of the participants' vocal thinking, notes taken by the observing researcher, participants' design outputs, participants' post-task self-ratings and commentary and experts' ratings on the criteria measured. Statistical analysis of the ratings provided by the study participants and expert judges carried out to identify any perceived differences in the study sessions. The results of these analyses are given below. This is followed by the findings from the analysis of the interviews and commentary provided by the study participants.

4.1 Statistical Analysis of Self-ratings and Experts' Ratings

Self-ratings. The same group of participants in given image material case and no image materials cases, whose self-evaluation of design results in the table below.

We collected 28 effective scores in the experiment without image materials and 27 effective scores in the experiments with image materials. As can be seen from the table, with the design of the image material, the participants are more satisfied with the design process and the work, meanwhile, the creative effect is better. Due to the different design topics of the two experiments, it may disturb the scorers. We did not conduct expert scoring on this control group. So the result of this analysis is based on the designer's perspective (Table 1).

Table 1. Statistical analysis of self-rating

	M	SD
Without stimulation	6.1786	1.6789
Stimulation	6.6111	1.3253

Experts' Ratings. We invited two industrial design teachers and three industrial design graduate students, composed an expert scoring team. They will rate the results of the second design task in study on the initial idea stage of design. Before scoring, we selected 27 effective sketches from the two sets of sketches, mixed and relabeled. The expert scoring team rated the designs using the grading sheet. The evaluation criteria are as follows (Fig. 8).

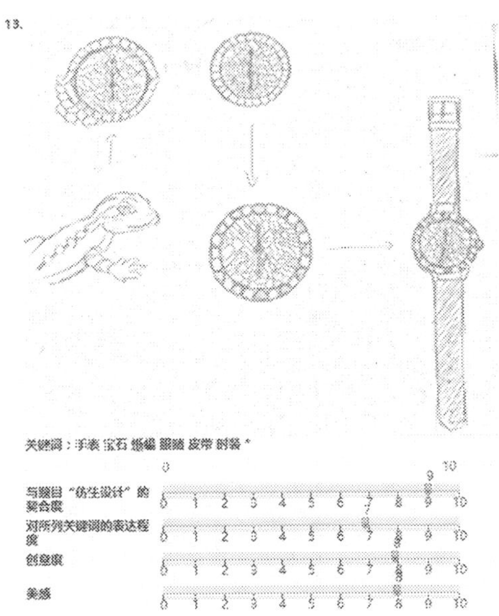

Fig. 8. Sample of the grading sheet

- Fit with the design theme (Bionic Design) or not.
- Express clearly or not (corresponding to the given keywords).
- Originality.
- Aesthetic feeling.

According to the numbers before scoring, the sketches that were stimulated at the beginning, were labeled as 1, 2, 5, 6, 9, 10, 13, 14, 17, 18, 21, 22 and 25; while, the sketches that were stimulated in the middle, were labeled as 3, 4, 7, 8, 11, 12, 15, 16, 19, 20, 23, 24, 26 and 27. Their average score of experts' rating are as follows.

From the Figs. 9 and 10, we can see that the group with the stimulating in the middle has the higher score in total, and the score distribution is relatively stable. Further, we calculated the average score of each group for comparative analysis in the table below. As we can see in the table, in the initial stages of generating ideas, applying image stimuli to different points can have different effects. The group with the stimulating in the middle fit with the theme better (M = 7.1429 vs. M = 7.0154), because it had better understanding in the design theme. In addition, this group is superior to the other in terms of expression (M = 6.3429 vs. M = 6.1538) and originality (M = 5.6429 vs. M = 5.4615). On the contrary, has more smooth lines and more beautiful shape (M = 5.7231 vs. M = 5.6286). The average total score of the group with the stimulating in the middle is 24.7571 (full score is 40), higher than that of the group with the stimulating at the beginning (24.3538).

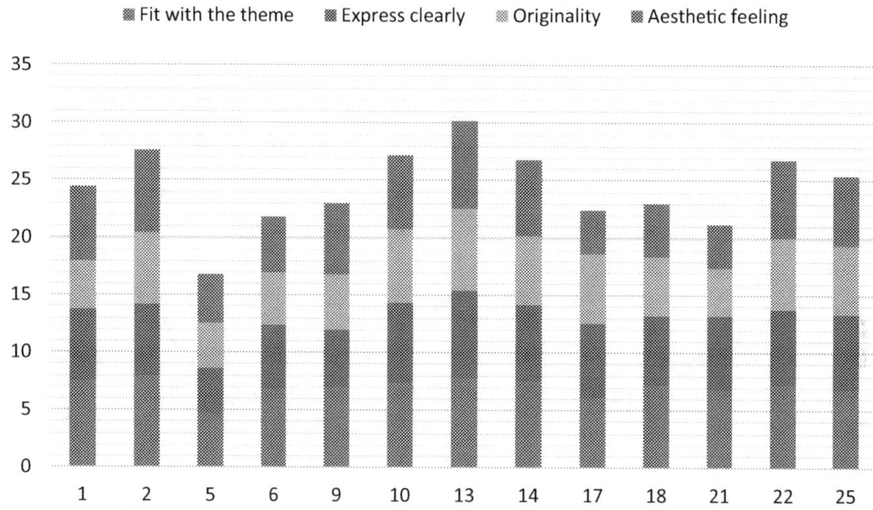

Fig. 9. Experts' rating for stimulation at the beginning

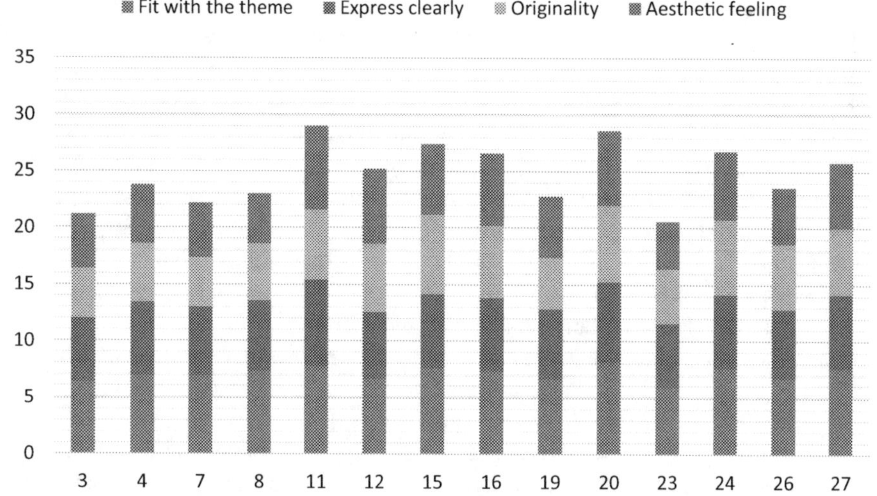

Fig. 10. Experts' rating for stimulation in the middle

Therefore, based on these analyses, thought there are some different effects when stimulate to different points, the difference is not very significant, applying stimulation in the middle of the process is slightly better (Table 2).

Table 2. Statistical analysis of experts' rating

	Fit with the theme	Express clearly	Originality	Aesthetic feeling	Total score
Stimulation at the beginning	7.0154	6.1538	5.4615	5.7231	24.3538
Stimulation in the middle	7.1429	6.3429	5.6429	5.6286	24.7571

4.2 Analysis of the Interviews and Commentary

In order to try and capture the participants' perspective on the design process, as well as their design output, short interviews were carried out with them at the conclusion of the first and second experiments during each study session.

After the first task, we choose three participants with three kinds of sketch thinking as interviewee, to find out how they think when sketching, what problems have they met and satisfied with their work or not. During the interview, comments were sought through open-ended questions, so that we can catch more valuable information. During the interview, we found that interviewees expressed their dissatisfaction with the design process and design work because of the lack of image materials. Some participants hope provided materials before designing, while, more participants expect shown the materials after giving them some time to think.

As with the first interview, once the participants had concluded their second design task, they were again invited to comment on the task they had just completed. They were also asked to review the effect of the image materials. During the interview, participants from stimulation in the middle stimuli group analyzed the image material more deeply.

From the interview result, we can also find that the creative design thinking can be divided into intuitive design thinking and logical design thinking. These two ways of thinking appear alternately in the design process. By analyzing the interviews and commentary of the first set of experiments, we got three kinds of the participants' design thinking (Fig. 6). We classify the thinking of starting from the shape as intuitive thinking, and classify the thinking of starting from the product and theory as logical thinking. Under the no stimulation situation, participants in intuitive thinking can design based on the accumulation of life, so their demand for image stimulation is not as strong as that of logical thinking. Besides, participants in both way of thinking said that the design process without image stimulation make their ideas dry up.

Design is a choice, and there are two space can offer for creativity. One is when we make a list of drastically different design drafts for selection, the other is when we definite the guidelines [13]. This is the thinking of divergence and divergence, each design phase is the process of divergence and convergence phases alternating [16]. By analyzing the initial idea stage of design, we find that two image stimuli act on the divergence and convergence stages of the design, respectively. So it led to the difference of divergence and convergence between the two groups of participants. The group that stimulated in the convergence phase had a higher evaluation of the effect of the image material and a lower evaluation of their work, and the group that stimulated in divergence phase was exactly the opposite.

5 Conclusions

In this paper we have described a study we conducted to better understand the influence of images on design creative stimulation. Although there are some limitations with our study, such as the small sample size (n = 28), and the fact that our participants were industrial design students rather than professional practitioners, nevertheless the study has provided us with a number of useful findings.

- At stage 1, by tracking designer's total design flow and in-depth interviews, we find that initial idea stage is a phase where a lot of ideas are generated. This phase demands image stimulus to stimulate creativity urgently.
- At stage 2, by analyzing the sketching patterns and the interviews of participants, we find that their thinking model can be divided into two types, i.e., intuitive design thinking and logical design thinking. Under different thinking model, the participants, however, report positive benefits from the availability of image stimulus during creative idea generated in different levels, and express higher satisfaction to the image-stimulating work as well. Besides, participants of logical thinking demand more for image stimulation. Although the actual effect is not as obvious as their subjective feelings.
- Further analysis on stage 2, we find that applying the stimulus in the middle stage of the sketch design, which is the convergent stage of thinking, can stimulate creativity more effectively and avoid the limitations of thinking from insufficient stimulus to some extent.

We are now in a better position to know more about where is the applicable point that image stimulus should be set and how to evaluate the effect of image stimuli. The next important step would be using eye movement, EEG and other method to choose creative stimulation images materials. This research can help us to establish the image material-driven creative design theory and method. Then lay the foundation for the iterative evolution and innovation of intelligent design that leads from creative thinking.

Acknowledgements. We would like to express our gratitude to the participants and review experts for their contributions to the study reported here.

References

1. Osborn, A.F.: Applied Imagination. Scribner, New York (1979)
2. Moon, H., Han, S.H.: A creative idea generation methodology by future envisioning from the user experience perspective. Int. J. Ind. Ergon. **56**, 84–96 (2016)
3. Hardy III, J.H.H., Ness, A.M., Mecca, J.: Outside the box: Epistemic curiosity as a predictor of creative problem solving and creative performance. Pers. Individ. Differ. **104**, 230–237 (2017)
4. Goncalves, M., Cardoso, C., Badke-Schaub, P.: What inspires designers? Preferences on inspirational approaches during idea generation. Des. Stud. **35**(1), 29–53 (2014)
5. Van Boeijen, A., Daalhuizen, J.: Delft design guide. TU Delft (2010)

6. Liu, Z., Sun, L.L., Lu, N.: Construction of process-oriented industrial design knowledge map. J. Mech. Eng. **46**(8), 181–187 (2010)
7. Al-Kazzaz, D.A., Bridges, A.H.: A framework for adaptation in shape grammars. Des. Stud. **33**(4), 342–356 (2012)
8. Cluzel, F., Yannou, B., Dihlmann, M.: Using evolutionary design to interactively sketch car silhouettes and stimulate designer's creativity. Eng. Appl. Artif. Intell. **25**(7), 1413–1424 (2012)
9. Sun, L., Xiang, W., Chai, C., Wang, C., Huang, Q.: Creative segment: a descriptive theory applied to computer-aided sketching. Des. Stud. **35**(1), 54–79 (2014)
10. Chen, A., Chen, R.: Design patent map: an innovative measure for corporative design strategies. Eng. Manag. J. **19**(3), 14–29 (2015)
11. Sarkar, P., Chakrabarti, A.: The effect of representation of triggers on design outcomes. Artif. Intell. Eng. Des. Anal. Manuf. **22**(2), 101–116 (2008)
12. Howard, T.J., Culley, S.J., Dekoninck, E.: Describing the creative design process by the integration of engineering design and cognitive psychology literature. Des. Stud. **29**(2), 160–180 (2008)
13. Greenberg, S., Carpendale, S., Marquardt, N., Buxton, B.: Sketching User Experiences. Morgan Kaufmann, San Francisco (2011)
14. Chen, S., Yang, Z.Y., Sun, L.Y., Lou, Y.: Research on design knowledge analytical method during sketching. Mod. Ind. Des. Inst. **49**(11), 2073–2082 (2015)
15. Laing, S., Masoodian, M.: A study of the influence of visual imagery on graphic design ideation. Des. Stud. **45**, 187–209 (2016)
16. Brown, T.: Change by Design: How Design Thinking Transforms Organizations and Inspires Innovation. Harper Business, New York (2009)

A Framework to Simplify Usability Analysis of Constraint Solvers

Broderick Crawford[1(✉)], Ricardo Soto[1], and Franklin Johnson[2(✉)]

[1] Pontificia Universidad Católica de Valparaíso, Valparaíso, Chile
{broderick.crawford,ricardo.soto}@pucv.cl
[2] Universidad de Playa Ancha, Valparaíso, Chile
franklin.johnson@upla.cl

Abstract. Currently, given the complexity of industrial problems, a powerful software is required to solve Constraint Satisfaction Problems. The constraint solvers are a kind of software that are based on a constraint approach. During the last years many constraint solvers have been created, some of them are intricate software and others are libraries to extend the features of a programming language. There are few efforts to have a framework that allows to compare a constraint system and less to allow the usability analysis of the solvers. In most cases, the users of these systems are more concerned about the number of enumeration and propagation strategies that can be used instead of the ease of use of constraint solvers. This paper presents a framework to compare and obtain a simple and objective analysis of the usability of these kind of systems. The paper shows that it is possible to establish comparison in terms of usability, allowing an analysis beyond the simple comparison of their internal strategies.

Keywords: Constraint programming · Constraint solvers · Usability

1 Introduction

The new industrial problems are increasingly difficult to solve, these problems use more complex models with more variables and data. Given the difficulty of these problems is not feasible to solve manually and it is necessary to use complex software to solve them. There is thus a strong need for powerful software tools using a simple user interface. This kind of complex problems are classified as combinatorial problems.

Constraint Programming (CP) [27] is a powerful programming paradigm devoted to the efficient resolution of combinatorial problems. Under this paradigm, a problem needs to be modelled as a Constraint Satisfaction Problem (CSP), which corresponds to a formal representation of the problem. The CSP mainly consist in a set of variables holding a domain and a set of constraints. CSPs are usually resolved by a constraint solver which has a powerful search engine. The search engine finds a proper solution by building and exploring a

© Springer International Publishing AG, part of Springer Nature 2018
G. Meiselwitz (Ed.): SCSM 2018, LNCS 10913, pp. 19–31, 2018.
https://doi.org/10.1007/978-3-319-91521-0_2

search tree. The constraint solver has two main process, the enumeration process and the propagation process. The enumeration process is responsible for assigning permitted values to the variables in order to generate partial solutions to be verified, while propagation aims at deleting from domains values that do not lead to any solution. The constraint solvers have different enumeration and propagation strategies, which are used in the resolution process of the problems [29].

During the last years many constraint solvers have been created [32], some of them are intricate software and others are libraries to extend the features of a programming language. Since different kinds of constraint solver are available, in some cases, it is difficult to objectively decide which constraint solver to use. A proper selection of a solver can be vital to a project. The developer must have a constraint solver which suits your needs. In some cases, these can be simple, using a constraint solver as a black box, in which only it is sufficient to enter and tune different parameters. But in other cases the developer will need a flexible system that allows him to develop more complex models, which is not available only by setting the solver.

Usability is defined as a quality attribute to measure the ease with which a user interacts with the system. The system users generally have different levels of expertise and experience. In software engineering, usability is the degree to which a software can be used by specified consumers to achieve quantified objectives with effectiveness, efficiency, and satisfaction in a quantified context of use [1]. A usability analysis may be conducted by a usability analyst. The usability includes methods of measuring usability, such as needs analysis and the study of the principles behind the perceived efficiency of an object. Usability differs from user satisfaction and user experience because usability does not directly consider usefulness or utility [22].

In the literature there are few works based on the usability of the Constraint Programming systems. In most cases, the studies are based on the performance and the number and kinds of strategies that the solvers implement [30] instead of the ease of use of the constraint solvers.

The main idea of this paper is to present a framework to compare and obtain a simple and objective analysis of the usability of the constraint solver. The proposed framework is based on the usability attributes proposed by Nielsen [22]. With this framework we try to measure attributes such as efficiency, ease-of-use, satisfaction, learnability, memorability. They are identified and related to usability scenarios. To test the simple usability analysis framework to constraint solver, we define the specific features needed, and then a heuristic evaluation can be performed using the proposed framework.

This work shows that it is possible to establish an adequate framework for comparing constraint Solvers in terms of usability, allowing an analysis beyond the simple comparison of their internal strategies.

This paper is organized as follows. Section 2 presents the constraint solvers, Sect. 3 presents some principles of usability the framework. Section 4 presents the framework to Simplify Usability Analysis of constraint solvers. The conclusions are outlined in Sect. 5.

2 The Constraint Solvers

A constraint solver is a Constraint Programming System that implements constraint programming to solve CSP [16]. These solvers can integrate a constraint logic language, a constraint programming libraries, and some languages that support constraint programming. The solvers have different enumeration and propagation strategies, which are used in the process of resolution of the problems.

A constraint solver implements an algorithm for solving allowed constraints in conformity with the constraint theory. The constraint solver collects the constraints that arrive. It puts them into the data structure for constraints (constraint store) and then it tests their satisfiability, simplifies and if possible solves them. When used from within a constraint programming language, a constraint solver should be able to perform the following reasoning services: Satisfiability test where evaluates whether it is feasible to satisfy a constraint. Simplification where tries to transform a given constraint into a logically equivalent, but simpler constraint. Determination where evaluates that a variable in a constraint can only take a unique value, Variable projection elimination where eliminates a variable by projecting a constraint onto all other variables [8].

2.1 Classification of CP Solvers

Traditionally, to modelling and solving CSPs logical languages have been used. The logical languages are declarative and efficient. Moreover, there are various efforts to solve CSP using other languages, for which they have implemented specialized library for the management of CP. These efforts to generate constraint programming systems commonly referred as Constraint solvers, have resulted in specialized compilers or libraries to implement Constraint Programming.

The classification of Constraint solvers can be performed by multiple criteria. In this case we classify according to: logic programming languages, libraries to other imperatives languages, or constructed as specific solvers [6].

Constraint Logic programming languages. A brief description of different logic programming language is presented. These languages are classified into two groups; The Glass-Box and Black-Box. So, first the features for each classification will be explained. The distinction between a Black Box and Glass Box is difficult to establish. The Glass-Box [13] languages provide very simple and primitive constraints, whose propagation scheme can be formally specified. The constraints are used to build high-level constraints, for each application. Moreover, the Black-Box languages provide a wide range of high-level constraints whose implementation is hidden from the user. These constraints perform specific tasks very efficiently. In these languages, it is difficult for a user to add new constraints, as these must be defined at a low level requiring a detailed knowledge of the implementation.

Glass-Box Languages: We can define two types of glass-box languages. These differ in the way that constraint propagation may be defined: either using a single form of relational construct called an indexical or by means of special Constraint Handling Rules (CHR) [7].

An indexical is a reactive functional rule of the form X in R where X is a domain variable. R is a set-valued range expression of the form $t_1 \ldots t_2$ in which terms t_1 and t_2 denote singleton ranges, parameters, integers, combinations of terms using arithmetical operators or indexical ranges.

This constraint can be seen as an abstract machine for propagation-based constraint solving. It is possible to directly encode most of the higher level FD (finite domain) constraints with this one basic constraint. Traditionally among these languages we can find clp(FD) [15], and SICtus [3].

On the other hand the Constraint Handling Rules is a declarative programming language extension introduced in 1991 [9] by Thom Frühwirth. Originally designed for developing (prototypes of) constraint programming systems, CHR is increasingly used as a high-level general-purpose programming language. A CHR languages can define simplification and propagation over user defined constraints.

The application of consecutively CHRs allows to solve the constraints defined by the user. Originally CHRs were created to simplify the constraint languages, but it has spread to build CP solvers for particular applications and domains.

Black-Box Languages: A Black Box is a system such that the user sees only its input and output data: its internal structure or mechanism is invisible to him [11]. This approach partially addresses the requirement for simplicity since the user does not have to be aware of (or modify or extend) embedded techniques and algorithms. However, a Black-Box constraint solver must have a default configuration that yields in most cases the best behaviour that could be obtained by fine tuning of available options. This can be achieved by making the solver robust. One of the most popular black-box languages are Eclipse [21], Oz [28], Ilog SOLVER [14], B-prolog [33].

Constraint programming libraries. A constraint programming library is a tool kit for developing constraint-based systems and applications. These libraries provide a constraint solver with all characteristics of an imperative language.

The constraint programming library differs from constraint logic programming systems like CHIP [5], Eclipse [2] or SICStus Prolog [3] in some topics such us imperative versus rule-based programming, stateful typed variables and objects versus logic variables and terms, no pre-defined search versus built-in depth-first search.

Constraint programming is often realized in imperative programming via a separate library. Some popular libraries for constraint programming are: Choco [4], Gecode [10], IBM ILOG CP [14], JaCop [17], OscaR [26] among others.

Specific solver systems. These correspond to black-box systems, these systems can be implemented using constraint logic programming languages or Constraint programming libraries. This specialized solvers are closed systems that aim to release the user from the complexity of the problem resolution, and only provide an interface to parameterize them. Some of these Constraint solvers can be Abscon [20], Mistral [12], CPHydra [25].

3 Principles of Usability

Usability refers to the user's experience when interacting with a system. A system with good usability is one that shows all the content in a clear and simple way to understand by the user, this is a fundamental aspect of the software. Jakob Nielsen [22], initially defined five basic attributes of usability:

1. Ease of learning: rapidity with which a user learns to use a system with which has not previously had contact (which user does in a simple, fast and intuitive way).
2. Efficiency: the user can achieve a high level of productivity by knowing how to use a system.
3. Retention in time: that the user easily remember how the system was used if he stops using it for a while.
4. User error rates: refers to the amount and severity of errors committed by the user. When committing a fault, the system must inform the user and help him solve it.
5. Subjective satisfaction: refers to whether users feel comfortable and satisfied using the system, that is, whether they like it or not (subjective impression).

Nielsen also defines ten principles of usability, which are useful and easy to verify.

1. System visibility: The system must keep users informed of what is happening, through reasonable periodic feedback.
2. Match between system and the real world: The systems must speak the language of the users, with words, phrases and familiar concepts for the user.
3. User control and freedom: Users often choose options by mistake and clearly need to indicate an exit for those unwanted situations without having to go through extensive dialogues.
4. Consistency and standards: Users do not have to guess that different words, situations or actions mean the same thing.
5. Error prevention: A careful design that prevents problems is better than good error messages.
6. Recognition rather than recall: Make objects, actions and options visible. The user does not have to remember information from one party to another. The instructions for using the system must be visible or easily recoverable.
7. Flexibility and efficiency of use. Design a system that can be used by a wide range of users. Provides instructions when necessary for new users without hindering the path of advanced users.
8. Aesthetic and minimalist design. Do not show information that is not relevant. Each piece of extra information competes with the important one and decreases its relative visibility.

9. Help users recognize, diagnose, and recover from errors. To help users, error messages should be written in simple languages, indicate the problem accurately and show a solution.
10. Help and documentation. The best system is the one that can be used without documentation, but always allows a help or documentation, this information should be easy to find, directed to the tasks of the users, list the concrete steps to do something and be brief.

Usability also used the heuristic evaluation (HE), HE is a usability engineering method "for finding usability problems in a user interface design by having a small set of evaluators examine the interface and judge its compliance with recognized usability principles (the "heuristics")". This method uses evaluators to find usability problems or violations that may have a deleterious effect on the ability of the user interact with the system. Typically, these evaluators are experts in usability principles, the domain of interest, or both. Nielsen and Molich [24] described the HE methodology as "cheap", "intuitive", "requires no advance planning", and finally, "can be used early on in the development process". Often this methodology can be used in conjunction with other usability methodologies to evaluate user interfaces [23].

4 The Proposed Framework

In the literature, there are not studies about usability in constraint solvers, just some works comparing constraint languages and constraint solvers [6,18,31] have been presented. But in all cases, there is an inherent difficulty in trying to compare different systems built it in different environments, languages, and paradigm. For this reason, we propose a framework to simplify usability analysis of constraint solvers and make an objective evaluation based on the usability attributes proposed by Nielsen.

To develop a general framework for different constraint solvers, we must establish some broad criteria, which are not subject to specific conditions. Furthermore comparing different solvers is subjected to factors such as differences in modelling for each solver, different settings, among others [19,32]. Thus we do not consider runtimes, or number of backtracks. we will only establish a simple and clear mechanism to measure constraint solver according to the specific usability features that the evaluator needs.

Our framework is based on the usability attribute proposed by Nielsen and the use of heuristic evaluation to test the usability using a standard test. This framework provides a methodology based on 2 phases; *Design phase*, at this stage the modelling of the test is carried out and the *Evaluation phase*, it is the application of heuristic evaluation. The framework is presented in Fig. 1.

The *Design phase* suggests focusing on starting by modelling the usability of the constraint solver, using the usability measurement model based on a three-level hierarchy. This model defines the usability of constraint solver in terms of: Criteria, metrics and attributes. This can be seen in Fig. 2.

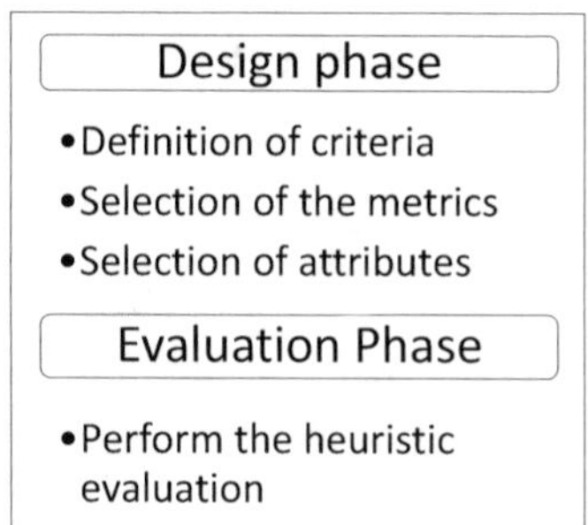

Fig. 1. Phases of the framework for CP solvers

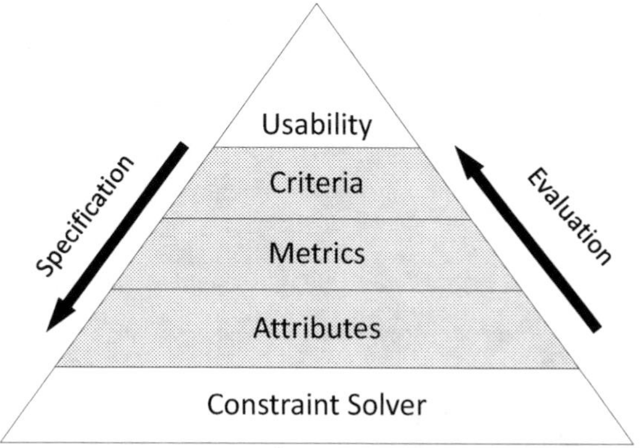

Fig. 2. Model based on a three-level hierarchy

First level: Definition of evaluation criteria. The criteria constitute the parameters for the evaluation of usability at the highest level (first level). The use of criteria refers to the use of a set of specific identifiers and primary characteristics, which allow a critical examination of a Constraint solver.

Second level: Definition of evaluation metrics. In this context, they are defined as two types of arguments; Attribute and measure of the attribute.

Third level: Definition of evaluation attributes. They are metrics that require the definition of attributes and must be declared qualitatively or quantitatively.

The Fig. 2 allows us to visualize the relationship between the levels and processes of usability evaluation. If a solver has more heuristic evaluations made using the three-level hierarchy model, it will have greater usability. And from this new specifications could be generated to modify the usability attributes of the solver. This last part is intended for future corrections and improvements that can be proposed to a solver.

4.1 Design Phase

In this phase, we determine the parameters for the measurement of usability. We have defined a set of criteria that allow us to evaluate the usability, for each criterion a metric is applied and for each metric an attribute is measured. For the specific case of constraint solver, we have taken some criteria defined in the previous section and adapted to be measured according to the features of the solvers. We have defined the following criteria; *Learning* (in Table 1), *Contents* (in Table 2), *Operability* (in Table 3), *Attractiveness* (in Table 4), and *Satisfaction* (in Table 5).

4.2 Evaluation Phase

At the end, of the design phase and once the evaluation guide has been defined, the heuristic evaluation can then be performed. In order to locate problems associated with usability, a heuristic evaluation can then be applied, which allows knowing in depth the constrain solver both functionally and its errors or possible improvements.

Table 1. List of metrics, and attributes associated to criteria Learning

Criteria	Metrics	Attribute
Learning	Ease of learning	Predictive
		Familiarization
		Synthetic
	Help	Consistency between the quality and quantity of help
		Context sensitive help
	Documentation	Access to documentation / tutorials
		Sufficiently explanatory and brief
	Effectiveness	Create a CSP model without help / documentation
		Solve a CSP without help / documentation
		Minimization of execution errors

Table 2. List of metrics, and attributes associated to criteria Contents

Criteria	Metrics	Attribute
Contents	Content to control the enumeration	Data type and data structures
		Variable selection heuristics
		Value selection heuristics
	Content to control the propagation	Definition of constraint
		Create propagators
	Content for cooperation	Integration and portability
		Input/output mechanisms

Table 3. List of metrics, and attributes associated to criteria Operability

Criteria	Metrics	Attribute
Operability	Modelling facility	Definition of constraint
		kind of constraint
		Facility of reification
		Facility to define propagators
		Facility to define value selection heuristics
		Facility to define variable selection heuristics
	Ease of running a model	By command line
		By code embedding
		By call of functions
	Easy to use	Ease of installation
		Simple and clear language
		Allows selection for operating parameters
	Error tolerance	Self-exploratory error messages
		Minimize recovery time
		Facilitates the correction to continue
		Detection and warning of entry errors
	Understanding	Interpretable interface functions
		Clear explanation of input/output actions
		Ease to understand the sequence of answers
		Short messages and simple language
		Clear functions that facilitate recall

Table 4. List of metrics, and attributes associated to criteria attractiveness

Criteria	Metrics	Attribute
Attractiveness	Attractiveness of the interface	Aesthetically pleasing
		Consistent presentation
		Presentation of results in text and graphics
		Combination of color/backgrounds
	Customize	Customization of elements for modelling
		Customizing elements to run CSP
		Changeable elements in the interface

Table 5. List of metrics, and attributes associated to criteria Satisfaction

Criteria	Metrics	Attribute
Satisfaction	Reliability	Texts and messages are easy to read
		Simple and pleasant overall appearance
		Allows access to help comfortably
		The results are clearly presented
	Acceptability	Update mechanisms
		Multiple functionalities
		On-line support

Heuristics evaluation tasks. To heuristic evaluation to be effective, that is, the greatest possible number of usability errors are found, it is recommended to perform a series of tasks, which are presented below:

- Study previously the constraint solver to familiarize yourself with it.
- Determine the usability parameters established in the design phase.
- Define, for each of the parameters, a series of questions to determine if they are met. Make an evaluation guide where each of the questions has the frequency with which the problem appears as well as its impact. The proposed criteria to estimate the impact of each of the questions is shown in the Table 6.
- Perform the heuristic evaluation of the tool using the guide.

Then it is proposed to make a selection of users. The selection of users is a fundamental element in the evaluation process. In the selection can be considered users with knowledge of the constraint solver, given the specific use of these software. Finally, in the framework a review and analysis of data is proposed. A systematic analysis of the data must be done in order to prepare a report detailing the problems and possible solutions applicable to solver.

Table 6. Definition of impact

Impact	Explanation
Low (1)	Although it is recommended that the statement be fulfilled, its non-compliance does not imply confusion or error in the user. It would not give important usability problems.
Medium (2)	Failure to comply can cause not very serious problems of usability although it is convenient to solve them since it would facilitate the operation of the system.
High (3)	Produces significant problems of understanding and functionality in the system so it is essential that the problem is solved. It can cause serious usability problems

5 Conclusion

Currently, there is a variety of constraint solvers. These can be of different types and cover different objectives. For this reason, it is difficult for a specialist to decide which solver to use in a particular project. On the other hand, there are no works that propose to evaluate any of these systems in terms of usability. For this reason, we propose a framework to simplify usability analysis of constraint solvers and make an objective evaluation based on the usability attributes proposed by Nielsen.

We presented a classification of the different constraint solvers. Later we presented the structure of the proposed framework. This framework is characterized by using two phases: a design phase, in which it is modelling the usability of the constraint solver, using the usability measurement model based on a three-level hierarchy. This model defines the usability of constraint solver in terms of: Criteria, metrics and attributes. Later it is defined the evaluation phase which consists in conduct the experimental evaluation, in this phase a heuristic evaluation is performed.

Although no more extensive work has been done, this framework aims to provide a general guide to analyse the usability of constraint solvers, delivering a series of criteria, metrics and attributes specially adapted for this type of system. In future work, the tests should be done and see if they are significant for the use of solvers.

Acknowledgments. Broderick Crawford is supported by Grant CONICYT/FONDECYT/REGULAR/1171243, Ricardo Soto is supported by Grant CONICYT/FONDECYT/REGULAR/1160455

References

1. ISO 9241–11: Ergonomic requirements for office work with visual display terminals (VDTs) - Part 11: Guidance on usability. International (1998)
2. Apt, K.R., Wallace, M.: Constraint Logic Programming Using Eclipse. Cambridge University Press, New York (2007)
3. Carlsson, M., Mildner, P.: Sicstus prolog - the first 25 years. CoRR abs/1011.5640 (2010)
4. choco Team. choco: an open source Java constraint programming library. Research report 10-02-INFO, École des Mines de Nantes (2010)
5. Dincbas, M., Van Hentenryck, P., Simonis, H., Aggoun, A., Herold, A.: The CHIP system: constraint handling in Prolog. In: Lusk, E., Overbeek, R. (eds.) CADE 1988. LNCS, vol. 310, pp. 774–775. Springer, Heidelberg (1988). https://doi.org/10.1007/BFb0012892
6. Fernández, A., Hill, P.: A comparative study of eight constraint programming languages over the boolean and finite domains. Constraints **5**(3), 275–301 (2000)
7. Fruhwirth, T.: Theory and practice of constraint handling rules. J. Logic Program. **37**(1–3), 95–138 (1998)
8. Frühwirth, T., Abdennadher, S.: Principles of constraint systems and constraint solvers (2005)

9. Frühwirth, T., Raiser, F. (eds.) Constraint Handling Rules: Compilation, Execution, and Analysis, March 2011
10. Gecode Team. Gecode: Generic constraint development environment (2006). http://www.gecode.org
11. Gent, I.P., Jefferson, C., Miguel, I.: MINION: a fast scalable constraint solver. In: Proceedings of ECAI 2006, Riva del Garda, pp. 98–102. IOS Press (2006)
12. Hebrard, E., Siala, M.: Mistral 2.0. LAAS-CNRS, Universite de Toulouse, CNRS, Toulouse, France, XCSP3 Competition (2007)
13. Hentenryck, P.V., Saraswat, V., Deville, Y.: Design, implementation, and evaluation of the constraint language cc(FD). J. Logic Program. **37**(1–3), 139–164 (1998)
14. IBM company: Ibm ilog cp (2006)
15. Jaffar, J., Michaylov, S., Stuckey, P.J., Yap, R.H.C.: The CLP(R) language and system. ACM Trans. Program. Lang. Syst. **14**(3), 339–395 (1992)
16. Mariott, K., Stuckey, P.: Programming with Constraints: An Introduction. MIT Press, London (1998)
17. Kuchcinski, K., Szymanek, R.: Jacop library user's guide (2010). http://jacopguide.osolpro.com/guideJaCoP.html
18. Lazaar, N., Gotlieb, A., Lebbah, Y.: A CP framework for testing CP. Constraints **17**(2), 123–147 (2012)
19. Lecoutre, C., Roussel, O., van Dongen, M.: Promoting robust black-box solvers through competitions. Constraints **15**(3), 317–326 (2010)
20. Lecoutre, C., Tabary, S.: Abscon 109 A generic CSP solver (2006)
21. Niederliński, A.: A gentle guide to constraint logic programming via eclipse (2012)
22. Nielsen, J.: Usability 101: Introduction to usability. Nielsen Norman Group, 4 January 2012
23. Nielsen, J., Molich, R.: Teaching user interface design based on usability engineering. SIGCHI Bull. **21**(1), 45–48 (1989)
24. Nielsen, J., Molich, R.: Heuristic evaluation of user interfaces. In: Proceedings of the SIGCHI Conference on Human Factors in Computing Systems, CHI 1990, pp. 249–256. ACM, New York (1990)
25. O'mahony, E., Hebrard, E., Holland, A., Nugent, C.: Using case-based reasoning in an algorithm portfolio for constraint solving. In: Iris Conference on Artificial Intelligence and Cognitive Science (2008)
26. OscaR Team. OscaR: Scala in OR (2012). https://bitbucket.org/oscarlib/oscar
27. Rossi, F., van Beek, P., Walsh, T.: Handbook of Constraint Programming. Elsevier, Amsterdam (2006)
28. Smolka, G.: The development of Oz and Mozart. In: Van Roy, P. (ed.) MOZ 2004. LNCS, vol. 3389, p. 1. Springer, Heidelberg (2005). https://doi.org/10.1007/978-3-540-31845-3_1
29. Soto, R., Crawford, B., Olivares, R., Galleguillos, C., Castro, C., Johnson, F., Paredes, F., Norero, E.: Using autonomous search for solving constraint satisfaction problems via new modern approaches. Swarm Evol. Computat. **30**, 64–77 (2016)
30. Soto, R., Crawford, B., Palma, W., Galleguillos, K., Castro, C., Monfroy, E., Johnson, F., Paredes, F.: Boosting autonomous search for CSPs via skylines. Inf. Sci. **308**, 38–48 (2015)
31. Tulácek, M.: Constraint solvers, bachelor thesis, Charles university in Prague (2009)

32. Wallace, M., Schimpf, J., Shen, K., Harvey, W.: On benchmarking constraint logic programming platforms. Response to Fernandez and Hill's "a comparative study of eight constraint programming languages over the boolean and finite domains". Constraints **9**(1), 5–34 (2004)
33. Zhou, N.-F.: The language features and architecture of B-Prolog. Theory Pract. Log. Program. **12**(1–2), 189–218 (2012)

Online Ethnography Studies in Computer Science: A Systematic Mapping

Andrei Garcia(✉), Bruna Pereira De Mattos(✉), and Milene Selbach Silveira(✉)

PUCRS, Faculdade de Informática, Porto Alegre, Brazil
{andrei.garcia, bruna.mattos}@acad.pucrs.br, milene.silveira@pucrs.br

Abstract. During the last two decades, online environments became rich grounds for ethnographic studies. In the same period, online communities have become a popular and broadly studied research topic. Along with online environments, the growth of online communities brought by the Computer-Mediated Communications created a solid research field for online ethnography studies. Online ethnography methods, such as virtual ethnography and netnography, are widely adopted for qualitative research. However, it is not clear how Computer Science field is using online ethnography for empirical studies. Thus, the main goal of this study is to present how online ethnographic studies have been performed in Computer Science. To accomplish this goal, we carried out a systematic mapping study regarding empirical studies on online environments. Through the analysis of 36 resulted papers, this systematic mapping provides a broad overview of existing online ethnography studies in Computer Science and by identifying how these studies have been performed considering adopted methods, collected and analyzed data, community characteristics, and researcher participation throughout these empirical studies.

Keywords: Online ethnography · Online communities · Systematic mapping

1 Introduction

The majority of ethnographic studies are related to direct observation. However, interviews, questionnaires, and studying artifacts used in activities also feature in ethnographic studies [1]. The basic tenets of ethnography are the recursive and inductive depth observation of a culture or a community as well as open-ended interviews designed to understand the perspectives of community's participant [2]. In order to help shape researchers' participant depth observation, some ethnographic procedures are used, such as making cultural entrée, gathering, and analyzing data, ensuring reliable interpretation, conducting ethical research, and providing an opportunity for member feedback. Furthermore, these procedures are completely known in ethnographies conducted in face-to-face situations [3].

During the last two decades, online environments became rich and vital grounds for ethnographic studies [2]. In the same period, online communities have become one of the most popular forms of online services [4]. Online communities are essentially forums for

© Springer International Publishing AG, part of Springer Nature 2018
G. Meiselwitz (Ed.): SCSM 2018, LNCS 10913, pp. 32–45, 2018.
https://doi.org/10.1007/978-3-319-91521-0_3

meeting and communicating with others [5], or in a more detailed definition, online communities are web-based online services with features that make the communication among members possible [4]. Along with online environments, the growth of online communities brought by the Computer-Mediated Communications (CMC) created a solid research field for online ethnography studies [6].

Online ethnography adopts principles of ethnographic research molded in offline environments and applies them to online environments with necessary adjustments [2]. According to Kozinets [7], online ethnography is a generic term for performing any ethnographic research by using some sort of digital or online environment. Methods such as netnography [8] and virtual ethnography [9] are widely adopted for qualitative research. However, it is not clear how Computer Science domain is using online ethnography for empirical studies. Thus, the main goal of this study is to present how online ethnographic studies have been performed in Computer Science.

To achieve this goal, we conducted a systematic mapping study regarding this methodological approach on Computer Science. In the mapping herein presented, we focused on empirical studies using any online ethnography method that were carried out on Computer Science discipline. In order to contextualize our findings, we present the background about online ethnography methods in the next section. Afterwards, we delineate the Research Method, including the research protocol. Next, in the Results section, we present the report and our results' analysis in order to answer our questions. Finally, we state our discussions and conclusion.

2 Background

Ethnography is a qualitative orientation to research, which emphasizes the detailed observation of people in natural environments. Ethnography seeks to present a picture of life seen and understood by those living and working within the domain in question, through direct involvement of the researcher in the environment under investigation [10]. The emergence of social media on the Internet provides qualitative researchers with a new window into people's outer and inner worlds, their experiences and their understanding of these [11]. The Internet has created many types of online communities that not only exist in cyberspace, but can also be studied through the internet itself. As collaboration and social activities became online, ethnographers adjusted their strategy to take into account computer-mediated communication [12]. This movement had several names, the most common being: Online ethnography [1], virtual ethnography [9], or netnography [13].

Developed by Robert Kozinets, netnography is a qualitative research methodology which adapts ethnography research processes to study cultures and communities that are emerging through CMC [3]. According to Kozinets [7], online ethnography is a generic term for performing any ethnographic research by using some sort of digital or online environment. As stated by Bengry-Howell et al. [14], the case study of netnography sits within a broader methodological context of online ethnography. Online ethnography encompasses approaches for conducting ethnographic studies of online communities. Normally, online ethnography includes observation of *postings* and *threads* within an online forum and interviews with an online community. However, it can implicate in data collection online as well as offline [9].

Another online ethnographic method is virtual ethnography [9]. Virtual ethnography is a form of ethnography for studying online communities based on textual data [15]. However, it appears to allow for a composition of online and offline ethnographic approaches to have an understanding of the online phenomena [16]. Meanwhile, netnography addresses online interactions and differ from other online ethnography methods by offering a more systematic, defined approach to addressing ethical, procedural and methodological issues specific to online research [17]. Nonetheless, both methods have been applied on computer science discipline.

One example of virtual ethnography application is Margaret and Walt's study [18]. In this research, the authors conducted an extensive virtual ethnography collecting data over a period of four years. Their goals were to deeper understand the ideology and work practices of free and open source software development, which is valuable to software developers and managers who wish to incorporate open source software into their companies. As an example of netnography, Di Guardo and Castriotta [19], applied an exploratory qualitative case study using the netnography method in order to analyze the open innovation experience and crowdsourcing of a large Italian company. Their results imply the effective use of collective knowledge in innovation processes. Besides software engineer applications, some studies also apply online ethnography for human-computer interaction domain, that is the case of Hussein, Mahmud, and Noor [20].The authors conducted netnographic approaches to investigate frustrations among practitioners while incorporating user experience design discipline in software development processes. Their findings provide insights to improve user experience design processes. Therefore, in order to present an overview of how online ethnographic studies have been performed in Computer Science, we conducted a systematic mapping that is detailed in the next section.

3 Research Method

This study was carried out by following the established guidelines for conducting Systematic Mapping Studies suggested by Petersen et al. [21]. A Systematic Mapping Study is a method designed to provide a wide overview of a research field by exploring the research data existence and by providing the amount and classification of such research data [22]. According to Petersen, the mapping process consists of planning, conducting, and reporting. Next sub-sections detail how each phase was performed from planning to conduction, delineating the research questions, search strategy, selection criteria, and data extraction strategy. In addition, the report is presented in the results section.

3.1 Planning

Before conducting the systematic mapping, we had forethought the research questions and establish the research protocol. The protocol was delineated considering the steps of search strategy, selection criteria, and data extraction strategy. As stated by Kitchenham

[22], a research protocol is essential for the sake of reducing chances of researcher's bias.

Research Questions. The main goal of this mapping is to present how online ethnographic studies have been performed in Computer Science. To accomplish this goal, we defined the following three research questions:

- RQ1 - Which areas of Computer Science have been using online ethnography research method?
- RQ2 - Where are online ethnography studies published?
- RQ3 - How are online ethnography studies performed?

By answering these research questions, this study provides an overview of how online ethnographic studies have been performed in Computer Science and we can understand where these studies are headed.

Search Strategy. Search strategy comprises the identification of search terms for querying applicable scientific databases. Seven relevant Computer Science databases were selected for the search: ACM Digital Library, EBSCO Host, Elsevier Science-Direct, IEEE Xplore, ProQuest, Springer Link, and Web of Science. The search string was composed based on well-known online ethnographic research methods such as netnography [13], virtual ethnography [9], webnography [23], and cyber-ethnography [24]. Therefore, in order to automate the search in the selected databases, we defined the following search string with their corresponding logical operators: "*online ethnography" OR netnography OR "virtual ethnography" OR webnography OR "cyber-ethnography*".

In addition, Springer and Web of Science databases provide a mechanism to filter by the discipline of Computer Science, which was helpful and returned more accurate results. For all other selected bases, the filter per discipline was performed manually since they do not provide an interface to refine the search considering the discipline. Furthermore, names of the computer science disciplines were not added as part of the search criteria in order to comprehend all possible computer science areas and avoid inaccurate results.

Selection Criteria. We assessed each publication returned from the automated search after selecting whether or not it should be included by considering the selection criteria. The selection criteria were composed by inclusion and exclusion criteria. In a first filter, we included/excluded papers based only on titles and abstracts. In a second filter, we ensured a full-text reading. Thus, the following inclusion criteria were applied in the first filter:

- Studies should be published in the computer science area.
- Studies should present reference(s) of use of online ethnography methods.

Publications that met at least one of the following exclusion criteria were removed:

- Books.
- Duplicated papers.
- Studies written in any other language other than English.
- Studies presenting summaries of tutorials, panels, poster sessions or workshops.

- Conference covers and table of content.

During the full-text reading stage, we analyzed all paper content. The goal of this stage was to select the studies according to the following inclusion criteria:

- Studies should present references of online ethnography methods application, being that a unique method or part of a mixed method.
- Studies should describe the methodology application.

Data Extraction Strategy. The data extraction strategy was based on defining a data set that should be collected in order to answer the research questions. RQ1 could be answered by defining the Computer Science area or sub-discipline which the study belongs to, such as User Interfaces and Human Computer Interaction, Software Engineering, and so on. RQ2 and RQ3 data set are composed of a conjunction of data as shown in Table 1.

Table 1. Data extraction for each research question.

Research question	Data set	Examples/Details
Which areas of Computer Science have been using online ethnography research method?	Computer Science areas	User Interfaces and Human Computer Interaction; Software Engineering; Database Management; ...
Where are online ethnography studies published?	Title Content Type Content Type name Year Author(s)	Study's title Journal or Conference Journal's or Conference's name Study's year Study's author(s)
How are online ethnography studies performed?	Research methodology Mixed Methods (if any) Application domain Number of communities Community size Timeframe Collected data Researcher involvement	Netnography, Virtual Ethnography, etc. Netnography + Survey + Interview, etc. Human Behavior, UX, Robotics, etc. Number of included online communities Number of community members Study's timeframe Text, Video, Image, etc. Active or Passive

3.2 Conduction

We searched for papers in the selected databases during April 2017. The first results led us to a set of 853 studies (Table 2). After the results' compilation, we applied the exclusion criteria, resulting in 762 publications. Afterwards, a total of 62 were selected in accordance with the inclusion criteria from the first stage, where only the title and abstract were considered. Finally, in the full-text reading stage, 36 publications were selected. The selection process is shown in Fig. 1.

Table 2. Number of publications per database

Database	Search	Inclusion/Exclusion criteria	Final set
ACM	17	7	3
EBSCO	89	6	3
Elsevier	362	17	6
IEEE	16	13	10
ProQuest	232	9	6
Springer	84	5	3
Web of Science	53	5	5
Total	853	62	36

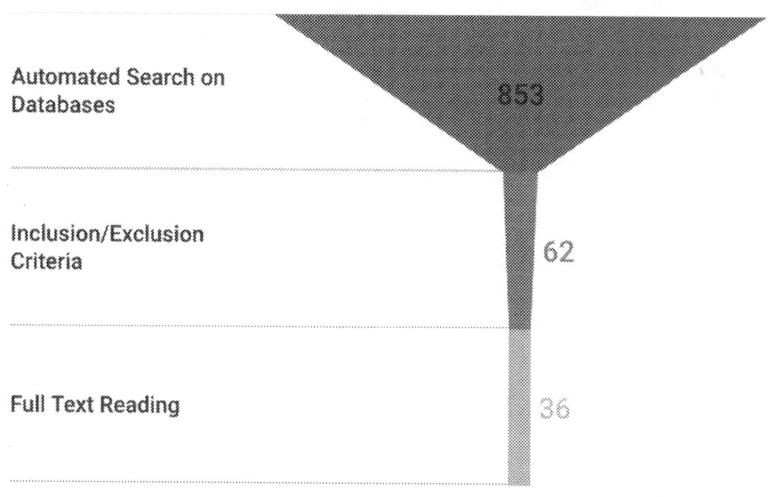

Fig. 1. Selection process

4 Results

4.1 Computer Science Areas Applying Online Ethnography Methods

The results for question RQ1 – Which areas of Computer Science have been using online ethnography research method? – revealed that 83% of result set studies applying

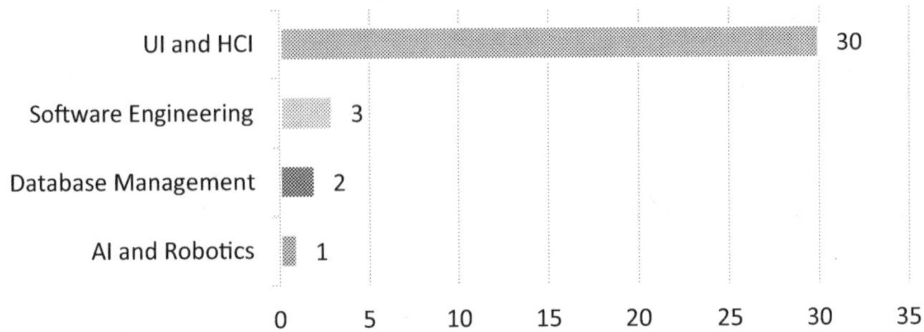

Fig. 2. Computer Science areas applying online ethnography methods

an online ethnographic method are classified in *User Interfaces and Human Computer Interaction* area. The remaining studies are categorized in *Software Engineering*, *Database Management*, and *Artificial Intelligence and Robotics* areas, as shown in Fig. 2.

4.2 Published Online Ethnography Studies

Results for question RQ2 – Where are online ethnography studies published? – revealed that 55.60% of result set studies are published as articles in journals and 44.40% are conference proceedings. All studies' references are shown in Table 3, which also shows the number of publications per year. The complete list of periodic and conference names is displayed in Table 4.

Table 3. Selected papers

Year	References
1999	[24]
2000	[5]
2008	[25, 18, 26]
2009	[27–29]
2010	[30]
2011	[31, 15]
2012	[32–35]
2013	[36, 19, 37–39]
2014	[40–46]
2015	[47–49, 16]
2016	[20, 50, 51]
2017	[52, 11]

Table 4. Periodic and Conference names where studies are published

Type	Name
Periodic	Bulletin of Science, Technology & Society
	Calico Journal
	Computers in Human Behavior
	Ethics and Information Technology
	Identity in the Information Society
	Information and Organization
	Information and Software Technology
	Information Systems Journal
	Information Systems Research
	Information Technology & People
	International Journal of Electronic Commerce Studies
	International Journal of Technology Management
	Journal of Computer-Mediated Communication
	Journal of Documentation
	Journal of Information Technology
	Journal of the Association of Information Systems
	Online Information Review
	Procedia Computer Science
	Procedia Technology
Conference	Computer Science and Electronic Engineering Conference
	Extended Abstracts on Human Factors in Computing Systems
	Hawaii International Conference on System Science
	International Conference in HCI and UX
	International Conference on Advanced Learning Technologies
	International Conference on Advances in Social Networks Analysis and Mining
	International Conference on Computing, Communication and Security
	International Conference on Well-Being in the Information Society
	International Multi-Conference on Society, Cybernetics, and Informatics
	International Professional Communication Conference
	International Scientific Conference eLearning and software for Education
	International Symposium on Open Collaboration
	International Symposium on Robot and Human Interactive Communication
	Panhellenic Conference on Informatics

4.3 How Online Ethnography Studies Are Performed

Considering the applied methodology, the results for RQ3 – How are online ethnography studies performed? – exposed that the majority of the studies on Computer Science (86.1%) followed virtual ethnography and netnography methods. Only one study adopted cyber-ethnography method and four studies called specifically online ethnography with no distinction for a specific method. Figure 3 shows the adopted methods on the selected set of studies.

Another outcome related to studies' methodology is that 15 studies employed mixed methods, using virtual ethnography or netnography plus interviews, surveys or experiments. Most of these studies applied two methods, except for Rozas' study [40],

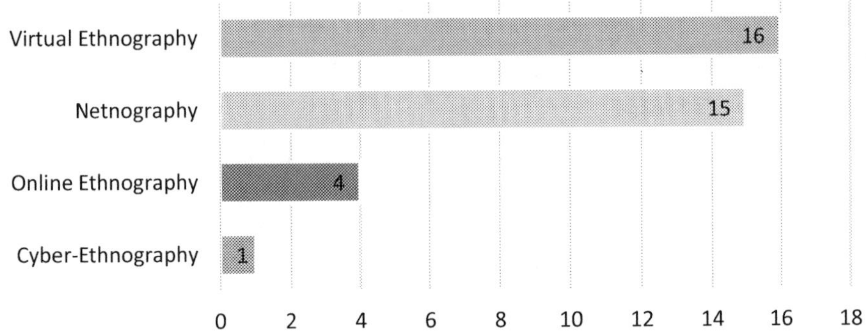

Fig. 3. Adopted methods

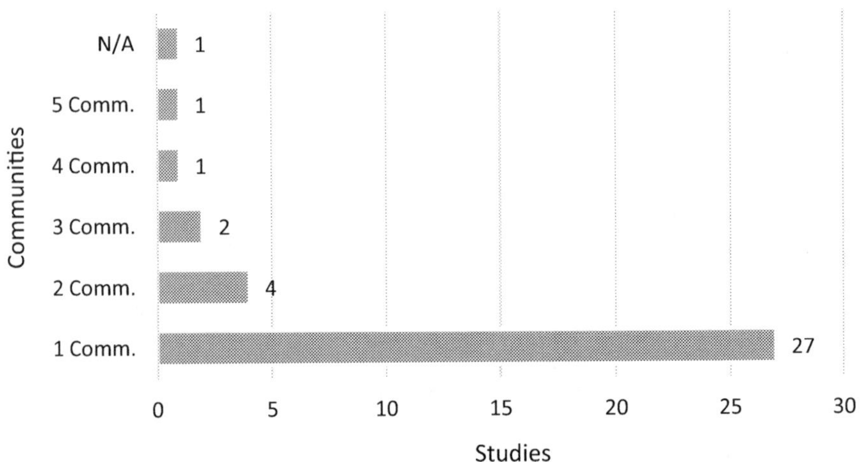

Fig. 4. Number of communities per study

which applied virtual ethnography, interviews, and survey, and Bauer, Franke, and Tuertscher's study [46] which applied netnography, survey, and experiment.

As stated by Bengry-Howell et al. [14] researchers have used online ethnography methods to study a particular online community, which is aligned with our mapping outcome. The majority of selected papers have used only one community to perform their studies. Only eight studies have adopted two or more communities to perform their studies, and one study has not provided this information. Figure 4 details the number of communities per study. In addition, the number of community's members, for those studies that informed this data, vary from a few members [33] to more than 1 million registered members [40].

The period performing an online ethnography study, for those studies that asserted this information, vary from 1 week [36] to 5 years [18]. One year or less is the most common period, stated in 17 studies. Two studies conducted a 2 years research, and other two studies conducted a 4 years research. While running an online ethnography research, the collected and analyzed data is mostly text-based. All studies have collected and analyzed text-based data. However, besides text, some studies also collected and analyzed videos [15, 31, 36] and images [11, 51, 46].

Depending on the participation of the researchers in the community, an online ethnography research can vary from non-participatory (passive) to participatory (active) [17]. The results from our mapping show that 58.7% of the researchers played as passive, while 41.3% participated as active.

5 Discussion

The primary goal of this systematic mapping study is to present how online ethnographic studies have been performed in Computer Science. Based on our analysis of 36 resultant papers, it is evident that Human Computer Interaction is the Computer Science area that most takes advantage of online ethnographic methods. Furthermore, these studies have been published in diverse conferences and periodic. However, the main thoughts and considerations were bounded around the online ethnographic methods approaches that have been used in Computer Science studies.

The majority of reviewed studies have adopted virtual ethnography or netnography methods to achieve their goals. For instance, Sigfridsson and Sheehan [15], used virtual ethnography method for studying free and open source software communities, which contributed assessing multiple and interlinked dimensions and interpreting the context of communities' activities. Another example is Synnott, Coulias, and Ioannou study [52], which applied virtual ethnography method as part of their multi-method approach to provide a case study analysis of a group of alleged Twitter trolls. In their case, the method provided the research engagement as observational and participatory in a specific online community. Additionally, Teixeira [45], has applied netnography method to delineate how patients use open source disease control software developed by other patients. Despite the fact that netnography has his roots in Marketing discipline, it has been adopted by other disciplines, including Computer Science.

Since online ethnography methods adopt principles of ethnographic research, such as user observation and researcher participation, they can easily be part of a mixed-method approach being used in combination with interviews, surveys or experiments for example. Online ethnography combined with interviews can provide a deep understanding of a specific raised theme.

Findings of this study show that most of the resulted publications focused to study a unique online community, depending on the particular researcher's interest and mainly on the research goal. However, there is no right or wrong regarding the number of communities included in a study, but it is important to bear in mind the criteria to select the appropriated community to perform the research. In general, as stated by Kozinets [3], online communities should be selected to have a focused topic relevant to the research question, higher number of posts and interactivity, heterogeneity, and rich in data.

The interactivity and number of posts commonly depend on the numbers of community's members. The number of members in a community vary from study to study and is related to the study goals and the selection criteria used to select the community. Another consideration is the period of time performing this sort of qualitative study. Such methods require researcher's immersion into the online community long enough to become familiar with the community's culture [53, 3].

After the researcher becomes familiar with community's culture, it is possible to begin collecting data. Since online communities data is predominantly text-based, researchers can benefit from the practically automatic transcription of gathered posts [3]. While all resulted studies from this mapping collected and analyzed text-based data, few studies additionally explored videos and images as part of their data collection and analysis. Furthermore, there are two important elements of data collection, which involves the straight gathered data from online communities members' communication, and the data the researchers address related to their participant observations and interactions with members' community [3].

Related to researchers' participation in the online communities, the applied methods can vary from passive participation to active participation. A passive participation means that the researcher is a member of the community but observes the group without interacting with people. On the other hand, an active participation implies that the researcher is actively engaged and involved in community's activities [54]. To conclude, active researcher participation aid to obtain rich data but it is not always an easy process.

6 Conclusion

In the research herein presented, we focused on how online ethnographic studies have been performed in Computer Science. Through a systematic mapping study about online ethnography methods, we deepen our understanding not only about the domain areas on Computer Science but also about the main processes applying these methods.

The mapping study presented that four Computer Science areas have been using some sort of online ethnography method, being them *User Interfaces and Human Computer Interaction*, *Software Engineering*, *Database Management*, and *Artificial Intelligence and Robotics* areas (with 30, 3, 2, and 1 citation(s), respectively). In addition, from the mapped online ethnography methods, we can highlight virtual ethnography and netnography, which also are often used in mixed-methods in combination with interviews, surveys, and experiments for example. Furthermore, the community selection is an important stage, where the researcher shall bear in mind its relevance to the research goals, activity, interactivity, heterogeneity, and rich in data. Additionally, the number of members in a community and period of time performing these qualitative methods vary according to each study. To complete, the researcher participation can be passive, when the researcher does not interact with the community, or active when the researcher interacts with the community members. For both researcher participation modes, the data collection and analysis are mostly grounded on text-based data, but it can also be supported by video and images.

The analyzed studies show us that the data analysis can be a challenge due to the large volume of data collected on online communities. Even when the researcher participates as passive, it is important to use a qualitative data analysis software to organize and filter the data. In addition, the use of online communities leads to ethical challenges for qualitative research, which is another perspective to be studied and we shall extend our understanding.

Acknowledgment. These results were achieved in cooperation with HP Brasil Indústria e Comércio de Equipamentos Eletrônicos LTDA. using incentives of Brazilian Informatics Law (Law no 8.2.48 of 1991).

References

1. Preece, J., Rogers, Y., Sharp, H.: Interaction Design: Beyond Human-Computer Interaction. Wiley, New York (2011)
2. Rotman, D., Preece, J., He, Y., Druin, A.: Extreme ethnography: challenges for research in large scale online environments. In: iCOnference, pp. 207–214 (2012)
3. Kozinets, R.V.: The field behind the screen: using netnography for marketing research in online communities. J. Mark. Res. **39**, 61–72 (2002)
4. Malinen, S.: Understanding user participation in online communities: a systematic literature review of empirical studies. Comput. Human Behav. **46**, 228–238 (2015)
5. Bakardjieva, M., Feenberg, A.: Involving the virtual subjects. Ethics Inf. Technol. **2**, 233–240 (2001)
6. Loanzon, E., Provenzola, J., Siriwannangkul, B., Al Mallak, M.: Netnography: evolution, trends, and implications as a fuzzy front end tool. In: 2013 Proceedings of the Technology Management in the IT-Driven Services-13, PICMET 2013, pp. 1572–1593 (2013)
7. Kozinets, R.V.: Netnography: Redefined. Sage Publications Inc., Thousand Oaks (2015)
8. Kozinets, R.V.: Netnography: Doing Ethnographic Research Online. SAGE Publications, Los Angeles (2010)
9. Hine, C.: Virtual Ethnography. SAGE Publications, London (2000)
10. Randall, D., Rouncefield, M.: The Encyclopedia of Human-Computer Interaction: Ethnography (2012)
11. McKenna, B., Myers, M.D., Newman, M.: Social media in qualitative research: challenges and recommendations. Inf. Organ. **27**, 87–99 (2017)
12. Dicks, B., Mason, B., Coffey, A., Atkinson, P.: Qualitative Research and Hypermedia: Ethnography for the Digital Age. SAGE Publications, London (2005)
13. Kozinets, R.V.: Netnography: Doing Ethnographic Research Online (2010)
14. Bengry-Howell, A., Wiles, R., Nind, M., Crow, G.: A Review of the Academic Impact of Three Methodological Innovations: Netnography, Child-Led Research and Creative Research Methods (2011)
15. Sigfridsson, A., Sheehan, A.: On qualitative methodologies and dispersed communities: reflections on the process of investigating an open source community. Inf. Softw. Technol. **53**, 981–993 (2011)
16. Kulavuz-Onal, D.: Using netnography to explore the culture of online language teaching communities. CALICO J. **32**, 426–448 (2015)
17. Costello, L., Mcdermott, M., Wallace, R.: Netnography: range of practices, misperceptions, and missed opportunities. Int. J. Qual. Methods. **16**, 1–12 (2017)

18. Elliott, M.S., Scacchi, W.: Mobilization of software developers: the free software movement (2008)
19. Di Guardo, M.C., Castriotta, M.: The challenge and opportunities of crowdsourcing web communities: an Italian case study. Int. J. Electron. Commer. Stud. **4**, 79–92 (2013)
20. Hussein, I., Mahmud, M., Noor, N.L.M.: Netnography approach for UX research. In: Proceedings of the CHIuXiD 2016, 2nd International Human Computer Interaction and User Experience Conference in Indonesia: Bridg. Gaps HCI UX World, pp. 120–124 (2016)
21. Petersen, K., Vakkalanka, S., Kuzniarz, L.: Guidelines for conducting systematic mapping studies in software engineering: an update. Inf. Softw. Technol. **64**, 1–18 (2015)
22. Kitchenham, B., Charters, S.: Guidelines for performing systematic literature reviews in software engineering version 2.3. Engineering **45**, 1051 (2007)
23. Horster, E., Gottschalk, C.: Computer-assisted Webnography: a new approach to online reputation management in tourism. J. Vacat. Mark. **18**, 229–238 (2012)
24. Ward, K.J.: Cyber-ethnography and the emergence of the virtually new community. J. Inf. Technol. **14**, 95–105 (1999)
25. Broillet, A., Dubosson, M., Trabichet, J.-P.: An Internet based distribution strategy of luxury products and services grounded on qualitative Web discourse analysis. In: 2008 IEEE International Professional Communication Conference (2008)
26. Fisher, W.: Digital ink technology for e-Assessment. In: Imsci 2008 2nd International Multi-Conference Society, Cybernetics and Informatics, Proceedings, vol. 1, pp. 4–5 (2008)
27. Banakou, D., Chorianopoulos, K., Anagnostou, K.: Avatars' appearance and social behavior in online virtual worlds. In: 2009 13th Panhellenic Conference on Informatics, vol. 5, pp. 207–211 (2009)
28. Baker, A.J.: Mick or Keith: blended identity of online rock fans. Identity Inf. Soc. **2**, 7–21 (2009)
29. Jacobsson, M.: Play, belief and stories about robots: a case study of a pleo blogging community. In: Proceedings IEEE International Workshop on Robot and Human Interactive Communication, pp. 232–237 (2009)
30. Bach, P.M., Carroll, J.M.: Characterizing the dynamics of open user experience design : the cases of firefox and OpenOffice. org. J. Assoc. Inf. Syst. **11**, 902–925 (2010)
31. Kongmee, I., Strachan, R., Pickard, A., Montgomery, C.: Moving between virtual and real worlds: second language learning through massively multiplayer online role playing games (MMORPGs). In: 2011 3rd Computer Science and Electronic Engineering Conference, CEEC 2011, pp. 13–18 (2011)
32. Lingel, J.: Ethics and dilemmas of online ethnography. In: Proceedings of the 2012 ACM Annual Conference Extended Abstracts on Human Factors in Computing Systems Extended Abstracts, CHI EA 2012, pp. 41–50 (2012)
33. Gray, K.L.: Diffusion of innovation theory and xbox live: examining minority gamers' responses and rate of adoption to changes in xbox live. Bull. Sci. Technol. Soc. **32**, 463–470 (2012)
34. Ferreira, A.O., Ferreira, S.B.L., Da Silveira, D.S.: Accessibility for people with cerebral palsy: the use of blogs as an agent of social inclusion. Procedia Comput. Sci. **14**, 245–253 (2012)
35. Ko, H.: Why are A-list bloggers continuously popular? Online Inf. Rev. **36**, 401–419 (2012)
36. Emad, S., Broillet, A., Halvorson, W., Dunwell, N.: The competency building process of human computer interaction in game-based teaching: adding the flexibility of an asynchronous format. IEEE Int. Prof. Commun, Conf (2013)
37. Manasia, L.: The impact of social software on developing communities of practice to enhance adult learning. In: The International Scientific Conference eLearning and Software for Education, pp. 598–603. "Carol I" National Defence University, Bucharest (2013)

38. Pihl, C., Sandström, C.: Value creation and appropriation in social media - the case of fashion bloggers in Sweden. Int. J. Technol. Manag. **61**, 309 (2013)
39. Marciano, A.: Living the VirtuReal: negotiating transgender identity in cyberspace. J. Comput. Commun. **19**, 824–838 (2014)
40. Rozas, D.: Drupal as a commons-based peer production community. In: Proceedings of the International Symposium on Open Collaboration, OpenSym 2014, pp. 1–2 (2014)
41. Burford, S., Park, S.: The impact of mobile tablet devices on human information behaviour. J. Doc. **70**, 622–639 (2014)
42. Marshall, A.: Sensemaking in second life. Procedia Technol. **13**, 107–111 (2014)
43. Ponti, M.: Hei mookie! Where do i start? The role of artifacts in an unmanned MOOC. In: Proceedings of the Annual Hawaii International Conference on System Science, pp. 1625–1634 (2014)
44. Rollins, M., Wei, J., Nickell, D.: Learning by blogging: understanding salespeople's learning experiences on social media. In: Proceedings of the Annual Hawaii International Conference on System Science, pp. 1656–1665 (2014)
45. Teixeira, J.: Patients using open-source disease control software developed by other patients. In: Saranto, K., Castrén, M., Kuusela, T., Hyrynsalmi, S., Ojala, S. (eds.) WIS 2014. CCIS, vol. 450, pp. 203–210. Springer, Cham (2014). https://doi.org/10.1007/978-3-319-10211-5_21
46. Bauer, J., Franke, N., Tuertscher, P.: Intellectual property norms in online communities: how user-organized intellectual property regulation supports innovation. Inf. Syst. Res. **27**, 724–750 (2016)
47. Peeroo, S., Jones, B., Samy, M.: Customer Engagement Manifestations on Facebook Pages of Tesco and Walmart (2015)
48. Sadovykh, V., Sundaram, D.: How Do online social networks support decision making? A pluralistic research agenda. In: Proceedings of the 2015 International Conference on Advances in Social Networks Analysis and Mining, Paris, France, 25–28 August 2015, pp. 787–794 (2015)
49. Mørch, A.I., Hartley, M.D., Caruso, V., Mørch, A.I., Hartley, M.D., Caruso, V.: Teaching Interpersonal Problem Solving Skills using Roleplay in a 3D Virtual World for Special Education: A Case Study in Second Life, pp. 464–468 (2015). https://doi.org/10.1109/icalt.2015.139teaching
50. Martinviita, A.: Online community and the personal diary: writing to connect at open diary. Comput. Hum. Behav. **63**, 672–682 (2016)
51. Germonprez, M., Hovorka, D.S.: Member engagement within digitally enabled social network communities: New methodological considerations. Inf. Syst. J. **23**, 525–549 (2013)
52. Synnott, J., Coulias, A., Ioannou, M.: Online trolling: the case of Madeleine McCann. Comput. Hum. Behav. **71**, 70–78 (2017)
53. Cherif, H., Miled, B.: Are brand communities influencing brands through co-creation? A cross-national example of the brand AXE: in France and in Tunisia. Int. Bus. Res. **6**, 14–29 (2013)
54. Kulavuz-Onal, D., Vásquez, C.: Reconceptualising fieldwork in a netnography of an online community of English language teachers. Ethnogr. Educ. **8**, 224–238 (2013)

Designing Training Mechanism for the Elderly to Use Social Media Mobile Apps – A Research Proposal

Abdulrahman Hafez(✉) and Yuanqiong (Kathy) Wang

Department of Computer and Information Sciences, Towson University, Towson, MD 21252, USA
abdul83hafez@gmail.com, ywangtu@gmail.com

Abstract. This is a research proposal to demonstrate a suggested training design that can potentially be suitable to training mild cognitively-impaired elderly to successfully use social media mobile apps. To test its success, the researchers propose a 3 by 1 experimental design involving three training groups receiving similar training treatments. Training sessions will be followed by observation meetings in two weeks period to conduct three measures: whether participants were able to repeat and complete tasks successfully; whether they were able to retain the information using the mobile tutoring app after a certain period of time; and whether the child narrator embedded in the mobile app design affects the learning process. The experimental design will be explained in greater detail later in this research proposal.

Keywords: Social media · Mobile tutoring app · Elderly · Cognitive impairment Social networking · Research proposal · Child narrator

1 Introduction

Preliminary results [1] indicate the need for social media mobile training among the elderly community to make them engage with technology. However, this research will particularly focus on the mild cognitively-impaired elderly group. Thus, a special training mechanism is needed to meet their cognitive disability. Generally, training plays a significant role in increasing technology awareness, and for the elderly to successfully use mobile social media applications on a regular basis. Therefore, this research investigates a blended training mechanism that can meet the elderly needs, while focusing on elderly with mild cognitive disability. This research will conduct a 3 by 1 experimental study to examine the effects of a blended learning approach to teach elderly with mild cognitive disabilities to successfully use social media mobile apps and be able to retain the information needed after certain period of time. This blended learning mechanism includes classroom training sessions with visual aids, flash cards, and mobile device, as well as a mobile tutoring app with a child narrator that will work as a supportive learning tool to help elderly with mild cognitive issues to retain the information they learned during classes. The mobile tutoring app will help participants during the self-study

© Springer International Publishing AG, part of Springer Nature 2018
G. Meiselwitz (Ed.): SCSM 2018, LNCS 10913, pp. 46–56, 2018.
https://doi.org/10.1007/978-3-319-91521-0_4

period particularly – two weeks the researchers will allow between the training phase and the observation phase. The mobile tutoring app will also include a child narrator presenting and singing some information to measure if singing will help participants remember short instructions. Creating an appropriate training mechanism will help elderly to engage with technology, and eventually, will allow them to successfully use social media mobile apps on a regular basis.

The mobile tutoring app will be given to the third group including the embedded child narrator, while the second group will receive the mobile tutoring app as instruction videos with no narration. The first experimental group will solely receive the classroom training includes visual aids, flash cards, and mobile devices but without the mobile tutoring app.

Instead of relying solely on Power Point presentations during classes (passive teaching style), elderly can receive a collection of flash card sets for each social media application addressing the necessary functions/features of a specific social media mobile app. These flash cards contain up-to-date user interface screen shots of app's main features and written how-to-accomplish-instructions of each task. Flash cards can work as a reference guide for elderly users if they are struggling with a certain functionality or feature, because they tend to be easier to point out information than written notebooks. Added user interface screen shots to the flash card set would be a big plus for elderly to be able to compare what they see in the flash card with what they have on their mobile devices.

Lastly, considering the idea of an app designed particularly to train the elderly how to use social media apps effectively, covering all possible activities a user may need to perform on those platforms. This mobile tutoring app should not terminate the face-to-face training sessions but can be an add-on to serve as a repetitive tool of what might be forgotten to remember in a later time to enhance the memory of mild cognitively-impaired elderly trainees.

2 Literature Review

2.1 Technological Impacts on Elderly

O'Connell [2] found that after participating in an educational computer-skills program, senior participants showed lower levels of physical difficulties, depression, and feelings of loneliness, as reported by several different psychological scales and checklists. They even reported improved life-satisfaction, sense of control, and quality of life. Increased participation in the Internet following a training program also leads to higher computer efficacy, lower levels of computer anxiety, more positive attitudes about aging, higher levels of perceived social support, and higher levels of connectivity with some seniors.

It can be a great motive for elderly who are apart from their children and grandchildren to learn to use social media networks and see their beloved ones every once in a while. There is a successful story of an elderly establishing her own blog to express her feelings and share her experiences; all started by participating in social media networks.

Paula Rice said, "I was dying of boredom" after she lost a husband, and while her children and grandchildren are way too far to visit. Yet, she spends around 14 h a day online communicating [3].

Research illustrates that the Internet has become a significant tool to exercise elderly's brains. Elderly who use social media, Skype, and email on daily basis tend to perform better cognitively and experience enhanced well-being [4]. A cognitive study was carried out in the UK and Italy of 120 elderly (65 years and older) for a duration of two years. Participants were provided with special computer training then were compared against a control group with no training. The results showed mental and physical improvements for those who received the training, while the control group members with no training showed a stable decline. A female participant stated feeling "invigorated" instead of "slipping into a slower pace" and become more concerned about her appearance and wants to lose weight. Overall, sustaining the basic social desires of elderly can have a positive impact on their overall health [4]. To sum up, if eagerness found in elderly to participate in social media, and a proper training mechanism was achieved, there shall be more elderly adults to use social media networking, which will ultimately enhance their well-being.

With the lack of social media network usage among elderly because of computer literacy, and with ascendancy of mobile devices against computers, the purpose of this research is to investigate a training mechanism that overcomes technology literacy among mild cognitively-impaired elderly and add great benefits to their lives. Participating in social media will not only make the elderly feel included within society, but will also improve their well-being and cognitive abilities [5, 6].

2.2 Cognitive Impairment Variations

Cognitive impairment is a syndrome that causes cognitive weakening for older people affecting their memory and education level but not necessarily collide with their activities of daily life. Studies show that between 3% to 19% in adults older than 65 years old experience mild cognitive problems. Some elderly with mild cognitive disability tend to stay stable or even improve over time, but over half worsen to dementia (severe cognitive impairment) within 5 years, which eventually leads to Alzheimer's disease [7].

Elderly with cognitive issues may suffer from mild to severe cognitive impairment. Mild cognitive disease causes a slight but noticeable and measurable decline in cognitive skills. Such declines are found in memory and thinking skills [8]. On the other hand, severe cognitive impairment can affect a person's ability to understand the meaning or the importance of objects and may affect the ability to talk or write, which prevents them from living independently [9]. Since this research will focus on mild cognitively-impaired user group, there is a need for a proper definition of cognitive impairments and how it can be measured in elderly in order to select the research experimental groups correctly. Cognitive ability measures will be explained in the next section of this research.

2.3 Cognitive Ability Measures

Cognitive abilities are usually measured using specific tests, in which they produce a score for each certain ability (e.g., numeric, verbal, reasoning), then the resulting scores represent measures of the specific mental abilities. Yet, one final score is also produced to assess the overall cognitive ability. Those tests are now designed as online tools that randomly generate questions include reasoning, perception, memory, verbal, language, mathematics, and problem-solving questions and allow a certain time for applicants to complete the test. Conventionally, the general trait assessed by cognitive ability exams is referred to as "intelligence" or "general mental ability." Nevertheless, an intelligence test is more of a specific mental abilities test that includes mathematical equations, verbal reasoning, comprehensive reading, number series completion, and spatial relations [10]. Traditional cognitive exams are consistent, include questions that are reliably scored, and can be assigned to large groups of applicants at once. Formats of cognitive ability tests include multiple choice, sentence completion, short answer, or true-false questions, and they are available either commercially or as free online tools. In this research, Wonderlic Test tool, a free online tool, will be used to measure cognitive abilities of potential candidates who will be recruited if eligible to participate in this research experiment.

The Wonderlic test is an intelligence test containing 50 questions designed to be taken in one set for applicants to complete in 12 min. The final score represents the questions that were successfully answered during the given time. The reason for the time restriction during the test is to produce stress among test takers which can help predict how applications perform under a certain amount of pressure. The scoring system of Wonderlic test is similar to the Stanford-Binet test that produce a Bell curve with results placed in the center of the curve [11].

2.4 Teaching Technology to Mild Cognitively-Impaired Elderly

Naumanen and Tukiainen [12] concluded that when elderly receive proper Information and Communication Technology (ICT) training, their ability increases to successfully engage with technology. In order to provide elderly with a proper training approach, we need to know how training can be effective for the elderly? Duay and Bryan [13] answered this question by stating that elderly adults need to feel they matter. They are in need for a recognition of their diverse experience and sociability which should be considered during the learning process. In other words, elderly do not enjoy the passive teaching style, they would rather be involved in a discussion and ask questions when they need to and be in an interactive learning environment. However, elderly's cognitive abilities can decline because of aging; such noticeable declines in reasoning, discourse comprehension, inference formations, reception of new information and its retrieval from memory [2].

Cognitive abilities are the key factors for fluid intelligence [14]. Successful completion of such a training approach this research is proposing demands for a wide range of perceptual, cognitive, and motor abilities. Similarly, Nair and colleagues [14] studied elderly performance carrying a computer-based study. They concluded that age crystallized intelligence, and fluid intelligence affected elderly early performance, as well as it affected the later performance that involved practice and experience. The impact of age on conceptual and motor abilities shows age-related declines. In addition, many cognitive abilities that are crucial to learning such as working memory, attentional processes, and spatial cognition, are weaken with aging. However, these factors do not necessarily halt elderly from learning a new skill set, even though it may take them longer to adapt to a new technology than younger adults [15]. Therefore, training mild cognitively-impaired elderly can be successful if it is done properly, while taking into account repetition and slow pace progression.

2.5 Child Narrator Impact on Elderly

Instruction videos – such as the mobile tutoring app this research is designing as part of its training mechanism – can increase elderly receptiveness of social media and boost their learning process. Elderly adults tend to accept video tutorials if they had prior background knowledge of contents, thus, the classroom teaching technique was thought of. A human narration incorporation that shows social presence can be highly effective in instruction videos to provide step-by-step details of usage. Existing findings recommend the inclusion of a human narrator in instruction videos as it is preferred by elderly over the ones without [16]. Specifically, elderly adults favored a child narrator in instruction videos over young adults and senior people; a sample of 124 elderly was collected by researchers [16]. Elderly adults are widely accepting a child narrator because they believe it conserves dignity. As elderly adults grow old, they have encountered vast amount of knowledge, therefore, it is hard for them to accept knowledge from others unless they are convinced what other presenting is true [17]. Hence, providing instructions to elderly should not violate dignity values for them to accept, which this research will take into consideration when designing material for the mobile tutoring app presented by a child narrator.

3 Research Questions

Examining such training technique is crucial in defining user engagement, their successful completion of given tasks, and their ability to remember these tasks after training is over. This can be achieved by designing appropriate training mechanism that is not only suitable for elderly but also meets their mild cognitive impairment needs. To address this training mechanism, this research will attempt to answer the following questions:

- To what extent the blended training approach of classroom and mobile tutoring app with child narrator is successful?
- Did it help the elderly users to use social media successfully? Are they able to successfully complete tasks after taking these training sessions?
- Does the designed mobile tutoring app help its users to retain information after certain period of time when completing their training sessions?

- Does child narrator embedded in the mobile tutoring app design have any effect in the learning process? Does the child singing information make any difference to remember short instructions?

The following research hypothesis will be examined in this study:

H1: The blended training mechanism used on group 3 (classroom and mobile tutoring app with child narrator) help elderly to successfully complete the given tasks, because it contains three repetitive treatments that work simultaneously.
H2: The mobile tutoring app helps its users to retain information after certain period of time since users can gain access to it anytime, especially during the self-study period.
H3: The child narrator included in the mobile tutoring app has a great impact on elderly learning process, as it has been proved by research [16].
H4: The child narrator songs included in the mobile tutoring app help elderly to remember short instructions.
H5: Songs are better remembered than words.

4 Experimental Design

The researchers will carry an experimental investigation involving a 3 by 1 design that is also called a between-group design, includes a training phase and an observation phase. During the training phase, the three examined groups will receive training as followed: the first group will receive a classroom training session including visual aids, flash cards, and mobile devices; the second group will receive classroom and mobile tutoring app training but will not experience the child narrator within the mobile tutoring app; the third group will receive a classroom training session and mobile tutoring app training including the child narrator. This will help distinguishing the effects of each method and will allow the researchers to draw a clear conclusion whether the premium treatment used on the third group (classroom, flash cards, mobile device, and mobile tutoring app with child narrator) is successful. 45 participants age between 65 and 80 will be recruited. They will be randomly assigned to one of the groups based on age. Therefore, each group will contain 15 participants of mild cognitively-impaired elderly participants age between 65 and 80 years old, and each class will have an equal distribution of participants' age range. Participants may have minimal to intermediate experience of social media applications. Additionally, cognitive ability will be measured during the recruiting process using Wonderlic Test; a free intelligent online tool [11]. In addition, other demographic data (e.g. age, sex, experience, etc.) will also be collected. The researchers will base recruiting decisions on candidates' final score after completing Wonderlic Test. Finally, the researchers will communicate with official institutes such as senior centers located in a Mid-Atlantic State for recruiting and will seek help from assisted living communities around the same location.

The researchers will allow a two-week-self-study period in between training phase and observation phase. Participants will be asked to freely use the course material and apply to a real-life social media mobile app interaction and note down their diaries. For instance, they write down the day and time they use social media and the activities or features they perform. This information will be utilized when running the statistical analysis later. Taking the self-study period into consideration will verify whether a participant performs successfully because of their treatment level or it was the effect of the two-week self-learning period, or both. Therefore, the researchers would not solely base results on participants' performance during classes but will also take into consideration their practices throughout the experiment including the self-study period.

Two weeks after completing phase one (the training sessions), participants will be called back to attend observation sessions (phase two). The second phase is designed to learn how much participants remember from the first phase and whether they will be able to retain the information and complete all tasks successfully. Participants will be asked to perform the same activities they accomplished during training classes to find out whether they are able to retain information and compare their performance levels within group. The results of each group will be compared to eventually decide which training mechanism works best for mild cognitively-impaired elderly.

The independent variable for this study is the training mechanism that includes the following values:

- Classroom only includes visual aids, flash cards, and mobile device;
- Classroom (includes visual aids, flash cards, and mobile device), and mobile tutoring app without narration;
- Classroom (includes visual aids, flash cards, and mobile device), and mobile tutoring app with the child narrator.

The dependent variables for this study are the effects as a result of the independent variable, such as success rate, task completion, and whether the social media app can be used without personal assistance during the self-study period.

4.1 Data Collection and Analysis

The researchers will run a factorial ANOVA statistical analysis to draw a conclusion of this experimental study. Factorial ANOVA or factorial Analysis of Variance method is used for empirical studies that embrace a between-group design to test two or greater values of an independent variable [18]. First of all, the researchers will conduct a preprocessing segment before analyzing collected data. This includes cleaning up data, which will be a significant step – especially for data recorded manually by participants – to avoid possible errors. This step will avoid negative impacts on results that human error can cause [18]. Next, collected data by the researchers will be coded to be used later by statistical analysis software. For instance, representing categorical variables of participants by codes "0" and "1" in order to be understood and run by the software [18]. The last stage of preprocessing data will be organizing collected data. The researchers will use SAS University Edition statistical analysis software [19] to compare a group of data collected for each group of participants. After initiating and finalizing the preprocessing stage, the researchers will conduct factorial ANOVA statistics.

Three factors that will be taking into consideration when measuring participants performance. These three factors are: successful completion of tasks, retain the information participants learned during the observation meetings, and measure the effect of the mobile tutoring app including the effect of the child narrator if any on mild cognitively-impaired elderly adults. The researchers will study if performance of the second and the third group will be any better than the first group to measure the mobile tutoring app effect, while taking into consideration how well participants practice on their own during the self-study period. Similarly, the researchers will compare performance of the second and the third group throughout, to measure the effect of the child narrator.

4.2 Participants

The researchers will recruit three groups of mild cognitively-impaired elderly adults to participate in two phases of this study. 10 participants for each group is needed. However, the researchers will plan to recruit 15 people for each group in case of drop offs. Therefore, the researchers aim to recruit a total of 45 participants; 15 participants per group.

The participants will be recruited through assisted living homes and Senior Activity Centers in a Mid-Atlantic State. Potential candidates will be asked to take the Wonderlic cognitive test [11] to measure their cognitive ability level. Then, the purpose of the study will be explained to all the potential candidates prior to recruiting. The researchers, if permitted, may use Senior Activity Center's computer classrooms, or will use the computer labs in the Department of Computer and Information Sciences in a Mid-Atlantic State. The instructor will explain to the participants that being part of the study is voluntary and will inform them of their rights to dismiss class at any point in time they do not feel comfortable completing the study. Then the participants in all three groups will be asked to sign the consent forms. After signing the consent forms a demographic information sheet will be distributed to class before training takes place.

4.3 Training Mechanism

The researchers will conduct a 3 by 1 experimental design consisting of two separate phases and three groups. The first phase will involve various training sessions of three groups of 15 participants each to train mild cognitively-impaired elderly, aging between 65–80 years old, who share similar experience level of social media mobile applications. Participants will then be asked to attend observation meetings (the second phase) after two weeks for the researchers to record how much information can be retained and how successful the participants will be in completing tasks. In addition, participants will be asked to write diaries of their social media activities during the self-study period that will be given between phases. During the second phase, contributors will be asked to repeat the activities were performed in class previously. The instructor will record if people completed tasks successfully as well as the tasks they needed help with, or uncompleted ones. The instructor will collect contributors' diaries at the beginning of phase two for data analysis later. Training classes and observation sessions will take place simultaneously to save time. This will allow accurate recordings and data analysis of all groups being close in time from each other.

4.4 Classroom Syllabus and Mobile Tutoring App Content

Each group of participants will have a chance to attend two classes for this study, a total of 6 classes for all three groups to cover the two phases. The first phase will be an introductory social media class that will run for 50 min for each group. The instructor will go over the importance of social media using Power Point slides to illustrate. Participants will be taught how to control what they share online so they feel confident to accept to use social media as privacy issues arose in literature. The instructor will go over the basics on how to use Facebook and will ask the participants to perform practically on their mobile devices. Participants will either have their own mobile device (smartphone or tablet) or will receive one from the instructor for the purpose of the study. The training session will be divided into multiple tasks that contributors will perform using their mobile devices. After each task, the instructor will allow some time for discussion. This will be a great opportunity for a collaborative class work where everyone shares their experiences, comments, and socialize.

The exercise will start by teaching participants how to download the Facebook mobile app from the App Store. Then participants will be asked to login or create new accounts if they never had one. Speaking of the third group of participants, the child narrator singing a login instruction will be presented via the mobile tutoring app at this point in time. The instructor will ask participants to help those who are struggling creating new accounts. A brief discussion session will take place to record participants' experiences with the login/create an account step. The instructor will show participants how to search and add a friend on Facebook and ask participants to invite each other to their friends lists. They will then be asked to post a text on a friend's timeline. In addition, they will be taught how to send a private message, how to post a picture or video on Facebook, how to like and share photos of others, and how to view friends' posts. Participants will learn how to read the news and other Facebook services. The last task the instructor will ask participants is to take a selfie using their mobile device front camera and try to include as many people in the photo as possible. They will then be asked to post this photo to their Facebook accounts, and start to tag the people who appear in it. The second and third group of participants will have a chance to review the mobile tutoring app as class progresses. There will be an instruction video of each task that participants will review after the instructor's explanation, and before they start performing the tasks on their own. The third group particularly will hear the child narrator speaking each activity when reviewing the mobile tutoring app.

Participants will be invited back to class after two weeks to repeat the same previously performed activities. The instructor will record performance as this class will be an observation session. Each group will attend independently, therefore, three observation meetings will take place during this phase. The instructor will videotape participants' performance to record data. Data collected on both sessions will be combined for later analysis.

Finally, the mobile tutoring app will include instruction videos of the main features that allow the user to successfully use the Facebook mobile app. Written instructions and a child narrator will also help explain each step the mobile tutoring app offers and provide a broad view of features and information of Facebook. The researchers will design two versions of the mobile tutorial app; one includes the child narrator and the other version without. The instructor will manage to present all app's content during classes to the second and the third group during the training phase. App content will be decided based on multiple interviews of randomly selected mild cognitively-impaired elderly, as well as previous literature review carried by the researchers.

5 Conclusion and Future Work

Preliminary research indicates elderly apprehension of anything new especially when involving new technology. One way to relieve the apprehension of the new technology is through proper training on mobile devices and social media applications [1]. To investigate how to best help the elderly overcome the apprehension, different training approaches are proposed in this research. Existing research recommends classroom training as an effective method that allow elderly to learn new technology, share their experiences, and socialize. However, the researchers are targeting mild cognitively-impaired elderly who may easily forget all the technical detail presented in class. Therefore, the researchers believe the mobile tutoring app that is especially designed with instruction videos and child narration will help boost the memory of elderly with mild cognitive disability, especially when training classes are over. Participants will have the opportunity to access the mobile tutoring app after class whenever they are struggling to perform a certain task.

In order to test the proposed premium training treatment (classroom, mobile tutoring app, and child narration), there is a need to compare the outcomes and performances of two more groups with less training treatment options. Therefore, the researchers decided to include a group of participants who will only receive a class training and mobile tutoring app without narration to measure effectiveness of the child narrator embedded in the app design when compared with the premium training treatment of the third group. In addition, the other group of participants will receive the class training only to measure the overall effectiveness of the mobile tutoring app when compared to the other two groups who will receive the mobile tutoring app as part of their training program.

Running factorial ANOVA statistical analysis will allow to test the three values of the independent variable this research is proposing; classroom only, classroom and mobile tutoring app without narration; and classroom and mobile tutoring app with the child narrator. The end results should draw a clear conclusion of which of these training approaches is the most effective that allow the mild cognitively-impaired elderly to successfully use social media mobile applications on a regular basis.

References

1. Hafez, A., Wang, Y.K., Arfaa, J.: An accessibility evaluation of social media through mobile device for elderly. In: Ahram, T., Falcão, C. (eds.) AHFE 2017. AISC, vol. 607, pp. 179–188. Springer, Cham (2018). https://doi.org/10.1007/978-3-319-60492-3_17
2. O'Connell, A.A.: Using instructions and behavioral skills training to teach Facebook skills to seniors (Doctoral dissertation, University of South Florida) (2016)

3. Clifford, S.: Online, 'a reason to keep on going'. The New York Times, p. 5 (2009)
4. Kamiel, A.: A Hot Trend: The Internet Social Media & The Elderly. The Huffington Post (2016)
5. Morris, S.: Study finds social media use beneficial to overall health of elderly, Media, The Guardian (2014). https://www.theguardian.com/media/2014/dec/12/study-finds-social-media-skype-facebook-use-beneficial-overall-health-elderly#. Accessed 8 Nov 2017
6. Marcelino, I., Laza, R., Pereira, A.: SSN: senior social network for improving quality of life. Int. J. Distrib. Sens. Netw. **12**(7), 2150734 (2016)
7. Gauthier, S., Reisberg, B., Zaudig, M., Petersen, R.C., Ritchie, K., Broich, K., Belleville, S., Brodaty, H., Bennett, D., Chertkow, H., Cummings, J.L., et al.: Mild cognitive impairment. Lancet **367**(9518), 1262–1270 (2006)
8. Albert, M.S., DeKosky, S.T., Dickson, D., Dubois, B., Feldman, H.H., Fox, N.C., Gamst, A., Holtzman, D.M., Jagust, W.J., Petersen, R.C., Snyder, P.J., et al.: The diagnosis of mild cognitive impairment due to Alzheimer's disease: recommendations from the National Institute on Aging-Alzheimer's Association workgroups on diagnostic guidelines for Alzheimer's disease. Alzheimer's Dement. **7**(3), 270–279 (2011)
9. Centers for Disease Control and Prevention: Cognitive impairment: A call for action, now. CDC, Atlanta, GA (2011)
10. OPM.gov. (n.d.). Cognitive Ability Tests. https://www.opm.gov/policy-data-oversight/assessment-and-selection/other-assessment-methods/cognitive-ability-tests/. Accessed 18 Dec 2017
11. What is the Wonderlic test? Sample Wonderlic Test. (n.d.). https://samplewonderlictest.com/what-is-the-wonderlic-test. Accessed 6 Dec 2017
12. Naumanen, M., Tukiainen, M.: Guiding the elderly into the use of computers and internet–lessons taught and learnt. In: Proceedings of Cognition and Exploratory Learning in Digital Age, pp. 19–27 (2007)
13. Duay, D.L., Bryan, V.C.: Senior adults' perceptions of successful aging. Educ. Gerontol. **32**(6), 423–445 (2006)
14. Nair, S.N., Czaja, S.J., Sharit, J.: A multilevel modeling approach to examining individual differences in skill acquisition for a computer-based task. J. Gerontol. Ser. B Psychol. Sci. Soc. Sci. **62**(Special Issue 1), 85–96 (2007)
15. Lee, C.C., Czaja, S.J., Sharit, J.: Training older workers for technology-based employment. Educ. Geront. **35**(1), 15–31 (2008)
16. Teh, P.-L., Phang, C.W., Ahmed, P.K., Cheong, S.-N., Yap, W.-J., Ma, Q., Chan, A.H.S.: Teaching older adults to use gerontechnology applications through instruction videos: human-element considerations. In: Rau, P.-L.P. (ed.) CCD 2017. LNCS, vol. 10281, pp. 582–591. Springer, Cham (2017). https://doi.org/10.1007/978-3-319-57931-3_46
17. Yardley, L., Donovan-Hall, M., Francis, K., Todd, C.: Older people's views of advice about falls prevention: a qualitative study. Health Educ. Res. **21**(4), 508–517 (2006)
18. Lazar, J., Feng, J.H., Hochheiser, H.: Research Methods in Human-Computer Interaction. Morgan Kaufmann, San Francisco (2017)
19. Free Statistical Software, SAS University Edition, SAS. (n.d.). https://www.sas.com/en_us/software/university-edition.html. Accessed 8 Dec 2017

Teaching Communication Strategies in Social Networks for Computer Science Students

Pamela Hermosilla(✉), Nicole Boye, and Silvana Roncagliolo

Pontificia Universidad Católica de Valparaíso, Valparaíso, Chile
{pamela.hermosilla,nicole.boye,
silvana.roncagliolo}@pucv.cl

Abstract. Nowadays, web technology has changed drastically by positioning itself as an interactive, dynamic and very important in people's life, which is why it has become a fundamental tool in social media and communication. As professors of Computer Science (CS) students, we have seen the need to encourage the development of "soft" skills in the learning process of future engineers, such as communication, which represents a major challenge in the field of engineering. The purpose of our study was to introduce to university students of CS, how to incorporate communication techniques when using social media networks as an effective communication tool. The study aims to show that it is not enough to know the current social media networks, but rather requires targeted communication strategies for this type of media. In this way, it involved CS undergraduate students who participate in an optional course called "Strategies of communication and diffusion in computer projects" and another group that have the same level of curriculum in the same career, but they have not internalized in the techniques of communication oriented to social media. As mentioned above, the opinion of both groups and the perception of how they can use properly communication skills for their future work are presented, and indicating the need to incorporate this ability as part of the learning process in CS students.

Keywords: Social media · Communication strategies
Computer Science (CS) students
Information and Communication Technologies (ICT)

1 Introduction

The impact generated by new technologies and the globalized world has led entrepreneurs to demand professionals with comprehensive conditions regarding not only their profession but also basic conditions of human interaction and the general understanding of what is social media. Therefore, nowadays a professional besides having a specific knowledge in his area must also fulfill competences such as communication, feedback, interaction among others. Nowadays key competencies are the essential competencies for effective participation in the emerging patterns of work and work organizations. They focus on the capacity to apply knowledge and skills in an integrated way in to work situations [1].

© Springer International Publishing AG, part of Springer Nature 2018
G. Meiselwitz (Ed.): SCSM 2018, LNCS 10913, pp. 57–66, 2018.
https://doi.org/10.1007/978-3-319-91521-0_5

Computer engineers are professionals who are at the forefront of what is today the impact of new technologies, but nowadays some students have more developed soft skills than others, so it is important for who don´t have those skills developed to offer during their studies the opportunity to learn new concepts and competences that will help to be integral professionals and that also generates a new attribute to their professional profile, "this implies a new conception of the engineer as an individual that surround a series of features not precisely of technological order, but rather committed to the search of problems solutions" [1]. Furthermore, understand users, know what they do, what they want, how and where; It is fundamental key for large companies, which is why it is a challenge for computer engineers to learn to understand individuals and how they use social networks. Whereby, we will begin to relate to basic concepts such as strategic communication, social media, computer science students (CS) and then relate them; and also make known how positive it is to have a professional with communication tools that make it an integral professional, for the global world in social media.

2 Using Communication Strategies in Social Media Interaction Trough Information and Communication Technologies (ICT)

Social media is fast, interactive, effective and inexpensive. Researches as Kaplan and Haenlein have given a perfect vision of how to perceive social media "what may be up-to-date today could have disappeared from virtual landscape tomorrow" [2]. Social media is not only the most efficient way to publicize a product or service, but it also helps users to understand any idea, product, to deliver a better service or create a new one, to give an opinion, etc. For this reason, social media is the most important thing in many agendas of business executives, or large companies that generate products or services. There are multiple applications where people can show themselves, make comments have an instant feedback or just look around for what they need. From the Facebook Boom applications have been appearing in addition to giving us more tools to get closer to everything or simply helps us to understand the global world; the way how we interact today is fast, simultaneous, just brilliant way! For this reason it is important to know how to use social networks effectively, and it is a fundamental complement for any type of company or microenterprise.

Today there are social needs that should be part of the training of today's world engineers. For this reason, the study plans must be designed based on the professional competences required by society. This is how each school must choose the competences that their graduates will possess at the end of their studies and design their curriculum based on these competences. These competences will define the professional profile of the graduates [3]. Authors such as Mayer [1] have classified competencies that are essential to be an integral professional, so he/she must:

- Collect, analyze and organize information
- Know how to communicate ideas and information
- Plan and organize activities
- Know how to work with others and as a team

- Use mathematical and technical ideas
- Solve problems
- Use technology

Within traditional education has always been learned to read and write, it has also been essential to learn mathematics as a capacity to analyze and reason for a solution to the problem, but when new technologies enter the world quickly is when begins to create a rethinking of how a professional should be trained, where in addition to learning the traditional must be impregnated with generic skills that help to respond flexibly and quickly to technological changes and their dynamism. Authors such as Huckin and Olsen have defined these skills as fundamental "… they are critical tools for success, even survival, in real world environments" [1]. Within these skills is the "communication" that is a fundamental tool to learn to communicate effectively.

There are studies carried out in scientific Universities and Schools of Engineering where it is demonstrated that communication is fundamental in the learning of engineers and that through this they generate other skills such as the ability to recognize and solve problems, then is when strategic communication is necessary as such. Therefore, developing communication skills should be an immediate objective in the academic and professional training of the students of this Century [1].

3 Strategic Communication Starts from the New Educational Paradigm Within Engineering

Globalization and technological changes have made traditional education systems begin to seek a change in the way courses and programs are taught and thus begin to develop in students skills and skills that will help them more effectively in the world of work. All this makes the new paradigm that shows the engineer as an integral professional who apart from learning specifically engineer skills must also have to learn to understand the real world in a practical, social, human way where there is a leadership and is able to face problems and to be able to solve them by interacting with others in an effective way and creating work groups [1]. This new paradigm is nothing more than a new vision of the profile of the engineer where new attributes and competences stand out, and we can state that this New comprehensive way to train professionals is a new educational culture.

"Man is immersed in his cultural context, from which he cannot isolate himself. Communication is, therefore, a permanent process that integrates multiple modes of behavior: words, gestures, looks, mimics, space management" [4].

The communication strategy is a series of programmed and planned actions that are implemented based on certain interests and needs, in a space of human interaction, in a wide variety of times. The strategy carries a principle of order, of selection, of intervention on an established situation [5].

Moreover, the communication strategy is the set of forms and modes of communication that aim to establish an effective communication of ideas, products or services with an implicit commitment of resources that help decision-making, and allow to achieve the organizational goals. However, the knowledge of communication strategies does not imply understanding how to use effectively the available tools in a suitable way through the diversity of current technological applications [6, 7], show in a following Chart 1:

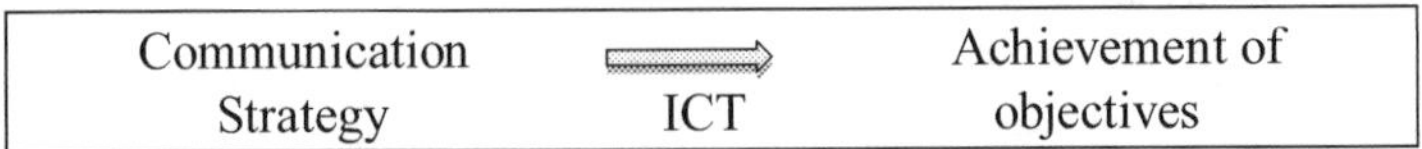

Chart 1. Interrelation of study

Based on the information presented above, the study seeks to answer two main questions:

- Do computer engineering students know communication strategies through the use of ICT applications available for communication and dissemination?
- Do the study programs of the area under study, should consider a course with this subject within the career training plan?

4 The Experiment

For the development of this study, two groups of students of the same career and of the same level of curricular advancement within their curriculum are considered. One of the groups has also completed an optional subject within the mesh, called "Strategies of communication and diffusion of computer projects", which has as main objectives.

Teach students communication strategies applying effectively in social media. And use strategic communication and personal marketing as a new trend to change the prototypes that exist today in professional environments. This way, students will be able to develop their qualities of communication to the maximum, helping them to reach an optimal performance in the personal life and in the work world. In addition to this students will be capable of use communication effectively and related to their professional field. *The communication strategy shows how effective communication can:*

- Help achieve those global objectives of the organization.
- Participate effectively with interested parties.
- Demonstrate and exhibit the success of their work.
- Ensure that people understand what we do.
- Change the behavior and perceptions of users.

In this way a data collection instrument was designed, which would allow to compare both groups of students, in order to verify if the students belonging to a curricular plan associated with the ICT area, know how to properly use the social networks in the field of business-level communication. The groups that participated in the study correspond to 19 students who studied the aforementioned elective (*Experimental Group*), and 22 students who have not studied said subject (*Control Group*), remembering that both groups are from the same career or are in the same year of study.

The instrument used considered 10 questions, grouped into three categories: general, specific knowledge and application, which together allowed investigating the subject. The evaluation scale used considers:

- Likert Scale: A measurement tool allows measuring attitudes and knowing the degree of compliance of the respondent. In this sense, the response categories will serve to capture the intensity of the respondent to certain statements. For this case scale with 4° of compliance is used.
- Dichotomous questions: with yes/no answer (Boolean).
- Selection of Options: identification of possible alternatives for a certain evaluated aspect.

Table 1 presents the questions according to category, dimension and evaluation scale considered:

Table 1. General category questions

Dimension	Questions	Scale Type
D1. Appreciation	Q1. Do you consider that as a computer scientist learning effective communication techniques to be used in social networks is important within the career	Likert scale levels: SA. Strongly Agree A. Agree D. Disagree SD. Strongly Disagree
D2. Importance	Q2. As a future Engineer, do you consider that social networks are important in the process of communication in the workplace	Likert scale levels: SA. Strongly Agree A. Agree D. Disagree SD. Strongly Disagree
D3. Preference	Q3. What social networks do you use most often? The three most significant responses are considered in the results of this question	Options selection: D3.1 Twitter D3.2 Facebook D3.3 Instagram D3.4 Tumblr D3.5 LinkedIn D3.6 MSN D3.7 Snapchat D3.8 Otro
D4. Tendency	Q4. In the field of the use of information technologies, in the process of the delivery and dissemination of information, indicate that you are aware of the following communicational tendencies:	Options selection: D4.1 Blogs Corporativo D4.2 Chatbots D4.3 Publicidad native D4.4 Marketing in moviles D4.5 Newsletter

5 Results and Discussion

The main results of the instrument used in this study are shown below. It will be determined *"Experimental Group"* to those students who attended the elective and *"Control Group"* to those who do not (Tables 2 and 3).

Table 2. Specific knowledge category questions

Dimension	Questions	Scale Type
D5. Knowledge	Q5. You know the concept of "Strategic communication"	Yes/No
D6. Identification	Q6. In the area of social networks in the communication process, "Digital Strategy" corresponds to the application of techniques that are carried out in the offline world and translated into the online world	Yes/No
D7. Classification	Q7. For the following statements, point out: 1: if corresponds to Operational Objectives 2: if it is related to communication objectives All the options are valid as an answer	Options selection: D7.1 Train the staff effectively to work with our clients D7.2 Ensure that customers perceive how cleanliness is a primary objective within the organization D7.3 Ensure that all employees know and understand the expected customer service standards D7.4 Keep the facilities clean and well maintained
D8. Asociation	Q8. For the following aspects: Point out: 1: if it corresponds to Premises to be considered in the digital co-communication strategy 2: if it is related to tools as a dissemination tool:	Options selection: D8.1 Search engine positioning D8.2 Flow D8.3 Feedback D8.4 Microsites D8.5 Link campaigns D8.6 Loyalty D8.7 Massive email D8.8 Contextual advertising D8.9 Functionality

The analysis of the results will be carried out for each of the identified categories, and the dimensions evaluated in them, comparing both groups of the study (Table 4).

- General Category Questions

In the case of questions of a general nature, it can be seen that perception (associated with D1 y D2) of the students of *Experimental Group* it reaches a greater acceptance in both dimensions (52.6%, 57.9%), which shows that this group shows a greater appropriation to the importance of effective communication techniques in social

Table 3. Application category questions

Dimension	Questions	Scale Type
D9. Evaluation	Q7. If you had to inform a product or service of your company, identify what aspects you would have in mind to ensure that your message is received as it is pursued Only the correct answers to the question are considered in the results	Options selection: D9.1 Target audiences D9.2 Keywords in message D9.3 Communication channels D9.4 Objective of the message D9.5 Business context
D10. Analysis	Q10. Which of the following types of study and/or analysis, would you use in the context of a project such as Electronic Voting in Chile? Only the correct answers to the question are considered in the results	Options selection: D10.1 The 4Ps (Product, Price, Place, Promotion) D10.2 5 Porter's five forces D10.3 SWOT (strengths, weaknesses, threats) D10.4 PEST (Political, Economic, Social, and Technological) D10.5 Competition analysis D10.6 CAME (Correct, Adapt, Maintain, Explore) D10.7 LRPD(Limitations, Risks, Potentialities, challenges)

Table 4. Percentage scores for dimensions (D1, D2)

	Dimension	(SA)	(A)	(D)	(SD)
Experimental group (19 participants)	D1	52.6	36.8	10.5	0
	D2	57.9	31.6	10.5	0
Control group (22 participants)	D1	47.8	34.8	17.4	0
	D2	34.8	47.8	17.4	0

networks as part of their future work performance, which is not possible to perceive in *Control Group* students, where relevance to what is considered in these dimensions they do not exceed 42% on average (Table 5).

Regarding the preferences of social networks and trends used, it is possible to appreciate that both groups maintain a high percentage for Facebook (D3.2) and assigns greater relevance to Mobile Marketing (D4.4), which responds to the wait-do,

Table 5. Percentage (D3) and average (D4) scores for dimensions

	D3. Preference			D4. Tendencies				
	D3.1	D3.2	D3.3	D4.1	D4.2	D4.3	D4.4	D4.5
Experimental group (19 participants)	21.1	**78. 9**	68.4	3.5	2.4	3.7	**4.6**	3.1
Control group (22 participants)	30.4	**87.0**	39.1	2.8	3.3	3.1	**4.3**	3.5

because it is about groups of students of the same career, of the same level of curricular advancement, and who are inherently inserted into the current information society, for them it is natural to live with virtual social networks and through mobile devices (Table 6).

Table 6. Percentage scores for dimensions (D5, D6, D7)

	Dimension	Yes	No	D7.1	D7.2	D7.3	D7.4
Experimental group (19 participants)	D5	57.9	42.1	**31.6**	**21.1**	**73. 7**	89.5
	D6	78.9	15.8				
Control group (22 participants)	D5	8.7	87.0	21.7	13.0	56.5	**95.7**
	D6	39.1	56.5				

- Specific Knowledge Category Questions

When reviewing the results of the dimensions associated with greater knowledge in the area of "Strategic Communication" and "Digital Strategies" (D5, D6) it is considerable to appreciate the lack of knowledge that the *Control Group* presents before the *Experimental Group*, where they average 23.9%, 68.4% respectively.

With respect to the classification of operational and communication objectives (D7) it is possible to appreciate that in 3 of 4 statements the *Experimental Group* overcomes the *Control Group* (Table 7).

Table 7. Percentage scores for dimension (D8)

	D8.1	D8.2.	D8.3	D8.4	D8.5	D8.6	D8.7	D8.8	D8.9
Experimental group (19 participants)	52.6	57.9	52.6	63.2	**73.7**	63.2	68.4	57.9	47.4
Control group (22 participants)	39.1	73.8	47.8	52.2	60.9	69.6	**73.9**	60.9	60.9

In the analysis of Association Dimension (D8), which looked for aspects related to whether it corresponds to Premises to be considered in the digital communication strategy and diffusion tool, it is observed that there is no clear relationship between both groups and that the statements: Link (D8.5) and Massive Email (D8.7), are the ones that obtain the best association in the *Experimental Group* and *Control Group* respectively, so it is not considered a relevant result in the study (Table 8).

- Application Category Questions

Table 8. Percentage scores for dimensions (D9, D10)

	D9.1	D9.3	D9.4	D10.1	D10.3	D10.4	D10.5
Experimental group (19 participants)	**100**	**89.5**	**57.9**	**73.7**	**100**	**26.3**	**36.8**
Control group (22 participants)	95.7	47.8	39.1	52.2	65.2	4.3	26.1

In the Evaluation Dimension, where the elements for reporting a product (D9) should be considered, the expected correct response corresponded to: Target audience (D9.1), Communication channels (D9.3) and Message objective (D9. 4). For the above, the *Experimental Group* presented a very good identification in the correct options, obtaining on average 82.5%, while the *Control Group* achieved a 60.8% assertiveness in this question. Respecting the ability to analyze the most appropriate studies for a particular case presented (D10), the correct options for this question were: 4P (D10.1), SWOT (D10.3), PEST (D10.4), Competition analysis (D10.5), for this case *Experimental Group* has in each of the correct options for this question a higher level of success than the *Control Group*. Although the *Experimental Group* had two very well evaluated options: "The 4P" (D10.1) and "SWOT" (D10.3) in conjunction obtained a 59.2% achievement, while the *Control Group* also obtained better results in (D10.1, D10.3) in total for this question only a 36.9% achievement, well below the *Experimental Group*.

6 Conclusions

Social networks are widely used today, in various areas of society, but this small sample made in computer science studentes makes it possible to demonstrate that it is not enough to know them, but that they need to be used in conjunction with communication strategies in order to enhance their effectiveness within the informative process.

For this study the dimensions (D1, D2) evaluated within the general category, results were obtained that show the relevance that the *Experimental Group* considers respect of the use of effective communication techniques in social networks, for a better professional performance. Regarding the preferences of social networks (D3) and trends for communication (D4), both groups agree on their results, which is within the expected, given that the groups of students have similar characteristics in their age range, and of studies achieved in the career.

For this study the dimensions (D1, D2) evaluated within the general category, results were obtained that show the relevance that the *Experimental Group* considers respect of the use of effective communication techniques in social networks, for a better professional performance. Regarding the preferences of social networks (D3) and trends for communication (D4), both groups agree on their results, which is within the expected, given that the groups of students have similar characteristics in their age range, and of studies achieved in the career.

In the last category considered in this study, that of Application, the *Experimental Group* obtained significantly greater achievements before two practical situations consulted: elements to inform products and studies appropriate for a particular case (D9, D10).

With the results obtained in this sample, it is possible to conclude that the CS students need to study effective communication strategies that can be used through social networks. It is not enough to just be part of a career linked to information technology, but the current demands and changes in communication require knowing the most effective way to reach the target public, so it is left in question the need to consider this topic as compulsory element within the training plan of a career related to Computer Science, and not only as part of an optional subject, such as this study account, which reveals the difference between both groups and what in the future may be considered a significant competitive advantage, considering an important soft skill such as communication.

As future work, it would be interesting to focus on those aspects that did not present a significant difference and that may require more detail for their study, as well as trying to measure the effectiveness of the communication process of a group that has knowledge of digital strategies.

Acknowledgments. We thank all the students involved in this case of study. They provided useful opinions that allowed us to prepare this article and could be the beginning of other related studies.

References

1. Kindelán, M.P., Martín, A.M.: Ingenieros del siglo XXI: Importancia de la comunicación y de la formación estratégica en la doble esfera educativa y profesional del ingeniero 732-733-734 (2008)
2. Kaplan, A.M., Haenlein, M.: Users of the world, unite! the challenges and opportunities of social media. Bus. Horiz. **53**(1), 59–68 (2010)
3. Martinez, A., Aluja, T., Sanchez, F.: Perfil profesional del ingeniero informático: diagnóstico basado en competencias XV JENUI. Barcelona, 8-10 de julio de 2009 (2009). http://jenui2009.fib.upc.edu/. Accessed 17 Dec 2017, ISBN: 978-84-692-2758-9
4. Ojalvo, V.: Componentes sociopsicológicos de la comunicación pedagógica y su importancia para el trabajo docente-educativo. CEPES, La Habana (1999)
5. López Viera, L.: Comunicación social. Editorial Félix Varela, La Habana 214 (2003)
6. Impacto de las tic en la comunicación corporativa e institucional Memoria para optar al grado de doctor Presentada por: Jaime de la Fuente Martínez, Madrid (2011). ISBN: 978-84-694-6644-5
7. López Jiménez, I.E.: El Impacto de la tecnología en la comunicación empresarial: Reflexiones y Análisis Razón y Palabra, Primera Revista Electrónica en América Latina Especializada en Comunicación. http://www.razonypalabra.org.mx. Accessed 7 Jan 2018

Posting Content, Collecting Points, Staying Anonymous: An Evaluation of Jodel

Philipp Nowak(✉), Karoline Jüttner, and Katsiaryna S. Baran

Department of Information Science, Heinrich Heine University, Universitätsstr. 1, 40225 Düsseldorf, Germany
{Philipp.Nowak,Karoline.Juettner, Katsiaryna.Baran}@hhu.de

Abstract. While most social networking services require a registration and a user profile, the student app Jodel works in a different way, as it is completely anonymous, like its American former counterpart Yik Yak. This article comprises a comprehensive evaluation of the app, leading to a final assessment of its quality as an information service. To conduct this evaluation, the Information Service Evaluation (ISE) model from Schumann and Stock [1] is used. The users' experiences were determined by an online survey with a total of 1,009 participants which therefore provided representative results. Furthermore, the expectations of the Jodel developers, regarding the user responses, were queried and the differences were determined according to Customers Value Research [2]. In addition, an expert interview with Jodel founder Alessio Avellan Borgmeyer was conducted. The survey showed that Jodel reached its target group, since 72% of the respondents are students. The users' satisfaction with the app is extremely high. 92% of the respondents are satisfied with the app while 97% said that they would recommend Jodel. The network effect is therefore particularly pronounced. The used elements of gamification were discreetly but effectively implemented. The results of Customers Value Research showed, on the one hand, the accuracy of the developers assessing the participants' responses and, on the other hand, that they are aware of possible vulnerabilities of the app. Overall, the evaluation showed that Jodel is a social information service of high quality.

Keywords: Usability · Social networking services · Anonymity
User experience · Jodel · Evaluation

1 Introduction

Nowadays social networking services appear in a variety of forms. For full use, most of them require a registration and thus a profile, while the social aspect lies in virtual friendships or followers. The student app Jodel, however, shows that a social network can be structured in a different way and work completely without registration, profiles and friendships. It emphasizes a free and creative exchange between users in the vicinity by granting anonymity. Therefore, Jodel resembles the US-American app Yik Yak which has been shut down in May 2017 [3]. On the one hand, Jodel is an app which should provide for the users' entertainment while on the other hand, it is a

© Springer International Publishing AG, part of Springer Nature 2018
G. Meiselwitz (Ed.): SCSM 2018, LNCS 10913, pp. 67–86, 2018.
https://doi.org/10.1007/978-3-319-91521-0_6

service for the informational exchange between people in the immediate vicinity. This study comprises a comprehensive evaluation of the app Jodel and leads to an assessment of its quality as an information service. In order to conduct this evaluation, the Information Service Evaluation (ISE) model from Schumann and Stock [1] is consulted and the three dimensions information service, information user and information acceptance are considered. Applied methods of this study are, besides a representative user survey, a developer survey and an expert interview with the founder of Jodel, Alessio Avellan Borgmeyer.

Jodel is a location-based social network for mobile devices. The app's users can anonymously send postings, so-called "Jodels", in the form of short texts or photos which are presented to other users in an area of about ten kilometers in a feed (see Fig. 1). These postings can be commented, shared, pinned or reported by others. Postings, as well as comments, can be either up- or downvoted (2) by every user inside the radius. A Jodel, which receives five downvotes, gets automatically deleted from the feed. Above the feed, the user sees his current location (8). Each posting contains its "age" and how far away from the user it was sent (1), as well as the number of received comments (4) and upvotes (2). There are three categories regarding the postings' order, "newest" (5), "most commented" (6) and "loudest" (7). In "loudest", the postings are ordered descending by the number of their upvotes. As an appeal for the app's use, users collect so-called "karma points" (3). They receive them, among other things, for own postings getting upvoted and active contribution to the community, e.g. voting other postings. The app also allows the use of hashtags, but a search for hashtags is not yet implemented. Furthermore, there are channels (9), which work like groups, in which users can exchange to a certain topic. Everyone can search for channels and see their content but has to join to interact. Channels are, however, not available for every location yet. The main aspect, in which Jodel distinguishes from Yik Yak, is the moderator system. An algorithm which regards aspects like positive contribution to the community chooses app users as moderators. An additional area is unlocked to these users where they can allow or block reported postings. Decisions are always made collectively by many moderators. As an addition, Jodel cooperates with the police when postings comprise crime acts or their announcement and offer information like location und IP address. These additional aspects are intended to ensure that the published content on Jodel remains within the law despite anonymity, as Yik Yak was repeatedly accused to have serious problems on this point [4, 5] which ultimately led to the app being shut down.

2 Methods

The approach of this evaluation follows the Information Service Evaluation (ISE) model from Schumann and Stock [1]. The model (see Fig. 2) combines the most important aspects of existing evaluation models from different academic disciplines to create a comprehensive holistic evaluation model which can be used for the assessment of every information service.

The three categories information service, information user and information acceptance were considered. The first dimension includes aspects like the perceived

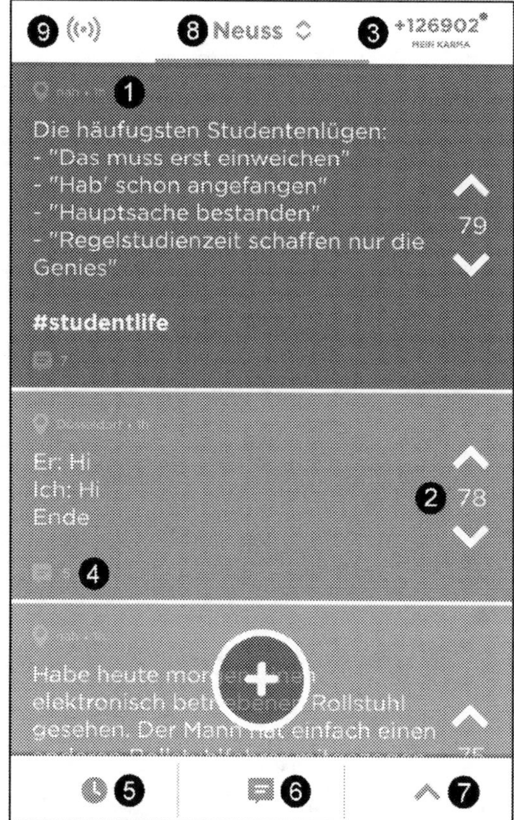

Fig. 1. Screenshot of the Jodel feed.

quality of the information service, the perceived quality of the content and the objective quality of the information service. The second dimension concentrates on the information user, in particular, on his information need and information behavior. The information service quality, as well as the aspects information need and information behavior, are decisive for the user's adaption of the service. Out of this adaption, a regular use can arise, depending on the user's satisfaction. Through a regular use, an influence on the user's information behavior can occur and the diffusion of the service can be driven forward, if the so-called network effect arises.

2.1 User Survey

With orientation towards the ISE model, a user survey was created which includes questions regarding the three dimensions information service, information user and information acceptance. Many aspects of the ISE model's dimensions walk along with the survey participants' perceptions. As a procedure to measure those perceptions, a Likert scale [6] was consulted. The survey consists of several sections which comprise

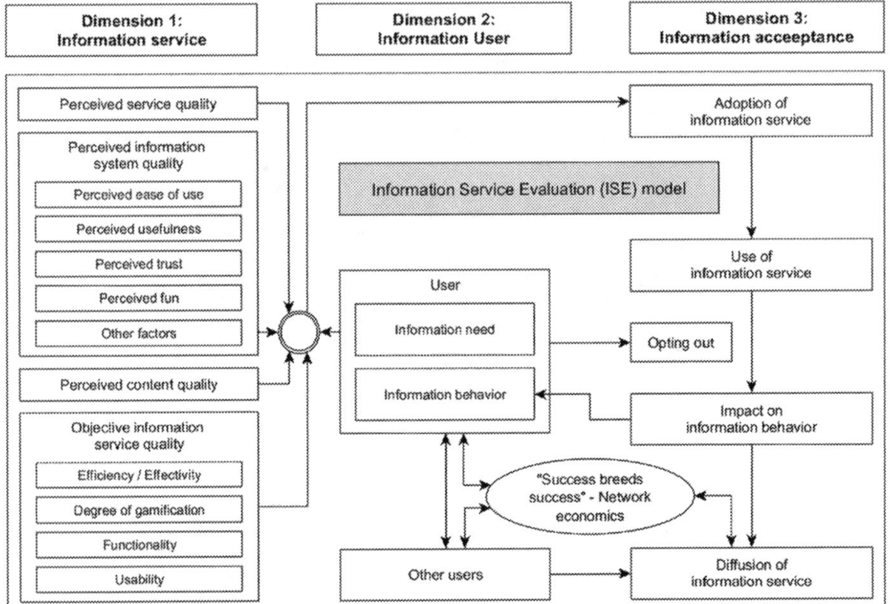

Fig. 2. The Information Service Evaluation (ISE) model based on Schumann and Stock [1].

different Likert items of a certain topic. The participants give their answers by means of a scale from 1 ("I totally disagree") to 7 ("I totally agree"). The number 4 represents abstention or indecision. After finishing and activating the online survey, it was distributed between October 2016 and March 2017, primarily over Jodel itself. The link to the survey was distributed in 16 larger German cities, like Cologne, Düsseldorf, Hamburg and Aachen. The survey was also spread over Facebook, because not only users were interesting for the evaluation, but also non-users and former users. Surveying the non-users should give an insight into aspects which stand against an adaption of Jodel while surveying the former users was important to figure out reasons for performing an opting-out. In total, 1,009 persons participated and finished the survey. As Jodel was the main distribution channel, the absolute majority is using Jodel, with a number of 877 participants, while 93 are non-users and 39 are former users.

For the evaluation of the online survey, it was mainly worked with the statistics software IBM SPSS Statistics and the spreadsheet program Microsoft Excel. To every topic that should be answered by means of scales of Likert-type it was computed the percentage distribution, the average and the standard deviation in SPSS Statistics. Furthermore, the survey results were tested for correlations. With the Pearson correlation coefficient r, it is possible to determine linear relationships between two variables. The accruing result is a value between −1 and +1, whereby a positive or negative correlation can be asserted. The more the value converges +1 or −1, the stronger is the effect. Although it is regarded controversially to equate ordinal scales such as the Likert scale with interval scales [7], in this study the distances between the values are considered exactly the same, following the opinion of Carifio and Perla [8].

Moreover, within the scope of the survey, the SERVQUAL method [9] was applied, where the users' expectations and perceptions are focused. The users' answers are obtained through Likert-type scales [6] which comprise seven gradations ("I totally disagree" (1) to "I totally agree" (7)) and divided into expectations and perceptions. Following, the difference between the two values is computed with a gap analysis. For the gap analysis, regarding to a question or statement, the perception value (P) is subtracted from the expectation value (E) to estimate the gap (Q). It follows: $Q = E - P$ [9].

2.2 Developer Survey

In addition to the user survey, inspired by the Customers Value Research method [2], the expectations of the developers were surveyed, to compare them to the users' actual perceptions. The aim was to find out, how the developers assume the users' answers to the user survey. For this purpose, an almost identical survey was created. The developers' answers were also given with scales of Likert-type. Their task was to assume, for every statement or question, how the participants of the user survey would answer in average. Afterwards, it should be determined, if the developers' expectations strongly differ from the participants' perceptions, with the mentioned gap analysis. The survey was answered by Alessio Avellan Borgmeyer, the app's founder, representatively for the developers, on November 30, 2016.

2.3 Expert Interview

To receive additional information about Jodel, an interview with the app's founder was conducted. The questions comprised topics like the foundation of Jodel, inappropriate behavior on Jodel, the moderator system and future plans. The questions were answered as a voice mail on November 30, 2016.

3 Results

3.1 User Survey

Out of 1,009 survey participants 87% are active users of Jodel. 9% do not use Jodel, whereas 4% of the participants are former users.

Demographic Data. Around 66% of the respondents are female, 34% are male. 95% of the survey participants were born between 1989 and 1999. The peak is at year 1997 with 15.3%. It can be summarized that most of the respondents are between 19 and 25 years old. Most of the respondents are students (72%). 10% are employed, 8% are training and 6% attend school.

Information Service. Jodel is very easy to handle (mean: 6.67) and is fun (6.24) for the users. Regarding usefulness (4.75) and trust in the promised anonymity (4.99), the results are mixed but with a positive tendency.

To examine the content quality on Jodel, the aspects orthography, currency, clearness and trust were considered, based on the suggestions from Parker, Moleshe, De la Harpe, and Wills [10]. For the clear majority of the survey participants it is important to pay attention to orthography (5.51). Also, the currency of topics (4.87) and the clearness of the postings' meaning (4.93) are relevant for the users. Furthermore, the users are skeptical regarding the veracity of user content in social networks generally (2.76). According to the respondents, it is rather not payed attention to orthography (3.62). The users are often skeptical regarding the veracity of user content on Jodel (2.90). On the other hand, the topics on Jodel seem to be quite relevant (4.87) and the clearness of the postings' meaning attains an intermediate level (4.26).

The users' expectations and experiences are contrasted by using the SERVQUAL method in Fig. 3. The results show that the expectation of the topics' currency is met exactly, both means yield 4.87. In turn, the clearness of the postings' meaning is on a lower level than expected. The difference is −0.67 with means consisting of 4.93 (expectation) and 4.26 (experience). On Jodel it is significantly cared less about orthography (3.62) than the users would have wished (5.51) which results in a profound difference of −1.89. The users evince increased trust in the veracity on Jodel compared to other social media, the difference is 0.14 with small means of 2.76 (expectation) and 2.90 (experience). However, the means are on a very low level.

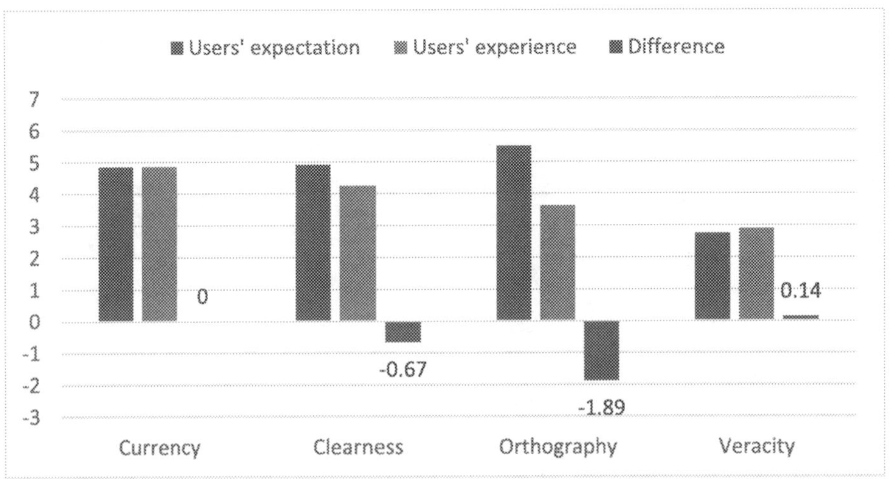

Fig. 3. Gap analysis of users' expectations and experiences with Jodel. N = 877, Scale: 1 (totally disagree) - 7 (totally agree).

Jodel is characterized by an intuitive usability (6.11) and a reliable handling (5.62). Technical problems occur rather rarely (5.40).

Within the Information Service Evaluation (ISE) model the aspect of gamification is also playing an important role. According to Knautz [11], gamification elements are implemented into "non-game contexts" [12] "to obtain an increase of the use motivation". Such elements can be, among others, a reward system, levels, virtual goods, leaderboards and so-called quests [13]. The grade of gamification is examined by counting the implemented gamification elements of the information service. Regarding Jodel, the karma point system is the most conspicuous element. It is intended to motivate the users to become active in a positive way, that means writing postings or answers which are appreciated by other users and using the voting system to receive karma points. Another element of gamification is the category "loudest". Here the current postings get sorted in descending order regarding their number of upvotes, hence the category can be considered as a sort of leaderboard. This can motivate the users to produce content which receives upvotes to position their postings as high as possible in the leaderboard. When a moderator successfully evaluated several postings, he gets rewarded with the "picture of the day". In the broadest sense, this feature also can be considered as an element of gamification, as reward or achievement which motivates the moderators to not abort the moderation session early. It can be doubted if the rewarding effect of the "picture of the day" suffices to apply it as an adequate element of gamification, because it is not really a game element. Therefore, we count 2.5 gamification elements.

Most of the survey participants would use Jodel even without the existence of karma points (5.99), whereas the voting function is considered as quite relevant ("usage without voting function": 3.57). The joy about the own posting being among the "loudest" is especially high (6.32). Thereby is no appreciable difference to the own posting becoming the "loudest" in the vicinity (6.33). The impact of karma points on voting is indicated as low (2.78). At the time of the survey, the users received karma points for both up- and downvoting. Since November 2017, downvotes "cost" the voting user karma points to restrict the so-called "Downvote mafia", which means users who vote other postings consequently down regardless of their content.

The number of required clicks to manage a specific task can be an indicator for the quality of the navigation elements [14]. The possible main activities include reading postings, voting postings, publishing short texts or photos (as posting or comment) and for several users moderating reported content. To read postings the app only needs to be opened since the start screen is already showing the "newest" feed. One click is necessary to switch between the categories "newest", "most commented" and "loudest". The voting function is usable directly in the feed by clicking the "arrow" up or down. To write a posting, one click on the big "plus" is required, for photos an additional click on the camera symbol. After writing a short text or taking a photo, it is required to click on "send". To moderate reported content, the user reaches the menu by one click on the karma points and selects the item "moderation". Due to these two clicks the user gets to the moderation screen which displays the first posting as well as the options "allow", "I don't know" and "block". Overall, not more than three clicks are necessary to perform the main activities.

Information User. Jodel is mainly used to read funny postings (5.94) and out of boredom (5.52). On the other hand, it is not a relevant issue to find a partner (1.42) or friends (1.76). Trolling, which means to provoke others aiming to cause damage [15], is also less pronounced (1.62). The incentive to collect karma points is with a mean of 3.13 rather weak, the high standard deviation of 1.88 indicates that the survey participants' motivations vary widely. There is a similar high standard deviation from 1.8

to 1.9 regarding the aspects "reach people in the vicinity" (3.51), "exchange with others" (4.01) and "ask questions anonymously" (3.81).

Information Acceptance. A vast majority (71.4%) uses Jodel several times a day, 11.1% even every hour. 9.5% of the respondents use the app once a day and 7% several times a week. The answer possibilities "once a week", "once a month" and "less than once a month" hardly attracted attention with summed about 1%.

Jodel is used 5 to 30 min at a stretch by 92.5% of the users, divided into three groups: 31.8% use Jodel around 5 min, 37.2% use the app circa 10 min at a stretch and 23.5% use Jodel 15 to 30 min. 4.3% use Jodel 30 to 45 min, whereas only a few users use the app under one minute (1.5%), one hour (1.3%) or even longer (0.5) at a stretch.

The users are satisfied with Jodel and widely agree on the grade of satisfaction: the mean of 5.80 is high and the standard deviation is with 0.99 at a low level. The survey participants averagely gained positive experience using Jodel: many users felt good after the use (4.52), encountered acceptance and appreciation (4.50) and received good advice on a topic (4.26). Apart from that, less users made friends on Jodel (2.10).

851 out of 877 active users would recommend Jodel or already did, which corresponds to 97%. Only 26 out of 877 (3%) respondents would not recommend the app to others.

Multiple answers were possible, regarding the question how the users encountered Jodel. With more than 76% the survey participants most commonly became aware of Jodel through friends. Facebook pages about Jodel aroused interest of 26%. Campus advertising almost reaches 11%, followed by discovering Jodel in the "App Store" or "Google Play Store" with more than 6% and through relatives with almost 5%. Print media has as good as no relevance on the adaption, only 1.7% of the respondents encountered Jodel this way. 1.9% of the survey participants gave an individual answer, primarily mentioning social networks like "Instagram", "YouTube" and "Twitter".

Jodel has the biggest impact on the daily smartphone usage (3.90). On social behavior (2.13), sleep (2.30) and leisure time (2.14) Jodel has less impact. Beyond that, the app most likely impacts the learning behavior (2.73).

Opting-out. The most distinct reason for active users to opt out is "too much distraction" (3.47). This is followed by "too many reposts" (3.17), that Jodel became boring or less interesting (2.98) and that too many inappropriate postings circulate (2.60). Less relevant are possible negative impacts on the user's behavior (1.73), friends or the environment not using the app anymore (1.54) or getting too old for Jodel (1.36).

The users also had bad experiences with Jodel. The respondents mostly bothered about baseless negative ratings regarding their own postings (4.37), followed by spotting racist, sexist or otherwise inappropriate content (3.97). The users were rather less offended or threatened in comments (2.38). After the usage hardly anyone felt bad (1.87). The very high standard deviations (circa 2) regarding the first three named experiences are striking: therefore, the extent of the negative experiences with Jodel were quite different.

Non-users. So far, 877 users of Jodel were examined, now the 93 non-users are considered. Main reason for not using Jodel is lack of interest (4.95). The second most important reason is that nobody in the vicinity uses the app (4.40). Also quite relevant are no awareness of the app (3.35) and the concern that Jodel would be too distracting (3.57). There is nearly no shortage of trust in the promised anonymity (2.89). Almost nobody was discouraged from using Jodel (1.69).

Former Users. Finally, the motives for the opting out of 39 former Jodel users are being investigated. The most important factor for ending the use of the app was that Jodel became boring or less interesting (4.97). Also quite relevant was that the app had caused too much distraction (3.92) and that in the opinion of some former users too many reposts were in circulation (3.59). Less relevant was that friends no longer used the app (3.08), too many "inappropriate postings" circulated (2.85), or negative impact on the users' behavior (2.36).

3.2 Correlations

Below, the results of our survey are examined on Pearson correlations regarding the answer options. Significant correlations (error probability below 5%) are marked with "*", very significant linear relationships (error probability below 1%) with "**". In the following, the focus is placed on correlations between usage reasons, usage time and usage frequency.

Usage Frequency, Usage Time and Usage Reasons. The more Jodel is used, the less the users do it out of boredom (−.068*), but rather due to productive reasons, led by collecting karma points (.190**) and asking questions anonymously (.185**). It is similar regarding the duration of use, except that there is no linear relationship between the length of the session and collecting karma points (.045) as well as between the length of the session and reading funny postings (.030). The more the app is used out of boredom, the less Jodel is used for other reasons. The most pronounced are the negative correlations with social interactions ("exchange with others": −.198**, "reach people in the vicinity": −.146**), whereas there is no connection between boredom and reading funny postings (.037) or collecting karma points (.063).

Anyone who uses Jodel primarily because of funny postings uses the app more often (.115**) and is less in search of friends (−.094**) or a partner (−.091**); however, the effect size here is low. To ask questions anonymously correlates with the exchange with others (.533**). Also, the linear relationship to reaching people in the vicinity is worth mentioning (.334**). In general, it can be stated that all social aspects of using Jodel correlate consistently with each other. Also worth mentioning in this context are the effect sizes of the correlations between finding friends and finding a partner (.546**), finding friends and reaching people in the vicinity (.410**) as well as exchanging with others and reaching people in the vicinity (.445**). Users who want to disturb the positively connotated construct of Jodel, prefer hating or trolling. The more such users do this, the more they want to earn karma points (.148**) and, interestingly, the more they are looking for a partner (.135**). There is also a negative correlation to reading funny postings (-.105**).

Usage Reasons and Perceived Quality. The higher the quality of Jodel is perceived, the more often the app is used. The usage frequency correlates most with the fun factor (.289**). Regarding the duration of use, the correlations are smaller, concerning the ease of use not even present anymore. Anyone who uses Jodel primarily out of boredom considers the quality slightly worse, except for the ease of use.

Particularly large are the effect sizes of the correlations between fun and reading funny postings (.318**), usefulness and asking questions anonymously (.286**), usefulness and the exchange with others (.319**), as well as usefulness and reaching people in the vicinity (.278**).

Usage Reasons and Gamification. The more often one uses Jodel, the greater is the joy of having a "loud" posting (.212**), and the more often the voting function will be used just to get karma points (.138**). In terms of usage time, there is no correlation with gamification incentives. The joy about a "loud" posting is especially high for those who want to collect karma points (.296**), compared to other usage reasons. The more these users want to earn the points, the less they would use Jodel without karma points (−.445**) and the more likely they would vote just to get karma points (.446**). The effect sizes here are noticeably high.

Usage reasons and positive experiences. Those who use Jodel more often gain more positive experiences and are more satisfied with the app (.140*). Those who use Jodel out of boredom feel less accepted (−.147**) and got rarely advised well (−.116**). On the other hand, users who use Jodel more for asking questions and exchanging information with others were able to find particularly good advice (.482** and .400**). If you want to find friends on Jodel, you will rather find them (.509**), even those who are originally looking for a partner (.347**). Above all, users who want to exchange with others (.292**) and reach people in the vicinity (.249**) felt good after using Jodel. The more you use the app because of funny postings, the more satisfied you are with Jodel (.244**). This is the only usage reason where there is a more pronounced correlation with user satisfaction. Haters and trolls are rather dissatisfied with the app (−.150**).

Usage Reasons and Change of Behavior. The more often or longer Jodel is used, the greater are the behavioral changes of the users. Especially the use of mobile phones is influenced by the frequency of use (.266**). While the use due to boredom and funny postings does not affect the users' behavior, it is, from a lesser to a moderate extent, the case due to social reasons. For example, the linear relationships between social behavior and the motivation to find friends (.238**) or a partner (.257**) or the correlation between sleep and asking questions anonymously (.202**) deserve special mention. The more users hate and troll, the greater is, at least to a lesser extent, the impact on their social behavior (.106**).

Usage Reasons and Possible Reasons to Opt Out. There are only a few linear relationships between the reasons for the use and possible reasons for opting out. Anyone who uses Jodel out of boredom more likely opts out than other users. Negative influences (.165**), distraction (.185**) and rising boredom (.203**) are the most important aspects.

Usage Reasons and Negative Experiences. Particularly negative experiences had the users who want to ask questions anonymously. Their postings were more often rated negatively for no reason (.251**) and they were more often insulted or threatened in the comments (.234**).

3.3 Customers Value Research

According to Customers Value Research [2], the expected values of Alessio Avellan Borgmeyer, representative for the Jodel developers, are compared with the results of the user survey and a gap analysis is used to calculate the difference between the values. Just like the users, Borgmeyer only had the opportunity to express his answers through integral numbers on a scale from 1 to 7. A deviation of less than 0.5 is therefore to be regarded as correctly expected, because in some cases a closer approximation was not possible.

Information Service. The usefulness of Jodel was rated much smaller by the developer than by the users. The difference is 1.75 (Fig. 4). Also, the confidence in the promised anonymity is greater among the users than Borgmeyer suspected. The ease of use and the high fun factor almost met the expectations of the developer.

With regard to the users' expectations about the clearness of a posting's meaning and respect for orthography, Borgmeyer's values are a bit lower (Fig. 5). The expectation regarding the currency of the topics was met. Concerning the trust in the veracity of content on social media in general, users are more critical than he expected.

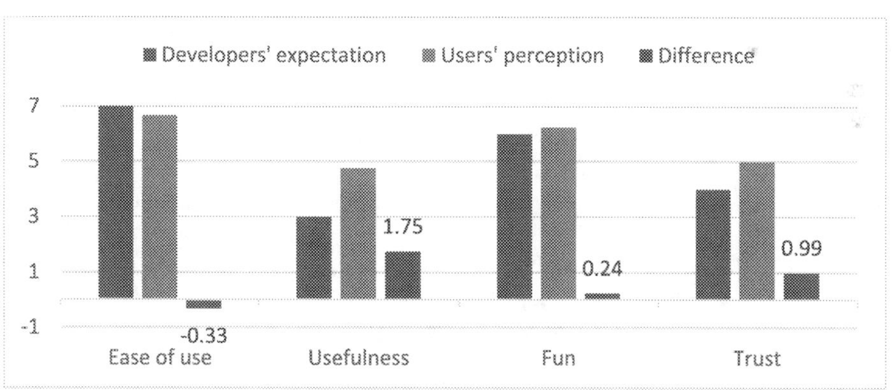

Fig. 4. Perceived quality of Jodel. N = 877, Scale: 1 (totally disagree) - 7 (totally agree).

On Jodel, respect for orthography and trust in the veracity are lower than Borgmeyer expected (Fig. 6). The currency of the topics and the clearness of the postings' meaning are in line with his assessment.

Users have far fewer technical problems than the developers suspect, with a gap value of 2.4 (Fig. 7). The level of immediate intuitive usability was rated a bit too high, but the reliable handling was expected correctly.

Borgmeyer overestimated the relevance of the voting function, the difference to the users' assessment is 1.57 (Fig. 8). The joy of seeing own postings among the "loudest"

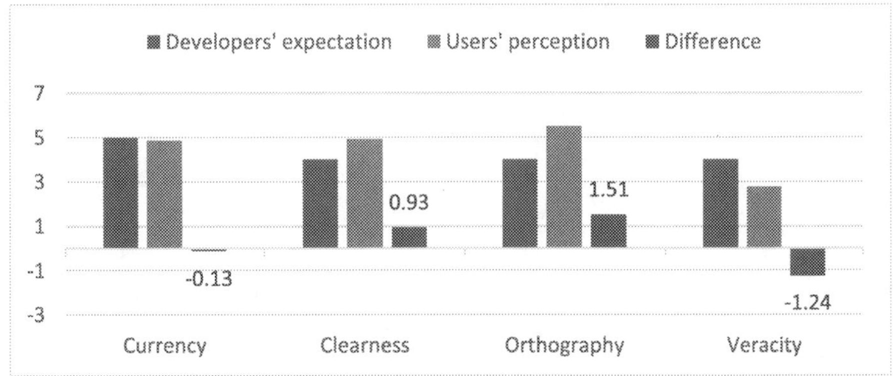

Fig. 5. Expected content quality of Jodel. N = 877, Scale: 1 (totally disagree) - 7 (totally agree).

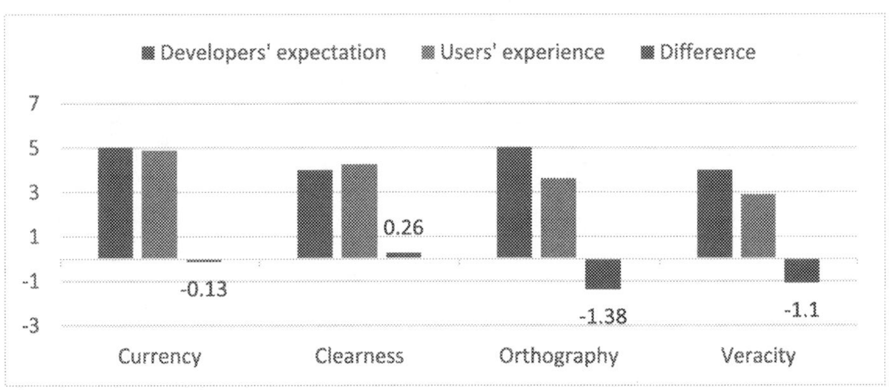

Fig. 6. Experienced content quality of Jodel. N = 877, Scale: 1 (totally disagree) - 7 (totally agree).

was correctly predicted, but he did not expect that there is almost no difference to reaching the first rank in the list. The influence of karma points on the voting activity was also overestimated. The developer correctly guessed that most respondents would use Jodel without karma points: the difference amounts to only −0.01.

Information User. With three exceptions, Borgmeyer was able to correctly assess the weighting of the different reasons for using Jodel (Fig. 9). He weighed the reasons to anonymously ask questions, to exchange with others and to reach people in the vicinity higher than the users of the survey did.

Information Acceptance. On the one hand, the users received less good advice and fewer friendships arose than Jodel's developer would have thought, but on the other hand, the user satisfaction is higher than expected (Fig. 10). Regarding the good experiences "encountered acceptance and appreciation" as well as "felt good after the use", Borgmeyer met the result of the user survey with a small deviation of −0.5.

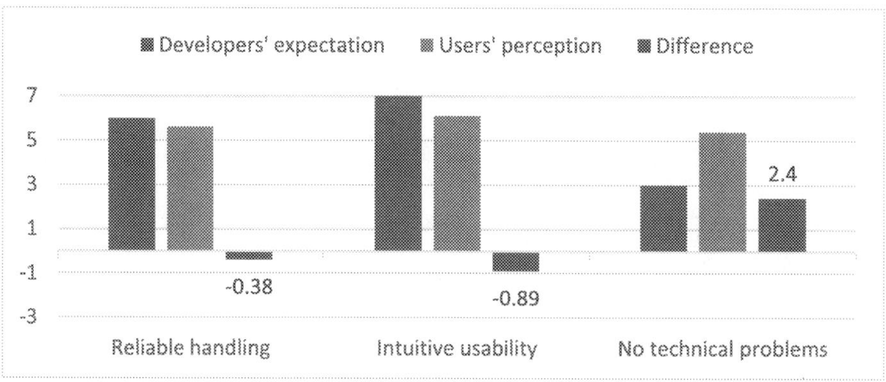

Fig. 7. Operability of Jodel. N = 877, Scale: 1 (totally disagree) - 7 (totally agree).

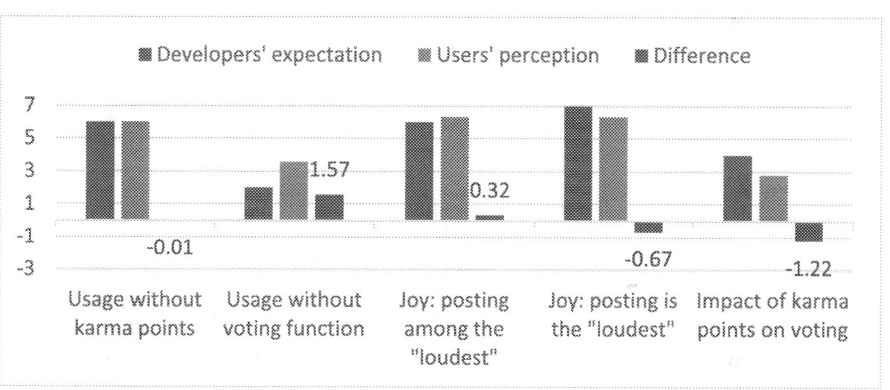

Fig. 8. Gamification and incentives. N = 877, Scale: 1 (totally disagree) - 7 (totally agree).

Opting-out. There is a large discrepancy between Borgmeyer's expected values and the users' assessment regarding possible reasons to opt out of Jodel (Fig. 11). There were consistently higher scale values assessed than it actually was the case. The biggest gap exists for the reason to have grown too old, with a value of −4.64. A difference of over −3 points exists for the motives that friends no longer use the app and that Jodel became boring. Also, the extent of too much distraction, inappropriate postings and too many reposts was overestimated. Only the low mean of the reason "negative influences on behavior" was correctly predicted.

Borgmeyer estimated the users' negative experiences with Jodel a bit too high in three out of four cases (Fig. 12). Self-written postings got less downvotes for no reason, not so much racist or sexist content was spotted, and the users felt less bad after use than the developer assumed. Borgmeyer expected correctly that users were not that often threatened or insulted in comments: the difference is only 0.38 here.

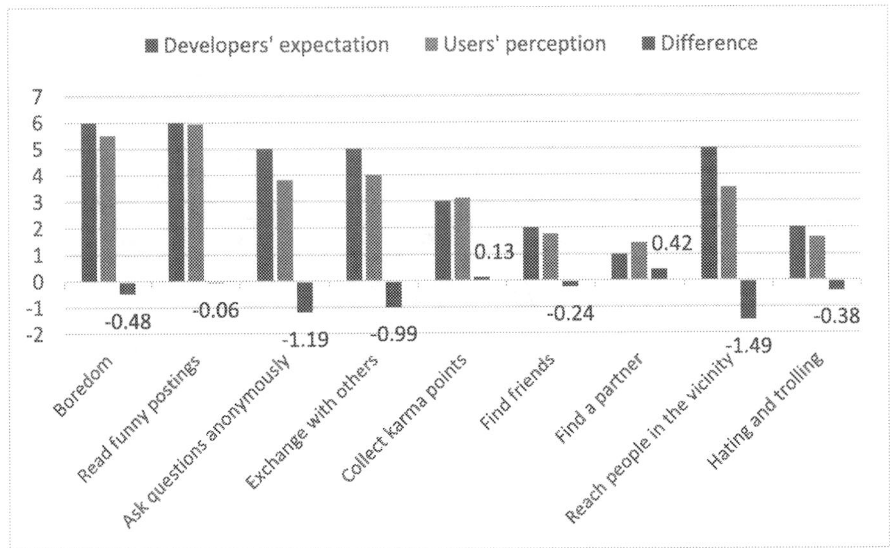

Fig. 9. Reasons for the use of Jodel. N = 877, Scale: 1 (totally disagree) - 7 (totally agree).

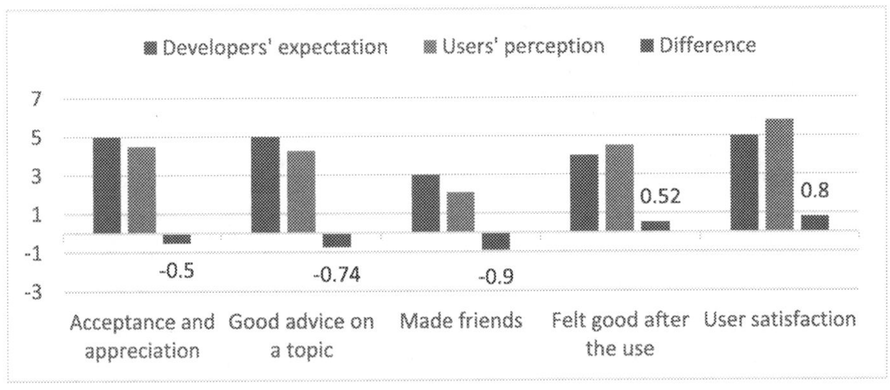

Fig. 10. Good experiences with Jodel and user satisfaction. N = 877, Scale: 1 (totally disagree) - 7 (totally agree).

3.4 Expert Interview

In addition to user and expert survey, an expert interview was conducted with the Jodel founder to get more detailed information about the app from the developers' site. Borgmeyer explained that the app launched on October 20, 2014 and that the diffusion nearly "exploded" for that time. Initially, the distribution took place mainly via campus advertising in form of flyers. He explained that at the time of the interview, Jodel was, besides Germany, successful in all the Nordic (European) countries, Austria, Switzerland, France and partly in Italy. The establishment in the Southern countries is

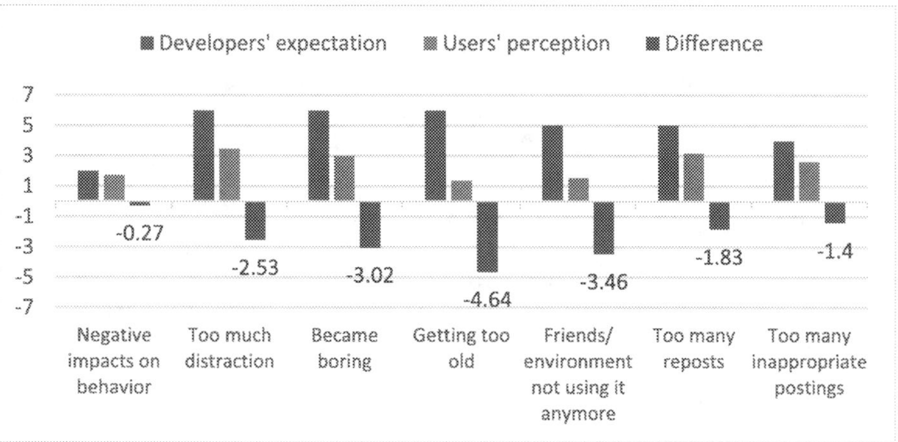

Fig. 11. Possible reasons for opting out. N = 800, Scale: 1 (totally disagree) - 7 (totally agree).

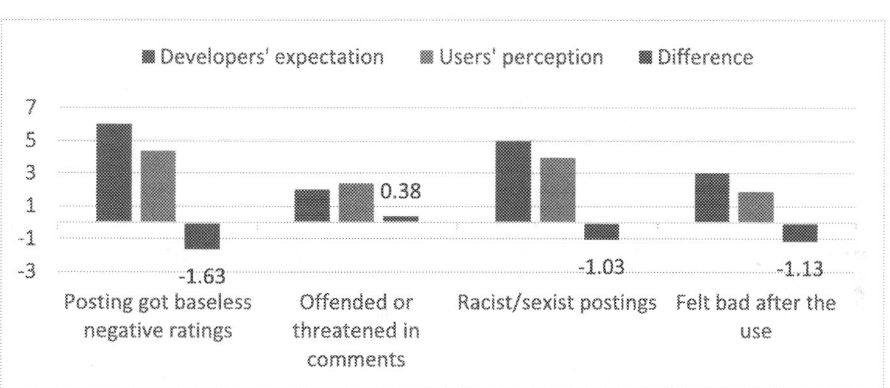

Fig. 12. Negative experiences with Jodel. N = 877, Scale: 1 (totally disagree) - 7 (totally agree).

harder, while especially in the Nordic countries, there are partly more downloads of the app than students, according to Borgmeyer. He explained, that they aim to make Jodel a must-have for every student and that they want to "instantly connect everyone sharing the same location". Furthermore, Borgmeyer stated that they want to reach this goal for students first and afterwards want to focus on other demographics, too. So far, no revenue could be generated with Jodel and the whole funding is currently held only by investors. According to Schlenk [16], Adam di Angelo, the first CTO of Facebook, is one of these investors. Borgmeyer said that Jodel's business model will most likely be ad-based in the future.

4 Discussion

From the answers of overall 1,009 participants it clearly arises that the student-app Jodel did not miss its target group. 72% of the participants are students. The users' age underlines this, because the majority is in the typical student age, between 19 and about 25 years. The participants' age indicates that they almost completely belong to the generation of "digital natives" and only a few exceptions can be associated to the "digital immigrants". Prensky [17] explains "The 'digital immigrant accent' can be seen in such things as turning to the Internet for information second rather than first". This statement emphasizes that a platform like Jodel is, especially for this reason, more interesting for those, who meet their information need, as well as their need to communicate, primarily online, which should be mainly the case for the "digital natives". As a border between the two generations, the birth year 1980 was considered, as suggested by Palfrey and Gasser [18].

Regarding the perceived quality of Jodel, the users agree that Jodel is easy to use, even at first use. This result is underlined by the analysis of the number of clicks which are necessary to perform certain tasks. All the main tasks, which are possible on Jodel, are reachable with at most three clicks. Jodel furthermore means a lot of fun for its users, a significant aspect in relation to the service quality. The usefulness of Jodel is regarded controversially, but the participants tend to agree with it, while it is even higher rated than the developers expected.

Referring to the content quality, it is noticeable that the currency of the topics is perceived just the same as expected. The perception of the orthography significantly sinks in contrast to the expectations of the participants and the developers. It is interesting that the participants tend to not believe what other users publish on Jodel, but still the veracity is perceived higher than in other social media, despite the anonymity.

From the sum of experiences, the participants made with Jodel, it is possible to gain an approximate impression of Jodel's objective quality. Overall, it is perceived as quite positive. Regarding the effectivity, the users agree that Jodel "always does, what it should do". Furthermore, the users tend to not have any technical problems with the app, even less than the developers expected. This can be seen positively for the effectivity, as well as for the efficiency of Jodel. The immediately intuitive use of Jodel, which represents the aspects ease of use, functionality and usability, also finds agreement among the participants. These positive experiences, especially at the first use, support the transition from the adaption to the use and significantly increase the probability that the users accept the service.

2.5 elements of gamification were counted: the karma points, the category with the "loudest" postings and the reward for the moderators with the "picture of the day". 2.5 gamification elements may seem less on the first impression, but if one regards the app's simplicity and that the "normal" use (without moderation) actually only comprises writing postings and voting, it can be seen that the two elements karma points and "loudest" category are used wisely and present to motivate the users to participate positively in the community. The survey results showed that karma points were not rated as very relevant for the use. A reason for this could be that it is neither possible to compare them to other users' points, nor to change them into something. But it is important to consider that a too strong incentive, like the possibility to use the karma points in online shops, would probably lead to misuse. The survey showed that the karma points, in the way they are implemented, are not having a big effect on the decision to vote a posting. A premium concept would probably change this and would show a negative influence on the honesty behind the votings. The fact, that the developers assessed the relevance of the karma points, with a deviation of 0.01, almost exactly like the users did, lets infer that the degree of motivation was not randomly estimated that low. The fact, that additionally the "picture of the day" was implemented for the moderators, shows, that there is at least one incentive for every (positive) action on Jodel. It is concluded that the selection of gamification elements was thought out and reasonable, especially regarding their degree of motivation, without overloading the app or causing misuse.

The survey showed very clearly in which time intervals the use of Jodel takes place. 92% of the participants use the app at least once a day, most of them even several times. This leads to the assumption that Jodel is not a service which is used in irregular time intervals, but that the app is more like an element of the users' everyday life.

Referring to the user satisfaction, Jodel performs very well. Almost 92% of the participants are rather up to totally satisfied. This satisfaction clearly reflects in the recommendations, as 97% said, they would recommend Jodel. That many users already did this is proven by 76% of the participants who stated they found out about Jodel through their friends. Only 11% took notice of Jodel through campus advertising. Within the interview, Borgmeyer said, this was the initial distribution method. It can be assumed that the campus advertising became redundant at the moment Jodel reached the critical mass and the network effect occurred, because the distribution hived off through recommendations.

It is striking that, despite the daily use of the most users, the participants rather not notice any (negative) influences on their behavior. For Jodel, this is quite positive, as it cannot occur a "pushback" [19] where users see negative influences on their behavior and reduce or even quit the use of the service.

To find out which aspects persuade the users to use Jodel regularly, they were asked about good experiences with the app. The main aspects were that the participants tend to feel good after the use, that they encountered acceptance and appreciation and that they found good advice on a topic. They felt even better than the developers expected. This is an evidence for a positive community which is also willing to share help, advice or information.

The possible reasons for an opting-out showed again that the users in average see no negative influences on their behavior. The highest rated reason to quit the use is distraction through the app. Too many reposts and inappropriate postings can be relevant factors, too. But overall, the participants in average did not agree with any of the reasons. This shows that Jodel has no certain weakness which would lead the users to opt out of Jodel. It seems more like the users see possible reasons, but none of them as critical. It is interesting that it seems like it is not playing any role if friends quit using Jodel. This presents clearly an advantage for the app, because therefore it cannot occur any "negative network effect" where users get enticed away from Jodel by their friends.

"Jodel became boring or less interesting" was rather not a reason to opt out for the surveyed users, however, it was the aspect which the former users rated as the most important reason for quitting the use of Jodel. As it was already found out that many users need Jodel as a remedy for boredom, it makes sense that they opt out when the app cannot serve this purpose anymore.

The strongest pronounced reason for not using Jodel is simply no interest. It is noticeable that the fact that no one in the personal environment uses the app is also not sensed as insignificant. According to this, the personal environment seems to play a role for the adaption of the service but gets irrelevant after passing to a regular use. The fewest participants got advised against using Jodel, what can be explained with the high degree of user satisfaction.

Bad experiences with Jodel were also examined to work out which aspects can provoke discontent. A bad experience which most of the participants had was receiving baseless downvotes for a posting. Regarding the joy, the users feel, when their postings make it into the category of the "loudest" Jodels, it is understandable that the anger about baseless downvotes is high. Racist or sexist postings are a sensitive topic on Jodel, too. To get insulted or threatened in the comments did either happen rarely or was perceived as less severe. However, it can be recorded that the respondents tend a lot more to feel well than bad after the use of Jodel and also less bad than the developers expected them to feel. In general, the users were able to have more positive than negative experiences. This is also reflected in the high user satisfaction and can be seen positively regarding the transition from adaption to regular use. A study from Johann, Wiedel et al. [20] underlines this, since they examined that only 16.7% of 2,542 regarded postings contain slightly toxic expressions, while 38% of all considered answers showed prosocial behavior, like empathy or advices for example.

All in all, the developers accurately assessed the users' answers, without overestimating the app's effect. Summed up, Borgmeyer, representatively for the developers, assessed 20 out of 49 answers "correctly" (with deviations up to 0.5). Eight more were only slightly off target, with deviations less than one point. Only five answers were strongly aside the users' answers, with more than two points of deviation. In those cases, the developers either underestimated positive or overestimated negative aspects of the app. This can be seen as a more conductive result than if the own service was consequently overestimated, because the critical attitude shows that the developers see problems, on which they have to work, even if they are not considered that distinct by the users.

This evaluation, based on the ISE model from Schumann and Stock [1], ensures a comprehensive investigation of the app, as it combines a user survey, a developer survey and an expert interview with the analysis methods correlation analysis, SERVQUAL [9] and Customers Value Research [2]. Arising from this evaluation, the final assessment can be derived that Jodel constitutes a social information service of high quality.

It will certainly be interesting to observe the further development of Jodel. Within the expert interview, Borgmeyer addressed a possible expansion of the user group, after the diffusion among students has been successfully performed. It can be interesting to observe, if such an expansion will actually occur and if it does, which impacts it will have on the communities' group identification if the homogeneity of the user group will be decreased. Furthermore, it will turn out if Jodel will be successful with its future business model and if the investments will finally pay off. On the one hand, Jodel is very suitable for precise local advertising, due to the restricted area, but on the other hand, students do not present the most profitable target group. According to Schlenk [16], an expansion of Jodel into the USA lies ahead which can be a relevant step regarding the further diffusion. It will be interesting to see if the establishment will succeed and if Jodel will be able to adopt the position of the shut-down app Yik Yak. If this case occurs, Jodel will certainly become the topic of many further researches.

References

1. Schumann, L., Stock, W.G.: The Information Service Evaluation (ISE) model. Webology **11** (1), 1–20 (2014). https://doi.org/10.3233/ISU-140759
2. McKnight, S.: Customers value research. In: Flaten, T.K. (ed.) Management, Marketing and Promotion of Library Services Based on Statistics, Analysis and Evaluation, pp. 206–216. K. G. Saur, Munich (2006)
3. Graham, J.: Yik Yak, The Once Popular and Controversial College Messaging App, Shuts Down. USA Today, 28 April 2017. https://www.usatoday.com/story/tech/talkingtech/2017/04/28/yik-yak-shut-down/101045670/. Accessed 31 Jan 2018
4. Lee, J.-S., Yang, S., Munson, A.L., Donzo, L.: What people do on Yik Yak: analyzing anonymous microblogging user behaviors. In: Meiselwitz, G. (ed.) SCSM 2017. LNCS, vol. 10283, pp. 416–428. Springer, Cham (2017). https://doi.org/10.1007/978-3-319-58562-8_32
5. McAuley, D.: Digital graffiti. Forum Mag. **2015**(1), 32–34 (2015)
6. Likert, R.: A technique for the measurement of attitudes. Arch. Psychol. **22**(140), 5–55 (1932)
7. Jamieson, S.: Likert scales: how to (Ab)use them. Med. Educ. **38**, 1212–1218 (2004). https://doi.org/10.1111/j.1365-2929.2004.02012.x
8. Carifio, J., Perla, R.J.: Ten common misunderstandings, misconceptions, persistent myths and urban legends about likert scales and likert response formats and their antidotes. J. Soc. Sci. **3**(3), 106–116 (2007). https://doi.org/10.3844/jssp.2007.106.116
9. Parasuraman, A., Zeithaml, V.A., Berry, L.L.: SERVQUAL: a multiple-item scale for measuring consumer perceptions of service quality. J. Retail. **64**(1), 12–40 (1988)
10. Parker, M.B., Moleshe, V., De la Harpe, R., Wills, G.B.: An evaluation of information quality frameworks for the world wide web. In: 8th Annual Conference on WWW Applications, 6–8 September 2006, Bloemfontein, Free State Province, South Africa (2006)
11. Knautz, K.: Gamification im Kontext der Vermittlung von Informationskompetenz. In: Gust von Loh, S., Stock, W.G. (eds.) Informationskompetenz in der Schule. Ein informationswissenschaftlicher Ansatz, pp. 223–257. De Gruyter Saur, Berlin, Boston (2012)
12. Deterding, S., Dixon, D., Khaled, R., Nacke, L.: From game design elements to gamefulness: defining "Gamification". In: Lugmayr, A., Franssila, H., Safran, C., Hammouda, I. (eds.) Proceedings of the 15th International Academic MindTrek Conference: Envisioning Future Media Environments, pp. 9–11. ACM, New York (2011). https://doi.org/10.1145/2181037.2181040
13. Knautz, K., Göretz, J., Wintermeyer, A.: "Gotta catch 'em all" - game design patterns for guild quests in higher education. In: Kindling, M., Greifeneder, E. (eds.) IConference 2014 Proceedings, pp. 690–699. iSchools, Illinois (2014). https://doi.org/10.9776/14010

14. Röttger, M., Stock, W.G.: Die mittlere Güte von Navigationssystemen. Ein Kennwert für komparative Analysen von Websites bei Usability-Nutzertests. Information - Wissenschaft und. Praxis **54**, 401–404 (2003)
15. Shachaf, P., Hara, N.: Beyond vandalism: wikipedia trolls. J. Inf. Sci. **36**(3), 357–370 (2010). https://doi.org/10.1177/0165551510365390
16. Schlenk, C.T.: Studenten-App Jodel erhält sechs Millionen – und will in die USA expandieren. Gründerszene, 7 June, 2017. https://www.gruenderszene.de/allgemein/jodel-studenten-app-usa-millionen. Accessed 31 Jan 2018
17. Prensky, M.: Digital natives, digital immigrants. Horizon **9**(5), 1–6 (2001)
18. Palfrey, J.G., Gasser, U.: Born Digital: Understanding the First Generation of Digital Natives. Basic Books, New York (2008)
19. Morrison, S., Gomez, R.: Pushback: the growth of expressions to resistance to constant online connectivity. In: Kindling, M., Greifeneder, E. (eds.) IConference 2014 Proceedings, pp. 1–15. iSchools, Illinois (2014). https://doi.org/10.9776/14010
20. Johann, M., Wiedel, F., Tonndorf, K., Windscheid, J.: Anonymous online communication between disinhibition, self-disclosure and social identity. a complementary mixed-method study. Presentation on the ICA Annual Conference, 9–13 June, 2016, Fukoka, Japan (2016)

MiGua! App for User Awareness Prior to Adopting Dogs in Urban Areas

Gerardo Real Flores(✉) and Rocio Abascal-Mena

Master in Design, Information and Communication (MADIC), Universidad Autónoma Metropolitana, Cuajimalpa, Mexico
gereflo@aol.com, mabascal@correo.cua.uam.mx

Abstract. This paper makes a proposal to reduce the abandonment of pets through a mobile application that makes the user aware before adopting a dog. This application is called MIGua! and the main function is to give some tasks to the user for some time so that he can feel the impact of having a pet before buying or adopting one. The developing process is based on the User-Centered Design methodology in order to include the user in all the stages and produce an app pertinent to the needs of our community. Paper and digital prototypes helped to evaluate the interface as the real impact on participants.

Keywords: Abandonment · Pet care · Usability
User-Centered Design · Protoyping · Evaluation

1 Introduction

According to some estimates, the current world population of domestic dogs Canis lupus familiaris may be as high as 500 million, of which a substantial although unknown proportion is poorly supervised or free-roaming [1]. Nowadays in the cities, the abandonment of pets has become a problem. According to data from the Ministry of Health of Mexico City every year 18,000 dogs are lost or abandoned by their owners [2]. Most street dogs may be abandoned by their owners because they are not previously aware of what their care means, such as the time, space and money invested in their adoption or purchase.

According to a study carried out by the Affinity Foundation in 2010, the main reasons for abandonment are: (1) unexpected litters (14%); (2)changes of address (13.7%); (3) economic problems (13.2%); (4) loss of interest (11.2%); (5) a problematic behavior of the pet (11%); (6) the end of the hunting season (10.2%); (7) family allergies (7.7%); (8) a newborn in the family (4%); (9) owner's death (3.5%); (10) vacations (2.6%) and (11) fear of contracting toxoplasmosis during pregnancy (2.4%) [3].

In order to try to solution this problem the first thing to do is to identify if a person meets the necessary requirements before adopting a pet. In this case, it does't not exist an established or official way to evaluate if the future owner can take responsibility of having a pet.

© Springer International Publishing AG, part of Springer Nature 2018
G. Meiselwitz (Ed.): SCSM 2018, LNCS 10913, pp. 87–96, 2018.
https://doi.org/10.1007/978-3-319-91521-0_7

Several studies have demonstrated that owning a dog has benefits for the owners' health. Dog owners are reported to visit the doctor less frequently [4] and to have lower health care costs than non-dog owners [5]. In addition to the health benefits of a pet, such as the dog, this can help create a better environment among the population. Having a companion animal can enhance social interactions between people, and this could lead to fewer depressive symptoms [6].

In the present article we propose a solution of awareness before the adoption of a pet through a mobile application designed with the User-Centered Design (UCD) methodology. The prototype, as a game, simulates the main responsibilities that must be taken when adopting a dog. And the end the user will be evaluated to determine if it is suitable for adoption.

The article is composed as following: in Sect. 2 is presented a state of the art of existing tools. Section 3 is dedicated to present the contextual study conducted to obtain and detect user's needs. The evaluation of the prototype is described in Sect. 4. Finally, conclusions and further work are stablished.

2 Background

Currently, there are different applications in the market that helps users with specific problems related to dogs, adoption and their care. For example, "Walk for a Dog"[1] is an application developed in New York that connects people with different shelters near them to walk dogs without their owner [7].

Another one is "Miwuki Pet Shelter"[2], which gives a list of protective associations, and a list of currently adoptable dogs and cats [8]. Also, it is possible to use "Dog care"[3], which is an application that helps and gives advice to dog owners for the care of the pet such as: basic care, toxic foods and first aid [9]. As well, "Dog walk-track"[4], is helpful for dog owners to keep track of the daily walks they have with their pets, and share photos of them on social networks during the walk [10]. As a closely related application, but in a complete virtual environment it exists "Dogotchi: Virtual Pet"[5], a video game that simulates the care, hygiene and feeding of a dog, but all the interaction is virtual and does not take any information from the phone sensors, nor change habits of the user in real life [11].

By analyzing and evaluating the previous applications we can notice that none of them were designed to raise awareness for dog care before the user makes the adoption of it. So, it is necessary to ensure the future owner about what means to have a pet so he can evaluate before adopting one by knowing that in his decision it is the life of a dog.

[1] http://www.wooftrax.com.
[2] https://www.miwuki.com.
[3] https://upstairs.co.in.
[4] https://tractive.com.
[5] http://mawges.com.

3 Contextual Study

Having the stray dog problem in mind, emerges the idea to encourage more people to adopt abandoned dogs by using technology. This could be achieved by generating a mobile application that could help user awareness to reduce abandonment rates. To identify the main needs of people who have adopted dogs, several interviews were carried out with different first-time owners, with no prior experience for adopting pets. From these interviews, it was possible to identify the main needs and motivations when adopting a dog. In this way, people need to be motivated in a good way by understanding the main responsibilities and work that has to be carried when a dog arrives to home. So, the questionnaire used to identify the needs of dog owners is presented below:

- What has been the biggest change you've made after adopting?
- What has been the biggest obstacle after adopting?
- Have you had bad experiences in your walks with your dog?
- What habits have you had to change after adoption?
- Have you experienced any bad experience when hiring a service for your pet?
- Would you adopt another dog?
- If you saw a case of animal abuse would you report it?

Through the surveys conducted, the various needs that users have were determined:

1. Time
2. Money
3. Security
4. Entertainment
5. Exercise
6. Responsibility

Under these variables, based on the people interviewed, it was conducted a modeling of different user's profiles. This, in order to stablish different communities of possible future owners and be able to canalize the prototype in an adequate way. Also, different ideal scenarios were built, where the main needs were met with the help of technology. In this way, it was possible to visualize the needs from a different point of view than the initial one.

3.1 User Study

The user profiles of the application must meet certain requirements, including: being a young person (between 18 and 35 years), having an Android mobile phone with Internet access, have a medium to high level of proficiency of the device, and to not have much experience in pet adoption. Below are the different profiles and scenarios created during the process of user-modeling.

Karla Pérez Age: 27

- Lives in the city.
- Has an apartment.
- Divorced.
- She has a toddler.
- She prefers dogs of small breeds.
- She found a dog in the street and decided to adopt it.
- She is often distracted.
- Before living on his own she never had a dog.
- Being a single woman with a small dog makes her more vulnerable because of the insecurity of the city.
- She does not have much free time.
- She prefers to take her dog to a canine groomer.

The scenario of this woman is the next: the young woman goes out to walk and she finds a dog in the street which she decides to adopt it. She easily finds what to do and which groomers and veterinarians are reliable to take the puppy and check its health. After adopting it, she can easily receive advice from other owners for the best care of her pet. Taking it for a walk makes her feel safe since she knows that there are dangerous places in her neighborhood (there is no public lighting, police, or there are cars at high speed) and she receives, constantly, reports or advices of other dog owners. Also, if she needs to do some shopping she can go to pet-friendly establishments.

Pedro and Carlos Average ages: 32.

- They live in the city.
- They prefer dogs of medium races.
- They do not have kids.
- They decided to adopt a dog from a shelter.
- They are usually organized.
- One of them already had experience with pets.
- In their neighborhood there are not many parks.
- They live in a neighborhood with a lot of people movement.
- Their free time is little.
- Sometimes they take their dog to canine groomers.

The scenario for the young couple living together is the next one: they decide to have a companion animal and so they adopt it in a shelter. Actually, they can locate nearby shelters and know what to do and what to change to be able to adopt. When they take a walk with their dog, they take a path in which the community has put bags to collect the dog's waste along the way (in case they need them) and know where they are. In addition they can see in their application which parks allow dogs (some parks are restricted in that area). They can also see which veterinarians and reliable hospitals are close to their home and their schedules.

Fernando Rodríguez Age: 24.

- Live in the city.
- He prefers cats.
- He decided to adopt because the life of the dog was in danger in his previous home.
- He is carefree.
- He is not used to take out his dog for a walk.
- Prefer to take care of his pet alone.
- Has little time.

The scenario for the young man is the next one: he adopts a puppy in danger but he also knows where there are reliable veterinarians to check the dog's health. Then goes for a walk with his dog and can plan a safe route in which he will take the right time to walk. Other dog owners help him by giving recommendations of how to take care of this dog.

By doing the profile modeling it was clear that the user's needs were different from those initially established. So, after discussing profiles and answers to the questionnary, the main needs identified are the time and money that an owner had to invest when adopting a dog. However, many of them did not know what kind of responsibilities it can take to care for a dog before having it. With this in mind, a brainstorming took place. In this brainstorm, it was identified that the problem of abandoning pets was not the motivation to adopt, but that the adopters are unaware of the responsibilities involved in adoption and cannot adequately meet them.

3.2 Development of User Interface

In the next step an inspiration-panel was developed in order to recognize what things or aspects have to be taken into account when developing the solution. This inspiration-panel includes words, colors and images that will help to give a clearer focus on what an ideal interface would look like. In addition to identifying what the application should be it also shows popular applications that are an inspiration because they have clear and simple interfaces. Some of the words that came into our mind when thinking in the final prototype were:

- Loneliness
- Tenderness
- Company
- Walk
- Happy
- Food
- Family
- Health
- Responsibility

At this step of the process, it was defined a point of view in order to make clearer what was the objective and the area of opportunity to be attacked. Our point of view devised for the app was: "being alone is very sad, what better company than a four-legged one!".

At this point of the research it was realized that the tool could focus on two different types of users: (1) first-time dog owners who needed a guide to understand the best way to care for their dogs (see Fig. 1) and (2) users who need to have awareness about what implies to adopt a pet (see Fig. 2). From these ideas, two scenarios were developed by using storyboards. This way, it was possible to decide in a more objective way the course that should be taken in the later development of the application. In addition, with the storyboards it is possible to communicate with the users and arrive to the solution of the problem. These ideas were presented to a group of people to know which possible solution would have more acceptance.

Fig. 1. Storyboard for the first-time owners. A girl is followed by a street dog and decides to adopt it even if she doesn't knows nothing about dog care. So she searches in Internet for an application that could help her about veterinarians, food for dogs, special care, etc. The dog care information from the app helps her a lot so right now she is very happy with her new dog.

Fig. 2. Storyboard for user awareness. A boy moves into his new apartment and feels lonely. So, he decides it would be a good idea to adopt a dog, but he doesn't know what are the main responsibilities he will carry by having a pet. He finds an application that simulates the care of a real dog and helps him to decide if he could be a good candidate to adopt a dog, or not.

3.3 Fast Prototype

It was decided that the line of prior awareness was the better one to work on because it solves the problem before it exists and prevents the abandonment of pets. After deciding this line of development, we started with the user interface design. For this we used the technique called "The Wizard of Oz" to increase the functionality and clarity of the application. "The Wizard of Oz" technique is performed with the participation of a human who is known as "Wizard" or "Accomplice", who simulates, without knowledge of the participant in the experiment, the role played by the computer during a Human-Computer Interaction [12]. This technique helps the creator to give an idea of how the interface will be developed, what are the possible errors, the different scenarios and the control of exceptions. In addition, it is possible to see in a real way if the interface created is simple and understandable for the profile of the user selected.

The prototype was tested on paper (see Fig. 3) by using the 10 Nielsen's heuristics [13]. From this first approximation, we found that the heuristics "family metaphors and language" have been violated, by having some words that were unclear. Also, the heuristic of "aesthetic and minimalist design" where the initial interface was composed of too many screens before reaching the solution.

Fig. 3. Example of paper prototypes before starting to develop the digital one.

4 Usability Test

The general idea of the application is to simulate a real dog care. After the instructions and presentation, it is possible to configure a virtual dog to take care of it. The dog can be small, medium or big. Then the dog will ask the user for food or exercise. The user needs to discount of his money the cost of the food and open the application to take a walk. After a few weeks, the virtual dog will say to the user if it was happy or unhappy with the care. If the results are poor because the user has forgotten to do some activities with the virtual dog then it will notify you. Indefinitely of the result it will disappear and the user can have another virtual dog and take another shoot to improve his skills in dog care (see Fig. 4).

Fig. 4. Cycle of Migua! app composed of main instructions, definition of the profile, reminders and final veredict.

After working on the violated heuristics, a digital prototype was created (see Fig. 5), which was already a more complete version than the paper version and closer to what would be the finished application. This tool has an ideal iconography as it respects the style of Android and the color palette.

Fig. 5. Digital prototype

This new prototype was tested with different users in order to verify if the tool was adequate. As in the previous steps, the 10 heuristics of Nielsen were used. In this second iteration of the evaluation it was detected the violation of the following heuristics "metaphors family and language", since some of the instructions did not seem to indicate that they were clear. For 100% of the users interviewed only half understood clearly the dynamics of the application. Also, "Memory Recognition", because some icons did not remind them clearly of their main purpose. However, the final prototype was generated (see Fig. 6) correcting the heuristics violated. Later on, it was shown again to the users in order to have their advice. This time, the modifications done allowed to have a better application and easy to use.

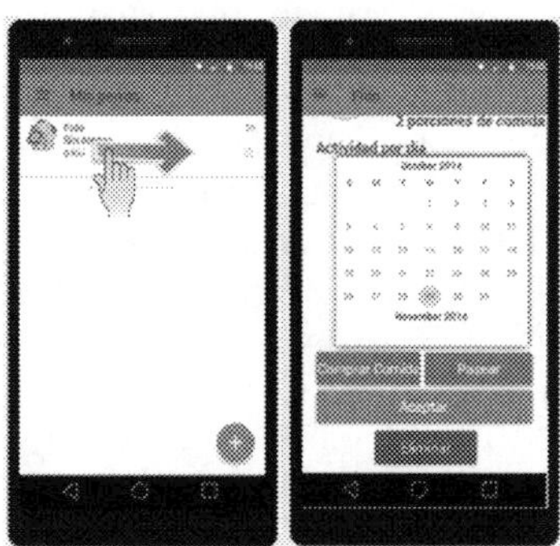

Fig. 6. Some screens of the last prototype generated.

5 Conclusion

Awareness is one of the most important factors to avoid the abandonment of pets. Therefore, motivating the population of cities is one of the first steps to reduce this problem. However, raising the awareness of the population is a major challenge that must be strongly taken by including a detailed process in order to offer to the users an application according to their profile. In this case, we have developed MiGua! to train users in having a pet. This app does not substitute pets, it only gives an advice about the responsibilities that a future owner has to have. That is why it must have a period of limit use, once that cycle has been completed the application will restart the progress, having to start again as shown in the first storyboard. This work was based on the methodology of User-Centered Design, which helped to check the reliability of MiGua! and how users are the most important part of all the development process. By taking into account our users, since the detection of needs, it is more easy to have a successful application at the end. Also, with the prototype it was proved that the interface is functional and works with the profile users that we have stablished. In future work, it is intended to implement the application with the rules that the Google store requires so general public can be able to download it. And finally, through the comments of users we are going to measure the impact of MiGua! in the resolution of animal abandonment.

References

1. Matter, H.C., Daniels, T.J.: Dog ecology and population biology. In: Dogs, Zoonoses and Public Health, pp. 17–62 (2000)
2. Morán Rodríguez, L.: Proponen solución al problema de los perros callejeros. Ciencia.unam.mx. http://ciencia.unam.mx/contenido/texto/leer/109/Proponen_solucion_al_problema_de_los_perros_callejeros. Accessed 3 Dec 2016
3. Las razones detrás del abandono de una mascota — Fundación Affinity. (2016). http://www.fundacion-affinity.org/perros-gatos-y-personas/busco-una-mascota/las-razones-detras-del-abandono-de-una-mascota. Fundacion-affinity.org. Accessed 30 Jan 2018
4. Siegel, J.M.: Stressful life events and use of physician services among the elderly: the moderating role of pet ownership. J. Pers. Soc. Psychol. **58**(6), 1081 (1990)
5. Clower, T.L.: The Health Care Cost Savings of Pet Ownership (2015)
6. Winefield, H.R., Black, A., Chur-Hansen, A.: Health effects of ownership of and attachment to companion animals in an older population. Int. J. Behav. Med. **15**(4), 303–310 (2008)
7. Walk for a Dog (2016). http://www.wooftrax.com/. Accessed 30 Jan 2018
8. Miwuki Pet Shelter (2018). https://www.miwuki.com/. Accessed 30 Jan 2018
9. Dog care (2018). https://upstairs.co.in/. Accessed 4 Jan 2018
10. Dog walk-track (2018). https://tractive.com. Accessed 4 Jan 2018
11. Dogotchi: Virtual Pet (2018). http://mawges.com/. Accessed 4 Jan 2018
12. Fraser, N.M., Gilbert, G.N.: Simulating speech systems. Comput. Speech Lang. **5**(1), 81–99 (1991)
13. Nielsen, J.: 10 usability heuristics for user interface design. Nielsen Norman Group, 1(1) (1995)

Approaches on User eXperience Assessment: User Tests, Communicability and Psychometrics

Virginia Zaraza Rusu[1(✉)], Daniela Quiñones[1], Cristian Rusu[1], Pablo Cáceres[1], Virginica Rusu[2], and Silvana Roncagliolo[1]

[1] Pontificia Universidad Católica de Valparaíso, Valparaíso, Chile
rvzaraza90@hotmail.com, danielacqo@gmail.com,
{cristian.rusu, silvana.roncagliolo}@pucv.cl,
pablo@psicometodos.com

[2] Universidad de Playa Ancha, Valparaíso, Chile
virginica.rusu@upla.cl

Abstract. Usability is a basic attribute in software quality. Its complex and evolving nature is hard to describe in a unique definition. Usability refers to ease of use and the way users can perform their tasks. User eXperience (UX) goes beyond the three generally accepted usability's dimensions: effectiveness, efficiency and satisfaction. UX covers all aspects of someone's interaction with a product, application, system and/or service including psychological ones. Psychometrics as a psychological assessment tool could be helpful in UX studies as a complement to usability evaluation methods. Communicability is a distinctive quality of interactive systems that effectively and efficiently communicate to the users the design intent and interactive principles. The paper explores how user testing (co-discovery), communicability evaluation, query techniques, and psychometrics (motivation scale) may complement each other when assessing UX. Empirical evidences are analyzed, using the World Digital Library (www.wdl.org) as a case study.

Keywords: User eXperience · Communicability · Psychometrics
User testing · Digital Library

1 Introduction

Usability is a basic attribute in software quality. Its complex and evolving nature is hard to describe in a unique definition. Usability refers to ease of use and the way users can perform their tasks. User eXperience (UX) goes beyond the three generally accepted usability's dimensions: effectiveness, efficiency and satisfaction. UX covers all aspects of someone's interaction with a product, application, system and/or service. UX takes a broader view, looking at the individual's entire interaction with the thing, as well as the thoughts, feelings and perceptions that result from that interaction [1]. As a psychological assessment tool, psychometrics could be helpful in UX studies as a complement to usability evaluation methods [2].

© Springer International Publishing AG, part of Springer Nature 2018
G. Meiselwitz (Ed.): SCSM 2018, LNCS 10913, pp. 97–111, 2018.
https://doi.org/10.1007/978-3-319-91521-0_8

The Semiotic Engineering views the use of interactive software systems as a computer-mediated communication between designers and users, at interaction time [3]. Semiotic engineering proposes two methods to evaluate the communicability: (1) the semiotic inspection and (2) the communicability evaluation. The latter explores the reception in the meta communication and tries to identify through observation empirical evidence of the effects produced by the designer's messages on the user as they appear during the interaction.

The paper explores how user testing (co-discovery), communicability evaluation, query techniques, and psychometrics (motivation scale) may complement each other when assessing UX. Empirical evidences are analyzed, using the World Digital Library (www.wdl.org) as a case study [4].

The paper is organized as follow: Sect. 2 explores the theoretical background. Section 3 presents the first experiment performed: co-discovery test, perception questionnaire and psychometric test. Section 4 presents the second experiment performed: the communicability test. Finally, the Sect. 5 shows conclusions and future work.

2 Theoretical Background

The ISO 9241-210 standard defines UX as a "person's perceptions and responses resulting from the use and/or anticipated use of a product, system or service" [5]. On the other hand the current ISO 9241 definition of usability refers to "the extent to which a system, product or service can be used by specified users to achieve specified goals with effectiveness, efficiency and satisfaction in a specified context of use" [5].

Usability evaluation methods are usually classified as: (1) empirical usability testing, based on users' participation [6], and (2) inspection methods, based on experts' judgment [7]. Evaluating UX is more challenging and arguably overwhelming for newcomers [8]. Almost 90 UX evaluation methods are described at http://www.allaboutux.org/ [9].

The "User Experience White Paper" [10] highlights the multidisciplinary nature of UX, which has led to several definitions of (and perspectives on) UX, each approaching the concept from a different point of view: from a psychological to a business perspective, and from quality centric to value centric.

It is important to mention that the ISO 9241-210 standard considers that UX "includes all the users' emotions, beliefs, preferences, perceptions, physical and psychological responses, behaviors and accomplishments that occur before, during and after use" [5]. Therefore, psychometrics as a psychological assessment tool that studies "the operations and procedures used to measure variability in behavior and to connect those measurements to psychological phenomena" [11] could be helpful in UX studies as well.

Motivation which can be understood as "the drive that produces goal-directed behavior" concerning the "initiation, direction, intensity, and persistence of behavior" [12] is a significant psychological concept in different life domains [13]. As a psychometric resource, the Multidimensional Work Motivation Scale (MWMS) seeks to provide information on the motivation of people with respect to their work [14].

As for the Semiotic Engineering, it views the use of interactive software systems as a computer-mediated communication between designers and users, at interaction time [3]. The system is therefore the designer's deputy, the artifact that transmits designer's intentions. Communicability is the attribute that defines the quality of the metacommunication ("communication about communication"). The semiotic engineering proposes two evaluation methods: (1) the semiotic inspection, and (2) the communicability evaluation method.

The communicability evaluation method analyzes the metacommunication. Evaluators observe of how a group of users interacts with a particular system identifying communicative breakdowns. Evaluators interpret the results and then prepare the semiotic profile and the meta-communicational message [3].

3 First Experiment: Co-discovery, Perception Questionnaire and Psychometric Test

The first experiment was carried out in the Usability Laboratory of the School of Informatics Engineering of the Pontificia Universidad Católica de Valparaíso, Chile, and was conducted by two experts from UX. Both have a Diploma in UX, but also one currently studies Psychology and has a degree in Architecture, and the other one has a Master Degree in Computer Science.

The World Digital Library (WDL) was evaluated, based on a set of predefined tasks. The participants explored the site in pairs, their comments and facial expressions were recorded with cameras and the screens of their computers were recorded. They were also observed by the evaluators through a polarized glass that allows one-way vision.

After the participants signed a confidentiality agreement, they were informed about the test conditions, about the website to evaluate, and about the different stages of the test.

The experiment consisted of 4 parts:

- (1) A pre-experiment questionnaire designed to individually identify the user profile and previous experience visiting portals similar to the evaluated product;
- (2) A co-discovery test in pairs, presenting to the participants a series of tasks to explore the site as a whole and comment out their opinions;
- (3) A post-experiment perception questionnaire that sought to know the different perceptions of each user regarding the site and the tasks;
- (4) A post-experiment psychometric test, to know the motivations of each participant to perform the requested tasks.

3.1 The Pre-experiment Questionnaire

In a first stage, each user had to complete a pre-experiment questionnaire to collect general information about their profile and experience in other Digital Libraries (DLs). Five questions were included regarding sex, age, level of education and information about previous visits in other DLs. There were 12 users, 9 men and 3 women, between 23 and 32 years old, all being graduate students in Computer Science. Only a quarter of the users (3) reported having experience in visiting DLs, but stated that they (almost) never visit these types of sites.

3.2 The Co-discovery Test

To perform this test belonging to the second stage, each pair of participants was provided with a list of predefined tasks to explore the site. In addition, they were asked to freely discuss and comment aloud their opinions regarding WDL, the tasks, and what they considered relevant.

The first task was to find certain items associated with historical events, using the "Timelines". Despite the fact that 83.3% of users completed the task, only 50% did it within the pre-established time (5 min). Difficulties arose to orient themselves within the different menus and to execute in a correct and efficient way the sequence of steps required to carry out the tasks.

The second task required opening a digital article, after placing it on the "Interactive maps" of the portal. 50% of the participants managed to do it and within the period of time assigned (5 min). The users had problems locating the different countries in the interactive map since their names did not appear, as well as difficulties to open the article since this option was only visualized when positioning the cursor over the main image. Therefore, the participants looked for alternative ways of doing the task or were distracted by other WDL contents.

The third task requested to find within the classification by places, articles associated to different geographical zones of Chile. 83.3% of users completed the task and did it within the time limit period (4 min). The rest showed a lack of attention to the instructions, looking for articles not associated with the geographical areas required. There was a tendency to look for articles in more direct ways, through the site's search engine.

The fourth task was to explore different types of articles, in order to play a movie. 100% of the participants could execute it within the time limit (4 min). Half of them used the portal search engine to find the articles more expeditiously.

The various tasks requested allowed to know certain difficulties presented by the experiment participants to orient themselves through WDL. Problems were highlighted in recognizing the navigation mode offered by certain sections and the functionality of different tools (for example regarding the "Timelines" articles) and to identify the content associated with different graphic symbols (for example for the geographical areas in "Interactive maps"). Due to this, there are functions offered by the site that users did not manage to use effectively and efficiently.

The users tended to repeat the sequences of steps requested, probably hoping to be successful on a new occasion in the face of the presented inconveniences. The use of alternative search routes also stands out, which increases towards the end of the experiment, evidencing perhaps the need for greater flexibility and immediacy in the use of WDL. The distraction in the participants with other site contents that where mentioned in the specified tasks, could eventually indicate a form of compensation for failing to execute those tasks or an authentic interest in the information offered by the portal.

There are no greater differences between the performances of users with previous experience in DLs than those who had not previously visited this type of site, except in the realization of the second task. This obtained the lowest performance (50% of achievement), even for users with previous experiences in DLs (66.7% of these did not manage to complete the task). This may suggest that WDL has significant usability problems compared to other similar websites, related to the lack of clarity in the functionality of some tools and the insufficient information associated with them.

3.3 The Post-experiment Perception Questionnaire

After completing the test, in a third stage, users had to respond individually to a post-experiment questionnaire based on the System Usability Scale (SUS) [15]. The aim was to identify the users' perceptions about the tasks' difficulty levels, the orientation in the site and the conformity and satisfaction with it. Five questions were used using a Likert scale of 5 points and 4 open questions.

Regarding the difficulty to complete the requested tasks, most of the participants indicated that they considered it easy to achieve (41.7%) or neutral (41.7%), while two users considered it difficult (8.3%) or very difficult (8.3%). Orientation within the portal was perceived as variable, with 41.7% feeling less oriented, 33.3% feeling neutral, and the rest feeling oriented (8.3%) or very oriented (16.7%). In relation to the degree of satisfaction with WDL, the majority found it satisfactory (41.7%) or neutral (41.7%), while 16.7% found it unsatisfactory. On the other hand, as to the information found on the site, the majority felt satisfied (58.3%), with one user (8.3%) who considered it very satisfactory, while 25% felt neutral and one user (8.3%) considered it unsatisfactory. Finally, most users express the intention to re-use the WDL, agreeing very much (25%), or agreeing (58.3%) while 25% of the users are neutral, and two users disagreed (8.3%) or strongly disagreed (8.3%).

Users with previous experience in DLs tended to perceive tasks with a lower degree of difficulty but with a varying degree of orientation within the site. This could suggest that although WDL navigation modes and interfaces are not necessarily easier and friendlier than in other similar portals, possibly familiarization with these allows developing greater intuition for the user regarding the portal's use. On the other hand, participants who have previously visited DLs tended to show satisfaction and intention of future WDL use, indicating probably a genuine attraction to the portal and towards this type of tasks, in comparison with other similar sites.

It should be noted that despite the users' overall perception seems neutral, with an average of 3.33 [2], analyzing each particular dimension allows to obtain a more precise understanding of the participant's perceptions and their experience. This results' description in conjunction with a more markedly quantitative reading allows us to observe that although task completion and orientation trough the website (with an average of 3) tend towards neutrality (with averages of 3.17 and 3 respectively) there is a majority who declare themselves satisfied with WDL (41.7%). There is also a tendency to express satisfaction about the information that WDL offers, and the intention of future use (with averages of 3.67 and 3.58 respectively). By complementing this type of reading, it can be assumed that despite the tasks' difficulty and the lack of orientation in the portal, users can feel challenged, interested in exploring the site and satisfied with it.

As for the aspects that most pleased the users, they rescue the site's content and the found information, its vastity and diversity, the graphic resources, the WDL organization and the user interface. These elements can explain the satisfaction, interest and intention of future use of the portal.

However, among the aspects that made it more difficult for the participants to navigate the site and that disliked, they themselves highlighted problems in identifying the navigation mode offered by some sections (for example in "Timelines"), in not being intuitive enough and assuming that the user has certain knowledge, and in some tools' functionality and problems to visualize and locate elements (as in the case of "Interactive maps"). These elements allow understanding also the failures in the performance of the requested tasks and the complications to be able to feel oriented within WDL. In addition, the users expressed that there is a lack of clarity in the use of search filters, many of which have gone to the use of these as an alternative search route.

Faced with the mentioned aspects, the participants pointed out the need for a guide to use the site, as well as a greater hierarchy and order, greater clarity and simplicity in navigation, and of a change in the portal's chromatic to be perceived as more attractive.

3.4 The Post-experiment Psychometric Test

The fourth part of the test consisted of a questionnaire about the motivation of each participant to perform the tasks, based on the Multidimensional Work Motivation Scale (MWMS) [14]. The scale was adapted to the academic context of the experiment and 19 questions were included using a Likert scale of 5 points. The questions were organized into 3 major categories, covering 6 dimensions. "Amotivation", "intrinsic motivation" and "extrinsic motivation" were the 3 major categories, and "extrinsic motivation" was divided into 4 dimensions: "external social regulation", "external material regulation", "introjected regulation" and "identified regulation". We analyzed preliminary findings in a previous study [2].

In the first category and dimension, the "amotivation" or absence of motivation, referring to a perceived waste of time, unworthy effort, and useless tasks, the scale was reverted, being 1 as "strong", and 5 as "lack of" amotivation, in order to be able to compare it with the rest of the dimensions. The majority of the participants stated that they did not have amotivation to perform the requested tasks (41.7% disagreed and 41.7% strongly disagreed with the presence of amotivation), while 16.7% showed neutrality. These results are understandable and expected since they freely volunteered in the experiment.

The second category belongs to "extrinsic motivation" or motivation based on winning rewards and avoiding punishments and includes 4 dimensions according to the type of regulation. With respect to the second dimension of "external social regulation" concerning other's approval, recognition, and criticism avoidance, this was denied by the majority (16.7% disagreed and 66.7% strongly disagreed) while 16.7% was neutral. In this way students reject as influencing factors the attitudes of others in their motivation. On the dimension of "external material regulation", referring to avoiding decreasing grades, getting academic rewards, and gaining experience, the majority of users were neutral (66.7%), 25% avoided recognizing their influence (being in disagreement), while one user (8.3%) agreed. Although the students do not deny that there may be material factors such as gaining experience (the reason that obtained the highest scores), those are probably not such determining factors in completing the requested tasks.

On the other hand, for the fourth dimension of "introjected regulation", related to demonstrate self-capability, feel proud of oneself, avoid dissatisfaction for not complying, half of the students were neutral (50%) while others confirmed their impact (8.3% agreed and 33.3% strongly agree), and one user denied it (8.3% disagreed). The fifth dimension of "identified regulation" regarding the importance, value and personal significance of putting effort into tasks was expressed by all the participants (41.7% agreed and 58.3% strongly agree). The obtained results could point out that users get involved in this type of activities with the same commitment, seriousness and motivation as they would in the case of interacting with a portal in which they need or wish to navigate in not only experimental contexts.

Finally, on the third category and sixth dimension, "intrinsic motivation", associated with doing inherently entertaining, interesting and challenging tasks and incitement to learn, this was expressed by all users (50% agreed and 50% strongly agree). This would allow reflecting once more on the authentic interest of the participants in carrying out this type of tasks and on the site, as they state in the perception questionnaire.

With respect to the users with previous experience in visiting DLs in comparison with the novice users of this type of portals, it should be noted that with respect to amotivation they tended towards greater neutrality. This is not necessarily surprising since they may have become accustomed to navigations and explorations in these types of sites. In relation to the external regulation dimensions, unlike the general trend, they showed in disagreement, but not in the rest of the dimensions, of greater internal regulation. This suggests once again that the repetitive navigation they have done on DLs probably comes from a real interest in this type of portals or in this type of exploration tasks.

It can be observed that there is a tendency for an increase in the motivation that the participants affirm as regards factors of a more internal, personal and subjective nature, such as their personal appreciations (identified regulation) and their innate attraction (intrinsic motivation) towards what do they do. The averages of the scores for each dimension (4.28 for "lack of" amotivation, from 1.39 for external social regulation to 4.44 in identified regulation, and 4.40 for intrinsic motivation) [2], also reflects this. Despite that from a quantitative view users' overall motivation seems neutral, with an average of 3.48 [2], this indicator does not turn out to be the most representative for an analysis around users' motivation. The applied psychometric test pretends to glimpse the presence of different dimensions involved in motivation, but these have different relevance, impact and weighting for the same participant, with amotivation possibly being the dimension that best integrates these aspects. The breakdown of each of these dimensions from a qualitative perspective is what allows a deeper and more complete understanding of the user's experience in relation to motivational aspects.

4 Second Experiment: Communicability Test

The second experiment was also performed in the Usability Laboratory of the School of Informatics Engineering of the Pontificia Universidad Católica de Valparaíso, Chile, a month after the first experiment. It was conducted by two UX experts. The same expert who studies Psychology participated, and the other is a PhD student in Informatics Engineering.

The WDL was once again evaluated, based on a set of predefined tasks. These tasks were different from those performed in the first experiment (Sect. 3). In the communicability test, the participants explored the website alone. Their facial expressions were recorded with cameras and the screens of their computers were recorded. They were also observed by the evaluators through a polarized glass that allows one-way vision.

The experiment involved 6 participants, all being graduate students in Computer Science. After the participants signed a confidentiality agreement, they were informed about the test conditions, about the website to evaluate, and about the different stages of the test.

The experiment consisted of 3 parts:

- (1) A pre-experiment questionnaire designed to identify the user profile and previous experience visiting portals similar to the evaluated product.
- (2) An individual communicability test, presenting to each participant a series of tasks to explore the website.
- (3) A post-experiment perception questionnaire that sought to know the user perceptions regarding the website and the tasks.

4.1 The Pre-experiment Questionnaire

Each user had to complete a pre-experiment questionnaire to collect general information about their profile and experience in other DLs. The same five questions included in the previous experiment were asked, adding a new one regarding the user's profession.

There were 6 participants, 4 men and 2 women, between 23 and 34 years old. 83.3% of the users (5) had already visited DLs before, but stated that they (almost) never visit these types of sites.

4.2 The Communicability Test

Each participant was provided with a list of predefined tasks to explore the website. The tasks were aimed to identify communicative breakdowns [3]. While the participants were accomplishing the tasks, the evaluators identified all signs of communicative breakdowns in the user's interaction. They took notes of these during the test.

The communicability test included 3 tasks:

1. T1: "*Search an article*". The first task was to search an article browsing the word "*mathematics*" in the main search engine of the website and applying different filters to select a specific language ("*Spanish*"), a specific place ("*Europe*") and to visualize the results in the form of a gallery.

2. T2: "*Read a book online*". The second task was to read the book "*Atlas of the Physical and Political History of Chile*", accessing to the "*Natural Sciences and Mathematics*" section, and then to the "*Animals*" section. The user should visualize the book in full screen, go to page 26 and write down the result of the animal that was shown on the screen ("*swan*").
3. T3: "*Find a museum on the map*". The third task was to visualize and search for a specific museum using the website map. After accessing to the "*Institution*" section, the user should filter the results by "museums" and select the museum "*Walters Art Museum*". The user should note in which country the museum is located and the number of related articles.

Table 1 shows the number of communicative breakdowns observed in the test for each user (U1–U6), considering all 3 tasks.

Table 1. Number of communicative breakdowns observed for each user.

Communicative breakdown	U1	U2	U3	U4	U5	U6	Total
Where is it?	5	1	2	4	3	6	21
What now?	2	0	1	2	0	2	7
What is this?	0	0	0	0	0	0	0
Oops!	1	3	3	2	0	2	11
Where am I?	1	1	0	3	1	2	8
I can't do it this way.	2	0	1	2	0	1	6
Why doesn't it?	2	1	1	0	0	1	5
What happened?	1	0	0	1	0	2	4
Help!	0	0	0	0	0	2	2
I can do otherwise.	2	0	0	2	2	2	8
Thanks, but no, thanks.	1	1	1	2	2	0	7
Looks fine to me.	5	1	2	1	2	1	12
I give up!	0	0	0	0	0	1	1

The communicative breakdowns with greater frequency were: "*Where is it?*" (21 communicative breakdowns) and "*Looks fine to me*" (12 communicative breakdowns). Users had difficulty finding the item (or information) they were looking for ("*Where is it*?"). In addition, users believed that they achieved their goal, however this did not happen. The user was not aware of the communicative breakdown ("*Looks fine to me*").

The communicative breakdowns with medium frequency were: "*Oops*" (11 communicative breakdowns), "*Where am I?*" (8 communicative breakdowns), and "*I can do otherwise*" (8 communicative breakdowns). The users made an error and immediately realized it. The users went back a step ("*Oops*"). In addition, users took actions that would be appropriate in another context. That is, they selected the wrong paths to achieve the task ("*Where am I?*"). Finally, users were not fully aware of the ways of action offered by the system to perform a task. Users chose to do something different than is expected, but achieved the same effect. That is, users achieved their goals by a non-optimal path ("*I can do otherwise*").

The communicative breakdown "*What is this?*" did not occur in any of the tasks (0 communicative breakdowns). This means that all the users understood the elements of the website.

Table 2 shows the number of communicative breakdowns observed for each task.

Table 2. Number of communicative breakdowns observed for each task.

Communicative breakdown	T1	T2	T3
Where is it?	5	10	6
What now?	1	2	4
What is this?	0	0	0
Oops!	6	2	3
Where am I?	1	2	5
I can't do it this way.	0	3	3
Why doesn't it?	0	1	4
What happened?	0	0	4
Help!	0	0	2
I can do otherwise.	0	2	6
Thanks, but no, thanks.	0	2	5
Looks fine to me.	6	4	2
I give up!	0	0	1
Total	19	28	45

As shown in Table 2, the task with the most communicative breakdowns was Task 3: "*Find a museum on the map*". The users could not properly filter the "*Institutions*" by "*museum*". This is because the website displays an unintuitive search filter for the user. The filter controls use a confusing color, so the user does not know when the filter is applied or not (see Figs. 1 and 2). Due to the difficulty in filtering the search by museum, users could not select the "*Walters Art Museum*" on the map. The communicative breakdowns that occurred most in task 3 were: "*Where is it?*", "*What now?*", "*Where am I?*", "*Why doesn't it?*", "*What happened?*", "*I can do otherwise*", and "*Thanks, but no, thanks*".

For Task 1: "*Search an article*", the communicative breakdowns that occurred most were: "*Oops!*" and "*Looks fine to me*" (both with 6 communicative breakdowns). Some users did the search without filtering by language and/or place, but they quickly realized the error and immediately corrected it ("*Oops!*"). Some users accessed the wrong book (they followed a different path) so they found the wrong animal. The users believed that they achieved the goal, but it was otherwise ("*Looks fine to me*").

For Task 2: "*Read a book online*", the communicative breakdown that occurred most was: "*Where is it?*", with 10 communicative breakdowns. Users could not find the option to read the book online. This was because the website did not have a clear button

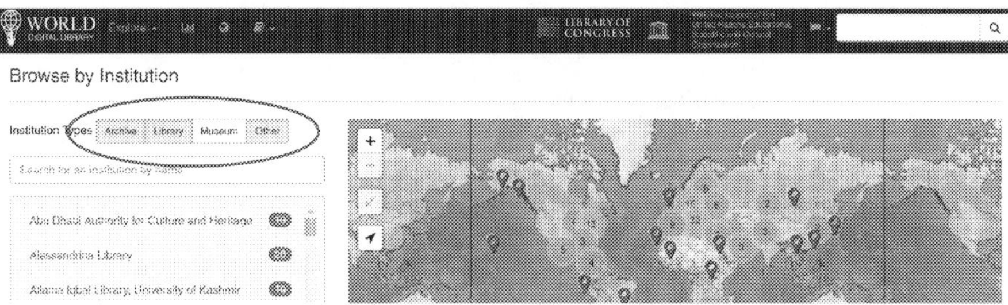

Fig. 1. The search was performed by the institutions types: “Archive”, “Library” and “Other”. “Museum” is not selected. The users thought that the white color indicated that “Museum” was selected, when it was otherwise.

Fig. 2. Correct filter applied. The search was performed by institution type: “Museum”. The gray color indicates that only “Museum” is selected. However, this was not clear to users.

to access that option. The users looked for the option in the page by minutes, until they realized that by positioning the mouse for a few seconds on the book image, it could be viewed to read.

Based on the results obtained in the communicability test, the evaluators explained the designer’s message [3]. To do this, the evaluators assumed the first person in discourse and spoke for the designer by answering the following questions:

- *Who do I think are the users of the product of my design?* They are users who access digital libraries very rarely but who have experience in the use of websites, so they know the meaning of the elements and symbols of a website.
- *What have I learned about these users’ wants and needs?* I have learned that users understand the elements of the interface without problems, but that some features of the website are not intuitive and easy to use (filters), which makes it difficult to use and search for information.
- *Which do I think are these users’ preferences with respect to their wants and needs, and why?* Users prefer an easy to use and navigate website, which is interactive and informative. Users prefer a website with intuitive filters that allow them to find what you want efficiently and quickly.
- *What system have I therefore designed for these users, and why?* I have designed a website that allows the user to acquire knowledge about different cultures of the

world. I have designed a confusing website for some users, because certain functionalities are unclear and do not help the user in their information search process.
- *What system have I therefore designed for these users, and how can or should they use it?* I have designed a digital library that allows users to access, free of charge and at any time, a wide variety of material about different cultures of the world. The user can search information by categories, timelines and interactive maps. In addition the articles can be reviewed using different multimedia elements.
- *What is my design vision?* My design vision is to distribute information to users about different cultures in different languages through the use of interactive elements, such as: audio, video, maps and online reading. My design vision is to present highlighted information on the home page of the website, allowing the user to search for content through a general search engine, categories, timelines and/or interactive maps.

Based on the semiotic profiling presented above, the evaluators identified the meta-communicational message using the template proposed by De Souza and Leitão [3]. The metacommunication template sums up what designers are communicating to the users through systems interfaces.

- *Here is my understanding of who you are, what I've learned you want or need to do, in which preferred ways, and why.* I think you are a user with or without experience in digital libraries. I think you are interested in learning about different cultures from different countries of the world. I think you would like to see digital articles. For this reason I have designed for you an easy-to-use and interactive website, with different articles, videos, audios and images that allow you to access information as if you were in a physical library. I have also designed the website with a structure that allows you to search for information by categories, sections, timelines and maps.
- *This is the system that I have therefore designed for you, and this is the way you can or should use it.* The system that I have designed for you is to obtain information of your interest. You can interact in different ways, using audio, video and/or interactive reading about articles from different cultures of the world. I have designed the website with different search methods, so you can access the content as best suits you. You can search for a particular concept; navigate through the different sections that I present to you; or interact with the world map to look for information.
- *In order to fulfill a range of purposes that falls within this vision.* The objectives that are within my vision are to promote the exchange of cultural knowledge in a global way; and expand the amount and variety of cultural content on the Internet. To do this, I present the content in a visual way and with appropriate sizes, allowing you to browse and search for information of interest through different mechanisms.

4.3 The Post-experiment Perception Questionnaire

After completing the test, in a third stage, users had to respond individually to a post-experiment questionnaire based on the System Usability Scale (SUS) [15]. The questionnaire applied was the same as in the first experiment (Sect. 3).

Regarding the difficulty to complete the requested tasks, half of the participants indicated that they considered it neutral (50%), while two users considered it easy to achieve (33.3%) and one user considered it difficult (16.7%). Orientation within the portal was perceived as variable. Half of the participants felt less oriented (50%), two users felt neutral (33.3%) and one user felt oriented (16.7%). In relation to the degree of satisfaction with WDL, two users found it satisfactory (33.3%); two users found it neutral (33.3%) and two users found it unsatisfactory (33.3%). With respect to the information found on the site, the majority felt satisfied (66.7%), with one user (16.7%) who considered it very satisfactory, while 33.3% felt neutral. Finally, half of the participants express the intention to re-use the WDL (50%), while 33.3% of the users are neutral, and one user disagreed (16.7%).

All users stated that the most difficult to use of the site were the search filters. Five users declared that after using the filter by "Institutions type" on the map for a while, they realized how it worked and managed to find the museum. However, they all declared that the use of the filter is not intuitive.

The users positively highlighted the website design, the variety and large number of articles presented, and the good images quality.

On the other hand, users stated that the website should correct certain elements, such as improving the use of filters (both the search filters of articles as the filter by institutions on the map) and clearly show the buttons of certain actions (e.g. reading articles online).

5 Conclusions

Users expressed similar concerns during the co-discovery experiment and the post-experiment perception questionnaire. They would like WDL to offer an intuitive and "user-friendly" navigation. They would also expect intuitive and easy to use functionalities. These are perceived as main obstacles in accomplishing the tasks that the experiment required. They had direct impact on UX, and also generated communicative breakdowns during the communicability test.

Users tried to find alternative ways to accomplish tasks. They wanted flexibility; they expected WDL to adapt to their working style and preferences. Flexibility could improve UX.

Users showed interest in WDL's content, and they hoped finding diverse information. The post-experiment psychometric test showed lack of amotivation, and more internal regulation (identified regulation, intrinsic motivation), than external regulation (extrinsic, material, and social). Results are consistent with the perception questionnaire's outcomes. It seems that a satisfactory UX would be influenced by accomplishing goals intuitively, and finding interesting content, that may generate enthusiasm.

Better motivation could lead to higher commitment in accomplishing tasks, better performance, better efficacy and efficient use of the tools that WDL offers. It would probably generate more sincere and authentic opinions on users' experience. This highlights the importance of psychological aspects on a positive UX, aspects that the interaction designer should always consider. The experiments that we made highlight, as expected, the importance and complementarity of quantitative and qualitative aspects. It is also relevant to break them down into each constituent aspect and angles, in order to understand users' experience in a more profound and holistic way.

Human-computer interaction is in fact a user-computer-designer interaction, a place where designer's intentions meet and intersect user's goals, and the way that there are accomplished. UX is the results of users' expectations, goals, beliefs, preferences, but also their emotions, perceptions, physical and psychological responses, behaviors and achievements, towards what the designer offers.

Successful user-computer-designer interaction has to conciliate user's goals and designer's intentions. As future work we intend to study modes to better conciliate both aspects, to collect and analyze more experimental data, and to address other psychological aspects.

References

1. Tullis, T., Albert, B.: Measuring the user experience: collecting, analyzing, and presenting usability metrics. Morgan Kaufmann, USA (2013)
2. Rusu V.Z., Rusu C., Cáceres P., Rusu V., Quiñones D., Muñoz P.: On user eXperience evaluation: combining user tests and psychometrics. In: Karwowski W., Ahram T. (eds). Intelligent Human Systems Integration. IHSI 2018, Advances in Intelligent Systems and Computing, pp. 626–632, Springer, Cham (2018)
3. De Souza, C.S., Leitão, C.F.: Semiotic engineering methods for scientific research in HCI. Synthesis Lectures on Human-Centered Informatics. Morgan & Claypool, USA (2009)
4. World Digital Library. http://www.wdl.org. Accessed 10 Jan 2018
5. ISO 9241-210: Ergonomics of human-system interaction — Part 210: Human-centered de-sign for interactive systems. International Organization for Standardization, Geneva (2010)
6. Dumas, J.S., Fox, J.E.: Usability testing: current practice and future directions. In Sears, A. L., Jacko, J.A. (eds). The Human-Computer Interaction Handbook: Fundamentals. Evolving Technologies and Emerging Applications, pp. 1129–1149. Taylor & Francis, USA (2008)
7. Cockton, G., Woolrych, A., Lavery, D.: Inspection-based methods. In Sears, A.L., Jacko, J. A. (eds). The Human-Computer Interaction Handbook: Fundamentals. Evolving Technologies and Emerging Applications, pp. 1171–1190. Taylor & Francis, USA (2008)
8. Rusu, C., Rusu, V., Roncagliolo, S., González, C.: Usability and user experience: what should we care about? Int. J. Inf. Technol. Syst. Approach **8**(2), 1–12 (2015)
9. All About UX. http://www.allaboutux.org. Accessed 10 Jan 2018
10. Roto, V., Law, E., Vermeeren, A., Hoonhout, J.: User experience white paper. bringing clarity to the concept of user experience (2011). http://www.allaboutux.org/uxwhitepaper. Accessed 18 Oct 2017
11. Furr, R.M., Bacharach, V.R.: Psychometrics. An Introduction. SAGE, Los Angeles (2014)
12. Strickland, B.R. (ed.): The Gale Encyclopedia of Psychology. Gale, Detroit (2001)
13. Deci, E.L., Ryan, R.M.: Facilitating optimal motivation and psychological well-being across life's domains. Can. Psychol. **49**(1), 14–23 (2008)

14. Gagné, M., Forest, J., Vansteenkiste, M., Crevier-Braud, L., van den Broeck, A., Aspeli, A. K., Bellerose, J., Benabou, C., Chemolli, E., Güntert, S.T., Halvari, H., Indiyastuti, D.L., Johnson, P.A., Molstad, M.H., Naudin, M., Ndao, A., Olafsen, A.H., Roussel, P., Wang, Z., Westbye, C.: The multidimensional work motivation scale: validation evidence in seven languages and nine countries. Eur. J. Work Organ. Psychol. **24**(2), 178–196 (2014)
15. Lewis, J., Sauro, J.: Revisiting the factor structure of the system usability scale. J. Usability Stud. **12**(4), 183–192 (2017)

An Online Travel Agency Comparative Study: Heuristic Evaluators Perception

Cristian Rusu[1(✉)], Federico Botella[2], Virginica Rusu[3], Silvana Roncagliolo[1], and Daniela Quiñones[1]

[1] Pontificia Universidad Católica de Valparaíso, Valparaíso, Chile
{cristian.rusu, silvana.roncagliolo}@pucv.cl, danielacqo@gmail.com

[2] Universidad Miguel Hernández de Elche, Elche, Spain
federico@umh.es

[3] Universidad de Playa Ancha, Valparaíso, Chile
virginica.rusu@upla.cl

Abstract. Forming usability professionals, particularly heuristic evaluators, is a challenging task. Heuristic evaluation is a well-known and widely employed usability evaluation method. A heuristic evaluation may be performed based on generic or specific heuristics. A key issue is how new heuristics are validated and/or evaluated; heuristic quality scales were proposed. The paper presents some recurrent problems when teaching the heuristic evaluation method. It also discusses novice evaluators' perception over Nielsen's usability heuristics, based on empirical data. The experiment that we made involved Computer Science graduate and undergraduate students, enrolled in a Human-Computer Interaction introductory course. 50 Chilean students and 18 Spanish students participated. The online travel agency Atrapalo.com was used as case study. We used a questionnaire that assesses evaluators' perception over a set of usability heuristics. It rates each heuristic individually (Utility, Clarity, Ease of use, Necessity of additional checklist), but also the set of heuristics as a whole (Easiness, Intention, Completeness).

Keywords: Usability · Heuristic evaluation · Usability heuristics Heuristic quality · Online travel agency

1 Introduction

The usability concept is known for decades and is still evolving. As there is still no general agreement on its definition, we prefer the one stated by the ISO 9241-210 standard: "the extent to which a system, product or service can be used by specified users to achieve specified goals with effectiveness, efficiency and satisfaction in a specified context of use" [1].

User eXperience (UX) extends usability concept beyond its three widely agreed dimensions: effectiveness, efficiency and satisfaction. The ISO 9241-210 standard defines UX as a "person's perceptions and responses resulting from the use and/or anticipated use of a product, system or service" [1].

© Springer International Publishing AG, part of Springer Nature 2018
G. Meiselwitz (Ed.): SCSM 2018, LNCS 10913, pp. 112–120, 2018.
https://doi.org/10.1007/978-3-319-91521-0_9

Several classifications are used for usability evaluation methods. Lewis identifies two approaches on usability evaluation: (1) summative, "measurement-based usability", and (2) formative, "diagnostic usability" [2].

For more than two decades heuristic evaluation (HE) has been proved that it is one of the most popular usability evaluation methods [3]. Even if HE is a formative or usability-oriented method, it identifies lots of issues that (potentially) affect a satisfactory UX. Therefore, even if HE does not "measure" UX, it may be considered as a UX method.

When performing a HE, generic or specific heuristics may be used. Nielsen's ten usability heuristics are well known, but many other sets of usability heuristics were proposed [4, 5]. A key issue is how new heuristics are validated and/or evaluated. A heuristic quality scale was proposed [6]. We developed a questionnaire with a similar purpose.

Forming usability/UX professionals is a challenging task [7, 8]. The paper presents some remarks on teaching the HE method. We also discuss the novice evaluators' perception over Nielsen's usability heuristics, based on an experiment that we made.

2 Introducing Heuristic Evaluation to Novices

An easy way to raise awareness about usability/UX topics among Computer Science (CS) students is to practice formal evaluations. Our students have to perform each semester at least one HE and one user test.

We are using Nielsen's protocol and students have to perform their first HE based on Nielsen's heuristics [9]. But teaching the HE method for more than a decade allowed us to highlight some pitfalls. Some recurrent problems occur and are described below. They express our teaching experience for almost two decades, but are also empirically supported by explicit students' comments during the experiment described in this paper.

It is quite difficult for CS students to identify usability and UX related issues. They usually focus on technical issues rather than putting themselves in users' shoes. It is challenging to make them forget about subjective judgment and to be sympathetic with potential users.

When students find usability problems, it is quite hard to relate them to usability heuristics. It is rather common to associate a usability problem to two or even three different heuristics. We have to emphasize that each heuristic has a different purpose and to correctly understand it needs (quite a lot of) practice. Linking usability problems to usability heuristics is particularly challenging when working with generic heuristics, as Nielsen's.

Heuristics should be used appropriately as usability issues' detection tool. However, novice evaluators (and sometimes even experienced ones) focus on identifying usability problems instead of heuristics compliance. This somehow explain why is so difficult to determine problem's nature and to associate it to specific heuristic(s).

Quantifying the severity of a usability problem is perceived as a rather easy task, but rating its frequency is much more difficult. Students tend to overrate the frequency. That means the criticality of the problem will also be overestimated, as criticality is the sum of severity as frequency. Usability problems' ranking will be affected. Evaluation's report may confuse the targeted audience. Subsequently, the effort of solving usability issues may be wrongly focused.

3 The Experiment

We systematically conduct studies on the perception of (novice) evaluators over generic and specific usability heuristics. All participants are asked to perform a HE of the same software product (case study). Then, all of them participate in a survey.

We developed a questionnaire that assesses evaluators' perception over a set of usability heuristics, concerning 4 dimensions and 3 questions:

- D1 – Utility: How useful the usability heuristic is.
- D2 – Clarity: How clear the usability heuristic is.
- D3 – Ease of use: How easy was to associate identified problems to the usability heuristic.
- D4 – Necessity of additional checklist: How necessary would be to complement the usability heuristic with a checklist.
- Q1 – Easiness: How easy was to perform the heuristic evaluation, based on the given set of usability heuristics?
- Q2 – Intention: Would you use the same set of usability heuristics when evaluating similar software product in the future?
- Q3 – Completeness: Do you think the set of usability heuristics covers all usability aspects for this kind of software product?

The set of usability heuristics is rated globally through the 3 questions (Q1 – Easiness, Q2 – Intention, Q3 – Completeness), but each heuristic is rated separately, on each one of the 4 dimensions (D1 – Utility, D2 – Clarity, D3 – Ease of use, D4 – Necessity of additional checklist). After performing the HE, all participants are asked to rate each usability heuristic, concerning each of the 4 dimensions, using a 5 points Likert scale. The 3 questions aim to evaluate the overall evaluators' perception; responses are also based on a 5 points Likert scale.

We made an experiment with CS graduate and undergraduate students enrolled in a Human-Computer Interaction introductory course in Chile (Pontificia Universidad Católica de Valparaíso, Valparaíso) and Spain (Universidad Miguel Hernández de Elche, Elche). 50 Chilean students (33 graduate and 17 undergraduate), and 18 Spanish undergraduate students participated; we did not select samples, all students enrolled in the HCI course were also participants in the experiment. All of them evaluated the online travel agency Atrapalo.com, based on Nielsen's 10 usability heuristics and following the same protocol. Actually, the Chilean students evaluated Atrapalo.cl and the Spanish students evaluated Atrapalo.es.

We chose Atrapalo.com as case study because it is a widely known online travel agency in Latin America and Spain; its Chilean and Spanish versions are very similar. Moreover, usability and UX in online travel agencies is one of the research topics that we have been working on for years.

After performing the HE all participants were asked to rate their experience, based on the questionnaire described above. Additionally, two open questions were included:

- OQ1: What did you perceive as most difficult to perform during the heuristic evaluation?
- OQ2: What domain-related (online travel agencies) aspects do you think Nielsen's heuristics do not cover?

4 Results and Discussion

Most of the answers to the open question OQ1 confirm the problems described in Sect. 2, in all three groups of students. The most recurrent comments were: "*it is difficult to criticize*", "*it is hard to identify usability-related problems, instead of technical problems*", "*it is hard to think as a user, not as a computer scientist*", "*it is challenging to think as a novice/expert user*", "*it is hard to imagine scenarios of use*", "*it is difficult to link problems to appropriate heuristics*", "*I know there is a problem, but I am not sure what kind of problem is*", "*it is hard to synthetize/specify the problem*", "*I am not sure how to evaluate frequency*". By far, the most recurrent perceived difficulty was to establish the association usability problem – usability heuristic(s). It worth mentioning that all students felt the need to express their thoughts, answering question OQ1.

Answers to open question OQ2 identify several domain-related (online travel agencies) aspects, which students think Nielsen's heuristics do not cover:

- Effective and efficient transactional process,
- Easy to perform transactional process,
- Clear information on how many steps the process includes,
- Slow query, slow response,
- Information of trust,
- Unexpected system behavior,
- Security-related issues,
- Privacy-related issues,
- Accessibility-related issues,
- Responsivity and adaptability.

All questionnaire items are based on a 5 points Likert scale. Observations' scale is ordinal, and no assumption of normality could be made. Therefore the survey results were analyzed using nonparametric statistics tests (Mann-Whitney U and Spearman ρ).

Table 1 presents the average scores for dimensions and questions. Chilean students had a better (rather positive) perception on Nielsen's heuristics. Dimension D3 (ease of use) got the lowest score, in all cases. Even if heuristics are perceived as useful and clear, students think they are not easy to apply in practice. Moreover, they feel the need for a more complete heuristics' specification (necessity of additional checklist).

Heuristics' perceived overall easiness is low, especially in the case of Spanish students. Heuristics' perceived overall completeness is more homogeneous for the three groups of students. In spite of the above, the intention of future use of Nielsen's heuristics,

Table 1. Average scores for dimensions and questions.

	D1 – Utility	D2 – Clarity	D3 – Ease of use	D4 – Necessity of additional checklist	Q1 – Easiness	Q2 – Intention	Q3 – Completeness
Spanish undergraduate students (18 participants)	3.83	3.43	3.30	3.60	2.78	3.89	3.33
Chilean undergraduate students (17 participants)	4.39	4.04	3.73	4.21	3.53	3.89	3.18
Chilean graduate students (33 participants)	4.39	4.19	3.75	4.27	3.12	4.42	3.60

when evaluating similar products, is rather high for undergraduate and remarkably high for graduate students.

Mann-Whitney U tests were performed to check the hypothesis:

- H_0: there are no significant differences between evaluators with different background,
- H_1: there are significant differences between evaluators with different background.

Spearman ρ tests were performed to check the hypothesis:

- H_0: $\rho = 0$, two dimensions/questions are independent,
- H_1: $\rho \neq 0$, two dimensions/questions are dependent.

In all Mann-Whitney U and Spearman ρ tests, p-value ≤ 0.05 was used as decision rule.

As Table 2 shows, there are significant differences between the perception of Spanish and (all) Chilean students in all cases, excepting question Q3 (Nielsen's heuristics completeness).

Table 2. Mann-Whitney U test for Spanish and (all) Chilean students.

	D1 – Utility	D2 – Clarity	D3 – Ease of use	D4 – Necessity of additional checklist	Q1 – Easiness	Q2 – Intention	Q3 – Completeness
p-value	0.000	0.000	0.003	0.001	0.022	0.028	0.551

When comparing only undergraduate (Chilean and Spanish) students (Table 3), there are significant differences in almost all cases, excepting questions Q2 (intention of future use), and Q3 (Nielsen's heuristics completeness).

On the contrary, the perception of Chilean graduate and undergraduate students is quite similar. There are significant differences only in answers to question Q2, regarding the intention of future use of Nielsen's heuristics (Table 4).

Table 3. Mann-Whitney U test for Spanish and Chilean undergraduate students.

	D1 – Utility	D2 – Clarity	D3 – Ease of use	D4 – Necessity of additional checklist	Q1 – Easiness	Q2 – Intention	Q3 – Completeness
p-value	0.005	0.001	0.022	0.021	0.006	0.666	0.902

Table 4. Mann-Whitney U test for Chilean graduate and undergraduate students.

	D1 – Utility	D2 – Clarity	D3 – Ease of use	D4 – Necessity of additional checklist	Q1 – Easiness	Q2 – Intention	Q3 – Completeness
p-value	0.992	0.185	0.735	0.788	0.091	0.045	0.294

Even if there are significant differences between the perception of Spanish and Chilean students in all dimensions and almost all questions, we do not suspect cultural or background-related issues as possible cause. All students analyzed the same product (Atrapalo.com), using the same set of heuristics (Nielsen's), and following the same protocol. However, some of the Spanish students reported difficulties when scheduling and coordinating their tasks. As this was the only observed difference between the two groups of students, it may somehow influence Spanish students' perception not only on how easy was to perform the HE, but also on the set of heuristics they used.

In the case of the Spanish students (Table 5) there are only two significant correlations:

Table 5. Spearman ρ test for Spanish undergraduate students.

	D1 – Utility	D2 – Clarity	D3 – Ease of use	D4 – Necessity of additional checklist	Q1 – Easiness	Q2 – Intention	Q3 – Completeness
D1	1	0.532	Independent	0.664	Independent	Independent	Independent
D2		1	Independent	Independent	Independent	Independent	Independent
D3			1	Independent	Independent	Independent	Independent
D4				1	Independent	Independent	Independent
Q1					1	Independent	Independent
Q2						1	Independent
Q3							1

- A strong one between D1 – D4. If heuristics are perceived as useful, the necessity of additional evaluation elements (checklist) is also perceived.
- A moderate one between D1 – D2. If heuristics are perceived as clear (easy to understand), they are also perceived as useful.

As Table 6 indicates, the same significant correlations identified for Spanish students also occur in the case of the Chilean students (a strong one between D1 – D4, and a moderate one between D1 – D2). But three other significant correlations occur:

- Two moderate correlations between D2 – D3 and Q2 – Q3. If heuristics are perceived as clear, they are also perceived as easy to use; when the set of heuristics is perceived as complete, there is an intention of future use.
- A weak correlation between D2 – D4. Even if heuristics are perceived as clear, evaluators think that additional checklist is necessary.
- Two weak negative correlations between D2 – Q3 and D3 – Q3. Even if heuristics are perceived as clear and easy to use, evaluators feel that Nielsen's set does not cover all usability aspects of an online travel agency.

Table 6. Spearman ρ test for all Chilean students.

	D1 – Utility	D2 – Clarity	D3 – Ease of use	D4 – Necessity of additional checklist	Q1 – Easiness	Q2 – Intention	Q3 – Completeness
D1	1	0.415	Independent	0.623	Independent	Independent	Independent
D2		1	0.479	0.329	Independent	Independent	−0.341
D3			1	Independent	Independent	Independent	−0.286
D4				1	Independent	Independent	Independent
Q1					1	Independent	Independent
Q2						1	0.416
Q3							1

When analyzing Chilean graduate students' perception, most of the significant correlations that occur for the whole group of Chilean students repeat (Table 7). There are four positive and one negative correlations:

- Two moderate positive correlations (D1 – D4, D2 – D3).
- Two weak positive correlations (D1 – D2, Q2 – Q3).
- A moderate negative correlation (D2 – Q3).

Table 7. Spearman ρ test for Chilean graduate students.

	D1 – Utility	D2 – Clarity	D3 – Ease of use	D4 – Necessity of additional checklist	Q1 – Easiness	Q2 – Intention	Q3 – Completeness
D1	1	0.384	Independent	0.500	Independent	Independent	Independent
D2		1	0.565	Independent	Independent	Independent	−0.527
D3			1	Independent	Independent	Independent	Independent
D4				1	Independent	Independent	Independent
Q1					1	Independent	Independent
Q2						1	0.390
Q3							1

Only two positive significant correlations occur when analyzing Chilean undergraduate students' perception (Table 8):

- A very strong positive correlation (D1 – D4).
- A moderate positive correlation (D3 – Q1).

Table 8. Spearman ρ test for Chilean undergraduate students.

	D1 – Utility	D2 – Clarity	D3 – Ease of use	D4 – Necessity of additional checklist	Q1 – Easiness	Q2 – Intention	Q3 – Completeness
D1	1	Independent	Independent	0.827	Independent	Independent	Independent
D2		1	Independent	Independent	Independent	Independent	Independent
D3			1	Independent	0.488	Independent	Independent
D4				1	Independent	Independent	Independent
Q1					1	Independent	Independent
Q2						1	Independent
Q3							1

The last one does not occur for any other group of evaluators, even if one would expect it. When heuristics are perceived as easy to use, the HE as a method is also perceived as easy to perform.

The only correlation that occurs for all groups of evaluators is D1 – D4. Other recurrent correlation is D1 – D2; it is absent only in the case of Chilean undergraduate students. For both Chilean and Spanish undergraduate students correlations are scarce; they occur only twice. On the contrary, they are relatively frequent correlations in the case of graduate (Chilean) students.

A previous study indicates that most correlations between dimensions occur in the case of evaluators with previous experience [10]. As all participants in our experiment were novice, fewer correlations were expected.

5 Conclusions

Heuristic evaluation is a well-known and arguably the most popular usability inspection method. But forming evaluators is not an easy task; some recurrent problems occur. As method's performance depends mostly on evaluators' skills, we are encouraging students to perform as much evaluations as possible.

We systematically conduct studies on the perception of (novice) evaluators over generic and specific usability heuristics. We developed a questionnaire that evaluates each heuristic individually (Utility, Clarity, Ease of use, Necessity of additional checklist), but also the set of heuristics as a whole (Easiness, Intention, Completeness).

In the comparative study that we have done there are significant differences between the perception of Chilean and Spanish CS students in almost all cases. The perception of Chilean graduate and undergraduate students is rather similar. We do not have evidences to suspect cultural or background-related issues as possible cause;

differences are more likely due to difficulties that some Spanish students reported, related to scheduling and coordinating their tasks.

As in previous studies, few correlations occur between dimensions/questions. Even expected correlations are scarce. The rather heterogeneous students' perception shows that usability heuristics are quite difficult to understand by novice.

The study of novice evaluators' perception helps us in at least two aspects. We better understand the challenges that students are facing; it help us to improve the teaching process, focusing on sensitive issues, explicitly stated in students' comments. They also help us to develop new set of heuristics, for specific domains. In this particular study students highlighted domain-related aspects that Nielsen's heuristics do not cover. Their comments are a valuable asset when designing/refining a set of usability heuristics for online travel agencies.

As future work we intend to analyze the perception of each heuristic individually. We will also analyze the usability problems that students identified during the experiment, the usability heuristic(s) they associated, and the way they rated problems' severity and frecuency.

Acknowledgments. We thank all the students involved in the experiment. They provided helpful opinions that allowed us to prepare this and (hopefully) further documents.

References

1. ISO 9241-210: Ergonomics of human-system interaction — Part 210: Human-centered de-sign for interactive systems. International Organization for Standardization, Geneva (2010)
2. Lewis, J.R.: Usability: lessons learned… and yet to be learned. Int. J. Hum-Comput. Interact. **30**(9), 663–684 (2014)
3. Nielsen, J., Mack, R.L.: Usability Inspection Methods. John Wiley & Sons, New York (1994)
4. Hermawati, S., Lawson, G.: Establishing usability heuristics for heuristics evaluation in a specific domain: is there a consensus? Appl. Ergon. **56**, 34–51 (2016)
5. Quiñones, D., Rusu, C.: How to develop usability heuristics: a systematic literature review. Comput. Stand. Interfaces **53**, 89–122 (2017)
6. Anganes, A., Pfaff, M.S., Drury, J.L., O'Toole, C.M.: The heuristic quality scale. Interact. Comput. **28**(5), 584–597 (2016)
7. Rusu, C., Rusu, V., Roncagliolo, S.: Usability practice: the appealing way to HCI. In: The First International Conference on Advances in Computer-Human Interactions (ACHI 2008) Proceedings, pp. 265–270. IEEE Computer Society Press (2008)
8. Rusu, C., Rusu, V., Roncagliolo, S., González, C.: Usability and user experience: what should we care about? Int. J. Inf. Technol. Syst. Approach **8**(2), 1–12 (2015)
9. Nielsen, J.: 10 Usability Heuristics for User Interface Design. http://www.nngroup.com/articles/ten-usability-heuristics/. Accessed 28 Dec 2017
10. Rusu, C., Rusu, V., Roncagliolo, S., Quiñones, D., Rusu, V.Z., Fardoun, H.M., Alghazzawi, D.M., Collazos, C.A.: Usability heuristics: reinventing the wheel? In: Meiselwitz, G. (ed.) SCSM 2016. LNCS, vol. 9742, pp. 59–70. Springer, Cham (2016). https://doi.org/10.1007/978-3-319-39910-2_6

Evaluating Online Travel Agencies' Usability: What Heuristics Should We Use?

Cristian Rusu[1(✉)], Virginica Rusu[2], Daniela Quiñones[1], Silvana Roncagliolo[1], and Virginia Zaraza Rusu[1]

[1] Pontificia Universidad Católica de Valparaíso, Valparaíso, Chile
{cristian.rusu, silvana.roncagliolo}@pucv.cl, danielacqo@gmail.com, rvzaraza90@hotmail.com
[2] Universidad de Playa Ancha, Valparaíso, Chile
virginica.rusu@upla.cl

Abstract. Online travel agencies' customers have nowadays a wide range of alternatives and are more demanding. Usability is a basic attribute in software quality. Heuristic evaluation is arguably the most popular usability inspection method, well-known and widely used. A heuristic evaluation may be performed based on generic or specific heuristics. Many sets of specific (usually domain-related) usability heuristics were published. Heuristic quality scales to validate and/or evaluate new heuristics were proposed. The paper analyzes evaluators' perception on three sets of usability heuristics, when evaluating the same product: Nielsen's generic heuristics, a set of cultural-oriented heuristics for e-Commerce, and a set of heuristics for smartphones applications (SMASH). We made an experiment with 38 Computer Science students, enrolled in a Human-Computer Interaction introductory course, using the online travel agency Expedia.com as case study; the web and mobile versions were evaluated. We assessed students' perception based on a questionnaire that rates each heuristic individually (Utility, Clarity, Ease of use, Necessity of additional checklist), but also the set of heuristics as a whole (Easiness, Intention of future use, Completeness).

Keywords: Online travel agency · Heuristic evaluation · Usability heuristics
Heuristic quality

1 Introduction

Usability is a basic attribute in software quality. The concept is known for decades and is still evolving. The ISO 9241-210 standard defines usability as "the extent to which a system, product or service can be used by specified users to achieve specified goals with effectiveness, efficiency and satisfaction in a specified context of use" [1].

Lewis identifies two approaches on usability evaluation: (1) summative, "measurement-based usability", and (2) formative, "diagnostic usability" [2]. Usability evaluation methods are usually classified in two categories: (1) usability testing, based on users' participation, and (2) inspection methods, based on experts' judgment.

© Springer International Publishing AG, part of Springer Nature 2018
G. Meiselwitz (Ed.): SCSM 2018, LNCS 10913, pp. 121–130, 2018.
https://doi.org/10.1007/978-3-319-91521-0_10

For more than two decades heuristic evaluation is arguably the most popular usability inspection method [3]. When performing a heuristic evaluation generic or specific heuristics may be used. Nielsen's ten usability heuristics are well known, but many sets of specific (usually domain-related) usability heuristics were also published [4, 5]. The results of a heuristic evaluation depend on several factors, but at least two of them are critical: (1) evaluators' expertise, and (2) the set of heuristics that are employed. Heuristic quality scales were proposed.

The paper presents a comparative study on the evaluators' perception over three sets of usability heuristics, when evaluating the same product: Nielsen's heuristics [6], a set of cultural-oriented heuristics for e-Commerce [7], and a set of heuristics for smart-phones applications - SMASH [8]. The experiment that we made involved 18 graduate and 20 undergraduate Computer Science students, enrolled in an Human-Computer Interaction (Usability and User eXperience oriented) course. We used Expedia.com as case study [9]; web and mobile versions were evaluated.

2 Comparing Three Sets of Usability Heuristics: A Case Study

We conducted studies on the perception of evaluators over generic and specific usability heuristics for several years [10–12]. All participants are asked to perform a heuristic evaluation of the same case study. All of them are then asked to participate in a post-experiment survey.

We developed a questionnaire that assesses evaluators' perception over a set of usability heuristics, concerning 4 dimensions and 3 questions:

- D1 – Utility: How useful the usability heuristic is.
- D2 – Clarity: How clear the usability heuristic is.
- D3 – Ease of use: How easy was to associate identified problems to the usability heuristic.
- D4 – Necessity of additional checklist: How necessary would be to complement the usability heuristic with a checklist.
- Q1 – Easiness: How easy was to perform the heuristic evaluation, based on the given set of usability heuristics?
- Q2 – Intention: Would you use the same set of usability heuristics when evaluating similar software product in the future?
- Q3 – Completeness: Do you think the set of usability heuristics covers all usability aspects for this kind of software product?

Each heuristic is rated individually, on 4 dimensions (D1 – Utility, D2 – Clarity, D3 – Ease of use, D4 – Necessity of additional checklist). The set of usability heuristics is also rated globally, through the 3 questions (Q1 – Easiness, Q2 – Intention, Q3 – Completeness). In all cases, we are using a 5 points Likert scale (from 1 – worst, to 5 – best).

We made an experiment with 38 Computer Science graduate and undergraduate students enrolled in a HCI introductory course at Pontificia Universidad Católica de

Valparaíso (Chile). We did not select samples; all students enrolled in the HCI course were also participants in the experiment. Group composition was as follows:

- 20 undergraduate students; 13 students without previous experience in heuristic evaluation, 7 students with some experience based on Nielsen's heuristics.
- 18 undergraduate students; 11 students without previous experience in heuristic evaluation, 7 students with some experience based on Nielsen's heuristics.

All students evaluated the online travel agency Expedia.com, following the same protocol, but based on three different sets of heuristics:

- Nielsen's 10 usability heuristics [6],
- A set of e-Commerce (cultural-oriented) heuristics [7],
- SMASH, a set of usability heuristics for smartphone applications [8].

They evaluated Expedia.com website (based on Nielsen's and e-Commerce heuristics), and also Expedia mobile application (based on Nielsen's, e-Commerce, and SMASH heuristics). After performing the heuristic evaluation all participants were asked to rate their experience, based on the standard questionnaire described above.

Additionally, two open questions were asked:

- OQ1: What did you perceive as most difficult to perform during the heuristic evaluation?
- OQ2: What domain-related (online travel agencies) aspects do you think the set of usability heuristics does not cover?

3 Results and Discussion

Table 1 presents the average scores for dimensions and questions. Results are presented globally, but also grouped by students' level (undergraduate or graduate), and level of expertise (with or without previous experience in heuristic evaluation).

Heuristics' perceived utility (D1) is high in all cases (average score over 4.00). In the case of students with previous experience, the perceived utility is slightly in favor of e-Commerce and SMASH heuristics, comparing to Nielsen's heuristics. In general, students with previous experience perceive e-Commerce and SMASH heuristics' utility better than their novice colleagues.

Students perceived e-Commerce and SMASH heuristics' clarity (D2) better than Nielsen heuristics' clarity. The perceived clarity is always better among students with previous experience.

Heuristics' perceived ease of use (D3) is lower than their perceived utility and clarity. Ease of use perception is always better in the case of students with previous experience. e-Commerce and SMASH heuristics are perceived as (slightly) easier to use than Nielsen's heuristics.

The perceived necessity for additional checklist (D4) is higher in the case of Nielsen's heuristics; it is still relatively high for e-Commerce and SMASH heuristics. It is generally higher for novices, comparing to their more experienced colleagues.

Table 1. Average scores for dimensions and questions.

Participants	Previous experience	D1 – Utility	D2 – Clarity	D3 – Ease of use	D4 – Necessity of additional checklist	Q1 – Easiness	Q2 – Intention	Q3 – Completeness
Nielsen's heuristics								
All students (38)		**4.25**	**3.97**	**3.67**	**4.24**	**3.05**	**4.05**	**2.76**
24 students	No	4.30	3.81	3.57	4.33	2.79	4.08	2.71
14 students	Yes	4.16	4.24	3.84	4.09	3.50	4.00	2.86
Undergraduate students (20)		**4.19**	**3.99**	**3.71**	**4.16**	**3.10**	**3.80**	**2.60**
13 students	No	4.25	3.88	3.71	4.32	2.85	4.00	2.92
7 students	Yes	4.09	4.19	3.71	3.87	3.57	3.43	2.00
Graduate students (18)		**4.32**	**3.95**	**3.62**	**4.32**	**3.00**	**4.33**	**2.94**
11 students	No	4.36	3.74	3.41	4.34	2.73	4.18	2.45
7 students	Yes	4.24	4.29	3.96	4.30	3.43	4.57	3.71
e-Commerce heuristics								
All students (38)		**4.19**	**4.09**	**3.84**	**4.05**	**3.13**	**4.08**	**3.63**
24 students	No	4.13	3.95	3.68	4.07	3.00	3.96	3.54
14 students	Yes	4.30	4.31	4.12	4.01	3.36	4.29	3.79
Undergraduate students (20)		**4.22**	**4.12**	**3.92**	**4.09**	**3.20**	**4.05**	**3.45**
13 students	No	4.19	3.96	3.77	4.26	3.08	3.92	3.46
7 students	Yes	4.27	4.42	4.20	3.76	3.43	4.29	3.43
Graduate students (18)		**4.16**	**4.05**	**3.75**	**4.00**	**3.06**	**4.11**	**3.83**
11 students	No	4.06	3.95	3.57	3.84	2.91	4.00	3.64
7 students	Yes	4.32	4.20	4.04	4.26	3.29	4.29	4.14
SMASH heuristics								
All students (38)		**4.21**	**4.05**	**3.77**	**4.04**	**3.26**	**4.05**	**3.61**
24 students	No	4.21	3.94	3.60	4.06	3.17	4.04	3.71
14 students	Yes	4.21	4.24	4.07	4.00	3.43	4.07	3.43
Undergraduate students (20)		**4.23**	**4.04**	**3.74**	**4.05**	**3.35**	**4.10**	**3.55**
13 students	No	4.30	3.95	3.56	4.22	3.31	4.15	3.77
7 students	Yes	4.11	4.20	4.06	3.75	3.43	4.00	3.14
Graduate students (18)		**4.19**	**4.07**	**3.81**	**4.02**	**3.17**	**4.00**	**3.67**
11 students	No	4.11	3.94	3.64	3.87	3.00	3.91	3.64
7 students	Yes	4.32	4.29	4.08	4.25	3.43	4.14	3.71

The overall perception on easiness (Q1, how easy was to perform the heuristic evaluation) is lower than heuristics' perceived utility, clarity, and ease of use. It is quite close to the neutral point of the scale (3). As expected, it is lower for novices than for more experienced students.

Even if the heuristic evaluation is not perceived as an easy task, the intention of future use (Q2) is remarkably high for the three sets of heuristics. It is slightly higher among graduate students, comparing to the undergraduate students.

As expected, students consider that e-Commerce and SMASH heuristics covers better than Nielsen's heuristics the usability aspects of online travel agencies (Q3). Their opinion is less favorable to Nielsen's heuristics in roughly 1 point.

The descriptive statistics presented above was complemented with inferential statistics. As mentioned, all questionnaire items are based on a 5 points Likert scale. Observations' scale is ordinal, and no assumption of normality could be made. Therefore the survey results were analyzed using nonparametric statistics tests (Mann-Whitney U, Friedman, and Spearman ρ).

As samples are independent, Mann-Whitney U tests were performed to check the hypothesis:

- H_0: there are no significant differences between the perceptions of students with different background,
- H_1: there are significant differences between the perceptions of students with different background.

As the same group of students evaluated three different sets of heuristics, Friedman test was performed to check the hypothesis:

- H_0: there are no significant differences between students' perception on Nielsen's, e-Commerce, and SMASH heuristics,
- H_1: there are significant differences between students' perception on Nielsen's, e-Commerce, and SMASH heuristics.

Spearman ρ tests were performed to check the hypothesis:

- H_0: $\rho = 0$, the dimensions/questions D/Qm and D/Qn are independent,
- H_1: $\rho \neq 0$, the dimensions/questions D/Qm and D/Qn are dependent.

In all tests p-value ≤ 0.05 was used as decision rule.

Table 2 shows Mann-Whitney U tests results when comparing students with and without previous experience. Significant differences occur in very few cases:

- In the case of Nielsen' heuristics, regarding Q1 (easiness) for undergraduate students, and regarding Q1 (easiness) and Q3 (completeness) for graduate students.
- In the case of e-Commerce heuristics, regarding none of the dimensions or questions.

Table 2. Mann-Whitney U tests results when comparing students with and without previous experience.

Set of heuristics	Students' level	p-value						
		D1 – Utility	D2 – Clarity	D3 – Ease of use	D4 – Necessity of additional checklist	Q1 – Easiness	Q2 – Intention	Q3 – Completeness
Nielsen	Undergraduate	0.632	0.321	0.936	0.376	0.036	0.243	0.101
	Graduate	0.785	0.186	0.102	0.681	0.015	0.541	0.015
e-Commerce	Undergraduate	0.662	0.110	0.112	0.260	0.237	0.366	0.966
	Graduate	0.524	0.467	0.237	0.187	0.232	0.328	0.130
SMASH	Undergraduate	0.358	0.404	0.032	0.295	0.599	0.354	0.123
	Graduate	0.645	0.340	0.146	0.169	0.235	0.678	0.689

- In the case of SMASH heuristics, regarding D3 (ease of use) for undergraduate students.

As Table 3 shows, there are no significant differences between undergraduate and graduate students, for none of the three sets of heuristics.

Table 3. Mann-Whitney U tests results when comparing undergraduate and graduate students.

Set of heuristics	p-value						
	D1 – Utility	D2 – Clarity	D3 – Ease of use	D4 – Necessity of additional checklist	Q1 – Easiness	Q2 – Intention	Q3 – Completeness
Nielsen	0.557	0.953	0.597	0.636	0.754	0.095	0.284
e-Commerce	0.682	0.849	0.349	0.713	0.481	0.946	0.127
SMASH	1.000	0.872	0.837	0.976	0.413	0.645	0.604

Table 4. Friedman test results when comparing students' perception on Nielsen's, e-Commerce, and SMASH heuristics.

Students' level	p-value						
	D1 – Utility	D2 – Clarity	D3 – Ease of use	D4 – Necessity of additional checklist	Q1 – Easiness	Q2 – Intention	Q3 – Completeness
Undergraduate	0.821	0.338	0.018	0.498	0.368	0.397	0.000
Graduate	0.559	0.808	0.087	0.692	0.607	0.122	0.001

Friedman test results (Table 4) show significant differences between students' perception on Nielsen's, e-Commerce, and SMASH heuristics in only three cases:

- D3 (ease of use) and Q3 (completeness) for undergraduate students.
- Q3 (completeness) for graduate students.

As Mann-Whitney U tests results show no significant differences between undergraduate and graduate students, and very few significant differences when comparing students with and without previous experience, Spearman ρ tests were performed for the whole group of 38 students.

In the case of Nielsen's heuristics, 8 correlations occur between dimensions/questions (Table 5):

- A strong correlation between D2–D3; if heuristics are perceived as clear, they are also perceived as easy to use.
- 4 moderate correlations between D1–D2, D3–Q1, D2–Q2, and Q2–Q3. If heuristics are perceived as clear, they are also perceived as useful, and there is also a declared intention of future use. If heuristics are perceived as easy to use, the whole heuristic evaluation is perceived as easy to perform. If the set of heuristics is perceived as complete, there is a declared intention of future use.

Table 5. Spearman ρ test for Nielsen's heuristics: correlations between dimensions and questions.

	D1 – Utility	D2 – Clarity	D3 – Ease of use	D4 – Necessity of additional checklist	Q1 – Easiness	Q2 – Intention	Q3 – Completeness
D1	1	0.575	0.352	Independent	Independent	0.357	Independent
D2		1	0.650	Independent	Independent	0.594	Independent
D3			1	Independent	0.424	0.348	Independent
D4				1	Independent	Independent	Independent
Q1					1	Independent	Independent
Q2						1	0.429
Q3							1

- 3 weak correlations between D1–D3, D1–Q2, and D3–Q2. If heuristics are perceived as useful, they are also perceived as easy to use and there is also a declared intention of future use. If heuristics are perceived as easy to use, there is a declared intention of future use.
- It worth mentioning that there is no correlation between D4 (necessity of additional checklist) and any other dimension or question.

As Table 6 indicates, 16 correlations occur in the case of e-Commerce heuristics:

Table 6. Spearman ρ test for e-Commerce heuristics: correlations between dimensions and questions.

	D1 – Utility	D2 – Clarity	D3 – Ease of use	D4 – Necessity of additional checklist	Q1 – Easiness	Q2 – Intention	Q3 – Completeness
D1	1	0.731	0.759	0.408	0.428	0.386	0.556
D2		1	0.800	0.377	Independent	0.541	0.413
D3			1	0.415	0.489	0.444	0.359
D4				1	Independent	Independent	Independent
Q1					1	Independent	0.396
Q2						1	0.336
Q3							1

- Dimension D1 is correlated with all others dimensions and questions. When heuristics are perceived as useful, they are also perceived as clear and easy to use (strong correlations D1–D2, D1–D3); however there is also a declared necessity for additional checklist (moderate correlation D1–D4). When heuristics are perceived as useful, the set of heuristics is perceived as complete, the whole heuristic evaluation is perceived as easy to perform (moderate correlations D1–Q1 and D1–Q3), and there is an intention of future use (weak correlation D1–Q2).

- Dimension D3 is also correlated with all others dimensions and questions. When heuristics are perceived as easy to use, they are also perceived as useful (strong correlation D1–D3), clear (very strong correlation D2–D3), and there is a perceived necessity for additional checklist (moderate correlation D3–D4). The whole evaluation is perceived as easy to perform, there is a declared intention of future use of e-Commerce heuristics, and the set of heuristics is perceived as complete (moderate correlations D3–Q1, D3–Q2, and weak correlation D3–Q3).
- Dimension D2 is correlated to all others dimensions and questions, excepting one (Q1). Correlations are very strong (D2–D3), strong (D1–D2), moderate (D2–Q2, D2–Q3), or weak (D2–D4).
- Question Q3 is also correlated to all others dimensions and questions, excepting one (D4). Correlations are moderate (D1–Q3, D2–Q3) or weak (D3–Q3, Q1–Q3, Q2–Q3).
- Question Q2 is correlated to all others dimensions and questions, excepting D4 and Q1. Correlations are moderate (D2–Q2, D3–Q2) or weak (D1–Q2, Q2–Q3).
- Fewer correlations occur for Q1 (moderate correlations with D1 and D3, weak correlation with Q3), and D4 (weak to moderate correlations with other dimensions, but not with questions).

Table 7 highlights 15 correlations in the case of SMASH heuristics:

Table 7. Spearman ρ test for SMASH heuristics: correlations between dimensions and questions.

	D1 – Utility	D2 – Clarity	D3 – Ease of use	D4 – Necessity of additional checklist	Q1 – Easiness	Q2 – Intention	Q3 – Completeness
D1	1	0.737	0.532	0.384	0.338	0.718	0.341
D2		1	0.820	0.486	0.414	0.651	Independent
D3			1	0.467	0.469	0.468	Independent
D4				1	Independent	0.365	Independent
Q1					1	Independent	Independent
Q2						1	0.419
Q3							1

- As in the case of e-Commerce heuristics, dimension D1 is correlated to all others dimensions and questions (strong correlations between D1–D2 and D1–Q2, moderate correlation between D1–D3, and weak correlations between D1–D4, D1–Q1, and D1–Q3).
- Dimensions D2 and D3 are correlated to all others dimensions and questions, excepting Q3. Correlations are strong or moderate. There is a very strong correlation between D2–D3.
- Dimension D4 is correlated to all others dimensions and questions, excepting Q1 and Q3. Correlations are moderate or weak.
- Question Q2 is correlated with all dimensions (weak to strong correlations), and with question Q3 (moderate correlation).

- Question Q1 is correlated only with dimensions D1 (weak correlation), D2 and D3 (moderate correlation).
- Question Q3 is correlated only with dimension D1 (weak correlation) and question Q2 (moderate correlation).

Correlations are fewer in the case of general (Nielsen's) heuristics (7) than in the case of specific heuristics (e-Commerce, 16, and SMASH, 15). When occur, all correlations are positive.

4 Conclusions

Heuristic evaluation is arguably the most popular usability inspection method. We systematically conduct studies on the perception of evaluators over generic and specific usability heuristics. We are using a questionnaire that evaluates each heuristic individually (Utility, Clarity, Ease of use, Necessity of additional checklist), but also the set of heuristics as a whole (Easiness, Intention, Completeness).

Performing heuristics evaluation based on Nielsen's heuristics is a standard practice when we are teaching Human-Computer Interaction courses. This time we asked students to evaluate the same product based on three sets of heuristics: Nielsen's heuristics, e-Commerce heuristics, and SMASH heuristics.

The experiment involved graduate and undergraduate students. There were no significant differences between undergraduate and graduate students' perception, for none of the three sets of heuristics. When comparing students with and without previous experience, significant differences occurred in very few cases. Friedman test results showed significant differences between students' perception on Nielsen's, e-Commerce, and SMASH heuristics in only three cases.

Correlations were fewer in the case of general (Nielsen's) heuristics (7) than in the case of specific heuristics (e-Commerce, 16, and SMASH, 15). When occurred, all correlations were positive.

In general, students' perception on specific (e-Commerce and SMASH) heuristics was slightly better than on generic heuristics (Nielsen). As expected, students consider that e-Commerce and SMASH heuristics covers better than Nielsen's heuristics the usability aspects of online travel agencies.

As future work we intend to analyze the perception of each heuristic individually. We will also analyze students' comments to open questions.

Acknowledgments. We thank all the students involved in the experiment. They provided helpful opinions that allowed us to prepare this and (hopefully) further documents.

References

1. ISO 9241-210: Ergonomics of human-system interaction—Part 210: Human-centered design for interactive systems. International Organization for Standardization, Geneva (2010)
2. Lewis, J.R.: Usability: lessons learned… and yet to be learned. Int. J. Hum.-Comput. Interact. **30**(9), 663–684 (2014)

3. Nielsen, J., Mack, R.L.: Usability Inspection Methods. Wiley, New York (1994)
4. Hermawati, S., Lawson, G.: Establishing usability heuristics for heuristics evaluation in a specific domain: is there a consensus? Appl. Ergon. **56**, 34–51 (2016)
5. Quiñones, D., Rusu, C.: How to develop usability heuristics: a systematic literature review. Comput. Stand. Interfaces **53**, 89–122 (2017)
6. Nielsen, J.: 10 Usability Heuristics for User Interface Design. http://www.nngroup.com/articles/ten-usability-heuristics/. Accessed 28 Dec 2017
7. Inostroza, R., Rusu, C., Roncagliolo, S., Rusu, V., Collazos, C.: Developing SMASH: a set of SMArtphone's uSability Heuristics. Comput. Stand. Interfaces **50**, 160–178 (2016)
8. Díaz, J., Rusu, C., Collazos, C.: Experimental validation of a set of cultural-oriented usability heuristics: e-commerce websites evaluation. Comput. Stand. Interfaces **53**, 89–122 (2017)
9. Expedia. http://www.expedia.com. Accessed 10 Jan 2018
10. Rusu, C., Rusu, V., Roncagliolo, S., Apablaza, J., Rusu, V.Z.: User experience evaluations: challenges for newcomers. In: Marcus, A. (ed.) DUXU 2015. LNCS, vol. 9186, pp. 237–246. Springer, Cham (2015). https://doi.org/10.1007/978-3-319-20886-2_23
11. Rusu, C., Rusu, V., Roncagliolo, S., Quiñones, D., Rusu, V.Z., Fardoun, H.M., Alghazzawi, D.M., Collazos, C.A.: Usability heuristics: reinventing the wheel? In: Meiselwitz, G. (ed.) SCSM 2016. LNCS, vol. 9742, pp. 59–70. Springer, Cham (2016). https://doi.org/10.1007/978-3-319-39910-2_6
12. Rusu, V., Rusu, C., Quiñones, D., Roncagliolo, S., Collazos, C.A.: What happens when evaluating social media's usability? In: Meiselwitz, G. (ed.) SCSM 2017. LNCS, vol. 10282, pp. 117–126. Springer, Cham (2017). https://doi.org/10.1007/978-3-319-58559-8_11

Evaluation of Store Layout Using Eye Tracking Data in Fashion Brand Store

Naoya Saijo[1(✉)], Taiki Tosu[2], Kei Morimura[2], Kohei Otake[3], and Takashi Namatame[2]

[1] Graduate School of Science and Engineering, Chuo University, 1-13-27, Kasuga, Bunkyo-ku, Tokyo 112-8551, Japan
al3.rx7m@g.chuo-u.ac.jp

[2] Faculty of Science and Engineering, Chuo University, 1-13-27, Kasuga, Bunkyo-ku, Tokyo 112-8551, Japan
{al4.cjbj,al4.ers7}@g.chuo-u.ac.jp, nama@indsys.chuo-u.ac.jp

[3] School of Information and Telecommunication Engineering, Tokai University, 2-3-23, Takanawa, Minato-ku, Tokyo 108-8619, Japan
otake@indsys.chuo-u.ac.jp

Abstract. In this study, we conducted purchasing simulation experiment using eye tracking device in fashion brand store. Using the gaze measurement data obtained through experiments, we conducted several analyses to evaluate the store layout. Firstly, we divided the inside of the store into several areas. We tried to identify the areas that can become areas that are easily visible (Golden Zone) by performing multiple comparison on visual time for each area. Through the result, we identify the area that could be Golden Zone. In addition, it became clarifying that the characteristics of the areas which can become Golden Zone. Secondly, we tried to clarify that relationship between good impression item and visual time. It is clarified that there had a positive correlation between "Purchasing time" and "The number of item held in hand." Moreover, "Purchasing time" and "The number of good impression item" also had a positive correlation. From the results, we proposed improvement plans for better store layout.

Keywords: Eye tracking · Gaze measurement · Purchasing behavior Store layout · Shelf arrangement · Fashion brand store

1 Introduction

Improving store layout in a real store is still one of important issues. In a real store, there are areas that are most eye-catching and are easily accessible to hands to the customer. These areas are called "Golden Zone". Traditionally, by utilizing Golden Zone, sellers gathered customers' attention and sold items. However, it is thought that Golden Zone differs depending on the type and quantity of items to be handled, as well as the customer base. Therefore, it is necessary to grasp Golden Zone correctly in each store. To achieve that, it is important to clarify the customer's behavior within a store.

Eye tracking is one of the methods to capture the customer's in-store search and Purchase behavior. Eye tracking is a method of tracking the movement of a person's

© Springer International Publishing AG, part of Springer Nature 2018
G. Meiselwitz (Ed.): SCSM 2018, LNCS 10913, pp. 131–145, 2018.
https://doi.org/10.1007/978-3-319-91521-0_11

line of sight and measuring how much and which place that person gaze. Since it can analyze human behavior and unconscious need from the movement of the line of sight, it is used for the display method of products and the usability investigation on the website. In recent years, various studies of consumer behavior using eye tracking has been carried out [1]. By clarifying what kinds of gaze consumers are searching for items, it is thought that useful suggestions can be obtained in improving the store layout.

2 Purpose of This Study

In this study, in order to understand detail purchasing behavior, we conduct purchasing simulation experiment using eye tracking device in a fashion brand store. It aims to evaluate the store layout by using the purchasing behavior and characteristics of customers obtained from the gaze measurement data. Specifically, visual time for each area is measured, and we tried to identify Golden Zone through the analysis of multiple comparison. Also, with regard to the item that the participant replied that it was a good impression, we measure the visual time of good impression items and its surrounding items. From these results, we try to clarify the relationship between good impression items and visual time.

3 Previous Studies

In this chapter, we describe previous studies using eye tracking in the retail store. Kitazume et al. investigated the relation between staying or visual time and rate of purchase in a real shop by observing consumer's line of sight and view point by using the Eye-Mark Recorder and analyzing the decision-making process [2]. Chandon et al. evaluated the effect of getting gaze by POP (Point of Purchase) advertisement based on eye movement data in real store [3]. Van der Lans et al. proposed a methodology to determine the competitive salience of brands, based on a model of visual search and eye-movement recordings collected during a brand search experiment [4].

In addition, there are some studies focused on store layout and shelf arrangement of items. Tetsuoka et al. observed the consumer behavior in the retail store and clarified the relationship between the display method of goods in one display and the purchasing behavior by using bayesian network and belief propagation [5]. Shirai et al. revealed the position (Golden Zone) which is easy to select in the ice cream showcase [6]. However, from the aspect of consumer's eye tracking, there has not been adequately studied about store layout.

4 Experiment Aimed at Gaze Measurement in Purchasing Behavior

4.1 Outline of Experiment

We conducted experiment of purchasing simulation in the fashion brand store to obtain gaze measurement data in purchasing behavior.

4.1.1 About Eye Tracking Device and Target Store

In this experiment, eye tracking device "Tobii Pro Glasses 2" and recording software "Tobii Pro Studio" are used to measure the gaze of participants [7]. These make it possible to accurately grasp the visual time of participants and the movement of eyes, and output the experiment result as numerical data. Eye tracking device is shown in Fig. 1.

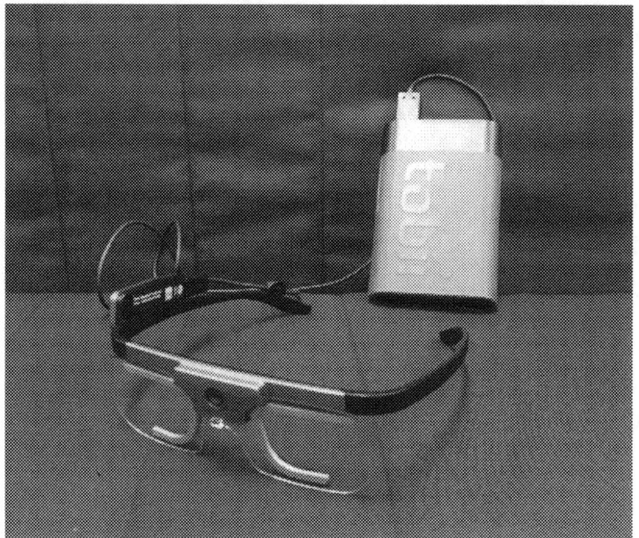

Fig. 1. Eye tracking device (Tobii Pro Glasses 2)

The target store for this study is a downtown street shop in Tokyo. Main customers are women in their twenties and thirties. The store treat handling bags, wallets, accessories, clothes, shoes and the other fashion items. The present layout is shown in Fig. 2.

The number of items laid out in each area and the size (height, width, depth) of the area are shown in Table 1. Also, the composition of item categories in each area is shown in Table 2.

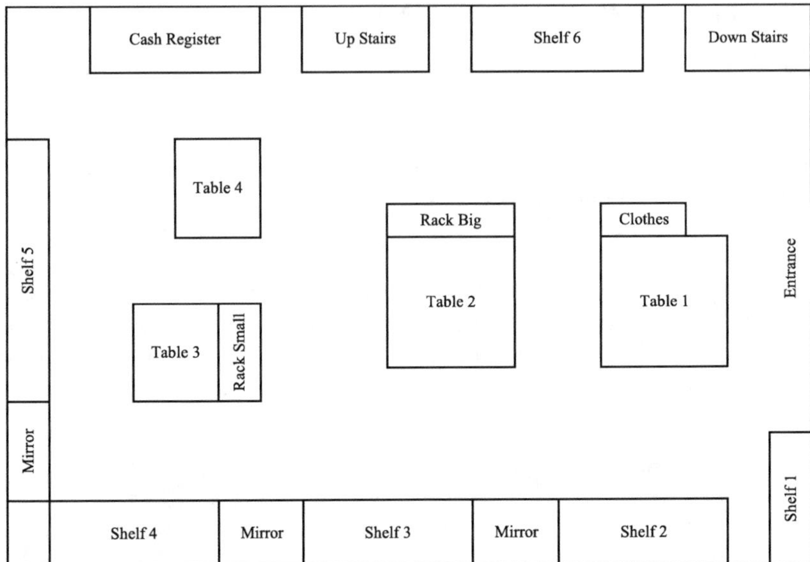

Fig. 2. In-store layout of the target store

Table 1. The number of items in each area and the size of each area

Area	The number of items	Height (cm)	Width (cm)	Depth (cm)
Shelf 1	9	231	117	
Shelf 2	26	280	280	
Shelf 3	31	280	280	
Shelf 4	60	250	200	
Shelf 5	22	220	220	
Shelf 6	29	291	312	
Table 1	14	63	122	304
Table 2	14	79	159	104
Table 3	26	92	44	150
Table 4	36	92	86	110
Rack Big	8	140	93	
Rack Small	5	140	80	
Cash Register	18	87	147	30
Clothes	8	117	150	

Table 2. The composition of item categories in each area

Area	Item categories
Shelf 1	Bag, key case, wallets, clothes, shoes (for men)
Shelf 2	Handbag, tote bags, shoes, wallet
Shelf 3	Handbag, tote bag, backpack, wallets, belt, pass case
Shelf 4	Handbags, wallets, pass case
Shelf 5	Handbag, shoes, wallet, key case
Shelf 6	Handbag, backpack, wallets, shoes, belt, pouch, pass case
Table 1	Handbag, shoes, belt, clothes, sunglasses
Table 2	Handbags, belt, bag charm
Table 3	Wallet, pass case
Table 4	Watch, bag charm
Rack Big	Tote bag
Rack Small	Shoulder bag
Cash Register	Earrings, bracelet, sunglasses, pouch
Clothes	Clothes

4.1.2 Experiment Participants

To gather the main target of this store, we selected 10 women in their twenties as participants. They are 4th grader at several university in Tokyo. First of all, we conducted questionnaire about sense of values in item purchasing and the target store. From the questionnaire, they have the following in common features.

- They purchase inexpensive items for themselves.
- They knew this shop from before.
- They rarely purchase items from this store.
- They rarely visit the EC site of this store.

4.2 Experiment Procedure

The specific procedure of the experiment is described.

1. Make participants gather in the room different from the store and explain the purpose and flow of the experiment.
2. Call the participant one by one outside the store, and the participant is attached the eye tracking device.
3. Calibrate the eye tracking device, and make it be accurate to collect recording data.
4. Record instore purchasing behavior of each participants from outside the store. At that time, 5 participants enter from the right side of the store, and the remained 5 participants enter from the left side of the store.
5. During the experiment, they can freely turn around the first floor of the store and pick up the items which they interest. Moreover, we observed the behavior of the participant and keep a brief note on the state of purchasing behavior. Expected time for purchasing behavior is about 5 min. When you finish watching them all, raise your hands and ask them to signal the end of the experiment.

6. After the purchasing behavior of the participant, we asked impressive items and items that they want as the after-questionnaire.

These experiments were repeated for 10 participants and gaze measurement data was acquired.

4.3 Measurement of the Visual Time Using AOI

We measure the visual time of each area and item from recorded data of purchasing simulation obtained by the experiment. To measure visual time, AOI (Aria Of Interest) installed in the recording software "Tobii Pro Studio" is used. By using AOI, it is possible to measure how much participants are viewing a certain areas or items. First, we specify the area to measure gaze time. Next, the recorded data is reproduced. If the area where we want to specify is shifted, move AOI or change its shape. When measurement is performed, the staying time of the line of sight is counted as 1 every 200 ms. By converting the measurement result at AOI to seconds, the visual time of each participant is clarified. The state of AOI is shown in Fig. 3.

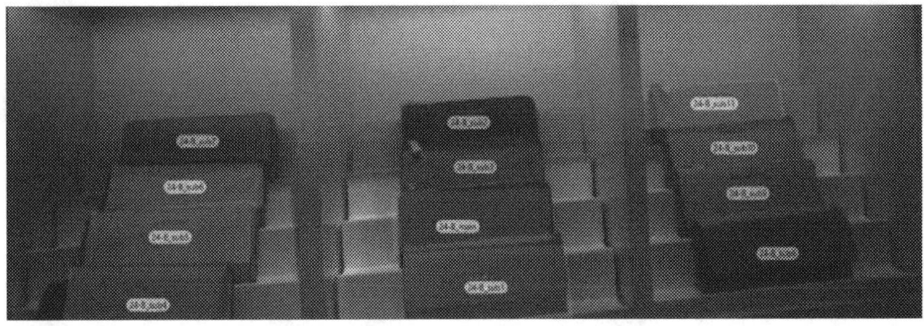

Fig. 3. The state of AOI

5 Analysis of the Visual Time of Each Shelf and Item

5.1 Multiple Comparison

In order to evaluate the difference in visual time in each area, we try to find dominant area by performing multiple comparison. In this case, since the sample size is small and a normal distribution cannot be assumed, the Steel-Dwass method is used [8]. The Steel-Dwass method is a multiple comparison using a rank order to simultaneously test all paired comparisons between groups for parameters representing the position of the distribution. The procedure of the Steel-Dwass method is as follows.

1. Make a combination of i and j $(i > j)$ for all groups.
2. Let the number of samples in both groups be $n_i, n_i(N = n_i + n_j)$.
3. The order of the i-th group and the j-th group is ranked, and the order of the k-th data of the i-th group is r_{ik}. Let the rank sum of the i-th group be $R_{ij} = r_{i1} + r_{i2} + \cdots + r_{in_i}$.
4. Calculate the expected value $E(R_{ij})$ and variance $V(R_{ij})$ under the null hypothesis.

$$E(R_{ij}) = \frac{n_i(N+1)}{2} \tag{1}$$

$$V(R_{ij}) = \frac{n_i n_j}{N(N-1)} \left\{ \sum\nolimits_{k=1}^{n_i} r_{ik}^2 + \sum\nolimits_{k=1}^{n_j} r_{jk}^2 - \frac{N(n_{ij}+1)^2}{4} \right\} \tag{2}$$

5. Calculate the test statistic t_{ij}.

$$t_{ij} = \frac{R_{ij} - E(R_{ij})}{\sqrt{V(R_{ij})}} \tag{3}$$

6. The p-value is calculated from the distribution of the studentized range of degrees of freedom ∞.

5.2 Identification of Golden Zone by Analyzing Visual Time

Firstly, the visual time of each area was measured from the recorded data using AOI. The visual time of each area for each participant is shown in Table 3. However, it is counted only when the line of sight stays for more than 1 s.

Table 3. The visual time of each area for each participant (s)

Area	Participant									
	A	B	C	D	E	F	G	H	I	J
Shelf 1	2	0	2	21	1	18	0	3	1	3
Shelf 2	25	33	70	53	18	7	4	9	96	53
Shelf 3	9	15	12	60	37	11	13	18	96	38
Shelf 4	19	74	26	77	14	4	65	77	110	8
Shelf 5	5	3	8	49	8	2	25	7	36	9
Shelf 6	32	0	10	4	42	81	0	24	7	6
Table 1	1	3	3	9	10	0	2	2	13	0

(continued)

Table 3. (*continued*)

Area	Participant									
	A	B	C	D	E	F	G	H	I	J
Table 2	0	14	0	3	24	10	0	3	32	0
Table 3	5	27	8	25	21	9	3	75	73	0
Table 4	16	38	25	16	36	27	8	31	23	0
Rack Big	25	15	4	0	0	0	0	99	0	0
Rack Small	0	15	0	5	14	0	2	29	48	0
Cash Register	0	0	0	5	0	0	8	0	0	2
Clothes	0	0	12	0	0	0	0	0	0	0

Secondly, a hypothesis test is performed by the Steel-Dwass method for the result of Table 3. By dividing by the visual time of all areas of each participant and the number of products in each area, the visual time for each area in "Table 3" is standardized. Visual time of "Shelf 1 to 4", "Table 3" and "Rack Small" is defined as "viewing time on the left side of the store as seen from the entrance" S_L. Visual time of "Table 4", "Rack Big", "Cash Register", "Shelf 6" and "Clothes" is defined as "viewing time on the right side of the store as seen from the entrance" S_R. How to divide the area is shown in Fig. 4.

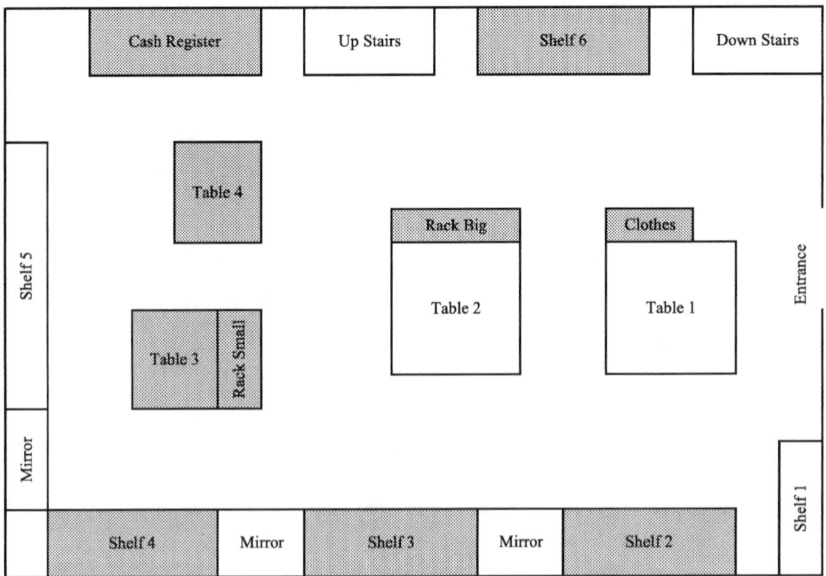

Fig. 4. How to divide the area

The hypothesis test at the significance level of 5% is performed with the following hypothesis.

$$H_0 : S_L = S_R$$

$$H_1 : S_L \neq S_R$$

As the result of the hypothesis test, the p-value is 0.00040. At the significance level of 5%, the null hypothesis is rejected and the alternative hypothesis is adopted. That is, the left side is seen more significantly from the right side.

Finally, the visual time of each area is standardized as $R_a(\text{Shelf 1}), \cdots, R_n(\text{Clothes})$. We make the following hypothesis for all combinations and perform a hypothesis test at a significance level of 5%.

$$\left.\begin{array}{l} H_0 : R_x = R_y \\ H_1 : R_x \neq R_y \end{array}\right\} (x, y) \in \{a, \ldots, n\}$$

The combination that became significant and its p-value is shown in Table 4.

Table 4. The combination that became significant and its p-value

Area		p-value
Shelf 2	Cash Register	0.034
Shelf 3	Cash Register	0.043

By combining the analysis results of both obtained above, it can be seen that "Shelf 2" and "Shelf 3", particularly on the left side of the store, can be candidates for Golden Zone.

5.3 Relationship Between Good Impression Item and Visual Time

Firstly, we summarize the results of purchasing simulation experiments of each participant. Specific value of each participant is shown in Table 5. "Purchasing time" represents the purchase simulation experiment time of each participant. Regarding "Enter position", it was defined as that 1 was entered from the left side of the store and 0 was entered from the right side. "View outside store" was set to 1 when the out-of-store window was visually observed, and to 0 when the out-of-store window was not visually observed. Also, "The number of good impression item" is counted from the after-questionnaire.

We clarify details about "The number of good impression item". The results of "The number of good impression item" for each area is shown in Table 6.

Table 5. Specific value of each participant

Participant	Purchasing time (sec)	Enter position (0: left, 1: right)	View outside store (0: No, 1: Yes)	The number of item held in hand	The number of good impression item
A	368	0	0	7	2
B	312	1	0	5	4
C	461	0	0	14	5
D	640	1	1	16	6
E	359	0	1	5	6
F	316	1	1	5	2
G	273	0	0	8	2
H	488	1	0	9	6
I	604	0	0	9	5
J	269	1	1	5	4

Table 6. The number of good impression items by area of each participant

Area	Participant									
	A	B	C	D	E	F	G	H	I	J
Shelf 1	0	0	0	0	0	0	0	0	0	1
Shelf 2	1	0	0	0	0	0	0	0	0	0
Shelf 3	0	0	1	0	2	0	1	0	1	1
Shelf 4	0	0	0	1	0	1	1	1	1	0
Shelf 5	0	1	1	2	0	0	0	0	0	0
Shelf 6	1	0	0	0	1	1	0	0	0	0
Table 1	0	0	0	1	0	0	0	0	1	0
Table 2	0	1	1	0	1	0	0	0	1	0
Table 3	0	1	0	0	1	0	0	3	1	0
Table 4	0	0	1	0	1	0	0	0	0	0
Rack big	0	0	0	2	0	0	0	2	0	0
Rack small	0	1	1	0	0	0	0	0	0	2
Cash register	0	0	0	0	0	0	0	0	0	0
Clothes	0	0	0	0	0	0	0	0	0	0
Outside store	0	0	0	0	0	0	0	0	0	2

Secondly, the visual time of each good impression item and surrounding items it can be measured from the recorded data using AOI. By comparing each visual time, it is possible to clarify the relationship between good impression item and visual time. For example, visual items and times of participant C on the area of "Rack small" is shown in Figs. 5 and 6.

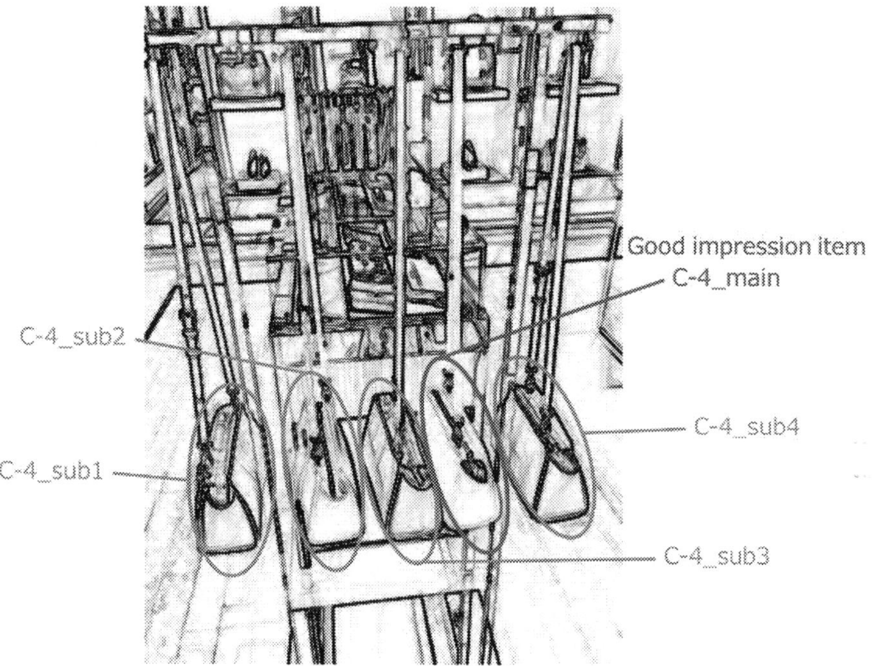

Fig. 5. Visual items of participant C on "Rack small"

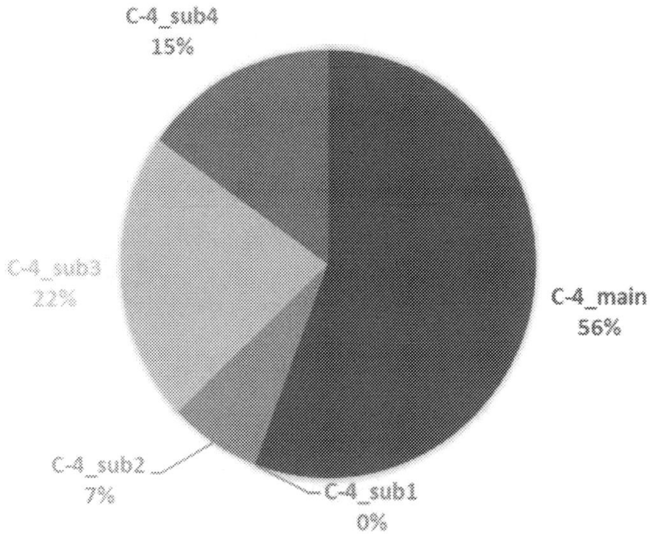

Fig. 6. Visual times of participant C on "Rack Small"

These results for each participant are shown in Table 7. Among the good impression items, the number of items with the longest visual time in the same area of the good impression items is defined as "The number of the longest visual time" in Table 7. Match ratio is defined as the ratio of "The number of the longest visual time" among "The number of good impression item" in Table 7.

Table 7. Relationship between good impression item and visual time

Participant	The number of good impression item	The number of the longest visual time	Match ratio
A	2	2	1.000
B	4	2	0.500
C	5	4	0.800
D	6	5	0.833
E	6	5	0.833
F	2	1	0.500
G	2	1	0.500
H	6	6	1.000
I	5	4	0.800
J	4	3	0.750

It clarifies that the items that remain in the good impression are seen long from the index of match ratio. Therefore, it is suggested that there is a relationship between good impression items and visual time.

Thirdly, we perform correlation analysis using each purchasing specific value. For correlation analysis, Pearson product-moment correlation coefficient is used. Pearson product-moment correlation coefficient is shown below.

$$r_{xy} = \frac{\sum_{i=1}^{n}(x_i - \bar{x})(y_i - \bar{y})}{\sqrt{\sum_{i=1}^{n}(x_i - \bar{x})^2}\sqrt{\sum_{i=1}^{n}(y_i - \bar{y})^2}} \quad (4)$$

Pearson's correlation coefficient of each purchasing specific value is shown in Table 8.

The bold letters in Table 8 are the correlation coefficients of $-1.0 < r < -0.6$ or $0.6 < r < 1.0$. This is a value that is generally thought to be a strong correlation. It can be seen that there is a positive correlation between "Purchasing time" and "The number of item held in hand" and between "Purchasing time" and "The number of good impression item".

Table 8. Correlation coefficient of each purchasing specific value

	Purchasing time	Enter position	View outside store	The number of item held in hand	The number of good impression item
Purchasing time	1.00				
Enter position	−0.03	1.00			
View outside store	−0.08	0.41	1.00		
The number of item held in hand	**0.77**	−0.08	−0.12	1.00	
The number of good impression item	**0.64**	0.13	0.15	0.44	1.00

6 Discussions and Evaluation of Store Layout Using Sales Ranking of Items

Based on the above analysis, we plan a better store layout. Firstly, we consider Golden Zone in the retail store. In this study, we divided the inside of the store into several areas and tried to identify Golden Zone by performing multiple comparisons. As the result, "Shelf 2" and "Shelf 3" are considered to be Golden Zone. Traditionally, it has been said that easiness for customers to see and to take in hand have a connection with Golden Zone. Therefore, it was assumed that the inner part of the store where customers can see the items slowly could be Golden Zone. That is, if utilizing existing knowledge at the target store, the areas such as "Shelf 4", "Shelf 5", "Table 3" and "Table 4" is defined as a Golden Zone. However, conventional studies are targeting daily necessities such as supermarkets.

In this experiment, there were store clerks and experiment record members at the "Cash Register" and "Down Stairs". In addition, the target store was a fashion brand store, which is expensive for supermarkets. As a result of after-questionnaire to participants, it was the first time that they visited this time, so they often checked the price. At that time, participants answered that they were concerned with the store clerk's gaze. Such a feeling is thought to be more likely to occur with higher price items. Due to these influences, in this study, it is considered that "Shelf 2" and "Shelf 3" were located away from the store clerk and the items can be seen slowly. Actually, in this store, store clerks often stay near the cash register during business hours, so "Shelf 2" and "Shelf 3" are considered to be Golden Zone.

Secondly, we consider relationship between good impression item and visual time. In this study, in many participants, the match rate between good impression items and the longest visual items in each area was high. Therefore, it is conceivable that the longer the viewing time of the item, the easier it is to have a good impression. As the result of correlation analysis, it was clarified that there is a positive correlation between "Purchasing time" and "The number of item held in hand" and between "Purchasing time" and "The number of good impression item". Therefore, it is considered that increasing the purchasing time and increasing the number of items to be taken by hand will lead to an increase in the number of good impression.

Thirdly, based on the sales ranking of the store targeted for this study, we compare visual time of popular items. Also, the popularity ranking is a ranking that integrates the qualitative evaluation based on the sales of the store and the subjective evaluation that the store clerks got through the customer service. The relationship between popular items and visual time is shown in Table 9.

Table 9. Relationship between popular items and visual time

Area	Item	The number of staying participants	Ratio of visual time in area
Shelf 3	Tote bag (black)	4	**0.15**
Shelf 6	Bag (black)	3	0.31
Rack Big	Tote bag (Upper right)	2	0.30
	Tote bag (Bottom right)	2	**0.14**
Rack Small	Bag (black)	2	0.26

As shown in "Ratio of visual time in area" in Table 9, tote bag which located in "Shelf 3" and bottom right of "Rack Big" are not seen much. Therefore, it is considered effective to place these items in a position more attracting attention.

Finally, we propose about store layout from these considerations. First of all, items that the store want to sell is arranged on "Shelf 2" and "Shelf 3". To make it stay longer in the store, the popular items is arranged in distributed fashion. Also, to get more items to be picked up, place the item in a position that is easy to pick up, or put the distance between the items wider. Considering the above proposal, it is thought that store layout can be more improved.

7 Conclusion and Future Works

In this study, purchasing simulation experiment was conducted to evaluate the store layout using eye tracking device in fashion brand store. By analyzing the gaze measurement data obtained in the experiment, we could identify the area that could be Golden Zone. Also, we analyzed relationship between good impression item and visual time, it was clarified that there is a positive correlation between "Purchasing time" and "The number of item held in hand" and between "Purchasing time" and "The number of good impression item".

As future works, the following can be considered. First of all, in this experiment, the number of participants was 10 and insufficient. In addition, all 10 participants were women in their twenties. By conducting experiments with participants in a wide range of age groups, it is thought that it is possible to clarify the difference in purchasing behavior by age. Also, in this experiment, we conducted experiments outside of business hours. For this reason, we do not consider the influence of other customers and service of clerks. By conducting experiments under conditions closer to actual purchasing behavior, it is thought that more accurate purchasing behavior can be clarified.

Acknowledgment. We are deeply grateful the target store for this study, employees of this store and participants of this experiment for providing experimental opportunity and useful comments.

References

1. Satomura, T.: Understanding consumer behavior by gaze measurement. Commun. Oper. Res. Soc. Jpn. **62**(12), 775–781 (2017). (in Japanese)
2. Kitazume, K., Yokouchi, T.: An analysis of relation between staying or visual time and rate of purchase in a department store. Trust Soc. **2**, 61–71 (2014). (in Japanese)
3. Chandon, P., Hutchinson, J.W., Bradlow, E.T., Young, S.H.: Measuring the value of point-of-purchase marketing with commercial eye-tracking data. In: Wedel, M., Pieters, R. (eds.) Visual Marketing: From Attention to Action, pp. 225–258. Psychology Press (2008)
4. van der Lans, R., Pieters, R., Wedel, M.: Research note: competitive brand salience. Mark. Sci. **27**, 922–931 (2008)
5. Tatsuoka, K., Yoshida, T., Munemoto, J.: Analysis on relationship between layout of display cases and purchase behavior by bayesian network. Trans. AIJ. J. Archit. Plan. Environ. Eng. **634**, 2633–2638 (2008). (in Japanese)
6. Shirai, A., Senba, K., Takagaki, A.: Influence of display position on product selection: search for golden zone in ice cream showcase. Bull. Fac. Soc. Stud. Kansai Univ. Jpn. **39**(3), 296–304 (2008). (in Japanese)
7. Tobii Pro Glasses 2. https://www.tobiipro.com/product-listing/tobii-pro-glasses-2/. Accessed 23 Feb 2018
8. Nagata, Y., Yosshida, M.: Basics of statistical multiple comparison method. Scientist Inc. (1997)

Evaluation of High Precision Map Creation System with Evaluation Items Unique to Each Feature Type

Masashi Watanabe, Takeo Sakairi(✉), and Ken Shimazaki(✉)

Mitsubishi Electric Corporation, 8-1-1 Tsukaguchi-Honmachi, Amagasaki, Hyogo, Japan
Watanabe.Masasahi@df.MitsubishiElectric.co.jp,
Sakairi.Takeo@db.MitsubishiElectric.co.jp,
Ken.Shimazaki@eb.MitsubishiElectric.co.jp

Abstract. Nowadays, autonomous driving systems are being developed and utilized. And high precision 3D map data is said to be necessary to realize autonomous driving. This data contains position data of features such as "road line", "road edge" and "road sign". This map data is created with point cloud data taken by Mobile Mapping System (MMS), and to reduce cost, it is desirable that the map creation work is automated, but in fact it is difficult to fully automate. So first, features detection process will be performed by an automatic process, and then a manual confirmation process will be performed. Therefore, in order to reduce cost, it is important to improve a manual confirmation process. In this paper, we propose a system that improves a manual confirmation process by displaying features in different color using evaluation items unique to each feature type.

In experiment, we used actual MMS data on Tokyo Japan, performed auto features detection process and got features data with evaluation item values. We investigated whether an appropriate low evaluation item values is given for error detection. As a result, we confirmed the effectiveness of this system.

Keywords: High precision map · Mobile mapping system · Point cloud

1 Introduction

Nowadays, autonomous driving systems are being developed and utilized. Information for automated driving is mainly provided by car equipped sensors such as cameras or radars, but detection range of these sensors are limited. Therefore, high precision 3D map data is said to be necessary to realize autonomous driving. This data contains position data of features such as "road line", "road edge" and "road sign" (see Fig. 1). Position data of road line can be used to recognize where to drive, it of road edge can be used to recognize where to escape in an emergency, it of road sign can be used to recognize self position precisely. Furthermore, to know road line shape over sensors range enable comfortable driving such as "changing lane for the destination in advance" or "decelerating before curved section".

© Springer International Publishing AG, part of Springer Nature 2018
G. Meiselwitz (Ed.): SCSM 2018, LNCS 10913, pp. 146–156, 2018.
https://doi.org/10.1007/978-3-319-91521-0_12

3D map data is created with point cloud data taken by Mobile Mapping System (MMS) (see Fig. 2) and map creation process should be automated to reduce cost. However, complete automation is much difficult because some road line may be faded, or some road edge may covered with grass, so manual confirmation process will be necessary. Therefore, in order to reduce cost, it is important to improve efficiency of manual confirmation process. In this paper, we develop system that can improve it.

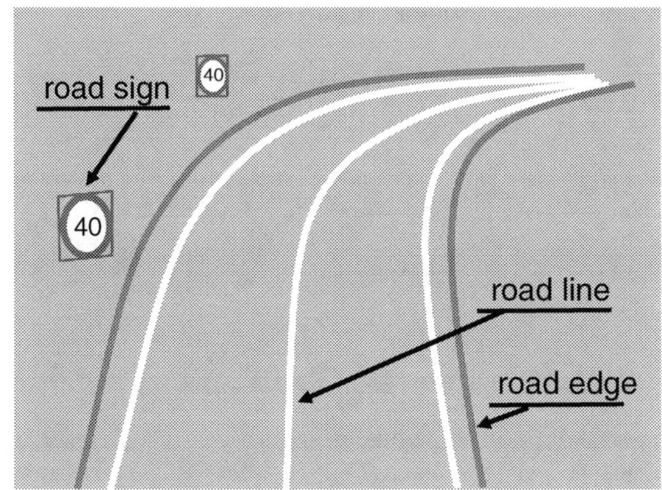

Fig. 1. Image of high precision map data

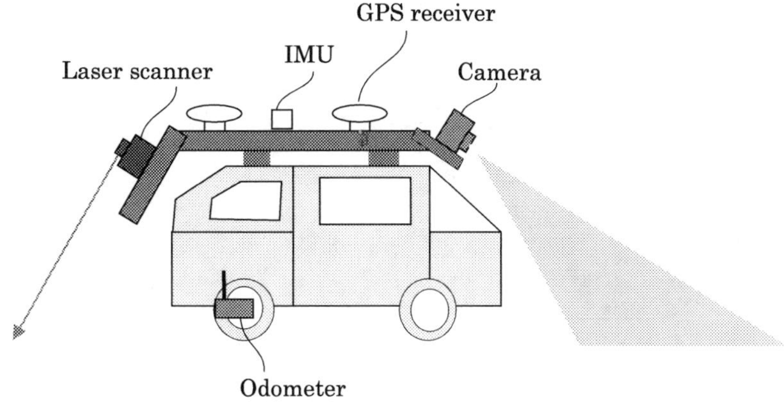

Fig. 2. Assumed components of MMS

2 High Precision Map Creation

In this paragraph, the overall flow of high precision map creation system is described. High precision map is generated through processing shown below.

- Getting measured data using MMS
- Automatic map creation processing by software
- Manual confirmation and modification processing

2.1 Getting Measured Data Using MMS

MMS is the car which has equipment for self-location grasping such as GPS antenna or IMU (Inertial measurement unit) and equipment for surrounding measurement such as radar or cameras. MMS can measure high precision 3D point cloud data and image data about the traveling road, although when a car is running by the side of MMS, MMS cannot get data about the back of it. And point data contains reflectance data as well as 3D position data.

2.2 Automatic Map Creation Processing by Software

Automatic map creation processing is the function that creates high precision positioning data such as road line, road edge and road sign from MMS data. For example, road line can be detected where reflectance is high, road edge can be detected where height difference is larger compared to the surroundings. These feature detection method is being studied in various way [1, 2, 3]. However, it may be difficult to detect the appropriate feature position fully, for example, due to faded and losing lines or road edge covered with grass.

2.3 Manual Confirmation and Modification Processing

Automatic map creation processing may cause over-detected failure or undetected failure (over-detected failure means detecting feature that should not be detected, undetected failure means not detecting feature that should be detected). So manual confirmation and modification processing must be performed. We should check failure detection and modify it correctly. The map data created through these processing can be actually used for autonomous driving.

3 Method of Calculating Reliable Value

In this paragraph, the method of calculating reliable value for road line, road edge and road sign is described. The reliable value is used to clearly indicate which features are highly necessary to modify, and it is derived from evaluation item values. So, first, we show the method of deriving the evaluation item values and then show the method of calculating reliable value using the evaluation item value.

3.1 Method of Deriving the Evaluation Item Values

We use different evaluation items depending on features, so we explain evaluation items for each feature.

Road Line

- Amount of change in reflectance.
  This item value indicates vividness of the road line, we consider vivid line is reliable. So if the line is faded, the evaluation value will drop. To calculate this value, we use points in the transverse direction from the point on road line, and we obtain the difference in reflectance between the point near line and the point with low reflectance (see Fig. 3).

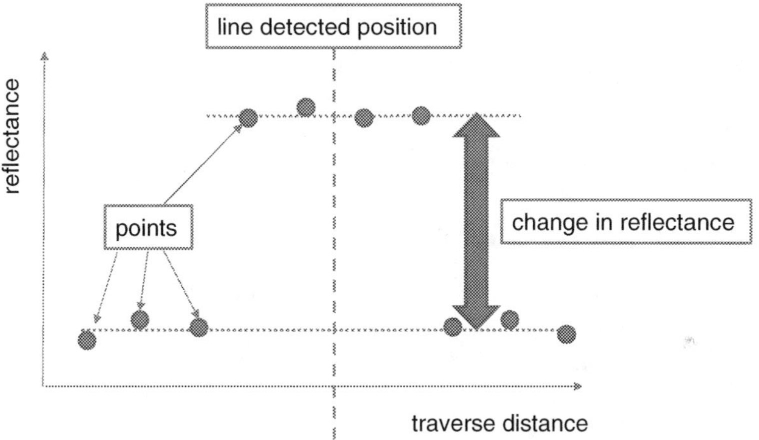

Fig. 3. How to get "Amount of change in reflectance"

- Linear shape
  This item value indicates whether the line shape seems good or not, we consider straight or clean curve line is reliable, so if the detected line is jagged, the evaluation value will drop. To calculate this value, we approximate a point sequence constituting a line to a function, and we obtain the sum of errors of approximation.
- Distance from MMS
  This item value indicates the distance from measuring car (MMS) to the detected line, point cloud is dense near MMS, and is sparse far from MMS, so we consider detected line near MMS is more reliable.

Road Edge

- Amount of change in height
  This item value indicates height of the road edge, we consider high edge is reliable. So if the edge is gentle, the evaluation value will drop. To calculate this value, we use points near the edge, and we obtain the height difference of these points.

- Linear shape
  Same as Linear shape in road line.
- Distance from MMS
  Same as Distance from MMS in road line.

Road Sign

- Direction of surface
  This item value indicates coincidence between direction of detected road sign surface and direction of MMS trajectory, we consider road sign which confronts to driver is reliable. To calculate this value, we create clusters by grouping nearby points, and fit clusters to a plane.
- Relationship between height and size
  This item value indicates properness of detected road sign size when considering its height, because the higher sign should be display larger in order to have the driver recognize the sign. To calculate this value, we create clusters and calculate its bounding box size.
- Distance from MMS
  Same as Distance from MMS in road line.

3.2 Method of Calculating Reliable Value Using the Evaluation Item Value

Reliable value is calculated by adding weighted evaluation item values. All evaluation item values range from 0 to 1. And in this system, all evaluation item values are equally weighted. To get better weighting or new function which derives better reliable value, it is possible to use machine learning and learn the correlation between features modified in manual processing and its evaluation item values.

4 Reliable Value Displaying System

In this paragraph, reliable value displaying system which uses detected feature data with reliable value is described. In this system, features are displayed in different colors, and point clouds data is also displayed on the background (see Fig. 4). A red feature indicates low reliability and a blue one indicates high reliability. The user can change view point position by mouse operation, and can check whether detections are correct.

Fig. 4. Image of reliable value displaying system. (Color figure online)

5 Experiment

In this paragraph, we perform experiment using reliable value displaying system. We used actual 35 km MMS data on highway in Tokyo Japan (details is shown in Table 1 and data image is shown in Fig. 5), performed auto features detection process and got features data with reliable value. We counted the number of wrong detections, wrong detections with reliable value lower than 0.8 and correct detections with reliable value lower than 0.8. Result of experiment is shown in Table 2, and example of detections with reliable value are shown in Figs. 6, 7, 8, 9, 10, 11 and 12.

Table 1. Details of point clouds data

Path	(start) Totsuka toll gate on Yokohama by-pass → Shin-hodogaya Interchange → Yokohama-Yokosuka Road → Kamariya Junction → Bayshore Route → Honmoku Junction → Kanagawa Route 3 Kariba Line → Ishikawacho Junction → Kanagawa Route 1Yokohane Line → Oi Parking Area (end)
Running Time of MMS	2300 s
Length	35 km

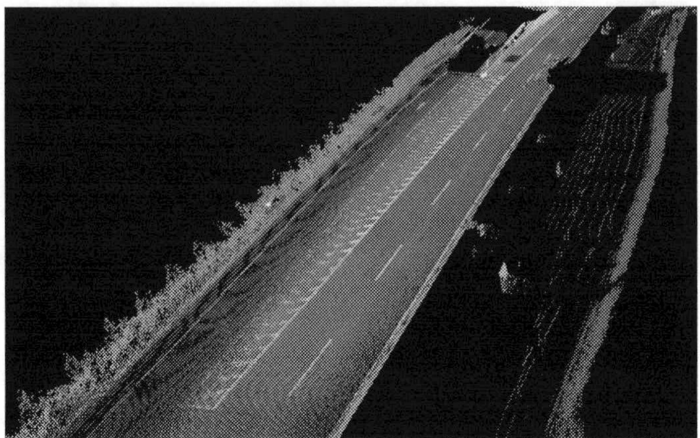

Fig. 5. Enlarged image of experiment use data

Table 2. Result of experiment

Feature	Line	Edge	Sign
Number of wrong detection (a)	275	235	389
Number of wrong detection with low reliable value (b)	183	232	184
Number of correct detection with low reliable value (c)	330	39	19
Low reliable value rate among wrong detection (b/a)	66.5%	98.7%	47.3%
Wrong detection rate among detection with low reliable value (b/(b + c))	35.7%	85.6%	90.6%

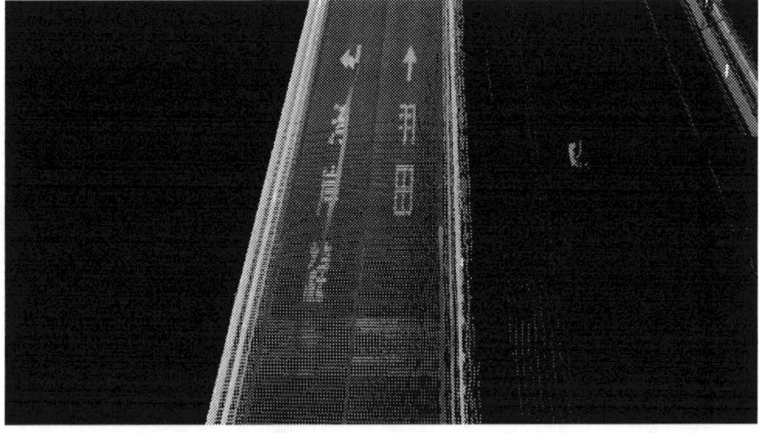

Fig. 6. Example of line wrong detection with low reliable value. (Color figure online)

Fig. 7. Example of line wrong detection with high reliable value. (Color figure online)

Fig. 8. Example of edge wrong detection with low reliable value. (Color figure online)

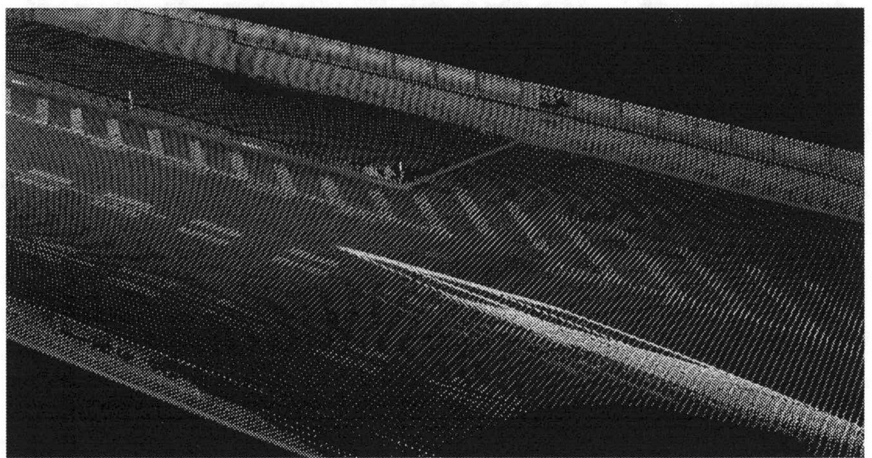

Fig. 9. Example of edge wrong detection with high reliable value. (Color figure online)

Fig. 10. Example of edge correct detection with high reliable value. (Color figure online)

Fig. 11. Example of sign wrong detection with low reliable value. (Color figure online)

Fig. 12. Example of sign wrong detection with high reliable value. (Color figure online)

Considerations on the results for each feature type are shown below.

Line:
Low reliable value rate among wrong detection (b/a) was not so high. Most of wrong detection were due to marks on road surface such as sign for destination or deceleration, and these marks were often arranged regularly. Therefore "Linear shape" value and reliable value became high. In order to set reliable value appropriately, it is necessary to use item for evaluating line spacing.

Wrong detection rate among detection with low reliable value (b/(b + c)) was also not so high. Because there were some faded line, and "Amount of change in reflectance" value and reliable value became low. But this evaluation item seems to be necessary, because line wrong detection may occur on the road surface.

Edge:
Both rate (b/a and b/(b + c)) were high. Most of wrong detection were due to shielding by parallel running vehicle and grass near the edge, and these were reflected in “Linear shape” value.

Sign:
Low reliable value rate among wrong detection (b/a) was not so high. Some wrong detection were due to the side of the bridge. In order to set reliable value appropriately, it is necessary to use item for Image recognition. MMS can get Image data, so using image data and image recognition, wrong detection on the side of the bridge can be set low reliable value.

6 Conclusion

To realize autonomous driving, high precision 3D map data is said to be indispensable, and it is important to reduce the cost of creating maps. We have proposed a novel system that displays features in a color according to reliable value, and enables to find suspicious map data quickly. Reliable value is calculated using evaluation item values unique to each feature type. Experiment result showed that the proposed system could recognize error detection to a certain extent, and it is assumed that more appropriate reliable value can be obtained by adding suitable evaluation items. Furthermore, we used reliable value calculated with the function which weights all evaluation item value equally, there will be a better function. In the near future, we will develop a better function using machine learning. And it is also necessary to investigate and improve the display method for making it easier for users to confirm error detection.

References

1. Guan, H., Li, J., Yu, Y., Wang, C., Chapman, M., Yang, B.: Using mobile laser scanning data for automated extraction of road markings. ISPRS J. Photogram. Remote Sens. **87**, 93–107 (2014)
2. Pu, S., Rutzinger, M., Vosselman, G., Elberink, S.: Recognizing basic structures from mobile laser scanning data for road inventory studies. ISPRS J. Photogram. Remote Sens. **66**, S28–S39 (2011)
3. Boyko, A., Funkhouser, T.: Extracting roads from dense point clouds in large scale urban environment. ISPRS J. Photogram. Remote Sens. **66**, S2–S12 (2011)
4. Liu, W., Anguelov, D., Erhan, D., Szegedy, C., Reed, S., Fu, C.-Y., Berg, A.C.: SSD: single shot MultiBox detector. In: Leibe, B., Matas, J., Sebe, N., Welling, M. (eds.) ECCV 2016. LNCS, vol. 9905, pp. 21–37. Springer, Cham (2016). https://doi.org/10.1007/978-3-319-46448-0_2

The Proposal of Cognitive Support for Driver by Voice Guide Using Soliloquy Expression

Takuya Yamawaki[1(✉)], Takayoshi Kitamura[1], Tomoko Izumi[2], and Yoshio Nakatani[1]

[1] Ritsumeikan University, Kusatsu, Siga, Japan
is0241sh@ed.ritsumei.ac.jp
[2] Osaka Institute of Technology, Hirakata, Osaka, Japan

Abstract. In a car navigation system, a voice guidance system is equipped to ensure the safety of driving. On the other hand, a driver requires a certain period of time to understanding instructions from the voice guidance and may misjudge them. To solve these problems, we propose a new expression (SVN) process of the voice guidance based on drivers' soliloquy. When drivers confirm a point to turn or a distance to the point based on the voice guidance, they quite often make correspondence the expressions of the guidance to their own expressions. Such the expression in the brain of a driver is expected to be similar to the soliloquy type expression of the driver. We assume that the soliloquy expression of the guidance will be understandable easier than the conventional expression. We conducted the experiment on the driving simulator to verify our hypothesis. The results suggest that SVN decreases the time for understanding the instruction and the frequency of misjudgment.

Keywords: Car navigation system · Human-machine-interface
Personal cognitive support · Voice navigation

1 Introduction

Today, the major car navigation system has adapted a display monitor to guide the route. However, it is pointed out that the gaze to the display during driving degrades the stability of the driving operation and prevents the driver from detecting a dangerous object [1]. Therefore, it is necessary to study a new route guidance process not using the display. One of the new route guidance processes is that using voice guidance only. This new processology is considered to solve the safety problem because drivers need not to gaze the display. However, using only voice guidance will cause a new problem. Freundshuh et al. [2] pointed out that the problem was caused by the high abstraction degree of voice information. That is, the disadvantage of the voice guidance is its low amount of information and its accuracy. As a result, a driver may take a time to understand the instruction and identify a different object, such as a corner or a landmark, as the target one in the instruction. Therefore, it is necessary to verify the expression process of voice guidance assuming the use only of voice guidance.

© Springer International Publishing AG, part of Springer Nature 2018
G. Meiselwitz (Ed.): SCSM 2018, LNCS 10913, pp. 157–172, 2018.
https://doi.org/10.1007/978-3-319-91521-0_13

2 Related Studies

Personalization is the concept of providing information according to the individual's ability, characteristics. Otani [3] indicated that personalized route guidance is effective in improving understandability, because individual differences exist in the environmental knowledge and the spatial cognitive ability. Therefore, by applying the concept of personalization to the voice guidance expression, it is possible to reduce the understanding time and improve the accuracy of decision. Audio information can be flexibly changed from the various points of views and have various expressions [4]. Therefore, it is considered that the voice guidance will be personalized in a relatively easy way. Kawai et al. [5] evaluated the personalized voice guidance based on a verbal navigation by participants. This evaluation revealed that there is a difference in preferred voice guidance expression depending on age in terms of the frequency of referring names of landmarks. However, in [5], they consider the situation where drivers use both of navigation from in-vehicle display and voice guidance. In the existing studies about the personalized voice guidance, the display guidance is also used, and there few studies focusing on using voice guidance only.

3 Proposal of New Voice Guidance Expression Process

In this research, we propose a new expression of voice guidance, called Soliloquy Voice Navigation (SVN), which uses soliloquy of a driver. We consider the situation where the navigation adopts the voice guidance only. In our work, we defined the "soliloquy of a driver during driving" as the monologue of the driver for confirmation by him/herself of a point of turning right or left. For example, "I wanna turn to the left a little further ahead…". The soliloquy during driving is caused when the divers confirms the routes or the target intersection to turn. Hence, it is considered that SVN, using the soliloquy is easy to understand for the driver.

4 Evaluation

4.1 Overviews

Our goal is to verify the hypothesis that the use of SVN is more effective than that of a general voice guidance (VN). To evaluate the efficiency, we focus on the time required for understanding the instruction content, called understanding time, and the accuracy of identifying the instruction, called the accuracy of decision. That is, we evaluate the understanding time and the accuracy of decision for SVN and VN in the experiment. To evaluate the efficiency of SVN, we compare the following evaluation indexes with VN:

1. The understanding time: It is the length of time from output of voice guidance to understanding of the instruction by a driver.
2. The accuracy of decision: It is the number of correctly decisions of the corner that the voice guidance instructed.

Furthermore, to reveal the suitable type of driver to SVN, we used the Driving Style Questionnaire (DSQ) which was proposed by Ishibashi et al. [3] and Workload Sensitivity Questionnaire (WSQ) which was proposed by Akamatsu et al. [4]. In this study, we do not focus on a process for construction of SVN. In the experiment, we made suitable SVN for each participant simply.

4.2 Participants

We focus on inexperienced drivers as participants. It is because that experienced drivers are accustomed to VN. Therefore, VN is a familiar voice guidance for experienced drivers, and then, such the experience will give some impact to evaluation results. In this research, the inexperienced driver is defined as a driver who driving frequency is less than once a month.

The details of the participants are shown in Table 1. The participants were selected from the university students and the graduate students majoring in the information science. All the participants were Japanese. We explained the contents in the informed consent, and all of the participants confirmed them in the writing. The previous research has reported that there are differences in cognitive ability during driving depending on gender [6]. On the other hand, other research pointed out that the differences between men and women is caused by the differences in their driving frequency [7]. In this experiment, since all the participants are inexperienced drivers, the gender of the participant does not affect the evaluation results. Figures 1 and 2 show the average scores about the results for DSQ and WSQ by the participants. The general drivers in these figures mean the average scores of general drivers in Japan obtained by the research of Ishibashi et al. [8] and Akamatsu et al. [9]. Compared with the general average, the average scores of the participants have the following features:

- The scores for “Anxiety about traffic accident” and “Hesitation for driving” are high.
- The scores for “Confidence in driving skill” is low.

It is considered that these results are caused by lacking driving experience of the participants. In addition, there is few difference between the participants and the general drivers excepting the points above. Therefore, in terms of the types of driving, it is said that there is few difference between the participants and the general driver.

Table 1. The details of the participants.

Item	Values
Number of participants	14
Gender	Male:10, Female:4
Age	19–25 years (average:21.7, SD:1.27)
Period of getting driving license	1–4 years
Driving frequency	Less than once a month

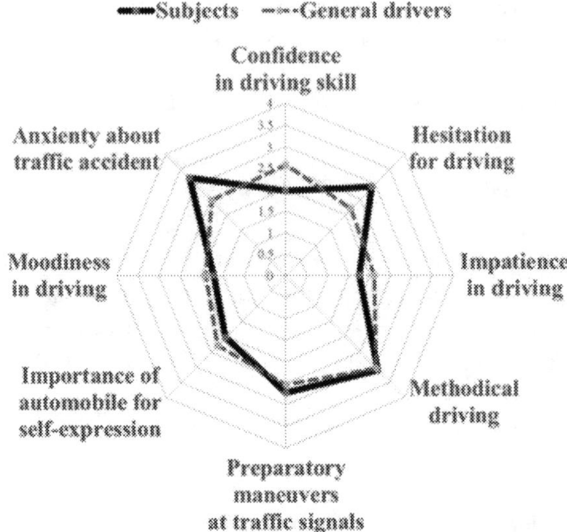

Fig. 1. The average scores about the results for driving style.

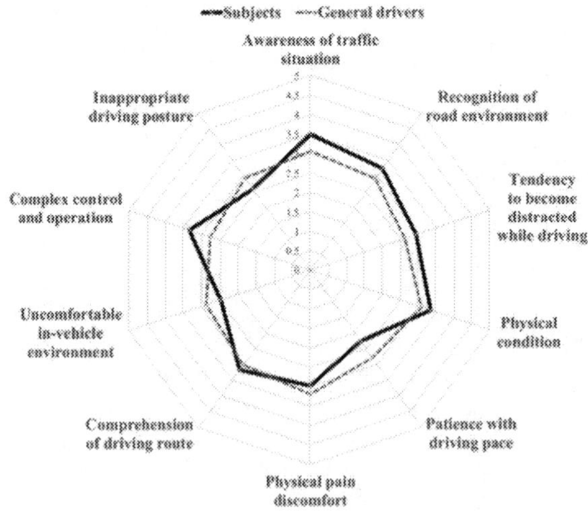

Fig. 2. The average scores about the results for workload sensitivity.

4.3 Procedure

Creating Soliloquy Voice Navigation

In the experiment, we create suitable SVN for each participant using the soliloquy of the In the experiment, we create suitable SVN for each participant using the soliloquy of the participant. This process is called the SVN create process hereinafter. This process uses three videos, called SVN create videos. Figure 3 shows the outline of the

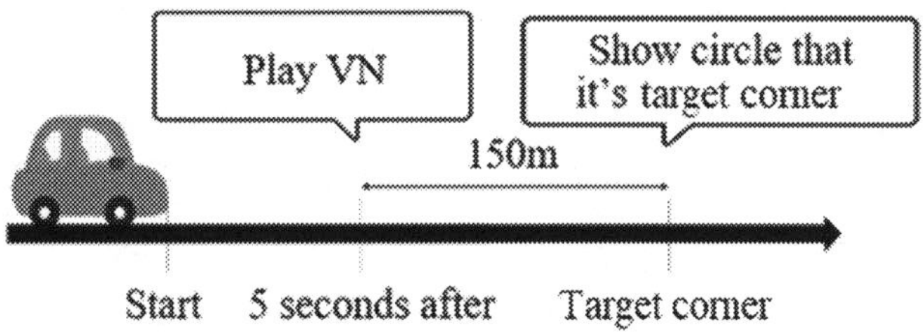

Fig. 3. The outline of SVN create video.

Fig. 4. The SVN create video capture. The red circle points out the target corner instructed by VN. (Color figure online)

SVN create video, 150 m version. At first, the VN starts at 5 s after the video starts. Next, the corner instructed by VN is pointed out by a red circle. Figure 4 shows one of the SVN create video capture, in which the corner is pointed out by the circle. In each video, the different contents of VN for the different driving roads are recorded. The SVN create videos were recorded the actual driving scenes from the driver's viewpoints in Nishi-oji-dori and Kita-oji-dori in Kyoto, Japan. We used GoProHERO4 camera to record. This camera was fixed to the installation position where does not cover the driver's operation. Regarding the parts of the recorded videos that were considered to have visibility problems due to the influence of sunlight, we adjusted the brightness so that it was in a general range during driving. The SVN create video was edited by adding VN at a point such that there was a corner at a distance 150 m, 250 m, or 325 m away from that point. The contents of three types of VN using in SVN create video are showed in Table 2. Each VN contents were created based on the voice guidance expression in the general car navigation systems in Japan.

Next, we explain the procedure of the SVN create process. The participants watch the three SVN create videos one by one. After watching one of the videos, we ask the soliloquy expression of the participants. More concretely, the procedure is as follows:

1. The participant watches the three SVN create videos.
2. The participant watches the SVN create video (150 m) three times.

Table 2. The contents of 3 types of VN using in SVN create video. The upper parts are the contents in Japanese, and the lower parts are the translated one in English.

Distance to a corner	Contents
150 m	"Mamonaku hidari houkou desu." (Please turn to the left soon.)
250 m	"Konosaki hidari houkou desu." (Please turn left in a little while.)
325 m	"Oyoso 300 m saki hidari houkou desu." (Please turn left after 300 meters away.)

3. We ask a question "How do you talk to yourself when the voice guidance was played?" and record the answer.
4. The participant watches the SVN create video (250 m) three times.
5. We ask the same question and record the answer.
6. The participant watches the SVN create video (325 m) three times.
7. We ask the same question and record the answer.
8. We generate the voices expressing the three answers above by the synthesis.

In this process, three types of SVN that is personalized to each participant are generated by using the SVN create videos.

At first, the participant watches all of the three SVN create videos. It is because that the participant is made understand the sense of distance to the corner instructed by the VN. Second, the participant watches SVN create video 150 m version. After that, we ask a question to get a soliloquy expression of the participant for the case of the preceding video.

The "talk to yourself" used for the question is synonymous with "soliloquy of the participant during driving", which is explained to the participants before the experiment. We record the answer to the question. The same process in the cases of the SVN create video (250 m) and the video (325 m) is performed. An example of the generated three types of soliloquy is showed in Table 3.

Table 3. The examples of the contents of SVN. The upper parts are the contents in Japanese, and the lower parts are the translated one in English.

Distance to corner	Contents
150 m	"Mousugu hidari houkou yana" (OK, turn left.)
250 m	"Kekkou saki susunndekara hidari (I'll turn left in a little bit.)
325 m	"Daibu sakimade susunndekara hidari" (After a time, I'll turn left.)

For video play, we used a system created using PsychoPy [10], which is an application for a psychological experiment environment. In addition, we used Rospeex API [11] to generate the synthetic voice. Moreover, It is considered that if the participants watch the video on the desktop display, the sense of distance will be changed from the feel of them. To solve this problem, we used Oculus Rift CV 1, which is a Head Mounted Display (HMD), for display the SVN create video.

Evaluation Experiment

We compared the efficiency of the SVN and the VN in terms of the understanding time and the accuracy of decision. We call this process the evaluation experiment. In the evaluation experiment, we used a system constructed using PhychoPy. The system outputs the videos, called the evaluation experiment videos. This video is created based on the actual driving scenes from the driver's view-points in Nishi-oji-dori and Kita-oji-dori in Kyoto, Japan. In each video, the different contents of VN for the different driving roads are recorded. The VN or SVN, indicating a corner 150 m ahead starts at 5 s after the video starts. The evaluation experiment video, 150 m version was edited by adding VN or SVN at a point such that there was a corner at a distance 150 m away from that point. In this video, the corner instructed by VN or SVN is not pointed out like the SVN create video. SVN was used synthetic voice, created by the SVN create process. Therefore, SVN contents was changed for each participant in this video. There are 20 types of the videos; we have 8 types of video for the SVN (150 m) or VN (150 m), 6 types for both of them (250 m), and 6 types for both of them (325 m).

In the videos, the color and the position of the circle for performing the tasks changes according to the driving behavior of the driver in the video:

- When the driver accelerates or drives at a constant speed, the color of the circle changes to blue.
- When the driver decelerates or is braking, the color of the circle changes to red.
- When the driver turns the wheel to the right, the circle moves to the right.
- When the driver turns the wheel to the left, the circle moves to the left.

Examples of changes of the circle are shown in Figs. 5 and 6.

Fig. 5. An example of changes the color of the circle when the driver is braking. (Color figure online)

Fig. 6. An example of changes of the color of the circle when the driver is accelerating. (Color figure online)

In the evaluation experiment, the participants watched the videos, and performed the following tasks:

- Driving task: It is the task for imposing the participants the work load like driving.
- Understanding task: It is the task for measuring the understanding time of the participants.
- Judgement task: It is the task for measuring the number of the correct decision of the target corner instructed by the guidance.

In the driving task, the participants performed the driving tasks like actual driving operations using the racing wheel, the brake and the accelerator pedal for games.

The movement of the wheel corresponds to the movement of the mouse cursor on the display. So, the participants can look the changes due to the wheel operation. The driving task is that the participant operates the steering wheel or the pedals according to the change of circle in the evaluation experiment videos. The operations performed by the participants are as follows:

- When the color of the circle is red, step on the brake pedal.
- When the color of the circle is blue, step on the accelerator pedal.
- When the position of the circle changes, operate the wheel so that move the mouse cursor to the center of the circle.

We conducted the preliminary experiments on 16 participants to evaluate the driving task. As a result, 11 participants evaluated that the task is similar to actual driving. Therefore, it is suggested that the task simulated the actual driving operation.

In the understanding task, the participants click the button on the wheel when they under-stand the instruction contents from the SVN or the VN. We record the length of time from the start of the evaluation experiment video to time the button pressed.

In the judgement task, the participants click another button on the wheel when they think the car in the video reached the target corner instructed by the SVN or the VN. We record the length of time from the start of the evaluation experiment video to the time the button pressed. To evaluate the accuracy of decision, we set a time range of

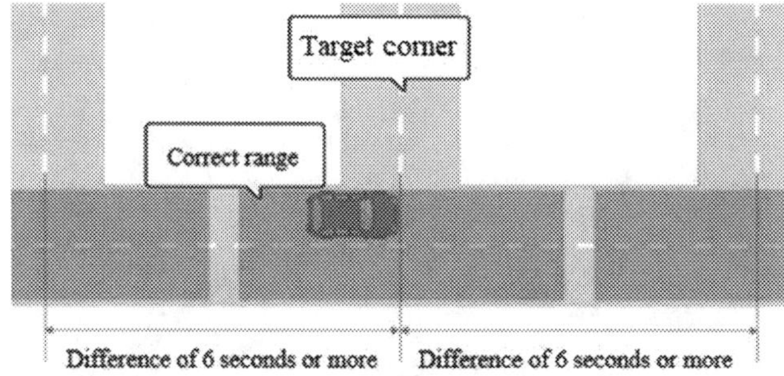

Fig. 7. The correct range in the judgement task.

correct answer for the understanding task. The correct answer range is set to 3 s before and after the time when the car reached the target corner. The correct range is shown in Fig. 7. The time length between all corners is more than 6 s in the videos. We set the correct answer range so that there is no overlap between the correct answer ranges and that there is no misjudgment. For the judgment task, we also conducted the preliminary experiment on 16 participants. As a result, we judged all of the indicated corners by the participants based on the recorded time in the judgement task. Therefore, there is no problem of the setting the correct answer range.

The outline of the evaluation experiment is shown in Fig. 8. At first, we separated the participants into group A and group B. After that, the groups A and B evaluated the SVN and the VN in a different order. In the evaluation experiment, we used the system for playing the evaluation experiment video and recording data about the tasks. At first, the system started playing the video and recording the data about the tasks. Next, the system started playing the SVN or the VN when 5 s after the video started playing. The participants performed the tasks while watching the videos and listening to the SVN or the VN. After repeating 5 times this flow, the type of voice guidance was changed. For example, at first, the group A evaluated the VN 5 times. after that, changed the type of voice guidance from the VN to the SVN. After that, the group A evaluated the SVN 5 times in the same way. It is because that avoid the influence of the order effect. By using Oculus Rift CV 1, the participant watched the videos and listened the SVN or the VN. The experimental environment is shown in Fig. 9. Finally, we applied the questionnaire about DSQ and WSQ to examine the participant's driving characteristics.

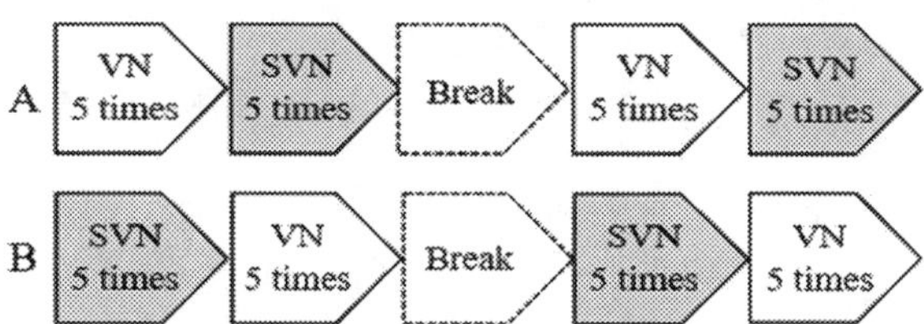

Fig. 8. The outline of the evaluation experiment.

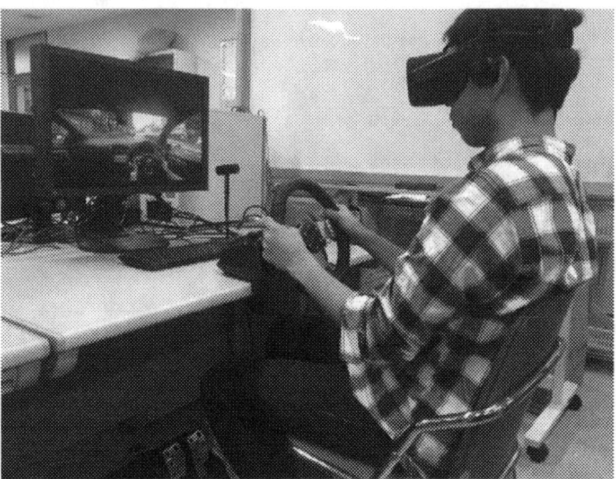

Fig. 9. The experimental environment.

4.4 Results

The results of the understanding tasks are shown in Table 4. The average of the understanding time when using the SVN was 4.71 s (SD: 1.38), and when using the VN was 5.16 s (SD: 1.40). The 13 participants of 14 participants had the shorter time for using the SVN than using the VN. In addition, in each participant, we analyzed the understanding time of the SVN and the time for the VN, using an unpaired t test. As a result, a significant difference was appeared for 4 participants ($p < 0.05$). The time for the SVN of all the participants who were recognized significant differences were shorter than the time for the VN.

Table 4. The results of the understanding tasks.

Participant	A	B	C	D	E	F	G	H	I	J	K	L	M	N
VN	3.97	5.05	5.44	4.47	5.54	5.52	4.84	5.06	4.98	4.55	9.25	5.73	5.8	5.07
SVN	3.37	3.93	5.56	3.59	5.21	4.64	3.75	3.69	3.84	4.44	9.17	4.95	5.24	4.25
t test	n.s.	**	n.s.	n.s.	n.s.	n.s.	**	**	**	n.s.	n.s.	n.s.	n.s.	n.s.

n.s.: Not significant **: $p < 0.05$

Furthermore, using the results of DSQ and WSQ, we compared the participants who were recognized significant differences to the other participants. The results of DSQ and WSQ are shown in Figs. 10 and 11. As the result, the participants with the significant differences have the following features:

- The scores for "Impatience in driving" is low.
- The scores for "Preparatory maneuvers at traffic signals" is low.
- The scores for "Patience with driving pace" is low.

The results of the numbers of the correct answers in the judgement task are shown in Table 5. The average number of correct answers in the judgement task when using

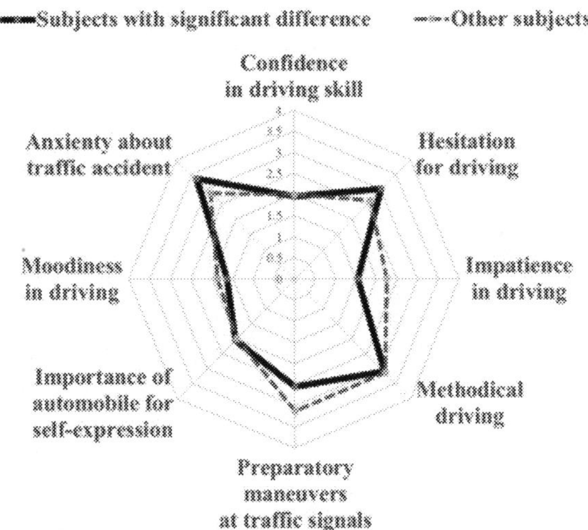

Fig. 10. The comparison of the results for DSQ between the participants with and without significant difference in the understanding time.

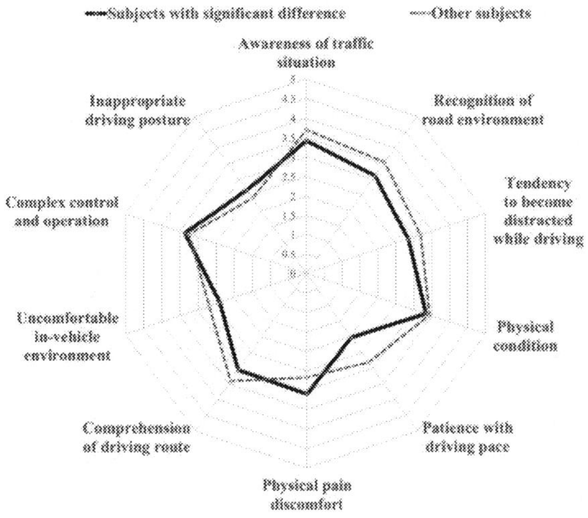

Fig. 11. The comparison of the results for WSQ between the participants with and without significant difference in the understanding time.

the SVN was 5.28 (SD: 1.94), and when using the VN was 4.21 (SD: 1.93). The 7 participants of 14 participants had the larger number of the correct answers for the SVN, 5 participants had the smaller number of the correct answers for the SVN, and 1 participants had the same number of the correct answers. In addition, in each participant, we analyzed the numbers of the correct answers in the judgement task for the SVN and the VN, using the $\chi 2$ test (Fisher two-sided test). As a result, a significant difference was appeared for 2 participants ($p < 0.10$). For all the participants who were recognized significant differences, the numbers of the correct answers for the SVN was larger than the numbers for the VN. In the results of the judgement task, there were only 2 participants who was recognized significant difference. For this reason, we didn't compare the results of DSQ and WSQ of the participants who were recognized significant difference to the other participants.

The contents used in the SVN as the soliloquy of the participants are shown in Table 6. From this result, we classified the types of the SVN as follows:

- Time (Abstract): It is the SVN using abstract time expression (e.g. "a little later").
- Time (Concrete): It is SVN using concrete time expression (e.g. "after 10 s").
- Distance (Abstract): It is SVN using abstract distance expression (e.g. "a bit far").
- Distance (Concrete): It is SVN using concrete distance expression (e.g. "300 meters away").
- Intersection: It is SVN using information of intersection (e.g. "the second intersection").
- Other: It is SVN not applicable to the above types.

Table 5. The numbers of the correct answers in the judgement tasks.

Participant	A	B	C	D	E	F	G	H	I	J	K	L	M	N
VN	7	5	4	4	4	2	6	4	5	3	7	0	6	2
SVN	3	6	7	7	4	7	5	9	3	7	4	2	4	6
x^2 test	n.s.	n.s.	n.s.	n.s.	n.s.	*	n.s.	*	n.s.	n.s.	n.s.	n.s.	n.s.	n.s.

n.s.: Not significant *: $p < 0.10$

The result of classifying the SVN is shown in Table 7. A feature of the SVN for 150 m distance is that many participants expressed the SVN of Time (Abstract) or Intersection type relatively. In addition, a feature of the SVN for 250 m is that most participants used Distance (Abstract) type. As a feature of the SVN for 325 m, there are various types of the SVN. In English, the participant F's SVN for 150 m means "It is about time I'll move to the left a bit.". The action of going to the left side is done before turning the corner to the left. Since this does not match to the above types, we classified it as Other. In addition, in English, the participant N's SVN for 325 m means "I will keep driving on this lane.". This does not directly represent the distance or time to the corner, and then we classified it as Other. The combination of types of the SVN which was used most frequently was "150 m: Time (Abstract), 250 m: Distance (Abstract), 325 m: Distance (Abstract)", and 4 participants used it.

4.5 Discussion

In the result of understanding task, we analyzed the understanding time of the SVN and the time for the VN, using an unpaired t test. As a result, a significant difference was appeared for 4 participants ($p < 0.05$). The time for the SVN of all the participants who were recognized significant differences were shorter than the time for the VN. That was, it was suggested that SVN is effective in reducing understanding time. Furthermore, from the

Table 6. The contents of all the SVN.

Participant	150 m	250 m	325 m
A	mousugu hidari houkou	mousukoshi saki de hidari houkou	mousukoshi tattara hidari houkou
B	soko hidari dayone	ato sukoshi de hidari houkou	2tume no shinngou wo hidari
C	tugi no magarikado wo hidari houkou	supu-do sonomama de 15byou teido	300m saki hidari kana
D	sukoshi saki hidari houkou	kono saki hidari houkou	15byougo hidari houkou
E	mousugu hidarihoukou yana	kekkou susunnde kara hidari	daibu saki made ittara hidari yana
F	sorosoro hidari yottkou kana-	kono saki hidari ikanakyana	moutyotto shitara hidari magannnakyana
G	mousukoshi de hidari	tyotto ittara hidari	kono saki moutyoi ittara hidari
H	tugikurai hidari	moutyotto saki hidari	30byou gurai ato hidari
I	sorosoro hidari yana	mousukoshi saki de hidari yana	ato mousukoshi hashittara hidari yana
J	2ko saki wo hidari houkou	100m saki wo hidari houkou	300m saki hidari yana-
K	mousugu hidari kana	mousukoshi saki hidari yana-	mousukoshi saki de hidari yana-
L	mousugu hidari kana	mousukoshi saki hidari yana-	tyotto ookime no kousatenn no tokoro wo hidari kana
M	2tume no tokoro hidari	moutyoi saki hidari	tyotto oodoori no tokoro wo hidari kana
N	shinngou no tugi no kado wo hidari yana	moutyotto saki yana	mada massu de ikka

results of the scores for DSQ and WSQ, the average score of the participants who had significant differences in their understanding time was smaller than one of the other participants in terms of "Impatience in driving", "Preparatory maneuvers at traffic signals", "Patience strength", "Patience with driving pace". Thus, it is considered that the SVN is especially effective for the drivers who prefer to drive at their own pace in the understanding time. However, in this experiment, since we focused only on the inexperienced drivers as the participants, it is not clear whether the same fact is confirmed for all common drivers.

In the result of the judgement task, we analyzed the average number of correct answers of the SVN and the number of the VN, using an unpaired t test. As a result, a significant difference was appeared for 2 participants ($p < 0.10$). For all the participants who were recognized significant differences, the numbers of the correct answers for the SVN was larger than the numbers for the VN. That was, it was suggested that SVN is effective in improving the accuracy of decisions. However, it is hard to conclude the effectiveness of the SVN because there are only a few participants who have significant differences. We expect that the reason of this result is the small count of trials. For example, in this

Table 7. The result of classifying SVN.

Participant	150 m	250 m	325 m
A	Time (Abstract)	Distance (Abstract)	Time (Abstract)
B	Distance (Abstract)	Time (Abstract)	Intersections
C	Intersections	Time (Concrete)	Distance (Concrete)
D	Distance (Abstract)	Distance (Abstract)	Time (Concrete)
E	Time (Abstract)	Distance (Abstract)	Distance (Abstract)
F	Other	Distance (Abstract)	Time (Abstract)
G	Time (Abstract)	Distance (Abstract)	Distance (Abstract)
H	Intersections	Distance (Abstract)	Time (Concrete)
I	Time (Abstract)	Distance (Abstract)	Distance (Abstract)
J	Interse tions	Distance (Concrete)	Distance (Concrete)
K	Time (Abstract)	Distance (Abstract)	Distance (Abstract)
L	Time (Abstract)	Distance (Abstract)	Intersections
M	Intersections	Distance (Abstract)	Intersections
N	Intersections	Distance (Abstract)	Other

experiment, the participant does not have a significant difference unless the difference between the correct answers of the SVN and the VN is 5 or more. That is, if the participant answers the target corners correctly more than half of trials in both cases of the VN and the SVN, there is no significant difference among them. Therefore, it is necessary to increase the number of trials and to evaluate them again. However, if the number of trials is increased, the participant will become familiar with the SVN and the VN. Thus, the appropriate number of trials should be set in the examination.

In addition, by analyzing the contents of the SVN, it was revealed that the tendency of the types of content varies depending on the distance to the instructed target corner. However, it is possible that the tendency was affected by the SVN create video. The contents of the VN was placed in the types of “150 m: Time (Abstract)”, “250 m: Distance (Abstract)”, “325 m: Distance (Concrete)”. Many participants answered their tweets categorized in “Distances (Abreast)” for the video of 250 m, and we generated the SVN (250 m) based on the tweets. Therefore, we need further consideration to yield any facts about the types of the SVN. On the other hand, there were the “Intersections” contents of the SVN, while the information of intersection is not used in the VN. The “Intersections” contents express the number of the intersections to the target corner in the SVN creation video. Such the number does not correspond to the number of the intersections in the evaluation experiment video. We explained this fact in the experiment, however, many participants used the expression using intersections. In addition, 6 participants of 7 participants using Intersection type had more number of the correct answers for the SVN than the VN. Furthermore, the contents for the SVN (250 m) of the participant J “100 m ahead to the left”. Even though the correct distances to the target corner and the distance in the SVN were different, the participant J had the more number of the correct answers for the SVN than the VN. Therefore, it is suggested that an exact expression of the voice guidance does not lead drivers to easy understanding. In the analyze of the types of the SVN, 2 contents were classified as Other. The content of the participant F for the SVN (150) means “It is about time I move to the left a bit.”. The action of going to the left side is done before turning the corner to the left. The expression about the action does not indicate the distance to the target corner clearly.

However, the participant F has the more number of the correct answers for the SVN than the VN, and the participant F was recognized significant difference. It seems that the distance to the target corner was recognized by imagining the action performed 150 m before. Since this result is seen only for the participant F, it is necessary to study in the future whether other drivers may have similar tendencies. The combination of types of the SVN which was used most frequently was “150 m: Time (Abstract), 250 m: Distance (Abstract), 325 m: Distance (Abstract)”, and 4 participants used it. Each of the other combination were used only by one participant. In the analysis, we classified the contents of SVN based on the expression about distance, time, or intersections excepting the endings of the tweets and dialect, the patterns of the SVN were dispersed. Therefore, it is said that SVN has so many patterns, and it is needed to further study and consideration to yield any findings about the patterns.

5 Conclusion

In this research, we proposed a new expression of voice guidance, called SVN, which uses soliloquy of a driver. We considered the situation where the navigation adopts the voice guidance only. Our goal was to verify the hypothesis that the use of SVN is more effective than that of a general voice guidance (VN). To evaluate the efficiency, we focused on the time required for understanding the instruction content, called understanding time, and the accuracy of identifying the instruction, called the accuracy of decision. That was, we evaluated the understanding time and the accuracy of decision for SVN and VN in the experiment. In addition, we showed the suitable type of driver to the SVN and the types of soliloquy during driving.

In each participant, we analyzed the understanding time of the SVN and the time for the VN, using an unpaired t test. As a result, a significant difference was appeared for 4 participants ($p < 0.05$). The time for the SVN of all the participants who were recognized significant differences were shorter than the time for the VN. From the result, it was suggested that SVN is effective in reducing understanding time. Furthermore, using the results of DSQ and WSQ, we compared the participants who were recognized significant differences to the other participants. From the results, it was speculated that SVN is especially effective for reducing understanding time of the drivers who prefer to drive at their own pace.

In each participant, we analyzed the numbers of the correct answers in the judgement task for the SVN and the VN, using the $\chi 2$ test (Fisher two-sided test). As a result, a significant difference was appeared for 2 participants ($p < 0.10$). For all the participants who were recognized significant differences, the numbers of the correct answers for the SVN was larger than the numbers for the VN. From the result, it was suggested that SVN is effective in improving the accuracy of decisions. In the results of the judgement task, there were only 2 participants who was recognized significant difference.

In addition, by analyzing the contents of the SVN, it was revealed that the tendency of the types of content varies depending on the distance to the instructed target corner. In addition, it is suggested that an exact expression of the voice guidance does not lead drivers to easy understanding.

References

1. Summala, H., Dave, L., Matti, L.: Driving experience and perception of the lead car's braking when looking at in-car targets. Accid. Anal. Prev. **30**, 401–407 (1998)
2. Freundschuh, S.M., Mercer, D.J.: Spatial cognitive representations of story worlds acquired from maps and narrative. Geogr. Syst. **2**, 217–233 (1995)
3. Otani, R., Kansaki, H.: The presentation of route guidance information and better wayfinding. Chukyo Univ. Bull. Psychol. **1**, 45–59 (2001)
4. Taylor, H., Tversky, B.: Assessing spatial representation using text. Geogr. Syst. **2**, 235–254 (1995)
5. Kawai, M., Kato, S., Minobe, N., Tsugawa, S.: Driver adaptive audio route guidance information for car navigation systems. Soc. Automot. Eng. Jpn. Acad. Lect. Meet. Prepr. **37**, 173–178 (2006)
6. Ishibashi, M., Okuwa, M., Doi, S., Akamatsu, M.: Indices for characterizing driving style and their relevance to car following behavior. In: SICE, 2007 Annual Conference, pp. 1132–1137 (2007)
7. Akamatsu, M., Kurahashi, T., Ishibashi, M.: Driver's status assessment using physical measures. In: XVth Triennial Congress of the International Ergonomics Association on Proceedings (2003)
8. Galea, L.A., Kimura, D.: Sex differences in route-learning. Pers. Individ. Differ. **14**, 53–65 (1993)
9. Tomita, Y., Kuriyagawa, Y.: Effect of sexuality, driving experience and habit on driving attitude and workload consciousness. Soc. Autom. Eng. Jpn. Acad. Lect. Meet. Prepr. 7–10 (2005)
10. Peirce, J.W.: PsychoPy - psychophysics software in Python. J. Neurosci. Methods **162**, 8–13 (2007)
11. Sugiura, K., Zettsu, K.: Rospeex a cloud robotics platform for human-robot spoken dialogues. In: Proceedings of IEEE/RSJ IROS. pp. 6155–6160 (2015)

Usability Analysis of the Novel Functions to Assist the Senior Customers in Online Shopping

Xinjia Yu[1(✉)], Lei Meng[1], Xiaohai Tian[1], Simon Fauvel[1], Bo Huang[1], Yunqing Guan[1], Zhiqi Shen[2], Chunyan Miao[1,2], and Cyril Leung[1,3]

[1] Joint NTU-UBC Research Centre of Excellence in Active Living for the Elderly, Singapore, Singapore
{XYU009, lmeng, xhtian, sfauvel, bo.huang, yunqing.guan, ASCYMIAO}@ntu.deu.sg

[2] School of Computer Science and Engineering, Nanyang Technological University, Singapore, Singapore
ZQShen@ntu.deu.sg

[3] Department of Electrical and Computer Engineering, The University of British Columbia, Vancouver, BC, Canada
cleung@ece.ubc.ca

Abstract. Online shopping provides a convenient and diverse shopping experience. However, elderly customers are unable to leverage such benefits due to their age-related impairments or lack of computer knowledges. To solve this problem, we extend our previous e-commerce website design with novel assistance functions including multimodal search and personalized speech feedback. In this paper, we evaluated the usability of these functions through a phenomenography based qualitative study. From the results, we found out several biases which affect senior users' interaction with the assistance functions. Firstly, there is a gap between the icon metaphor and the senior users' real world experience. Secondly, consistency is more important than flexibility in e-commerce website design for the elderly. Thirdly, senior users tend to show less interest to explore the website than younger ones. These findings and considerations will guide us in the following rounds of age-friendly assistance function designs to improve the senior user's online shopping experience.

Keywords: E-commerce · Multimodal search · Speech feedback Age-friendly · Usability

1 Introduction

Shopping through websites has been a popular life style in the recent years. It improves the shopping experience with convenience, diversity and richness. However, while the younger generation enjoys the benefits from their online shopping experience, the elderly customers show less interest due to their age-related impairments or lack of familiarity with computers. In the past decades, the world population has been aging at

© Springer International Publishing AG, part of Springer Nature 2018
G. Meiselwitz (Ed.): SCSM 2018, LNCS 10913, pp. 173–185, 2018.
https://doi.org/10.1007/978-3-319-91521-0_14

an unprecedented rate. According to the report of the United Nations [19], the number of older people aged over 60 has reached 1 billion in 2017. The aging population will continue to grow in the following decades. By 2050, this number will expand to 2.1 billion, which is more than double the size of the current elderly population in 2017. Hence, how to adapt the E-commerce systems for the elderly to improve their quality of life has become an important research topic.

To solve this problem, we need to provide assistance functions to the senior users on the pain points during their online shopping experience. Our previous work [12] developed the following novel assistance functions [18]: (1) a multimodal search engine. The searching engine accepts image, speech, text and the combination of them as inputs to help the elderly find products easily and accurately. This function assists the elderly with problems in literacy or typing. (2) A personalized speech feedback engine with the aim of reducing the elderly's visual burden when browsing the website.

The interface of our website is shown in Fig. 1. To provide a better experience to the senior users, bigger font sizes with high-contrast colors (i.e. white text on dark backgrounds) are chosen to improve the visibility of our web browser, and fewer items are listed in a single page for simplified layouts.

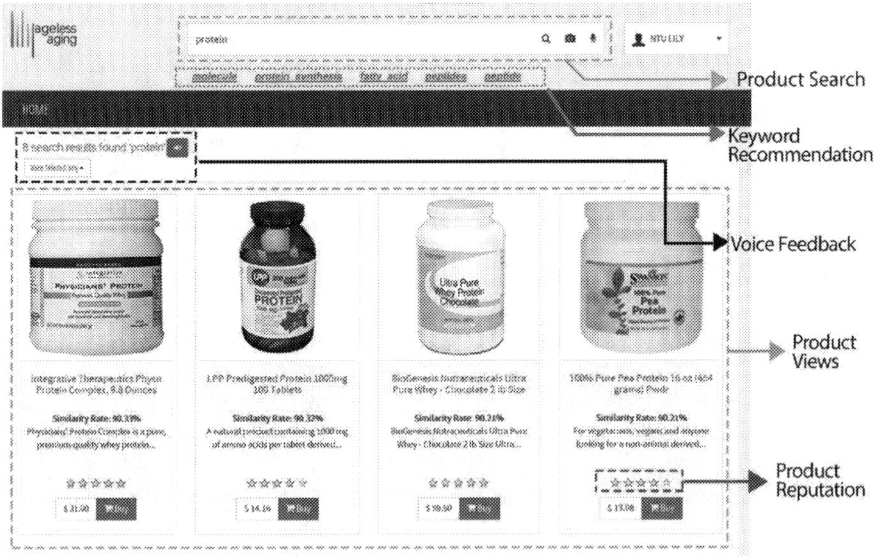

Fig. 1. The interface of our targeted website in this study

Additionally, to make the UI of our system more intuitive use, we chose icons to indicate the multiple search functions. In our system, a user can search by using a part of speech, an image from a camera or the Internet. As such, the user is able to find the desired products more accurately with comparable time of simple search function.

Furthermore, a personalized voice can be generated by the engine for the old users to better understand the speech feedback. In practice, the inexperienced users may miss some important information about the search results. In order to access the full information of search results for these users, a personalized speech feedback engine is developed. When the product search is finished, the summary of the search results will be presented to the user through voice. A personalized speech engine is built on voice conversion technology, which can transform one speaker's speech as if it was uttered by another speaker with limited training data. Although text-to-speech (TTS) can also realize the similar function with better speech quality, long recording of the target speaker is required to build such a system. Hence, voice conversion is a more cost-efficient way to achieve this goal. After the product searching, a summary of search result is showed in the web page. In the example shown in Fig. 2, there are 77 results found. However, the inexperienced users, especially the elderly, may believe the products are all shown on the existing screen without the mental model to roll down the progress bar or turn to the next page. This speech feedback function can help them to avoid this situation. Moreover, the elderly can also choose or create the voice they prefer for better understanding. They can also replay the speech by clicking on the audio button.

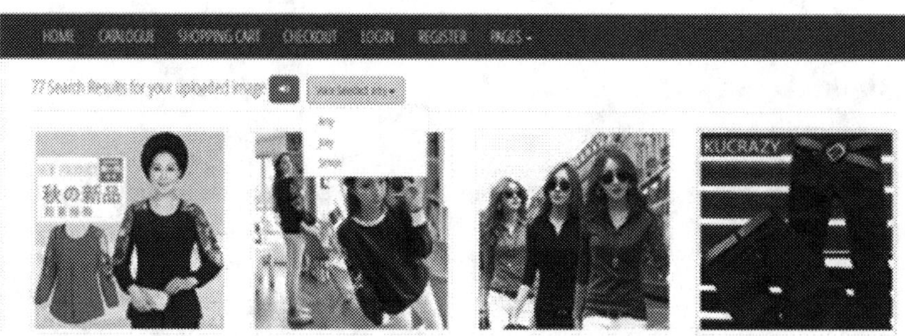

Fig. 2. The interface of the personalized speech feedback function

Do these functions really meet the age-friendly goals? In this paper, we use a lab experiment based qualitative study to analyze the usability of these novel assistances as a follow-up study.

2 Related Work

There are several barriers preventing the elderly from using E-commerce. One of the most important is the declining physical and cognitive functions of the elderly [1]. A number of previous studies are focus on the evaluation and development of age-friendly browser designs [1, 4, 5, 8–11, 14, 20]. In [9], the age-friendly design principles in terms of the usability were introduced. In [7], 36 websites were evaluated by older adults in items of 25 "senior-friendly" guidelines recommended by the National Institute of Aging.

Some other barriers root from the inexperience of the senior users in online shopping. To solve this problem, some researchers and developers developed age-friendly functions to assist the elderly in their interaction with the E-commerce websites [3]. In [2, 16], speech technology was adopted for products search. In [7], a study on using voice commands for the elderly when browsing websites was presented. In [12], our team designed three age-friendly functions, crowd-improved speech recognition, multimodal search and personalized speech feedback, to improve the elderly's online shopping experience.

Based on these studies, we extend our previous work in [12, 18] with the following aspects: (1) An age-friendly UI following the human factor designs for the elderly, such as simplicity and intuitiveness, is designed for our E-commerce system; (2) we integrate the previous developed functions, namely multimodal search engine to assist the senior users with problems in visual or typing; (3) a personalized speech feedback is also imbedded into our proposed system to help the inexperienced users. As well as its assistance effect, this function can also enhance the senior users' psychological experience by providing familiarity in voice. With these integrated functional modules, our age-friendly E-commerce system provides the users with less computer or online shopping experience a more understandable online shopping platform.

3 Methods

In this study, we use phenomenography as a qualitative method to analyze the participants' experience with the assistance functions based on the e-commerce website. Phenomenography is "a research method for mapping the qualitatively different ways in which people experience, conceptualize, perceive, and understand various aspects of, and phenomena in, the world around them" [13]. It is a methodology with the aim of seeking and describing the variations in the ways people experience. In this study, we observe and analyze the experiential differences of the interaction experience with our online shopping platform between the senior users and young users. Besides the performance observation, we also use semi-structured interviews and comments recording to collect data.

Participants

According to Jakob Nielsen [15], 5 users are enough to find about 80% of potential issues while with 10–12 users, we are able to find out 100% of the problems. In this study, we recruited 5 senior participants aged over 60 and 11 younger participants under the age of 40. The younger participants are all Chinese while the senior ones include 3 Chinese, 1 Malaysian Singaporean, and 1 Caucasian from the U.S. All of them can speak English. The details of the participants are displayed in Table 1. The participants' self-rated computer and online shopping experiences can help us to understand the reasons behind their various performances by cross comparison analysis. These participants are volunteers of this study. Ethical clearance to conduct this study was obtained from the Nanyang Technological University Institute of Research Board.

Table 1. Participant profile

No.	Age group	Gender	Computer using experience	Online shopping experience
1	30–39	F	3	4
2	20–29	M	4	5
3	20–29	F	5	5
4	>60	M	1	3
5	20–29	F	5	5
6	20–29	M	5	4
7	30–39	M	4	4
8	30–39	F	2	3
9	20–29	F	3	5
10	20–29	F	5	5
11	20–29	F	3	5
12	20–29	F	4	5
13	>60	M	5	1
14	>60	F	4	5
15	>60	F	3	2
16	>60	M	5	4

Table 2. Pre-experiment interview questions

1. Which year were you born in?		
2. Are you retired?	If "Yes"	May I know what your occupation was before retiring?
	If "No"	May I know what your current occupation is?
3. Who are you currently living with?		
4. Do you use a personal computer in your daily life?	If "Yes"	How often do you use a computer?
		What do you usually do with your computer?
		How do you rate your computer skill with the scale 1–5
5. Have you ever shopped online?	If "Yes"	Did anyone help you with it? If so, who?
		How frequently do you shop online?
		How do you rate your online shopping experience with the scale 1–5
		Which online shopping websites have you used?
		Do you like online shopping, and why?
	If "No"	What are the major obstacles that stop you from online shopping?
		If you could shop online, what types of goods would you want to buy online?

Procedure

In this study, each participant used the targeted online shopping website individually under the guide of our researchers. Before the experiment, they were interviewed by in a semi-structured interview with the questions listed in Table 2. Then, the participants experience the platform for about 15 min. Firstly, they browsed the website randomly as their previous online shopping habit. Secondly, they performed a task to search a targeted product by the multimodal search functions. At last, they tried the personalized speech feedback function. During the experiment, the participants were encouraged to think out aloud while their comments and feedbacks were recorded by the researcher. They can also seek help from the researcher any time during the study. After their experience, the participants were invited in a post-experiment interview about their feelings and preferences.

The experiment was conducted on a laptop running Windows 10, and viewed at a 1280 × 1024 pixel using interactive prototype version published in May, 2017.

Data collection

The following data were collected from this study:

- Interview: The two parts of interviews were both conducted individually. The pre-experiment interview helped us to collect the participants' demographic information, computer knowledge and online shopping information. While the post-experiment interview aimed to collect information about participants' subjective perceptions.
- Phenomenon observation: During the experiment, our researchers were observing the participants' performance and recording their comments. Each participant's task finishing speed, error rate, and operation behaviors were recorded.
- Comment: The participants were encouraged to think out aloud. Their oral comments were recorded by the researcher during the experiment.

Data Analysis

- Thematic coding: we used the thematic coding method to analyze the text material collected from the interview replies and comments as well as the observation results of the participants' performance.
- Phenomena categorized: the themes generated from the thematic coding analysis provides categories. These efforts helped us to understand the variety phenomena between different age groups.

4 Results

Based on the observation and interview results, we found that though both the younger group and the senior one met some problems during their operations, their reactions and subsequent behaviors differed. The error of mis-clicking buttons happens equally in the two groups. However, the majority (7 of 11) of the young participants can fix their mistakes by multiple tries independently. At the same time, the senior participants were swamped in the usability biases, especially in their interactions with the assistance functions. None of the 5 of our senior participants finished the tasks by themselves for the first time. This phenomenon also happened on 4 of the younger participants with little computer knowledge.

Based on the thematic coding analysis of the comments we collected during the experiment, we revealed the following biases of the elderly which affect the effectiveness of the assistance functions:

(1) Incomprehension of the icon.

In this study, the website used the icons of camera and microphone to represent the image search and voice functions. The aim of this design is to assist the novice users with little computer experience in input process.

In the experiment, each participant was asked to "search products by the given image". The majority of the younger participants finished this task without problem. However the senior ones met difficult from the beginning. They asked questions like "what should I do?", "Where should I start?" and "How can I put an image into the system?" In this situation, the researcher showed them the image search function step by step from clicking the camera icon on the searching bar. After the demonstration, 4 of our senior participants finished their tasks following the demo way. The last 1 (Participant NO.15) met other problems in uploading an image. At last, she finished the task under the researcher's oral guide. Since all of the participants browsed the platform before this task, the researcher asked whether they noticed the camera icon during their browsing. All of them answered "yes", but none of the senior participants realized that the icon was related to image searching. They even did not realize that they can click it.

Following the image searching task, the participants were asked to search products by speech searching. At this time, both the senior and the younger participants noticed that they should click the microphone icon. To the senior participants, a mental model about icons is established.

There is a gap between the icon design and the users' real world experiences. The metaphor of icons cannot match the senior users' real world experience. However, the effect of this metaphor critically depends on the users' computer culture mental model. The end users, such as the elderly who really need this help, can hardly understand these iconic communications.

(2) Confusion of the multiple paths.

In the task of searching a specific product, at the first time, the researcher introduced the task involving all the possible ways to achieve the goal. We told the participants "please find some pants on the platform by searching. You can use key word searching, image searching, or speech searching." Nine of the younger participants understood this idea immediately and started their journey. Two of them lost in the requirement with all the 5 senior participants. They asked for a repeat with a common theme: confusion over information. Their responses conclude the following keys: "lost", "not catch up", "fail to follow", and "no idea about what should I do." When be asked about the reason, their answers flocked around the keys like "too much information" and "too many searching words". One of the participants mentioned that he "tried to distinguish and understand the XX searching and XX searching and missed the following sentence." Four of the senior participants mentioned "separately" or "one by one" in the conversations when they wanted the researcher to repeat the requirement.

To solve this problem, we split the task introduction into three parts as "Please find some pants by key word searching", "please find some pants by image searching", and "please find some pants by voice searching". Each requirement was thrown out after the participant finish the previous one. The 5 senior participants and the 2 younger ones who failed in the first round understood their tasks at this time. One of the senior participants showed a little confusion and asked "why do I need to finish the same task three times using different methods?"

To the senior users, consistency is more important than flexibility. Many website designers provide multiple routes to achieve one single goal to enrich the users' personalized experience. This design philosophy is demonstrated to be powerful to benefit the skilled users. However, for the novice users such as the elderly, they feel confused and lost facing so many operational options.

(3) Ignore of the assistance function.

Comparing with the younger generation, the senior ones tend to show less impulse to explore the website. In this case, the website provides a voice feedback to assist the users who have visual bias during browsing. The users can activate this function easily by clicking the icon. However, the senior participants ignored the icon and the function. Unlike the younger participants who showed curiosity to each button and could hardly stop from clicking randomly, the elderly followed the researcher's guide strictly for they are afraid of damaging the system by wrong operations.

We asked the senior participants whether they noticed the button to activate the speech feedback. Three of them answered "no". Their reasons include "my attention was fully on your words", "I was attracted by the product list", and "I did not think of activating anything manually".

5 Discussion

In this study, we found several obstacles in the novel assistance functions designed for the elderly to make their online shopping experience convenient. These findings show us several directions to improve the usability of assistance functions in e-commerce website design.

(1) The senior users need more clues other than a simple metaphor icon. In the instance of image searching, we can use an image as a key clue through the whole interface design. When the functions such as product category are all displayed based on images, it will help the users to establish a mental model and lead them to accept searching products by an image. When the users are "thinking with images", they can find the path to search by images easily. We call this principle "image thinking" design. Furthermore, we can change the position of the traditional searching bar to produce a consistent idea bridge from image browsing to image searching. Figure 3 shows a proposed interface of the "image thinking" design. Besides the existing searching bar design, we display the image searching

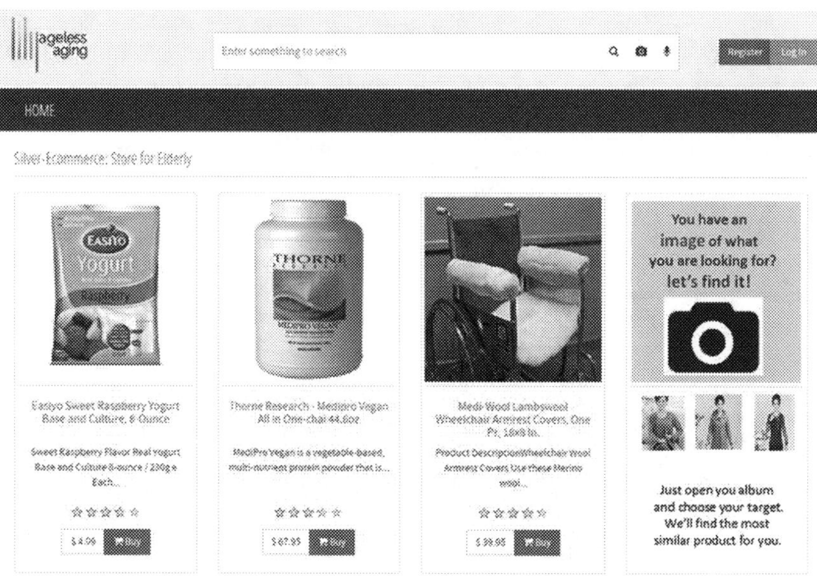

Fig. 3. The proposed image search interface

icons in the cluster of product images. Text introduction and image recommendation are leading the users further as clues. This design idea can also apply to the voice searching function. If we display the voice searching icon and guide under an environment with system voice feedback, the user may learn to use it quickly.

(2) The simpler the better is an important principle. For the senior users with less familiarity of the computer and weaker memory, a single path to a specific goal is an easier concept to be accepted in control designs. When the system guides the users to achieve a specific goal, the existing personalized design always provide several paths to enhance the users' self-control experience. The introduction is always like "to achieve this goal, we provide Option A, Option B, and Option C, you can choose anyone you like based on your personalized consideration." The process in Fig. 4 is showing an example of this multiple path task introduction.

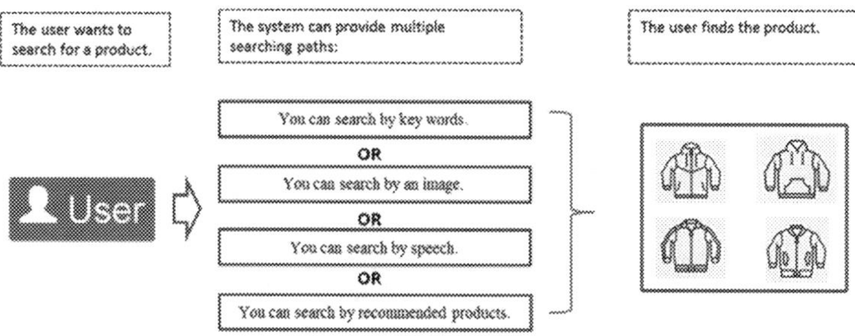

Fig. 4. The process of multiple path task introduction

However, from this study, we found out that for the inexperienced users especially the senior ones, this considerable design lead to confusion in both learnability and memorability. The users may confound various paths together into a way heading nowhere. This confusion also enhances the users' belief of task difficulty level and weakens the users' self-confidence at the same time. To solve this problem, a single path flow may introduce the task better as shown in Fig. 5.

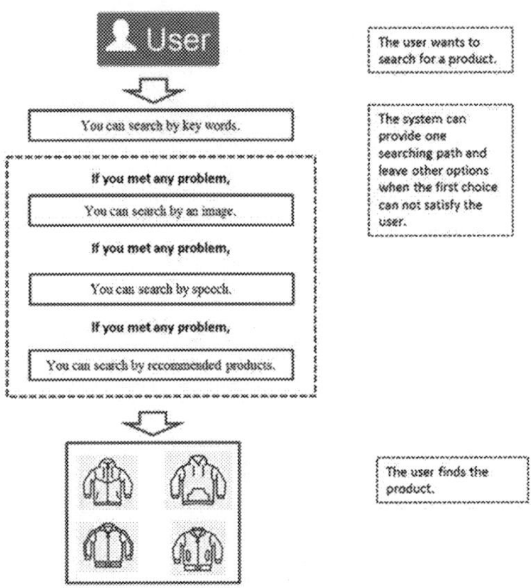

Fig. 5. The process of single path task introduction

Under the single path task introduction flow, the experienced users can still choose personalized task achievement routes by trimming the flow in their own way. Meanwhile, the inexperienced users can learn one way to finish this task quickly without many distractions.

(3) We should take the initiative to show every function, but not depend on the users' exploration. The traditional HCI design tends to leave the users the authority to turn on non-necessary functions. However, in this study we found that the senior users are too easy to be satisfied by the basic system which leads to overlooking the novel functions. In this case, we consider turning on every function as default and leaving the option to turn them off for the senior users instead of the opposite way for the younger ones.

(4) Age vs. mental model, which is the most critical obstacle in the senior users' interaction experience with online shopping? In this study, the younger participants with less computer experience met the same problems as the senior ones. This situation highlights the critical idea about various mental model considerations in online shopping website design. The decline of cognition ability caused by age and the mental model lack of computer experience are both preventing the

elderly from enjoying online shopping. However, since the mental model problem is not as visible as the former one, it is easy to be ignored in HCI design. Just like the findings in this study, sometimes, the assistance functions may cause usability problems because of the misunderstanding by the senior users.

(5) The elderly may overrate their computer knowledge and experience. This is a new problem cropping up from the analysis unexpectedly. Two of our senior participants marked their computer experience as 5 which was the highest score. They explained that they worked on personal computer everyday with the Microsoft Office and other software. Meanwhile, the younger participants with the same computer operation level only mark their ability as 3 or 4. Only the developers and students major in computer science marked their computer experience as 5. Based on the concept understanding range in both breadth and depth, the senior participants look more optimistic about their ability. This phenomenon inspires that in user study involving various age groups, a simple self-rated experience, knowledge, or ability mark may not reveal the real situation. More detailed information is needed to assess their self-evaluation.

6 Conclusion

Providing novel functions in the e-commerce website to improve the senior users' online shopping experience is a popular attempt in both website design and usability study. However, the lack of computer and online shopping experience by the elderly may weaken the effect of these novel functions. In this study, we found out several obstacles which affect the senior users' online shopping experience with the assistance functions. Firstly, the senior users have problems understanding the widely accepted metaphor icons. Helping the users to establish a "thinking environment" may bridge this gap. Secondly, consistent introduction or guide is more convenient to the elderly than flexible ones. In task introduction design, we should provide one way to finish the task and leave other personalized options for later, but not to push them all together at the same time in front of the user, especially at the first time. Thirdly, the elderly show less curiosity in exploring the system than the younger generation. If the system wants to persuade them to use any function, it should activate it as a default but leave the option to turn it off. These findings and considerations will guide us in the following age-friendly assistance function designs to improve the senior users' online shopping experiences.

Acknowledgments. This research is supported by the Interdisciplinary Graduate School, Nanyang Technological University, Singapore; and the National Research Foundation, Prime Minister's Office, Singapore under its IDM Futures Funding Initiative.

References

1. Becker, S.A.: A study of web usability for older adults seeking online health resources. ACM Trans. Comput.-Hum. Interact. (TOCHI) **11**(4), 387–406 (2004)
2. Demirbilek, O., Demirkan, H.: Universal product design involving elderly users: a participatory design model. Appl. Ergon. **35**(4), 473–478 (2004)
3. Chen, L., Gillenson, M.L., Sherrell, D.L.: Consumer acceptance of virtual stores: a theoretical model and critical success factors for virtual stores. ACM Sigmis Database **35**(2), 8–31 (2004)
4. Dickinson, A., Gregor, P., McIver, L., Hill, R., Milne, S.: The non browser: helping older novice computer users to access the web. In: Proceedings of the 2005 International Conference on Accessible Design in the Digital World, p. 18. British Computer Society (2005)
5. Dickinson, A., Smith, M.J., Arnott, J.L., Newell, A.F., Hill, R.L.: Approaches to web search and navigation for older computer novices. In: Proceedings of the SIGCHI Conference on Human Factors in Computing Systems, pp. 281–290. ACM (2007)
6. Fisk, A., Rogers, W., Charness, N., Czaja, S., Sharit, J.: Designing for Older Adults: Principles and Creative Human Factors Approaches. CRC Press, Baco Raton (2009)
7. Hanson, V.L., Richards, J.T., Lee, C.C.: Web access for older adults: voice browsing? In: Stephanidis, C. (ed.) UAHCI 2007. LNCS, vol. 4554, pp. 904–913. Springer, Heidelberg (2007). https://doi.org/10.1007/978-3-540-73279-2_101
8. Hart, T.A., Chaparro, B.: Evaluation of websites for older adults: how "senior friendly" are they? Usability News **6**(1), 12 (2004)
9. Hollinworth, N., Hwang, F.: Investigating familiar interactions to help older adults learn computer applications more easily. In: Proceedings of the 25th BCS Conference on Human-Computer Interaction. pp. 473–478. British Computer Society (2011)
10. Holzinger, A., Searle, G., Kleinberger, T., Seffah, A., Javahery, H.: Investigating usability metrics for the design and development of applications for the elderly. In: Miesenberger, K., Klaus, J., Zagler, W., Karshmer, A. (eds.) ICCHP 2008. LNCS, vol. 5105, pp. 98–105. Springer, Heidelberg (2008). https://doi.org/10.1007/978-3-540-70540-6_13
11. Kang, L., Dong, H.: B2C websites' usability for chinese senior citizens. In: Kurosu, M. (ed.) HCI 2014. LNCS, vol. 8512, pp. 13–20. Springer, Cham (2014). https://doi.org/10.1007/978-3-319-07227-2_2
12. Meng, L., Nguyen, Q., Tian, X., Shen, Z., Chng, E., Guan, F., Miao, C., Leung, C.: Towards age-friendly E-commerce through crowd-improved speech recognition, multimodal search, and personalized speech feedback. In: International Conference on Crowd Science and Engineering, pp. 1–8 (2016)
13. Marton, F.: Phenomenography: a research approach to investigating different understandings of reality. In: Sherman, R.R., Webb, R.B. (eds.) Qualitative Research in 245 Education: Focus & Methods, pp. 141–161. RoutledgeFalmer, London (2001)
14. Niehaves, B., Plattfaut, R.: Internet adoption by the elderly: employing is technology acceptance theories for understanding the age-related digital divide. Eur. J. Inf. Syst. **23**(6), 708–726 (2014)
15. Nielsen, J., Landauer, T.K.: A mathematical model of the finding of usability problems. In: Proceedings of ACM INTERCHI 1993 Conference, Amsterdam, The Netherlands, pp. 206–213, 24–29 April 1993
16. Schalkwyk, J., Beeferman, D., Beaufays, F., Byrne, B., Chelba, C., Cohen, M., Garret, M., Strope, B.: Google search by voice: a case study (2010)

17. Stroud, D., Walker, K.: Marketing to the Ageing Consumer. The Secrets to Building an Age-Friendly Business. Palgrave Macmillan UK, London (2013)
18. Tian, X., Meng, L., Liu, S., Shen, Z., Chung, E., Leung, C., Guan, F., Miao, C.: Novel functional technologies towards age-friendly E-commerce. In: proceedings of HCI 2017 ITAP: International Conference on Human Aspects of IT for the Aged, pp. 150–158 (2017)
19. United Nations Department of Public Information, population division: World population prospects- The 2017 revision, Ageing population, p. 1 (2017)
20. Wang, D., Yu, X., Fauvel, S., Tan, A., Miao, C.: Elderly friendliness evaluation of mobile assistants. In: IEEE International Conference on Agents (ICA), pp. 115–120 (2017)

Individual and Social Behaviour in Social Media

Political Opinions of Us and Them and the Influence of Digital Media Usage

André Calero Valdez(✉), Laura Burbach, and Martina Ziefle

Human-Computer Interaction Center, RWTH Aachen University,
Campus-Boulevard 57, Aachen, Germany
{calero-valdez,burbach,ziefle}@comm.rwth-aachen.de

Abstract. Democracies in the late 2010s are threatened by political movements from the borders of the political spectrum. Right-wing populist parties increasingly find agreement in larger parts of the population. How are these people convinced to these political beliefs? One explanation can be seen in polarization and the phenomena that arise from it such as the spiral of silence. In this article we empirically investigate how digital media usage influences the perception of polarization in Germany using a survey with 179 respondents. We use polarized opinions and measure agreement from two perspectives with them. We find an influence of social media usage on the perception of polarization in our sample. Further, polarization seems to be perceived differently depending on the topic. The results contribute to an understanding of how to adequately design presentation of sensitive or controversial topics in digital social media and could be utilized in student eduction to sensitize social media users to the effect of polarization of opinions.

Keywords: Opinion forming · Fake news · Polarization
Social media use

1 Introduction

Polarization is the social process of diverging opinions forming in social groups in a society. An example for a topic for which polarization can be observed in the United States of America is gun-control. There are at least two groups of people, whose opinions seem to continuously diverge. One group strongly advocates stricter gun-control, the other argues against gun-control. Independently of which opinion one might agree with, the process of polarization can be observed in the news streams and comments sections of social media.

Polarization [1] poses a considerable risk for the stability of societies, as they promote the perception of sub-groups with strong within-group coherence and strong out-group rejection—the perception of *us vs. them*. Once polarized opinions have formed it becomes increasingly difficult to find compromise on middle ground which is necessary in democratic societies that need to be flexible enough to react to changes [2].

© Springer International Publishing AG, part of Springer Nature 2018
G. Meiselwitz (Ed.): SCSM 2018, LNCS 10913, pp. 189–202, 2018.
https://doi.org/10.1007/978-3-319-91521-0_15

A core aspect of polarization are the perceptions of within-group and out-group opinions. Before the rise of social media, these perceptions where limited and cultivated by exposure in mass media such as TV, radio, and newspapers [3]. With the increasing spread and use of social media, it is possible to be exposed to the opinion of everyone everywhere, given that the algorithm that controls your news-feed presents it to you. Pariser proposed the phenomenon the *Filter Bubble* [4], referring to the positive-feedback loop of preferential media consumption and algorithmic presentation. You read what you like, the algorithm behind the news feed, presents more of what you like. More complicated, when non-factual news (or Fake News) are mixed into the equation, the feedback-loop could increase the believability of Fake News, as they match the overall impression of the news feed [5].

The question remains how much does the usage of social media influence the perceptions of opinions within a population? In Germany, the strongest polarization can currently be observed in politically hot topics such as immigration, refugees, and the right-wing party AFD (Alternative für Deutschland, transl. Alternative for Germany). But how pervasive are these perceptions, and even more importantly, how pervasive is the perception of polarization regarding these topics? Further, does the media a person uses influence the strength of perceived polarization in society?

In this paper we empirically investigate the perceptions of opinions in Germany regarding immigration, the AFD party, and some control-opinions to investigate how the selection of different media influences the perception of polarization.

2 Related Work

To understand how opinion forming in different sub-groups of society works and how the resulting perception of polarization is determined by media consumption we have to look into several fields of related work.

2.1 Opinion Forming

The study of how people form their opinion has been heavily investigated since the sixties of the last century when first efforts were made to understand how opinion leaders influence their social circle and how they can be identified [6]. First, with the purpose to understand what products reach market saturation quickly and then to understand how political opinion is shaped by opinion leaders. One aim of the early research was to identify who these leaders are and how they can be characterized [7]. Opinion leaders have high domain knowledge, they are highly educated, are strongly integrated in their social network, and are extroverted in their nature.

In the seventies the effect of media on political opinion forming was studied. In particular, it was explored to what extent the media contribute to determining what topics are part of the public discourse and what topics are non-relevant.

By putting attention towards a topic, the media decides what is on the public agenda [8] and what is not. Media thus shows an indirect effect on opinion forming, by shaping what topics influence political decision making. This is particularly interesting when the media portraits some opinions to a larger extent than they are actually present in society. The availability heuristic [9] influences how humans estimate the importance of something in society. As the human brain was designed by evolution to deal with tribal life, anything that is immediately experienced or by hearsay is recognized as rather important. The media changes what is perceived as important or pervasive, by selectively exposing its consumers to highly emotional content that happens rarely (e.g. plane crashes). This biases to believe that such events are more present, important, or pervasive than they actually are. People estimate the frequency of events by the frequency of their exposure. You believe what you see often, to happen often.

The culmination of this effect is the so-called *spiral of silence* [10]. This phenomenon describes the feedback-loop of the availability heuristic. An opinion that is not being reported on in the media, is perceived to be less pervasive in society. This leads to less people holding this opinion to speak their mind, reducing the presence of this particular opinion even further—until a majority of people believes that their opinion is shared by only a select few. Similar phenomena have been observed in social media as well [11]. This complex mechanism can be seen as a direct consequence of the network effects of micro- and macro-structures in opinion networks [12]. Slight changes in the micro-structure of opinions can lead to large changes in the macro-structure of opinions.

From a social science perspective, one question that is important to understand regarding opinion forming is when do people actually change their opinion [13] regarding political topics. First insights indicate that deliberation, the rational *personal* discussion of political arguments, can lead to opinion changes at least in certain subgroups of society. The presence of heterogeneity in such discussions increases the likelihood of opinion changes. No such effect has been found in political deliberation in online media.

2.2 Polarization

If feedback mechanisms in opinion networks lead to macro-structure changes, the next question one might ask is, what are stable configurations of such structures. These depend on the heterogeneity of the underlying network structure [1]. Opinions may diverge into two or multiple separate clusters that show little common ground. The opinions have polarized. In other cases opinions converge on compromises. But what structures lead to what outcomes?

Simulation models such as the Shelling model of segregation [14] try to understand such processes from first principles. In the case of the shelling model: Does segregation occur from two simple rules? First rule – stay if more than x% of my neighbours are similar to myself. Second rule – move otherwise. Strong Segregation does occur if x is larger than 40%. Independently from individual differences, some phenomena occur predictably from structure alone [15].

When looking at opinion forming, the influence of fake news on polarization has been investigated [5]. Fake news are mostly believed if they confirm the presuppositions of the reader and thus reinforce preexisting beliefs. This concordance between fake news and own beliefs increases the positive feedback-loop and, as a consequence, may increase the speed and the extent of polarization of opinions. This is further complicated by the algorithms underlying social media that select what a user sees.

2.3 Selective Exposure and the Social Web

To improve customer time and to control customer attention on a social media web site, companies optimize and customize content for the individual user. The aim is to keep the user on the website longer, and to increase page-impressions of commercials. The underlying algorithms used to customize content are so-called recommender systems [16]. Content that is liked or frequently interacted with is compared to the content that other users like. Similar content is presented to the user to keep them interacting with the social media site. This leads to the so-called *Filter Bubble* effect proposed by [4] in 2011.

But how does the filter bubble affect political opinion making? The effect of customization and selective exposure on users has been recently investigated [17]. Dylko et al. found that the system-immanent customization features have the strongest effect on political opinion forming. Further, the effect is strongest in groups with ideologically moderate individuals and occurs most strongly with news that run against the beliefs of the user.

These mechanisms can be exploited by political campaign makers as supposedly during the Donald Trump presidential campaign in 2016. Here, possibly undecided voters were micro-targeted by analyzing personality from social media profiles and presenting them customized campaign ads [18]. Overall, it is yet insufficiently understood how the use of digital media affects political knowledge and participation in political deliberation and opinion forming [19]. Moreover, the effect media usage has on the perception of polarization has not been studied sufficiently. It is unknown what user diversity factors influence the perception of polarization.

3 Method

In order to study how media usage influences the perception of polarization we conducted an online survey. The survey was conducted in December 2016 in Germany. The survey structure is depicted in Fig. 1. We assessed the following variables.

Demographics. In the survey we asked for the participants' age, gender, and level of education.

Media Usage. We also measured how frequently participants used a set of media. The set was *facebook*, *social media* in general, *newspapers*, *tv*, *radio*, and the *internet*. Usage frequency (UF) was measured on a six-point Likert scale

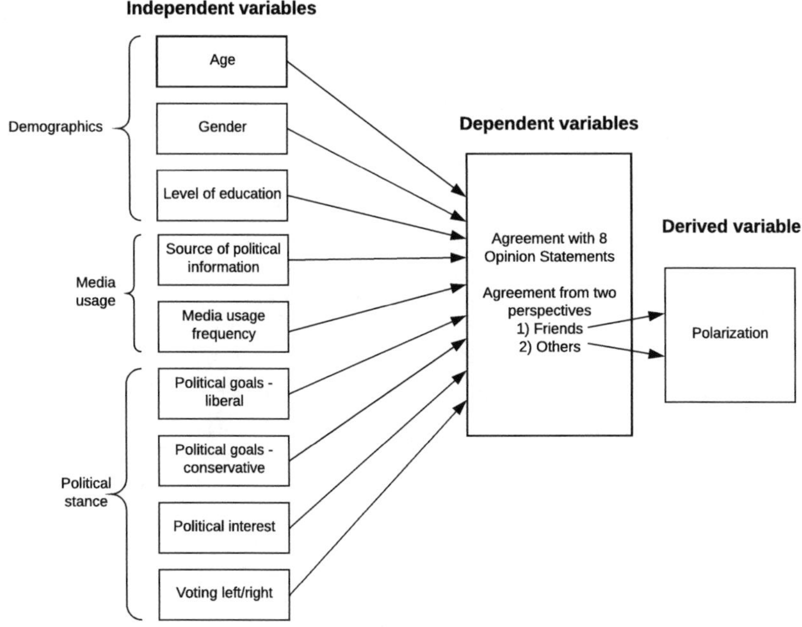

Fig. 1. Model of the investigated variables

(1 - vary rarely, 6 - daily). We assessed to which of these media participants relied on as their source for political information (SPI) and added the category *friends* to assess non-media channels for political information. SPI was also measured on a six-point Likert scale (1 - not at all important, 6 - very important).

Political Stance. We further asked participants to rate a set of 10 items on their political opinion. These items were used in a principal component analysis which yielded two factors, one factor for the agreement with liberal political goals and one factor for a conservative political goals (see Table 1). Interest in politics was measured using four items (i.e. *I am interested in politics.*, *I am interested in political events*, *I am interested in politics globally.*, *I am interested in politics in Europe.*). All measurements were conducted on six-points Likert scales (1 = low confirmation, 6 = high confirmation). Further we asked whether someone would vote for both right and left-wing parties.

In order to measure the difference in opinions in one owns social circle and the perception of agreement with these opinions in the general population, we presented the following opinions (see Table 2). All opinions were presented in a facebook-like comment as shown in Fig. 2.

Table 1. Item texts for political goales. The liberal scale showed a Cronbach's α of .794, the conservative scale one of .803.

Variable	I agree with the following goal
Liberal 1	Improvement of social security systems
Liberal 2	More social Justice
Liberal 3	More environmental protection
Liberal 4	More gender equality
Liberal 5	Reduction of poverty
Conservative 1	More security and order
Conservative 2	More political stability and continuity
Conservative 3	More flexibility in the job market
Conservative 4	More support for top performers
Conservative 5	More national pride

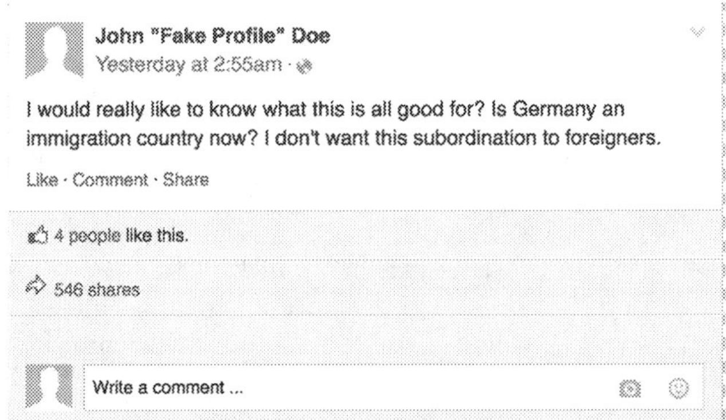

Fig. 2. Example forged facebook post

We then asked participants to rate on a scale of 0 to 100, how large the percentage of people is that would agree with such a statement. We explicitly instructed participants to estimate the real percentage and not the one present to them on social media. These topics were selected as some of them relate to polarizing topics, namely immigration and voting for the AFD. Lastly, we asked participants whether or not they could imagine voting for the most left-wing party in the German party system (i.e. Die Linke) and the most right-wing party (i.e. AFD) on a six-point Likert scale (1 = very unlikely, 6 = very likely).

All participants were instructed that they were doing the survey on a voluntary basis and that no identifying data would be stored. We explicitly informed participants that we were going to ask sensitive topics, and that they should answer honestly without thinking what was "right" or "wrong". They should focus on their political opinion.

Table 2. Text used in individual opinion posts.

Variable	Opinion text
Womens' rights	Where are all the womens' rights organizations now? Women's dignity is being mistreated and nobody seems to care
Immigration	I would really like to know what this is all good for. Is Germany an immigration country now? I don't want this subordination to foreigners
Russia	Trump is going to approach Russia. This will make the world a safer place. Clinton threatened war with Russia, this was way too aggressive
afd	The leftwing parties plan to not coalize with the CDU[a] and plan the downfall of our nation. The CDU should consider coalizing with the AFD
afd2	CDU with Merkel has become left and green. The only conservative party remaining is the AFD. The quiet majority will cause a political earthquake in the next polls
demosoc	We aim for a concrete goal. We fight for a society, with no kids in poverty, for self-determined peace, dignity, and social security, where we can construct a democratic society. We need a different economic system: a democratic socialism
leftdisconnect	All these complains lead nowhere. Nobody wants TTIP or CETA. If there is going to be a surge in right wing parties, it's the leftwing governments fault.[b]
immigration2	I give away half of income for high social and educational standards, but when our government floods our country with countless, uneducated religiously fanatic, and aggressive economic refugees that want to exploit our country, I say stop! Stop destroying our children's future

[a]Christliche Demokratische Union, a right-centric party in Germany that the majority of voters had voted in the last elections. Angela Merkel is also a member of the CDU.
[b]It is important to note, that currently a coalition of right-centrist and left-centrist parties governs Germany.

Participants were acquired by posting to various facebook groups. This convenience sampling method yields a high social media usage bias, which must be integrated when analyzing findings.

3.1 Statistical Methods

We analyzed all data using *R* version 3.3.2 using RStudio. We conducted correlation analysis using the *corrplot* package. Principal component analysis and reliability analysis was conducted using the *jmv* package. Likert data was analyzed using the *likert* package. For the principal component analysis we use the KMO and Bartlett criterion to test for sampling adequacy and sphericity/homogeneity of variances. We further use the simulation of the *jmv* package and the eigenvalue screeplot to determine factor count. All items with cross-loadings of more than .4 were removed. These items are no longer reported here. We use Cronbach's α to measure internal reliability and only use scales that are larger than $\alpha > .7$, indicating good internal reliability.

All data is reported using 95% confidence intervals. For null-hypothesis significance testing we set the level of significance to $\alpha = .01$. This means that when we find a correlation or difference in means, only 1 out of 100 samples would show a result as ours, even if no correlation or difference in means existed in reality. Given our sample size of 179 people we achieve a 95% power $(1 - \beta)$ for correlations larger than $r > .239$, and differences in means for within-subjects comparisons that are larger than $D > .271$ (Cohen's D). This means if an effect is present in reality there is a 95% chance that with a sample of our size the effects larger than these thresholds would be present in the sample, given it were a truly random sample. We use non-parametric correlations (Spearman's ρ) if ordinal scales are used, otherwise Pearson's r is reported.

4 Results

We first look at the results from a descriptive point of view to understand how our sample looks like. From our 179 respondents 63 were male 116 were female. This ratio indicates a strong over-representation of female participants. The mean age of participants was 28.5 years with a standard deviation of 9.4 years.

Our participants reported to use the Internet on a daily basis, similarly social media in general. Facebook was used multiple times per week and more traditional media such as tv, radio and newspapers were used only a few times per week (see Fig. 3). Newspapers are used least frequently.

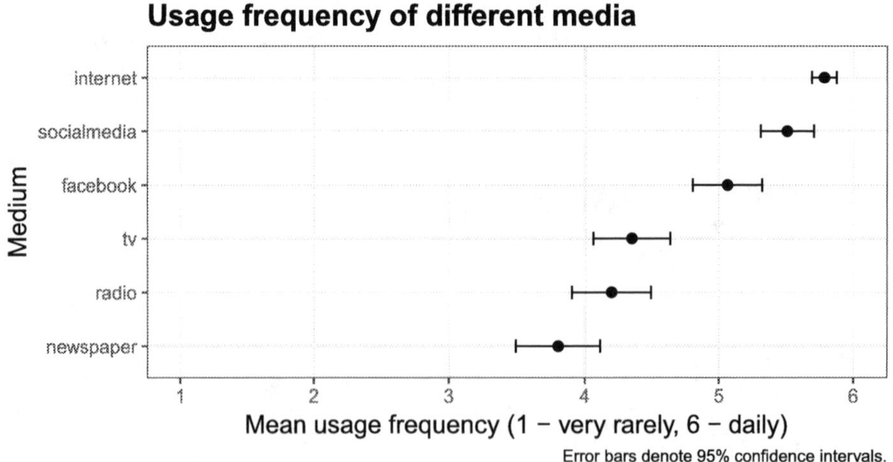

Fig. 3. Users report to use the Internet most frequently and the newspaper least frequently.

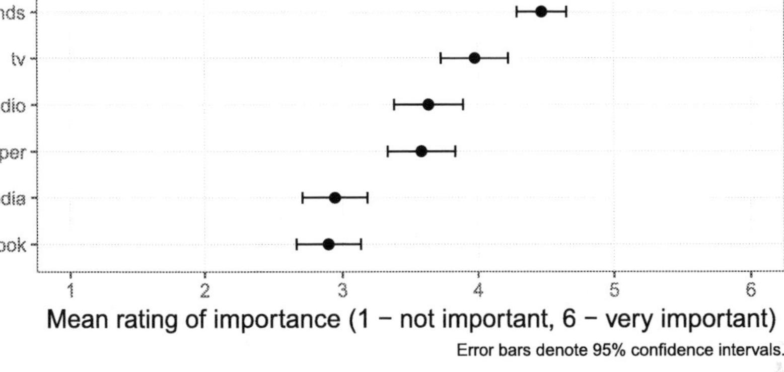

Fig. 4. Where do our participants get their information about political events?

When asked where the participants go for political information a different picture unearths. Participants do report to consult the Internet as a source of political information most importantly, but immediately afterwards real social connections—namely friends—are placed. This is followed closely by the TV, radio and newspapers, while social media and facebook are considered least important as a source of political information (see Fig. 4).

Facebook users tend to also be social media users in general (Pearson's $r =$ .66, $p < .001$), and radio listeners also watch TV ($r = .31$, $p < .001$). The more frequently persons use the Internet the more often they read newspapers ($r = .2$, $p < .01$) and the more often they listen to the radio ($r = .21$, $p < .01$).

When asked about how strong the political interest is present using our four item scale we find a mean of $M = 4.63$ (SD = 1.08), thus a rather high reported political interest. Participants showed a relatively high agreement with politically conservative goals ($M = 3.98$, $SD = 0.92$), and an even stronger agreement with politically liberal goals ($M = 4.81$, $SD = 0.86$).

We find an effect of gender on some of these variables. Women report to have a lower political interest than men ($t(155) = 2.68$, $p < .01$, $D = 0.442$). They further show higher agreement with liberal ($t(150) = -3.14$, $p < .01$, $D = -0.522$) but not conservative goals ($t(150) = -2.22$, $p = .028$, n.s.). They are also, on average, 3.8 years younger than the men in our sample ($t(177) = 2.77$, $p < .01$, $D = 0.434$). Age and the agreement with conservative political goals correlates ($r = .21, p < .01$) positively. Older participants do agree more strongly with conservative goals. When looking at reporting voting behavior voting left-wing correlates with political interest ($r = .31$, $p < .01$), liberal political goals ($r = .37$, $p < .01$) and negatively with conservative political goals ($r = -.26$, $p < .01$) and age ($r = -.21, p < .01$). Voting right-wing only correlates negatively with liberal political goals ($r = -.47$, $p < .001$). Interestingly the more one agrees with liberal political goals, the more they rely on friends as their source of political information ($r = .29$, $p < .001$).

4.1 Evaluating Opinions

Next we look at how participants rated the opinions presented to them using our forged facebook posts. The highest personal agreement is seen for the item *demosoc*, which measures whether a person agrees with the opinion that democracy should be social (or even socialist). The next strongest agreement is given for the perception that the government and their leftwing orientation have disconnected from what people really want (see Fig. 5). Approaching Russia and enforcing women's rights follow and opinions that criticizes immigration and propagate voting for the AFD are the last on the list, when participants are asked, how much they agree with these opinions.

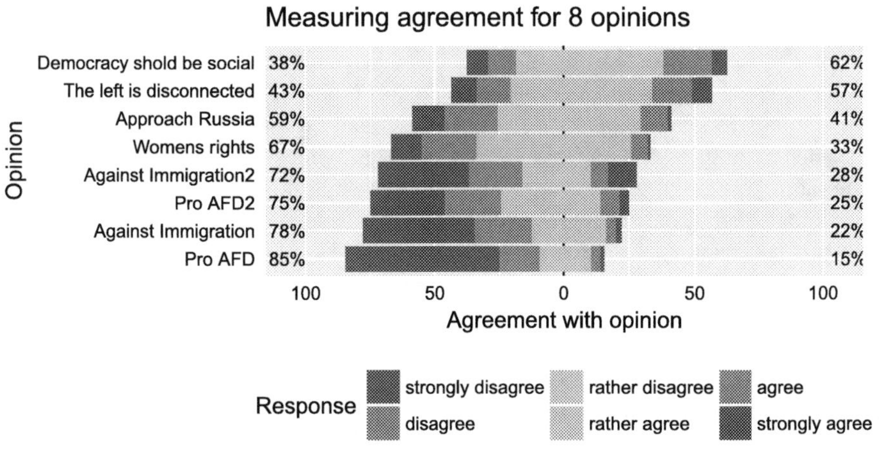

Fig. 5. Comparison of agreement with 8 opinions as seen for ones friends and other people in society.

More interestingly, when comparing how participants perceive these opinions to be pervasive in either their friends or in society in general, an interesting picture appears (see Fig. 6). For almost all topics the pervasiveness of an opinion is seen more strongly in society than in the individuals' friends group. However, only for the topic of immigration, AFD and the disconnect of the left-wing government do these differences become significant (within-subject t-tests: $p < .001$). When using within differences of means we can derive a score of perceived polarization.

Perceived polarization refers to the extent that a person perceives an opinion to be diverging from society in general and his own peer group. For example, if I believe that the average citizen is very strongly against gun-control, but me and my friends are very strongly advocating gun-control, it can be said, that I have a

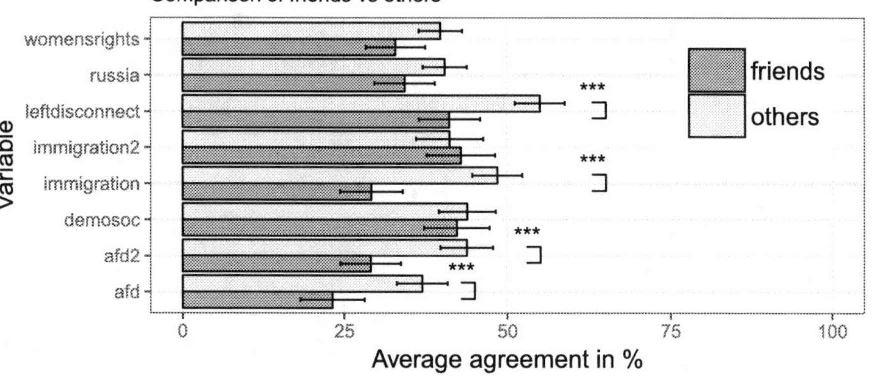

Fig. 6. Comparison of agreement with 8 opinions as seen for ones friends and other people in society.

perception of polarization for the topic of gun-control. We now look at polarization for the individual opinions in the study (see Fig. 7. We can see that the strongest polarization can be seen for the topics of immigration, voting for the AFD, and the disconnect of the left-wing government—as previously shown by t-tests.

Next it is interesting to see, which of the independent variables influences polarization. When using principal component analysis to analyze the factorial structure of polarization a single factor solution becomes apparent. When dropping the item *immigration2* a single scale with a reliability of Cronbach's $\alpha = .81$ results.

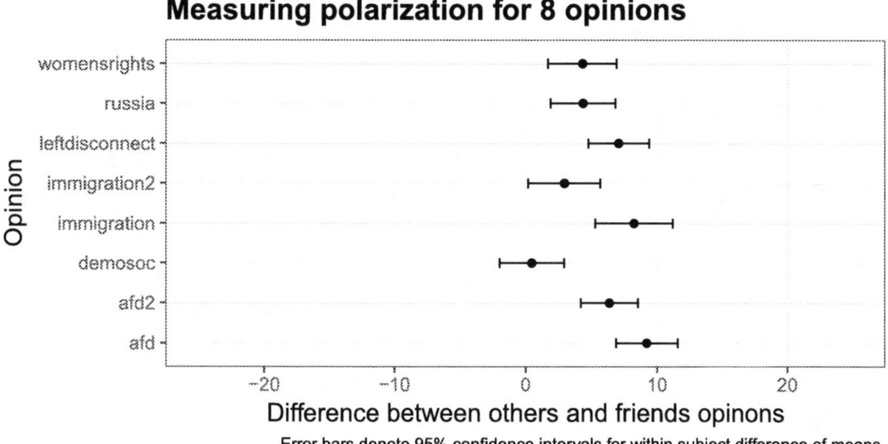

Fig. 7. Comparison of agreement with 8 opinions as seen for ones friends and other people in society.

When running correlation analysis on this new polarization scale, we find only two variables correlate with it. First, usage frequency of social media ($r = .22$, $p < .01$) and second, usage frequency of facebook ($r = .23$, $p < .01$). This indicates that the perception of polarization increases with the use of social media and in particular with the use of facebook. It is independent of the agreement with either political goals, or political interest, or even age and gender. There is a difference in means between sexes ($D = 0.435$), however it is not statistically significant ($p = .012$) on the significance level set.

5 Discussion

The results from our study indicate that the perception of polarization of opinions is existent in the sample and has associations with the use of social media. Simply put, people who use social media and facebook in particular more often, tend to see polarization more strongly than those who use it less frequently. It is interesting to note, that not the reported source of political information has an influence on polarization, but the mere usage frequency of social media. It is the overall amount of social media usage that is predictive of perceptions of polarization, not the explicit search for political information in social media. It seems to be that an (possibly subconscious) exposure effect might affect how polarization is at work. The perception of how much an opinion is shared in the general population typically has no "ground truth". So the estimation of this is typically conducted using mental heuristics. For example: "How often do I see or hear about this opinion" is a proxy for "How many people have this opinion". From a social science perspective, this is interesting, as a similar heuristic is at play when risk judgments are made. The availability heuristic, which is fine-tuned to tribal life, where every meaningful event is either perceivable directly or by hearsay, fails to adjust for both mass and digital social media. The frequency of public opinion forming is heavily biased towards the polarized states, as people with centric views, rarely rally in social media crying out for a less heated debate.

As taking part in on online discussion requires users to overcome a motivational threshold—"This is important enough for me to actually type in something"—no opinion of little affective value will induce pages of comments. This "natural" state of social participation in social networks leads to a more polarized state in social media. How much the opinion space in social media actually diverges from the opinions held by the general public needs to be established, yet. Also whether users actually compensate for this distortion cognitively must be investigated in future research. It could be that users are aware of this phenomenon consciously or subconsciously and only use social media as an indirect indicator. One could argue that effects such as the *anchoring effect* to still impact to what extent polarization is perceived, yet similar things could be said about more traditional media such as TV and radio. It could also be that for these media a better understanding of distortion is present in the general public, so that news are filtered and corrected for.

The question of what the *true* proportion of people with a certain opinion is irrelevant for the questions asked in our study. We focus on the perception of polarization. It could be interesting to investigate the true proportion using indirect means such as social media mining and direct means using representative surveys. But, recent events such as the opinion polls about Brexit have shown, that even representative sampling does not immunize against the high volatility and dynamic of opinion shifts. Opinion forming itself can be considered a chaotic system, as the individual parts (i.e. the people) adjust opinion depending on their belief about opinions. Thus, feedback loops are unavoidable. The opinion poll itself becomes the reason for opinion changes.

5.1 Limitations and Future Work

The study we conducted was performed using convenience sampling and yielded a heavily biased result. The findings must be interpreted in the light of this bias. First, the sample over-represents younger females with high education and confounds age and education. In our sample these are negatively correlated. When the young in our sample are also the educated, findings relating age and education might be inverse to what typically is present in a representative sample. This must be considered when generalizing our results. Nevertheless, the findings indicate that a skewed perception of polarization in society correlates with social media usage. Future work will have to investigate how this bias translates to other social groups and other types of media.

The opinions we used were taken from real discussions on facebook and then anonymized and without changing the wording reduced to possibly singular topics or domains. However, we still think that some of the opinions do "crossload" to other opinions. In future studies we aim for less realism and would try to create opinions that stem from more singular domains. In order to determine domains, it would be helpful to use topic modeling approaches of social media posts (e.g. latent dirichlet allocation) and then manually construct opinions that only load on singular topics.

5.2 Conclusion

In this paper we investigated the effect of media usage and user diversity factors on agreement with political opinions and the perception of polarization between the individual's peer group and the general population in Germany. We found that the perception of polarization was most prevalent in topics with right-wing political agendas. The strongest correlate with perceptions of polarization was the usage frequency of social media and facebook. Our results indicate that polarization is domain-specific, user dependent, and possibly not symmetrical. Further studies will have to investigate how these findings translate to an improved understanding of opinion forming in digital media communication.

Acknowledgments. The authors thank all participants for their openness to share their personal view on a sensitive topic. This work was funded by the State of

North Rhine-Westphalia under the grant number 005-1709-0006, project "Digitale Mündigkeit" and project-number 1706dgn017. We also thank Karina Herdt, Jens Keulen, Natia-Marta Tsikelashvili, Ceren Yilmaz and Victoria Yuryeva for setting up the survey and collecting the data.

References

1. Lee, J.K., Choi, J., Kim, C., Kim, Y.: Social media, network heterogeneity, and opinion polarization. J. Commun. **64**(4), 702–722 (2014)
2. Calero Valdez, A., Kluge, J., Ziefle, M.: You gotta fight for your right - of opinion leadership, distrust in elites, political efficacy, and the willingness to protest. Energy Res. Soc. Sci. **SI**(Special Issue on Populism) (submitted)
3. Gerbner, G., Gross, L., Morgan, M., Signorielli, N., Shanahan, J.: Growing up with television: cultivation processes. Media Eff. Adv. Theory Res. **2**, 43–67 (2002)
4. Pariser, E.: The Filter Bubble: What the Internet is Hiding from You. Penguin, New York (2011)
5. Spohr, D.: Fake news and ideological polarization: filter bubbles and selective exposure on social media. Bus. Inf. Rev. **34**(3), 150–160 (2017)
6. Rogers, E.M., Cartano, D.G.: Methods of measuring opinion leadership. Public Opin. Q. **26**, 435–441 (1962)
7. Myers, J.H., Robertson, T.S.: Dimensions of opinion leadership. J. Mark. Res. **9**, 41–46 (1972)
8. McCombs, M.E., Shaw, D.L.: The agenda-setting function of mass media. Public Opin. Q. **36**(2), 176–187 (1972)
9. Schwarz, N., Bless, H., Strack, F., Klumpp, G., Rittenauer-Schatka, H., Simons, A.: Ease of retrieval as information: another look at the availability heuristic. J. Pers. Soc. Psychol. **61**(2), 195 (1991)
10. Noelle-Neumann, E.: Die Schweigespirale. Piper (1980)
11. Gearhart, S., Zhang, W.: 'Was it something i said?' 'No, it was something you posted!' A study of the spiral of silence theory in social media contexts. Cyberpsychology Behav. Soc. Netw. **18**(4), 208–213 (2015)
12. Clemm von Hohenberg, B., Maes, M., Pradelski, B.S.: Micro influence and macro dynamics of opinions. SSRN (2017)
13. Suiter, J., Farrell, D.M., O'Malley, E.: When do deliberative citizens change their opinions? Evidence from the Irish citizens' assembly. Int. Polit. Sci. Rev. **37**(2), 198–212 (2016)
14. Schelling, T.C.: Dynamic models of segregation. J. Math. Sociol. **1**(2), 143–186 (1971)
15. Mark, N.: Beyond individual differences: social differentiation from first principles. Am. Soc. Rev. **63**, 309–330 (1998)
16. Resnick, P., Varian, H.R.: Recommender systems. Commun. ACM **40**(3), 56–58 (1997)
17. Dylko, I., Dolgov, I., Hoffman, W., Eckhart, N., Molina, M., Aaziz, O.: The dark side of technology: an experimental investigation of the influence of customizability technology on online political selective exposure. Comput. Hum. Behav. **73**, 181–190 (2017)
18. González, R.J.: Hacking the citizenry?: Personality profiling, 'big data' and the election of Donald Trump. Anthropol. Today **33**(3), 9–12 (2017)
19. Dimitrova, D.V., Shehata, A., Strömbäck, J., Nord, L.W.: The effects of digital media on political knowledge and participation in election campaigns: evidence from panel data. Commun. Res. **41**(1), 95–118 (2014)

Leadership and Social Media or About Hubs and Connectors: Useful Information and Meanings in the Selection Process of Potential Leaders

Adela Coman(✉) and Ana-Maria Grigore

The University of Bucharest, Bucharest, Romania
adela.coman@faa.unibuc.ro, anagrig27@gmail.com

Abstract. Social media and social network sites (SNSs) in particular are a response to the ever-changing, increasingly connected world – a world that needs more and more learning and collaboration to solve complex problems. In this context, the role played by leaders in organizations also changes. Despite the many studies published about leaderships, too little is known about the way we can use social networks in discovering/identifying potential leaders. So far we have been able to classify networks (Borgatti and Foster 2003; Plastrik and Taylor 2006), to define leadership networks and to see how they work (Hoppe and Reinelt 2010), to discuss about the social influence of leaders on followers, as well as the active way in which the followers, in their turn, influence leaders, particularly their behavior (Burak and Bashshur 2013). Much has been discussed about the skills leaders need to have (Mumford et al. 2007) on various hierarchical levels within the organization, but there is still no study on how we could identify these leadership skills by using social network sites (SNSs).

Our research is qualitative. We aim to analyze the skills leaders need – cognitive, interpersonal, entrepreneurial and strategic skills – and the way these can be identified in social networking, mainly using observation and surveys as methods of research. The paper is organized as follows: in the first part we discuss the basic concepts of the network theory (Barabasi 2002), leadership (Maxwell 1991), skills (Mumford et al. 2007) and influence (Cocheci 2017). In the second part, we present four concrete cases of identifying leadership skills within and with the help of social network sites, namely of the information gathered and interpreted by us, according to the specialized literature. The subjects we chose are leaders of four large companies in Romania who allowed us to access their SNSs and answered our questions during interviews organized on this occasion. The purpose of these interviews was to outline some types of desirable/undesirable behaviors in specific situations (access to and distribution of information; direct or indirect interaction with third parties; attract material and/or financial resources; formulate a vision; identify problems and consequences, objective assessment of situations and people).

The outcome of the whole work could be a model that can form the basis of a useful methodology for human resource departments, as well as for the head-hunting companies interested in finding people with leadership skills and potential.

© Springer International Publishing AG, part of Springer Nature 2018
G. Meiselwitz (Ed.): SCSM 2018, LNCS 10913, pp. 203–220, 2018.
https://doi.org/10.1007/978-3-319-91521-0_16

Keywords: Leadership · Social media · Skills · Hubs · Connectors

1 Introduction

Why is the digital world a new context, with implications on leadership?

There is neither a definition of globalization in a universally accepted form, nor a definitive one probably. The reason resides in the fact that globalization sub-includes a multitude of complex processes with variable dynamics, reaching a variety of fields in society. It can be a phenomenon, an ideology, a strategy and all of these at the same time (https://ro.wikipedia.org/wiki/Globalizare).

A key aspect in globalization is the change in technology and innovation. The occurrence of digital technologies is one of the biggest challenges the companies have to fight today. There is no organization that is immune to the assault of the digital. However, the question that is asked is how companies should use digital transformation and how they can make a competitive advantage out of it. If they want to progress, companies must think of strategies that take into consideration the opportunities offered by the new technologies and their applications. Whereas the transformation is not only digital, it cannot take place without the digital (Capgemini Consulting 2011).

On the other hand, the digital world models the battlefields in all sectors. Data show a widening of the gap between innovative companies that quickly learn how to use the tools of digital technology and those that choose not to do so. The more a company bases on the digital, the bigger the gap that separates it from the rest of the competitors. However, the digital phenomenon does not stop at the gates of traditional sectors. The digital technology permits a better targeted approach to business, a process that is more scientifically oriented by taking decisions and a new type of relations with the customers. Therefore, the companies that are part of all sectors of activity have to master the digital tools.

This assault of the digital revolution on companies and people does not diminish the importance of human initiative and responsibility, but on the contrary: it is more important than it has ever been to acquire the necessary abilities and to place them strategically to support transformation programs of the companies.

In this new context (globalization and digitalization), recruitment and the selection of leaders can often become a difficult task. According to May et al. (2003), when economy goes well, almost every type of leader is perceived as being good/efficient. In reality, the truly good/efficient leaders are rare, and the identification of a right leader for a company may be a long and expensive process. In order for the organizations to remain competitive, they need leaders and superior leadership. During hard times, a great leader can make the difference between the significant economic increase and downturn (the loss of a part of the market share). Because of this, identifying potentially talented leaders becomes the equivalent of having a key to success – for the organizations that wish to remain relevant on the market. However, research shows that only 30% of the employees with high performance have increased potential for leadership, and 90% of these people will encounter the problem at the next level once they are promoted (Balan 2017). What is interesting is that specialists involved in the evaluation process of talents (high potentials) perceive the process of their identification as having a success prediction rate of only 50%. The problem resides in the fact that the personal ability of ascending quickly on the hierarchic scale does not always predict the performance in the new role of leadership. In this case, performance is more about establishing and maintaining motivated and efficient teams, with an optimal/high level of engagement and productivity, rather than about the abilities a leader should possess.

Some of the most advanced techniques of recruitment available for professionals in human resources for identification, recruitment and selection of leaders include the recommendations of candidates (made by colleagues or directors), professional societies in which they operate, but also social networks. The use of social networks for recruitment and the selection of human resources constitute a relatively recent trend. For instance, a study from 2013 showed that 20% of the organizations that were part of the study used SNSs for the screening of candidates, whereas 12% of these planned on using SNSs for screening (Matei 2014). In 2017, 41% of women-leaders and 46% of men-leaders used SNSs for professional purposes (Roseti 2017).

It seems that employers notice quickly enough the SNSs potential as a monitoring tool of less "orthodox" behaviors in potential candidates. As monitoring becomes increasingly common, questions about the type of candidate the employer looks for start to appear when he/she eliminates right from the beginning those candidates that show an undesirable/inacceptable behavior on social networks.

The aim of the present study is that of researching whether critical abilities of leadership (Mumford et al. 2017), the way of thinking, attitudes, behaviors and an individual's actions may be captured on SNSs, and of analyzing to what extent these abilities (older and newer) are a precious indicator in the process of recruitment and selection of people with potential leadership skills. For this, we try to answer the following questions: why should we use social networks to identify potential leaders? What does leadership mean on SNSs? What are the older and the newer abilities of potential leaders, identifiable by using SNS? In the second section of the paper, we present the research methodology we used, the outcomes of our analysis being then discussed into detail. The model we propose at the end of the paper is meant to highlight a new perspective on the new model in which we should look at the potential leaders – as people being in a process of transformation of their cognitive, behavioral and emotional processes – this process being carried out with and through the use of SNSs.

2 Why Should We Use Social Networks to Identify Potential Leaders? A Literature Review

Maxwell (2002): "spread through all sectors of life…there is a handful of people with an extraordinary ability of making new friends and meeting new people. They are the connectors". The connectors are an extremely important part in our social network. They launch trends and fashion, have important business affairs, they create an uproar or help in launching a restaurant. They are the binding agents of society, who with disarming ease manage to bring together various groups, people with different origins and levels of education. The connectors – the nodes/the people – with an abnormal number of connections, from the economy to the cell; they present a fundamental feature in the majority of networks, a fact that arouses scientists' curiosity in various disciplines: the capacity of creating connections.

Cybernetics permits, among others, the freedom of extreme/total expression. Some like it; some feel threatened, but the content of a web page is difficult to be censured. Once posted, the message becomes accessible for hundreds of million people. This right of expression without precedent, together with the costs of a reduced publication, turns the web network into the supreme forum of democracy: everyone's voices can be heard and everyone's chances are equal. The logical question that may be asked is: if you put information on the network, will someone notice it?

In order to be read, you need visibility. On the network, the measure of visibility is given by the number of connections. The more input connections you have to your page, the more visible it will be. If every document on the network had a connection towards my page, everybody would know what I have to say in a short time. However, in reality, the possibility for a typical document to send to my page is almost non-existent.

Likewise, in society where some connectors know an abnormal number of people, the global architecture of the network is dominated by several nodes with a great number of connections: the so-called hubs. The hubs, like Yahoo or Amazon are extremely visible. No matter where you are navigating on the internet, you will find a connection towards them. In the network behind www, there are a lot of unpopular or rarely noticed nodes, with a small number of connections that are kept together with these pages that are very connected (Barabasi 2002).

On a smaller scale, we may say that hubs like Facebook, Twitter or LinkedIn are created somehow in a collective way, by all the people who create an account on the mentioned networks. These are special and function like some miniature worlds. The hubs are those that create shortcuts between any two nodes from the system. Therefore, even if the average distance between two randomly chosen people on Earth is of 6 steps, the distance between anybody and a connector is the address of only one or two steps. On the social networks – that we consider real hubs, there are people who play the role of connectors, that is, they have an extremely great number of connections with people from the most various fields, with different levels of education and different origins. Many of these connector-people exercise a particular influence on the social networks, through the messages they post, through the content they publish and through the actions they undertake and in which they manage to involve an impressive number of people, mainly due to the visibility they enjoy. In this paper, we will try to demonstrate that these connector-people are people with leadership features and abilities whom organizations should take into consideration for recruitment and selection for one's employment in a leading position.

2.1 What Does Leadership of Social Networks Mean?

A study carried out in 2016 shows that over half (53%) of the leaders of Romanian companies are present on SNSs. The most popular social network for these people is LinkedIn (67% of the respondents), followed by Facebook (61%) and then, at a considerable distance, Twitter (16%). The respondents give higher credibility to CEO's comments quoted online by the media (38%) and CEO's posts on the company's website (35%). Among the main audiences for the CEO's posts on networks, the majority are customers (74%), investors (44%), general public (43%) and their own employees (43%). 61% of the respondents have declared that the leaders post on the networks contents related to business issues (sales problems, management, HR, feedback, investment opportunities), 48% say that the leaders post information related to the company, and 33% state that the leaders post leadership content (https://rbd.doingbusiness.ro/).

Therefore, we may say that social networks help leaders in at least 3 ways: to accumulate and filter information; to communicate better/more efficiently; to organize activities faster.

The power and influence in leadership have always been seen as central elements (Maxwell 1991). The power is a fundamental force both in formal and social relations. Nevertheless, the digitizing and social media change the balance between leader-follower from the point of view of leadership development. Thus, an essential question occurs: how does the leader exercise their influence in the digital era?

In 2013, Bennis said that, if leaders at every level do not understand how to use the digital world and …" if they do not understand the power that it has in their relationships with their stakeholders, then they will seriously be left behind" (p. 7).

The (apparent) loss of power and influence can be re-established by using social media where the quality of interactions and the reach of the message transcend time and space. Deiser and Newton (2013) stated this thing in an expressive way in a McKinsey article, showing that: "social media encourages horizontal collaboration and unscripted conversations that travel random paths across management hierarchies. It thereby short-circuits established power dynamics and traditional lines of communication". This thing was fully demonstrated at the presidential elections in the USA, when president Trump favorably avoided social media on the purpose of not delivering his message directly to all the Americans.

Many platforms on social media, especially Twitter, give access for users to actively listen to everything followers, employees, customers, as well as the competitors have to say about products, services and the leadership of their own companies or the leadership within their country/nation. Therefore, leaders need to actively hire these stakeholders and to establish a communication network interwoven with them in order to influence conversations/discussions, to extend their social power and to build/to consolidate trust.

The requests concerning leadership in a digital world bring new challenges whereas actual approaches offer only fragmented explanations. A relatively recent approach of leadership, that has not received sufficient attention yet, is the one based on the L-A-P model (leader-as-practice). In 2016, Hibbert and Cunliffe stated in their paper that practitioners/exercise leaders want to gain an (a more) intimate awareness of their practice/experience on the purpose of browsing better in the future. From this results that the (experience) practice of the individual will dictate the type of leader he will become, and the interactions of the persons with the other people will shape his leadership style. The L-A-P model integrates theory with practice within a holistic framework of leadership and learning. Using social media for its own advantage, the different relationship between power and follower may be established on the basis of active listening, engagement and reciprocal trust regarding collaborative learning and personal development (Hibbert and Cunliffe 2016).

Leadership in the digital era has acquired dimensions that have not been sufficiently studied or understood yet. In the new context, the well-known leadership models seem to be old-fashioned and/or inappropriate today.

A new species of leaders is necessary in order to deal with a future where digitizing, continuous learning and change, critical and creative thinking, adaptability – will be the key attributes for the management of a diverse and complicated reality.

Actual leadership studies offer various "recipes" and approaches in order to become a successful leader: we talk about transformational leadership, servant leadership, authentic and ethical leadership, contingency leadership, etc. – all of these models start from an approach of leadership based on either behaviors or values (Griffith et al. 2015; Keller 2006; Marta et al. 2005).

The notion of power acquires a particular connotation in the digital world, governed by networks in which social platforms (SNSs) get all the attention. Power, in the new context, seems to migrate towards the extremes, i.e. in the direction of the one who knows how to operate with knowledge/information/cognition and towards the consumer. This made the occurrence of a more "agile" and flexible organization possible.

2.2 What Are the Necessary Abilities for Leaders? Older and Newer Abilities

Previous research concerning leadership abilities (Minzberg 1973; Zaccaro 2001); Mumford et al. 2001) classify them in 4 large categories: (1) cognitive abilities; (2) interpersonal abilities; (3) business (entrepreneurial) abilities; (4) strategic abilities. We will present them briefly:

(1) **Cognitive abilities – are considered to be basic leadership abilities.** They refer essentially to: collect, process and disseminate information (Zaccaro 2001) and to the capacity of learning (Mahoney and Barthel 1965).

An important cognitive ability is also the capacity of adaptation. This is favored by the existence of active learning abilities, skills that allow leaders to work with new information and to notice the implications of the newly appeared information. Thus, leaders can adapt their behaviors and strategies in order to deal with dynamic and/or unusual elements that appear within their job (Kanungo and Misra 1992).

Critical thinking is also part of the cognitive abilities category (Gillen and Carroll 1985) – extremely important for leaders that have to use their logic in order to analyze strong and weak points of various variants/scenarios of work.

(2) **Interpersonal abilities.** These refer to interpersonal and social abilities necessary for a leader in order to interact with other people and influence them (Mumford et al. 2000). Part of this category is social receptivity (Yukl 1989) that allows the leader to realize the reactions of the others and to understand the reasons why they react in the way they do.

Also part of the category of interpersonal abilities are the following: abilities of coordinating one's personal actions and the actions of the others (Mumford et al. 2000); negotiation abilities – for the reconciliation of the differences between individuals (Minzberg 1973); and persuasion abilities – to influence the others to achieve their objectives proposed at the level of the organization (Yukl 1989).

(3) **Business (entrepreneurial) abilities.** These refer to the abilities that contribute to the creation of the context in which the leaders work (Connelly et al. 2000). These include: abilities of material resource management – important for the management of the patrimony and technology the organization has at its disposal (Katz 1974); human resource management – for individuals' identification, motivation and promotion at their job (Kristof 1996).

(4) **Strategic abilities.** These are necessary for a leader because they allow him to understand the complexity and the ambiguity of the system/organization and to exercise his influence (Zaccaro 2001). Part of the strategic abilities category are: the abilities of systemic perception and formulation of a vision (Mumford 2000) supposing that the leader knows how to articulate an image of the environment in which the system/organization should advance, to decide if it is necessary (or not) to make changes in the organization and when these should be made.

The systemic perception and the capacity of creating a vision are connected to the ability to identify causes and consequences of an action (Mumford et al. 2000). According to specialized literature (Yukl 1989), the identification of causal connections between events allows the leader to build a sort of mental map of events and relations between them in the interior and exterior of the organization. The identification of the components of the mental map helps the leader recognize the relation between the identified problem and the possible solution/opportunity and to project an appropriate strategy in order to solve the problem.

Hence, strategic abilities also include an important component on solving problems. Therefore, leaders must possess the capacity of identifying and solving problems (Yukl 1989), but also abilities of (objective) evaluation of the variants on how a problem can be solved. (Mumford et al. 2000).

2.3 Specific Abilities for the New Context

A leader in the digital world always has in mind the overview, but at the same time, he can clearly see the steps necessary to be taken so that he carries out the objectives of the organization. The leader knows that the whole activity is not only a final point/an objective that needs to be achieved, but also a journey from which he has to learn continuously (Dicu 2015).

The authentic leader primarily thinks about the people he works with. He has the capacity to channel energy, by using positive emotions like trust and gratitude to his or the team's advantage. He creates an environment of acceptance and listens to the other people's opinion, (almost) all the time and everywhere (face-to-face or online). An authentic leader directs on the wish of serving the others. The authentic leader treats crises that appear as opportunities, learning primarily from mistakes (his personal mistakes or the other people's). In situations of crises, by adopting an open attitude, he transmits the message that every problem has solutions, and these can be discovered by using a common effort.

On Responsibility. The word "responsibility" is used in various situations with different meanings. We may use the word in order to assign an event to a cause or to assign a task to a certain role (acquired by a person). Responsibility has also been associated with: duty (towards someone), moral obligations, trust and support (Winter 1992). The most common meaning to responsibility, derived from legislation, is based on the model provided by Young (2011) according to which "a person assigns responsibility to individual entities that prove to be causally connected to the circumstances for which they look for responsibility" (p. 97).

The theory of responsibility, like social connection, assumes that the agents can be made responsible for their actions not only in the case when a direct causal connection between an action and a result can be established, but also in the case when connections/liaisons are indirect. As Young shows (2011), those who "contribute by their actions to the structural processes that produce injustice have responsibilities to work to remedy these injustices" (p. 137). In other words, Young asks the individuals/agents who have the resources and the power to correct social injustices, to assume their responsibility for these. Supporting the same idea, Maak and Pless (2009) showed that leaders from the business world have a higher responsibility (than common people) to involve in solving social and environmental problems because they were privileged: they have the power and the potential to make changes.

Even if Young's perspective, when he defines responsibility – is a global one – we believe that the elements identified by him as being essential for the concept of responsibility – can also be transferred to the microeconomic level of leadership, namely: (1) the leader's responsibility (in the digital world) is not about him acting in an isolated manner (on his own); (2) responsibility means that the leader critically evaluates the norms and the basic regulation and knows the people's opinions that are situated at a distance of one or two clicks; (3) the leader's responsibility marks the fact that he looks ahead (long term) rather than back (short/past term); (4) the leader's responsibility in the digital world is divide (between him and the others) and needs collective actions when it comes to solving problem(s).

The integration of responsibility in the concept of leadership has important implications. Solving problems becomes, hence, a problematic based on the dialogue with the stakeholders. In this sense, assuming responsibility does not become visible until the moment when the leader of the company and the stakeholders do not begin to communicate with one another. Especially when they try to solve some serious social problems (pollution, poverty, health), the stakeholders can also contribute with their knowledge and specific skills, assuming therefore, roles of leadership.

For instance, the NGO representative can operate as expert whereas the leader of the company can be the initiator or moderator of the dialogue with the expert, but with other stakeholders, as well. The leader, thus, goes from the exclusive quality of "leader" to divided leadership, that is, a process of (management) and sharing of meaning between the actors (Tourish 2014).

On Wisdom. Wisdom supposes taking informed decisions, thinking at the consequences on both short and long terms of each follower and stakeholder. Rooney et al. (2010) have articulated the concept of vision and long term perspective that incorporate objectives of prolonged impact. This type of perspective can be contradicted with the notion of efficient leadership that is based on objective measurements and tangible benefits.

Wisdom entails to get over the level of the proper decisions and to reach higher ideals of improvement and support of the human condition, to protect resources and environment. Wisdom also marks the understanding of how complex systems work, permanently looking for that knowledge that can help you distinguish important problems and act on them. Likewise, wisdom entails the ability to unify various interesting parts (in finding solutions for a problem) – in a winning coalition (Krahnke et al. 2014).

What is to be noticed is that specialized literature does not make any expressive reference to the fact that context changes also impose changes at the level of leader's abilities. Moreover, specialized studies do not mention personal and professional transformations that the leader experiments with the help of social media, and in some cases, even because of it. Hence, I started from the idea that especially SNSs can represent a valuable tool when attempting to highlight these transformations. Briefly, we try to demonstrate that the potential of leadership can be identified and tested by using networks.

3 Methodology

We chose to present four concrete cases of identifying leadership skills within and with the help of social network sites, namely of the information gathered and interpreted by us, according to the specialized literature. Our subjects (**to see Appendix**) are leaders of four large companies in Romania who allowed us to access their SNSs and answered our questions during interviews organized on this occasion. The purpose of these interviews was to outline some types of desirable behaviors in specific situations (access to and distribution of information; direct or indirect interaction with third parties; attract material and/or financial resources; formulate a vision; identify problems and consequences, objective assessment of situations and people).

For this, we used a questionnaire with questions meant to highlight the leaders' activity on social networks. Then, the four leaders granted us an interview based on the observations formulated by us, followed by the analysis of the answers to the questions on the questionnaire. What must be mentioned is that, during this whole process (interview), we had access to the open accounts of these people of SNSs, so we could directly identify the type of interactions these leaders had with the people in their contact list.

Also, we must highlight the fact that the four leaders work in "critical" domains in Romania: health and culture. The reasons we have chosen these two fields are the following:

(1) In Romania, we talk a lot about health, but we do too little. The last Report of the European Commission (Neagu 2017) places Romania on the last place in the EU from the point of view of the "performance" of the system, namely: we have the worst infrastructure in the EU (poor/inappropriate equipment in the medical system), the worst health services in the EU (because of the corruption), as well as a high rate of mortality and morbidity increased by cancers, hurt and lung diseases.

Also, we talk a lot about culture, field in which we do too little, again. Therefore, Romania can be "proud" of: a cultural heritage that is in an advanced degradation state; buildings – historic monuments restoration projects that take a very long time, sometimes even 10 years; high risk of alteration/demolition/destruction of buildings which are part of our cultural heritage because they are situated on attractive terrains from the point of view of real estates. What is to be mentioned is that both activity sectors are underfinanced by the state.

(2) In education and culture, especially, we need leadership because "without a sane and educated person, you cannot build the Romania of tomorrow" (Manoilescu 1933). Both fields contribute to the outline of people's identity – physical and mental. Furthermore, health and culture have been characterized through often changes at the top of the hierarchy, many hospitals and cultural institutions getting no benefits from professionals/skilled people in the field.

Given the transformations through which mankind faces – globalization and digitalization – we believe that this new context needs, more than ever, new people with abilities that permit them to cope with revolutions they confront at both macroeconomic and microeconomic levels. Thus, we tried to identify what are the abilities that had recommended our respondents to occupy their present jobs and to obtain great results.

Studying the way our respondents use social networks to reach other people, we tried to look for the motivation behind the connection/connections (attitudes and thinking), the identified problems and the ways of finding a solution (behavior and actions), as well as the way they react on the emotional plan (expressed feelings, empathy, solidarity, etc.) when it comes to challenges.

We mention that our respondents work both in the public sector (2) and the state sector (2), and they are between 37 and 60 years old. Every person selected has a vast experience in their field activity and has, at least, 10 years of leadership.

4 Results and Discussion

On the question about the networks on which they created accounts, the responses were, in 3 cases, Facebook, Twitter and LinkedIn. Only one respondent declared that he created accounts on Google+ and Instagram, as well. Moreover, the companies they direct have accounts on the same SNSs.

We wanted to observe if our respondents make the distinction between the personal and professional networks. All the respondents expressed themselves in the sense that, from their point of view, Facebook and Twitter are both personal and professional networks. This things has two possible meanings: (1) in the digital world, the leaders consider that there is no boundary between public and private; (2) the leaders think that the personal and the professional plans interweave or, as one of them said: "It does not matter what network you use as long as you can achieve your purpose".

To test the respondents' familiarity degree with SNSs, we asked them what methods they used when they applied for the current position. Two of the respondents mentioned the professional sites (LinkedIn) and/or the employer's web page, and the other two mentioned the predecessor's death, respectively the previous collaborations with the organization.

All the respondents post content on the SNSs, and their reasons for doing so are the following (in the order of importance): they want to express their point of view regarded what is right/wrong in the society; they want to bring into attention a certain problem; they want to mobilize people in a cause in which they believe. One respondent placed on the first place (main reason) the need to generate a current of favorable/unfavorable opinion among consumers.

This hierarchy of responses demonstrates that the leaders are responsible people, involved in problems appeared at the level of organization and/or community (local or national). Given the big number of friends (Facebook), followers (Twitter) or contacts (LinkedIn) that these people have on their SNSs, their opinions, comments, posted and shared articles on the network have a high potential of exercising a meaningful influence on perception, attitudes, and behaviors for those present in the virtual environment. One of the respondents mentioned that: "The article I wrote some time ago on the increasing incidence of mental disorders because of poverty in my community generated stormy discussions concerning the causes that can provoke these diseases. Then, the idea of building a hospital/a community center for treating patients with these diseases, coming from poor environments, was born."

The social networks are, for all respondents, useful tools that they use both in professional purposes and as employers. Hence, 3 out of 4 respondents have stated that they use SNSs to get in touch with professionals from their activity fields and to discuss certain tendencies, and a respondent indicated the taking of the pulse of the economic market and finding details about the important names in the field as aims they follow when they create connections in the virtual environment. From our point of view, reporting to other professionals in the field indicates, at least, the following: (1) the respondents want to learn from those who are authorized in the field; (2) they try to be as well informed as possible regarding novelties; (3) they gain visibility in their field and increase their chances of being perceived as skilled creative people, oriented towards performance.

As employers, the respondents have stated that they find it normal to use SNSs in order to check the applicants' profile. What 3 out of 4 respondents look for within the process of recruitment and selection of new candidates is: (1) the candidate's profile on SNSs; (2) the résumé and the life experience; (3) critical thinking. However, one of the respondents mentioned that he was interested in the way the candidate spent his free time, his critical thinking and his IQ. According to specialized literature, responsibility, but also wisdom is associated with a higher experience (Mumford et al. 2017), which is mutually available. Then, we can suppose that, if by checking the candidates' profiles on SNSs, the employers look for: (1) attitudes; (2) behaviors; (3) language, etc., by analyzing the résumé and life experience, they want to find out what is relevant for an individual from a professional and human point of view. Hence, one of the respondents made the following remark: "Of course, the candidate can show a certain behavior on the SNSs that has no link with what we can find written in his résumé. This thing automatically disqualifies him from the competition of getting the job. Or, it can show a false profile. Or, he can lie about his abilities. This is still a lie. Everything is uncovered at the interview. The good part is that you are prepared for this in the moment you view the profile and then, you meet the person face-to-face…".

According to specialized literature (Mumford et al. 2000), the leaders operate with determination in order to achieve their proposed aims and to transpose their vision into reality. Observing the action taken by our respondents on SNSs, we concluded the following: (1) they defend their position on SNSs when their right as consumers are violated; (2) two of the respondents are coordinators of some NGO; (3) all respondents are actively involved as volunteers in actions happening at a local or national level, using SNSs for these. In other words, they act and react differently towards ordinary people, being real connectors, and the hubs created by them (NGOs) include an impressive number of volunteer-participants in the most various activities: environment protection, fundraising for a social cause; education, etc.

Maxwell (1991) says that leadership is influence, nothing more, nothing less, highlighting the fact that, if you want to test a person's leadership abilities, all you have to do is to tell him to found and direct an NGO. If he manages to do this, then he is a leader in the true sense of the word. Why? Simply because at your job people follow you as the leader because "they must". In the case of an NGO, people follow you because they want and because they feel that you care. They follow you because they believe in you and they respect you. In this context, our respondents demonstrate that they are true leaders. Moreover, the actions undertaken by them emanate optimism, positivity, responsibility, faith that they can change the world, and by changing the world, they change themselves. Therefore, the leaders also go through a process of transformation and personal development, ceaselessly learning (from their mistakes and from the other people's), accumulating knowledge, experience and wisdom. It is a long term process that supposes a strategy. By paraphrasing Ma (2018), the founder of Alibaba, a good leader knows how the world will look in 50 years and he knows, at the same time, what he DOES NOT want to do – a dimension of strategic management that all the successful leaders possess.

Through the following 3 questions we wanted to identify cognitive abilities (the capacity of judging in an objective manner a person/situation; the capacity of defining the problem, of analyzing the objectives, the constraints and risks); creative thinking (the capacity of recognizing and evaluating valuable ideas); strategic abilities (the capacity of formulating a vision and giving sense to actions); interpersonal abilities (to listen and understand the other; to recognize the validity of another point of view when it is the case). Thus, the respondents appreciated that: (1) they try to be objective and they are willing to openly recognize when the interlocutor has a different opinion from theirs; (2) at the end of every year they usually make the lucid analysis of the causes of the present situation, of the constraints that have generated it; (3) they present their vision without diminishing the risks, considering that the elaboration of the vision regarding the organization is very important; (4) when they discover a valuable idea, they make efforts to put it into work for the benefit of the organization, even if they are or not the authors of the idea; (5) they usually talk about achievements at first and make optimistic projects regarding their future.

A prominent figure in the business world has compared management with the conducting a symphonic orchestra, and leadership, with conducting a jazz band. We have also asked our leaders if they associate the directing of an organization with conducting a symphonic or a jazz orchestra. The answer was: "jazz orchestra", and by making this, they consider themselves free to manifest their entrepreneurial abilities. Jazz is a form of "chaos" with structure (even if it might seem contradictory) whereas the interpretation of a symphony can only succeed if every musician perfectly knows his orchestral part and executes it flawlessly. However, even in Jazz, there is a melody that is apparently insane. Leadership, like Jazz, can be studied, refined, aside the inborn qualities (sometimes in spite of these). "Champions do not become champions in the arena – they only get the recognition there", says an old dictum. Even if a person has inborn talent, he or she still has to prepare and exercise in order to be successful. Leadership is developed and requires perseverance.

One of the 21 rules of leadership (Maxwell 1991) says: "people buy into the leader, and then the vision". In other words, the followers appreciate the human being/the leader at first; afterwards, his vision. According to Maxwell, good leaders see the problems, build a strategy in order to solve them, appreciate the diversity (of people, of opinions and ideas) and schedule thoroughly the way they have to take. The perspective is, in all cases, on a long term.

In Fig. 1, we tried to highlight leadership as a process, as a result to cognitive, behavioral and emotional transformations imposed by the new context of globalization and digitalization.

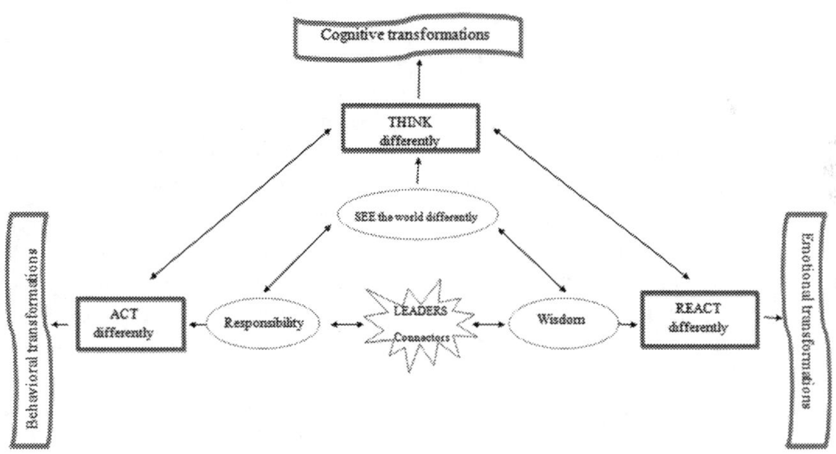

Fig. 1. Leaders in a digital world (new context)

The model proposed by us sends to the older and newer abilities every leader should possess in the digital world. The leaders should operate as connectors, that is, by means of digital tools, to be able to create connections with a number of people as many as possible. The leader needs information and can take good decisions, even in the conditions in which the information is not complete or is not sufficient – this is why they have to keep in touch with everyone that can help them with information in due time.

Leaders See the (Digital) World in a Different Way Than the Others. They have the capacity of seeing the world in its assembly (at a high scale), they can observe the world within perspective (long term) and because they have access to information, they know how the world will look in 50 years and they prepare for it. The leaders have the capacity of seeing the world in a different way and thus, they can also see it on a microeconomic level (organizational level). Because they know to define problems, to analyze causes, to analyze objectives (and to select only the relevant ones for the organization), but also constraints, they manage to infer possibilities where common people only see "a problem to be solved". These are all the old abilities, but they are also the ones that – if valorized in the next context of the digital world – allow leaders to manipulate complexity, based not only on instinct, but also on the identification of new ways of doing things. It is about inborn cognitive abilities (the instinct) and acquired cognitive abilities (through continuous learning and training) that transform into mental models under the impact of contact and (almost) permanent communication with professionals from the field, with people having the experience and with those having the ideas. Cognitive transformations and mental models lead to an unconventional thinking, a more agile one, more sensitive and more connected to changes of substance, but also of nuance in the exterior and interior environment of the organization.

Leaders Operate/Act in a Different Way in the Digital World. They use informational technologies in order to get involved in the identification of new solutions for old problems, such as: poverty, health, culture and environment. Their behaviors are different from those of common managers because leaders know how to easily collaborate with various teams; they hire more intelligent people than them and give these people the power to take important decisions; the leaders value the contribution of new partners (experts, NGO directors or other stakeholders) because they are aware of the fact that solving some complex problems requires a common effort; they invest time and energy in order to make things right and they learn from their mistakes; they plan and forecast on long terms, thinking thoroughly, at the same time, on the paces they have to take/on the details. Every taken decision is a manifestation of responsibility: responsibility towards followers and their future, responsibility towards resources that the leaders administer, ad everything they do is in accordance with the system of values on the basis of which they build their vision. Their different behavior, in their case, marks briefly the idea that you should let yourself guided by responsibility and you should attribute a sense to every operation you undertake.

Leaders React in a Different Way in the Digital World. Beyond the emotional abilities Goleman talks about, and those that determine him to report them to the people around in terms of empathy and preoccupation for the human being, the leader in the digital world discovers better and more practical solutions due to the easy access to information. He is convinced that he can change the world and, because of this, he involves himself in volunteer operations. Furthermore, he founds NGOs in which he attracts people with passions and beliefs similar to those that he has. These poles of action, often born by using SNSs, get to grow in time when more and more people resonate with the mission/purpose/operation taken into account. Leaders' opinions and comments can bring birth to passions, and passions can also transform in ideals. According to specialized literature, the leader's wisdom is associated to some social judgments: he wants to change the world not because this must be changed, but because he feels that he can do it; he believes in himself and in his own life experience; he sees the world within perspective and always thinks on long terms; he believes in his power of creating alliances to achieve a superior purpose.

5 Conclusions

The development – hard to foresee two or less than two decades ago – of channels and ways of communication at a global level, took to an explosion of information dissemination in all the areas of human activity and all social strata.

General and immediate access to information smoothed many subjective bumps and oiled the wheels of all processes, taking to their acceleration without precedent. This meant a powerful development and significant, fundamental change. Highly important changes in both areas of knowledge and economy impose serious changes of mentality and behavior at the leadership level. The assault of digital revolution on companies and people does not diminish the importance of human initiative and responsibility, but on the contrary: it is more important than ever to acquire the necessary abilities and to place them strategically in order to support transformation programs of companies.

Leadership is both a process and a property. As a process – focusing on what leaders actually do – leadership is the use of non-coercive influence to shape the group or organization's goals, motivate behavior toward the achievement of those goals, and help define group or organization culture. As a property, leadership is the set of characteristics attributed to individuals who are perceived to be leaders (Griffin 2006). In addition – almost all the qualities and features might help, but what really matters is the way they combine together in order to efficiently answer the requirements of the situation in which he has to operate as the leader. In order to truly become the leader of a nation or of a company, you must understand the personal and emotional concerns and the preoccupations of the others, entering, thus, in the personal world of those people (Garelli 2017). Ideally, a manager whom people must follow should have the talent of making people want to follow him. Force someone to get out of his comfort zone and you will get a rebel.

The limits of our study refer to several important elements, namely: the small number of participants/leaders, respondents of our questions; the specific fields of their activity; the impossibility of mathematically quantifying the abilities we analyzed in the present paper. Nonetheless, we hope that the our proposed model open a new research field to those who look for talented people, with leadership potential, by using for this, as a complementary tool, social networks.

Appendix

A. M. is the director of the Bucharest Municipality Museum – an institution that has under its control 12 museums and memorial houses. He is a historian, anthropologist, journalist and associated professor for two higher education institutions. Under his control, the Bucharest Municipality Museum has continuously prospered with newer and newer exhibitions, as well as with a multitude of programs and educations activities destined both for children and adults.

I. M. N. is a film critic, a radio and television show producer with a vast journalistic and translator activity (during the communist era, she was the voice of over 3000 movies on video tapes she had translated). She was part of SOROS Foundation Board of Directors in Romania. She is the honorific president of Vodafone Foundation today.

A. P. is a specialist psychiatrist and Ph.D. in Medical Sciences. He had the Psychiatrist Hospital in Campulung Moldovenesc in his charge, and today he owns a private mini clinic. He is the president of Social Psychiatry Association in Romania, affiliated to the World Association of Social Psychiatry, representing an active presence in the Romanian and international psychiatry field.

C. T. is a physician specialized in kidney diseases. He worked as a specialist physician at Fundeni Hospital in Bucharest. Today, he is the director of a private clinic, branch of a company that owes specialized companies in a number of countries all around the world.

References

Balan, M.: Organizatiile cu lideri eficiente au sanse de 13 ori mai mari sa depaseasca rezultatele competitorilor: Oameni si bani. RevistaCariere, p. 15, martie 2017. www.cariereonline.ro/leadership/

Barabasi, A.-L.: Linked: The New Science of Networks. Perseus Publishing, Cambridge (2002)

Bennis, W.: On Becoming a Leader. Basic Books (2013)

Borgatti, S.P., Foster, P.C.: The network paradigm in organizational research: a review and typology. J. Manag. **29**(6), 991–1013 (2003). https://doi.org/10.1016/50149-2063(03)00087-4

Burak, O.C., Bashshur, M.R.: Followership, leadership and social influence. Leadersh. Q. **24**(6), 919–934 (2013)

Capgemini Consulting: Digital Transformation Review no.1/2011 (2011). https://capgemini.com/consulting

Cocheci, H.: Poate HR sa influenteze cu adevarat leadership-ul organizatiei? Revista HR Manager, September 2017. https://hrmanageronline.ro/

Connelly, M.S., Gilbert, J.A., Zaccaro, S.J., Threlfall, K., Marks, M.A., Mumford, M.D.: Exploring the relationships of leadership skills and knowledge to leader performance. Leadersh. Q. **11**, 65–86 (2000)

Deiser, R., Newton, S.: Six Social Media Skills Every Leader Needs. McKinsey Q. (2013). https://mckinsey.com/industries/high-tech/

Dicu, L.: Surprinzator sau nu, liderii de succes fac lucrurile diferit. Revista Cariere, ianuarie (2015). https://www.cariereonline.ro/leadership/surprinzator-sau-nu-liderii-de-succes-fac-lucrurile-diferit

Garelli, S.: Un líder adevarat intra in lumea interioara a celor pe care ii conduce. Rev. Cariere **241**, 72–73 (2017)

Gillen, D.J., Carroll, S.J.: Relationships of managerial ability to unit effectiveness in more organic versus more mechanistic departments. J. Manag. Stud. **22**, 668–676 (1985)

Gladwell, M.: The Tipping Point: How Little Things Can Make a Big Difference. Back Bay Books, New York (2002)

Goleman, D.: Emotional Intelligence. Bantam Books, New York (1995)

Griffith, J., Connelly, S., Thiel, C., Johnson, G.: How outstanding leaders lead with affect: an examination of charismatic, ideological, and pragmatic leaders. Leadersh. Q. **26**, 502–517 (2015)

Griffin, R.W.: Fundamentals of Management. Houghton Mifflin, New York (2006)

Hibbert, P., Cunliffe, A.L.: The philosophical basis of leadership-as-practice from a hermeneutical perspective. In: Raelin, J.A. (ed.) Leadership as Practice: Theory and Application. Routledge/Taylor and Francis, New York (2016)

Hoppe, B., Reinelt, C.: Social network analysis and the evaluation of leadership networks. Leadersh. Q. **21**(4), 600–619 (2010)

Kanungo, R.N., Misra, S.: Managerial resourcefulness: a reconceptualization of management skills. Hum. Relat. **45**, 1311–1331 (1992)

Katz, R.L.: Skills of an effective administrator. Harv. Bus. Rev. **52**, 90–102 (1974)

Keller, R.T.: Transformational leadership, initiating structure and substitutes for leadership: a longitudinal study of research and development project team performance. J. Appl. Psychol. **81**, 202–210 (2006)

Krahnke, K., Clinebell, S., Wanasika, I.: Wisdom of a Leader. In: Marques, J., Dhiman, S. (eds.) Leading Spiritually. Palgrave Macmillan, New York (2014)

Kristof, A.L.: Person-organization fit: an integrative review of its conceptualizations, measurement and implications. Pers. Psychol. **49**, 1–49 (1996)

Maak, T., Pless, N.M.: Business leaders as citizens of the world. advancing humanism on a global scale. J. Bus. Ethics **88**(3), 537–550 (2009)

Marta, S., Leritz, L.E., Mumford, M.D.: Leadership skills and group performance: situational demands, behavioral requirements and planning. Leadersh. Q. **16**, 97–120 (2005)

Milward, H., Brinton, P., Provan, S., Keith, G.: A Manager's Guide to Choosing and Using Collaborative Networks. IBM Center for the Business of Government (2000)

Ma, J.: Jack Ma's Keys to Success: Technology, Women, Peace and Never Complain (2018). https://www.youtube.com/watch?v=-nSbkywGf-E

Mahoney, F.I., Barthel, D.W.: Functional evaluation: the Barthel index. Md State Med. J. **14**, 61–65 (1965)

Manoilescu, M.: Tendintele tinerei generatii. Bucuresti: revista Lumea noua (1933)

Marks, M.A., Mathieu, J.E., Zaccaro, S.J.: A temporally based framework and taxonomy of team processes. Acad. Manag. Rev. **26**(3), 356–376 (2001). https://doi.org/10.2307/259182

Matei, M.: Antreprenorii vorbesc. Barometrul antreprenoriatului romanesc 2013 (2014). https://www.eyromania.ro/2014/

Maxwell, J.C.: Cele 21 de legi fundamentale ale leadership-ului, Almateea, Bucuresti (1991)

Maxwell, J.C.: The 21 Irrefutable Laws of Leadership. Amalteea, Bucharest (2002)

May, D.R., Chan, A.Y.L., Hodges, T.D., Avolio, B.J.: Developing the moral component of authentic leadership. Organ. Dyn. **32**, 247–260 (2003)

Minzberg, H.: The Nature of Managerial Work. Harper & Row, New York (1973)

Mumford, M.D., Schultz, R.A., Van Doorn, J.R.: Performance in planning: processess, requirements and errors. Rev. Central Psychol. **5**, 213 (2001)

Mumford, M.D., Marks, M.A., Connelly, M.S., Zaccaro, S.J., Reiter-Palmon, R.: Development of leadership skills: experience and timing. Leadersh. Q. **11**, 87–114 (2000)

Mumford, M.D., Troy, V., Campion, M.A., Morgeson, F.P.: The leadership skills strataplex: leadership skill requirements across organizational levels. Leadersh. Q. **18**, 154–166 (2007). https://doi.org/10.1016/j.leaqua.2007.01.005

Mumford, M.D., Todd, E.M., Higgs, C., McIntosh, T.: Cognitive skills and leadership performance: the nine critical skills. Leadersh. Q. **28**, 24–39 (2017)

Neagu, A.: Raport al Comisiei Europene: Romania, pe ultimul loc in UE la cheltuielile pentru sanatate pe cap de locuitor (2017). https://www.hotnews.ro/stiri-esential-22141266

Plastrik, P., Taylor, M.: Net Gains: A Handbook for Network Builders Seeking Social Change. Version I.0 (2006). https://networkimpact.org/

Roseti, R.: Intre consum si comportament: Situatia dezvoltarii domeniului digital in Romania. Revista Cariere, octombrie 2017

Rooney, D., McKenna, B., Liesch, P.: Wisdom and Management in the Knowledge Economy. Routledge, New York (2010)

Studiu EY: 53% dintre liderii companiilor din Romania sunt prezenti pe retelele de socializare, iar cei mai multi prefera platforma Linkedin (2015). https://rbd.doingbusiness.ro/

Tourish, D.: Leadership, more or less? A processual, communication perspective on the role of agency in leadership theory. Leadership **10**, 79–98 (2014)

Winter, D.G.: Responsibility. In: Smith, C.P. (ed.) Motivation and Personality: Handbook of Thematic Content Analysis, pp. 500–511. Cambridge University Press, Cambridge (1992)

Young, I.M.: Responsibility and global justice: a social connection model. Soc. Philos. Policy **23**, 102–130 (2006)

Young, I.M.: Responsibility for Justice. Oxford University Press, New York (2011)

Yukl, G.: Managerial leadership: a review of theory and research. J. Manag. **15**, 251–289 (1989). https://doi.org/10.1177/014920638901500207

Zaccaro, S.J., Rittman, A.L., Marks, M.A.: Team leadership. Leadersh. Q. **12**, 451–483 (2001)

This Is How We Do It: Untangling Patterns of Super Successful Social Media Activities

Tobias T. Eismann(✉), Timm F. Wagner, Christian V. Baccarella, and Kai-Ingo Voigt

School of Business and Economics, Friedrich-Alexander-Universität Erlangen-Nürnberg, Lange Gasse 20, 90403 Nürnberg, Germany
{tobias.eismann,timm.wagner,christian.baccarella, kai-ingo.voigt}@fau.de

Abstract. Online social media plays an important role in the marketing communications mix of many companies. Thus, scholars have recently tried to uncover patterns that have a positive impact on the effectiveness of social media communication, predominantly focusing on message characteristics. Although a lot of valuable insights have been generated, it remains unclear what the drivers of 'super successful posts' (SSP) are. Therefore, the purpose of this paper is to reveal why a very small proportion of social media posts significantly outperform the majority of other posts. For this purpose, we employed case evidence from the automotive industry and collected 2,000 Facebook posts. In regard to the numbers of likes, comments, and shares, the 20 most successful posts each were selected. After removing the duplicates, a final sample of 42 SSP remained. With an explorative multi-level approach, including two focus group sessions, an in-depth analysis was conducted for every post. Aiming to capture a comprehensive picture, we also investigated the context of each post beyond the online environment. With our analysis, we reveal five typical patterns of social media excellence (co-branding, wow effect, cognitive task, timing, and campaign). In addition, we further elaborate on four selected SSP to enhance the understanding of underlying mechanisms. Among other things, our findings encourage practitioners to employ a broader view when planning social media posts. Thus, the understanding about the five patterns of SSP may support practitioners in enhancing the popularity of their future posts.

Keywords: Social media · Social networking site · Facebook · Social media posts Marketing communications · Social media excellence

1 The Value and Challenge of User Engagement

Social media is crucial in the mix of the marketing communications of many companies and represents an essential new way of fostering relationships and interactions with and among customers [1–3]. On Facebook, for example, companies can create corporate social networking site (SNS) identities in the form of brand fan pages [4, 5], enabling them to spread influential marketing messages [6–9]. Typically, users engage with the brand through 'following' the brand fan page and interact by liking, sharing, or

© Springer International Publishing AG, part of Springer Nature 2018
G. Meiselwitz (Ed.): SCSM 2018, LNCS 10913, pp. 221–239, 2018.
https://doi.org/10.1007/978-3-319-91521-0_17

commenting the brand posts [10–13]. These actions of user engagement mostly indicate agreement with or affinity for a brand, a product, or just the message in the post [14]. Additionally, Wallace et al. [15] found that fans with a higher level of engagement regarding social media posts tend to develop closer relationships with the respective brand. Further findings even indicate that customers who become a follower of these brand fan pages tend to visit the stores more often, generate more positive word-of-mouth, are more emotionally attached to the brand than non-brand fans [16], and tend to be increasingly loyal to the brand and willing to receive additional advertising information [17]. That is why user engagement is critical for SNS post success. The fact that marketing managers spent more than $4.3 billion worldwide on SNS advertising [18] shows that they have recognized the value of maximizing user engagement on SNS [19]. However, along with the obvious advantages of engaging users on social media, there is a severe need for better understanding the underlying mechanisms of user engagement in order to fully exploit the benefits [2, 20–22].

Several scholars have already started to uncover the drivers of user engagement on SNS. These studies have predominantly focused on certain message strategies regarding the content or design of a post, revealing various valuable insights [e.g., 23–25]. It has been shown, for instance, that posts containing images attract more user engagement in the form of likes and comments, or that posts which have been published during business hours were more likely to be commented on [23]. Besides these rather simple influencing factors, users may also be willing to engage more when a specific post touches them on a more emotional level [26]. However, it is still not fully understood how these different levels of post characteristics influence each other, and, more importantly, how they finally influence user engagement [23].

Interestingly, although most brands have already reached a certain professional level of social media expertise, there is a surprising lack of knowledge as to why some of the social media posts significantly outperform the majority of all other posts. Thus, it is of particular importance that both researchers and practitioners understand the peculiarities of these 'super successful posts' (SSP) on social media. Although a significant amount of research has unveiled individual antecedents of SNS post success, an in-depth investigation on those SSP is still lacking. We argue that an SSP consists of an intelligent interplay of many contextual and individual characteristics that make all the difference.

Against this background, the aim of this study is to analyze the SSP of brands in order to untangle their underlying patterns. Thereby, we contribute to the literature on social media research and, more particularly, expand the understanding of SNS user engagement by introducing a new perspective. For practitioners, we generate valuable insights that will enable them to reflect critically on their current social media strategy, aiming to enhance communication effectiveness.

Our article is structured as follows. First, we explain the applied methodology, which we performed to collect and analyze the data. Then, we present five typical patterns of social media excellence, followed by further explanations on four selected examples of posts. Our article concludes by stating implications for practitioners and scholars, as well as by outlining limitations and future research avenues.

2 Methodology

2.1 Data Collection

The sample used in this paper is the same as the one used by Wagner et al. [22]. As described in Wagner et al.'s paper, 2,000 Facebook posts from automotive brands were collected. We used Facebook as the research object because it is the most popular SNS with more than one billion active users per day [27]. We focused on one single industry, aiming to avoid distortions related to industry-specific circumstances [28]. All posts have been collected from the ten most valuable car brands, compiled by Millward Brown [29]. Accordingly, we included posts by Audi, BMW, Chevrolet, Ford, Honda, Hyundai, Mercedes-Benz, Nissan, Toyota, and VW (Volkswagen). In December 2014, 200 Facebook posts from each brand were collected and manually saved as screenshots. Between the final post to be collected and the first day of data collection, there was a time span of around five weeks. In accordance with Sabate et al. [23], we assumed that, after those five weeks, additional user engagement on a post is negligible. Since photo album posts are automatically re-posted when new photos are added to the album, these posts may go through several rounds of user engagement. We therefore excluded those 52 posts. Our data set then comprised 1,948 SNS posts.

After data collection, the number of likes, comments, and shares of each post was assessed. We then calculated relative performance measures for each post by dividing the number of likes, comments, and shares by the number of respective Facebook page followers. On average, the Facebook pages of brands had 7.14 million (SD = 6.00 million) followers. The BMW page had the most followers (19.03 million), and Hyundai had the smallest number of followers (1.76 million). Finally, for clarity in presentation, we multiplied the calculated measures by one million. Thereby, we generated performance measures (adjusted user engagement) which indicate how much user engagement (likes, comments, shares) a post received among one million page followers. We consider this user engagement measurement as more useful than directly looking at the number of likes, comments, and shares, regardless of the number of page followers.

During the next step, the most successful posts were identified. In particular, we collected the top 20 posts for (adjusted) likes, comments, and shares across all brands. Because some of the posts were in more than one of the three top 20 lists, the final sample of SSP comprised 42 posts.

2.2 Data Analysis

A thorough in-depth analysis of each of the final 42 SSP was conducted. For this purpose, our research team consisted of two senior researchers and two doctoral students with appropriate experience in the social media research field. The analysis followed a twofold explorative approach inspired by the qualitative case study method [30]. First, we comprehensively examined each post to understand and determine similarities across the sample. Second, we investigated the online and offline context of each post. This step involved extensive online research on the related topics of each post during the time the post was published. The research included brand websites, press websites, websites mentioned in the post, search engines, and other SNS brand fan pages. The goal was to reveal the strategy behind each post and how it was embedded in the media ecosystem and marketing activities of the respective brand. Overall, a holistic, in-depth analysis of each post was conducted.

Subsequently, two focus group sessions were performed—one with six social media practitioners, and one with seven marketing students. During these sessions, in regard to every single post, the participants were asked to discuss possible reasons for its popularity. At this stage, we provided necessary background information on each post that we had gathered beforehand in our context analysis. After the focus group sessions were completed, we aggregated the results in our research team. A final discussion clearly led to five typical patterns of social media excellence, which are presented in the following figure.

3 The Five Patterns of Social Media Excellence

The framework we use (see Fig. 1) consists of five distinct patterns that were identified among SSP: co-branding, wow effect, cognitive task, timing, and campaign. The five patterns are neither mutually exclusive nor do they all have to be implemented in an SNS post. Our sample showed several posts containing more than one of our five patterns. In the following paragraphs, we introduce the five patterns one by one and describe the underlying mechanisms that foster user engagement.

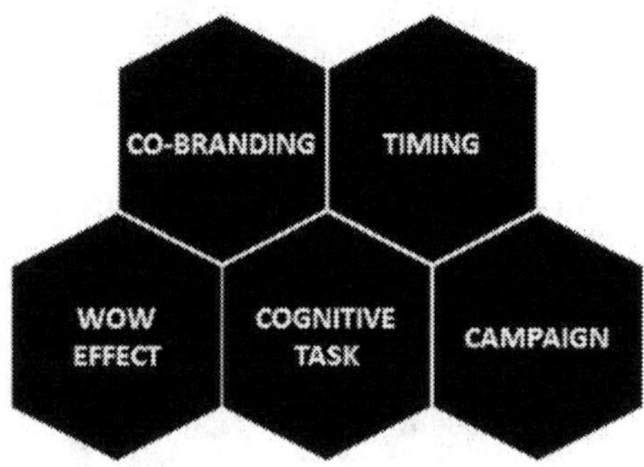

Fig. 1. The five patterns of social media excellence (Source: own illustration)

3.1 Pattern 1: Co-branding

The first identified pattern refers to co-branding. In fact, many of the SSP had in common that they utilized at least one more brand to maximize user engagement. This additional brand can be any player on social media, such as a classical product brand, a celebrity, a professional athlete, or a newspaper (e.g., The New York Times). The approach of bundling different brands in one post is aligned with a co-branding strategy as a means to gain more media exposure [31].

Joint advertising as a specific strategy in co-branding was also used, for example, the Apple Macintosh PowerBook campaign that featured the movie "Mission Impossible" as the second brand [33]. The concept of co-branding and its advantages and disadvantages are well known [e.g., 34–36]. An important goal of co-branding in social media posts is tapping into the community of a second brand, in addition to the already existing own followers. This increases the potential reach of the social media post and may therefore lead to increased user engagement. As Sabate et al. [23, 37] prove, the bigger the follower-base of a brand on SNS is, the more user engagement of the respective social media post is attracted.

There are two ways to take advantage of another brand. First, just mentioning the other brand's (or celebrity's) name can help to draw more attention towards the brand's own post. That, in turn, may lead to more user engagement. Nissan, for instance, posted a photo of their new pickup truck with a built-in diesel engine by Cummins Inc., which is a US-based company selling diesel engines worldwide. Although there is no link to the Cummins Inc. fan page (with almost 400,000 SNS followers at the time of data collection), this strategy enabled Nissan to leverage Cummins' brand awareness as well as their fan base.

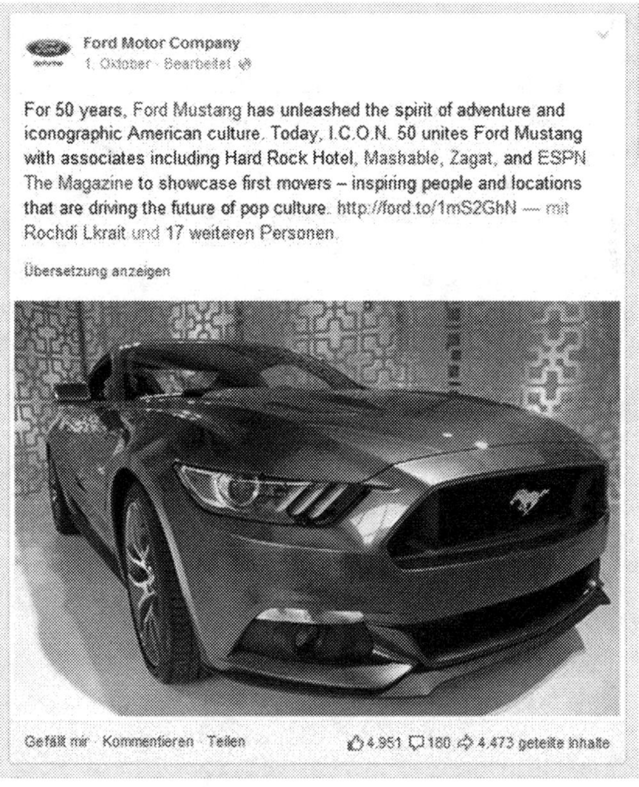

Fig. 2. Ford Mustang's 50th anniversary (Source: Ford Motor Company, 2014 [32])

Second, it is possible to integrate a direct link to another brand or celebrity fan page or website into the post. For example, Nissan posted a press review of the Nissan GT-R and integrated a link to the New York Daily News fan page that has more than 2.1 million Facebook followers. Ford linked press websites like Mashable (www.mashable.com) or ESPN Magazine (www.espn.com/magazine) with their own posts (see Fig. 2). Honda included a link to the fan page of the entertainer and celebrity Nick Cannon who has more than 2.7 million followers on Facebook. Toyota also tried to harness the network of the celebrity Oprah Winfrey in a social media post by advertising and pointing out that the company was a financial supporter of the Life-You-Want-Campaign. As a result, the main effect of the co-branding pattern is to maximize the potential audience for a post.

3.2 Pattern 2: Wow Effect

The Wow Effect is the second identified pattern derived from our SSP analysis. Especially in the social media sphere, users are confronted with a large amount of information. However, consumers are not able to process all available information [38]. Against this background, social media managers face intense competition for the limited attention of social media users. This leads to a situation where brands try to generate a high level of user engagement through differentiating themselves from the mass of other posts. In our sample, it is clearly noticeable that SSP often create a moment of surprise and astonishment for the users (the Wow Effect) that leads to greatly enhanced user interaction. Studies about the effect of surprise state that attention is predominantly drawn towards completely new and unseen content or content that is in some way special and does not belong to the particular situation [39].

In order to achieve this Wow Effect, brands in our sample followed two approaches. First, in order to create a post that contains something that deviates from a standard solution, SSP strongly emphasize aspects such as luxury, high-performance, futuristic design, or a product's innovative technology features. Mercedes-Benz, for instance, posted videos of a highly sophisticated future truck (see Fig. 3) or of future LED headlamps. Nissan focused on high-performance features by posting pictures of their sports car, the Nissan GT-R, and the potential successor, the Nissan Concept 2020 Vision GT. Honda published a post about the first ever built-in car vacuum cleaner in the new Honda Odyssey to capture the attention of their followers. Second, building on these emphasized features, user engagement increases even more if the post unveils a completely new product that has never been shown before, thus creating some sort of exclusivity. Mercedes-Benz, for example, published the first pictures of their new Mercedes-AMG GT. The post contained a strong focus on high-performance and luxury and additionally included on the SNS the first ever published picture of this new car.

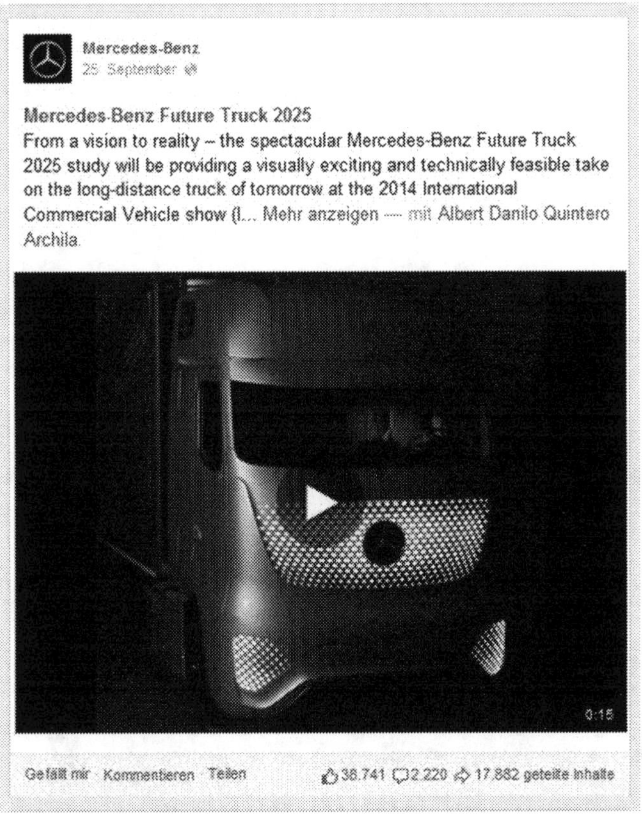

Fig. 3. Mercedes-Benz Future Truck 2025 (Source: Mercedes Benz, 2014 [40])

3.3 Pattern 3: Cognitive Task

The third pattern of SSP is related to the cognitive effort when working on a task. Posts that draw on that pattern raise a question, demanding a direct user response and, thus, provoke user engagement, mostly in form of comments. Brand posts by Ford, for example, asked the SNS user about their favorite Ford color and about the first Ford model they drove.

Another driver of user engagement in this pattern might be the fact that posts containing a cognitive task stimulate curiosity. According to human epistemology theory, curious, strange, surprising, or puzzling information arouses curiosity. This triggers the user to spend more time on the puzzling information in order to create meaning from it or to relate unknown information to something familiar [41]. Ford, for instance, published a post showing a photo of the parking area in a Ford manufacturing site in 1965 with many different Ford cars. The SNS user was requested to name the different Ford models portrayed in the picture, leading to a variety of user comments. Another example of a Ford post shows a macro photograph (extreme close-up shot) of a car part, asking the user to post the correct name of the part. Similarly, Volkswagen posted a picture of a Volkswagen Golf GTI filled with soccer balls and asked the users to guess how many soccer balls can fit into the car to encourage users to post their guesses in the comments (see Fig. 4). The best performing post of this category, and a prime example of the cognitive task pattern, was published by Toyota. The post shows a picture puzzle which had to be solved. We elaborate on this example later. In sum, the cognitive task pattern increases user engagement by direct calls for interaction with the post and it stimulates the inherent curiosity of SNS users.

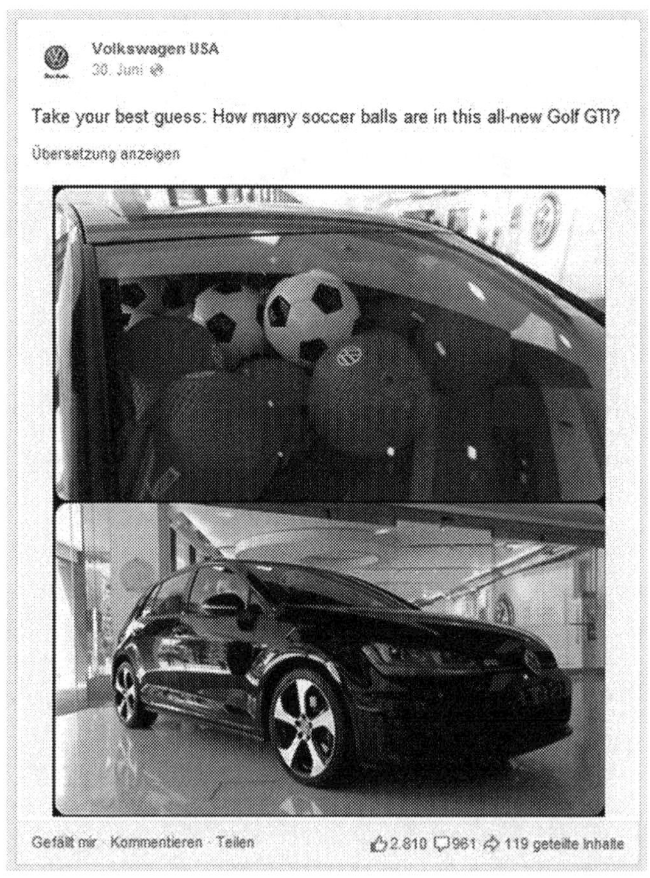

Fig. 4. Volkswagen soccer ball challenge (Source: Volkswagen, 2014 [42])

3.4 Pattern 4: Timing

Our results reveal that many SSP have in common the fact that they come with intelligent publication timing. These posts take advantage of certain events that already attract a lot of attention. Agenda-setting theory supports this notion by indicating that an audience perceives issues as more important when more emphasis is placed on them by mass media [43]. The more the media cover an event, the more attention it gets and, therefore, the more user engagement is generated on SNS. This means that publishing a post related to a specific event at a time when the attention for this event peaks, helps other brands to reach out to new audiences.

Two types of events could be identified that especially drove user engagement. First, non-brand-related events such as the FIFA World Cup, Mother's Day, Father's Day, or Independence Day are used in this context. Of course, each event can become a brand-related event through sponsorship, but here we mean events that are initially not directly related to a brand. For example, Volkswagen posted a video showing VW Golfs playing soccer in the national colors of Germany and Argentina, celebrating the win of the FIFA World Cup of the German national team in 2014. The video was posted right after the end of the match. This is an interesting approach intended to benefit from the attention of that event without being an official sponsor and without having paid for media coverage. Another post by Volkswagen celebrates Independence Day in the US by publishing a photo of the VW Beetle in a scene with fireworks and the text: 'Independence is something we love to celebrate'. This post attracts the attention of people celebrating Independence Day even though they might not be directly interested in the Volkswagen brand. Ford marked Mother's Day and Father's Day by posting emotional stories and by linking them to automobiles. On Father's Day, for example, they published a video telling the story of a father trying to comfort his daughter because she had failed the driver's test the second time (see Fig. 5). Second, brands also align their posts with brand-related events, such as anniversaries. Ford, for example, published posts to celebrate the anniversaries of the first ever Ford Model A or Ford Mustang sold

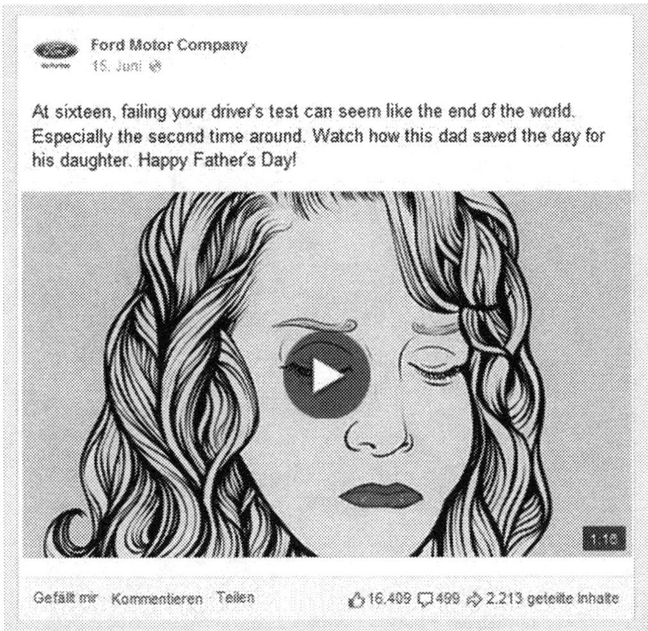

Fig. 5. Ford celebrating Father's Day (Source: Ford Motor Company, 2014 [44])

(see also Fig. 2). Honda posted content about sports events such as Honda winning the famous Indianapolis 500 race. Furthermore, Volkswagen announced their participation in the “Global Rallycross” race in Las Vegas. All in all, the timing pattern shows that marketing managers can increase user engagement by intelligently choosing the timing of their SNS posts.

3.5 Pattern 5: Campaign

The last identified pattern is based on the fact that brands implement their social media posts in a larger cross-media marketing campaign. In this case, a SNS post represents a smaller element of a much larger campaign, which often includes television, newspaper, radio, internet or other advertising activities. The “mere-exposure” effect can provide one explanation regarding the positive impact of this approach. Repeated exposure of a communication message to an audience induces a more positive attitude towards a message [45]. This positive attitude might lead to a higher likelihood of user engagement with the initial social media post. The fact that a user has already seen the campaign on other media channels can lead to a higher likelihood of engagement with the social media post.

Fig. 6. BMW i8 post as part of product launch campaign (Source: BMW, 2014 [48])

For instance, BMW's SSP concerning the new BMW i8 model (see Fig. 6) was part of a much broader cross-media product launch campaign. In addition to their social media activities, the campaign included seven different print advertisements and three TV spots that were shown worldwide on TV, in cinemas, and on their corporate website [46]. Similarly, for the launch of the 2013 Honda Civic, the brand created a television advertisement, and, in addition, they focused on a digital advertising campaign, including two Facebook posts, which turned out to be highly successful [47].

4 Four SSP Examples and Underlying Mechanisms

Our findings show that creating an SSP requires an appropriate understanding of the media ecosystem in which the post is situated. In order to understand the underlying mechanisms better, we elaborate further on four SSPs in the following section to provide a more nuanced picture of the corresponding influencing factors. Again, great emphasis was placed on a holistic approach, taking into consideration not only the post itself, but also its online and offline context.

4.1 Introduction of the new BMW i8

The first example and best-performing post in our sample in terms of the number of likes is concerned with the introduction of the new BMW i8 hybrid car in 2014 (see Fig. 7).

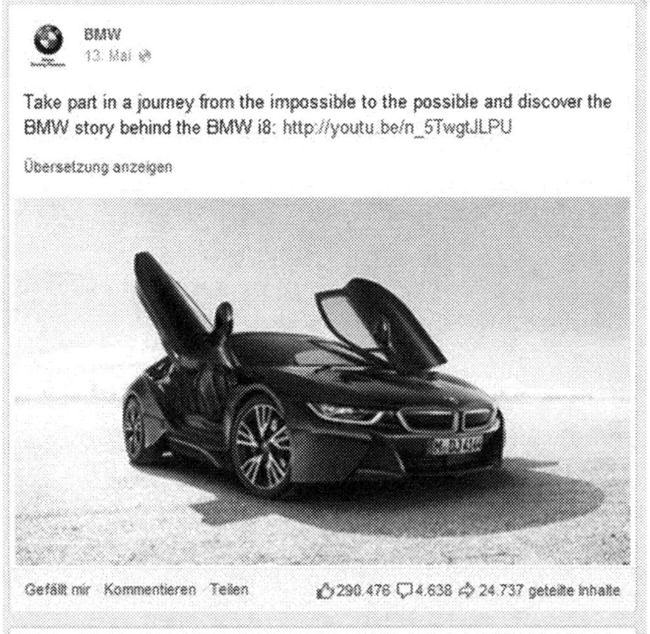

Fig. 7. Introduction of the new BMW i8 (Source: BMW, 2014 [49])

The post includes a picture of the new car and the claim: “Take part in a journey from the impossible to the possible”. Additionally, a link to a video about the story of the car is provided.

First of all, and as mentioned above, the social media post was part of a much larger cross-media product launch campaign. The marketing campaign started in May 9, 2014, and the post was published four days later, on May 13, 2014. The first German TV spot aired on May 10, 2014 and was repeated worldwide over several weeks [46]. During this time, the audience was able to learn about the new car and to discover the car’s features and technology aspects. Thereby, BMW was able to increase the awareness and attention for their new car. The campaign also included online advertising on various influential business, automotive, and lifestyle websites. Advertisements in print magazines and national daily newspapers had already started earlier in late April, accompanied by billboard advertising, projections on buildings, and video installations in selected areas. The increased exposure of the BMW i8 created not only familiarity and attention towards the product, but also a more positive attitude towards the product. A possible explanation for this might again be the already mentioned “mere-exposure” effect [45]. Because the BMW i8 was brand new at that time and because of the emphasis on high-performance and the futuristic features of the car, the post also corresponded to the previously mentioned Wow Effect pattern. The high degree of novelty of the car attracted additional attention on the social media post. People who got in contact with the BMW social media post were thus more likely to engage because they either had already seen content from the cross-media campaign or were intrigued by this so far unseen new product.

4.2 Honda’s “Best Yourself” Campaign

The SSP that received the second most likes in our sample is a post from the automobile manufacturer Honda (see Fig. 8). The post contains the slogan “Challenge yourself – push yourself – best yourself”. At first glance, the post and content seem to be very ordinary. Nevertheless, it received a high level of user engagement. Honda successfully set up a highly emotional cross-media campaign for the launch of the new product, including the social media post. This approach belongs to the above mentioned campaign pattern. Honda’s campaign called “Best Yourself” was generally about one’s motivation to achieve success through non-traditional paths. By celebrating individual achievements towards personal growth, the campaign perfectly matched the mindset of millennials. The desire to belong to something special helped to foster user engagement because people were more easily immersed in the campaign. The campaign further comprised a TV commercial placed in the commercial break of the popular TV show “America’s Got Talent”, which fits with the idea of reaching achievements through continuous personal improvement.

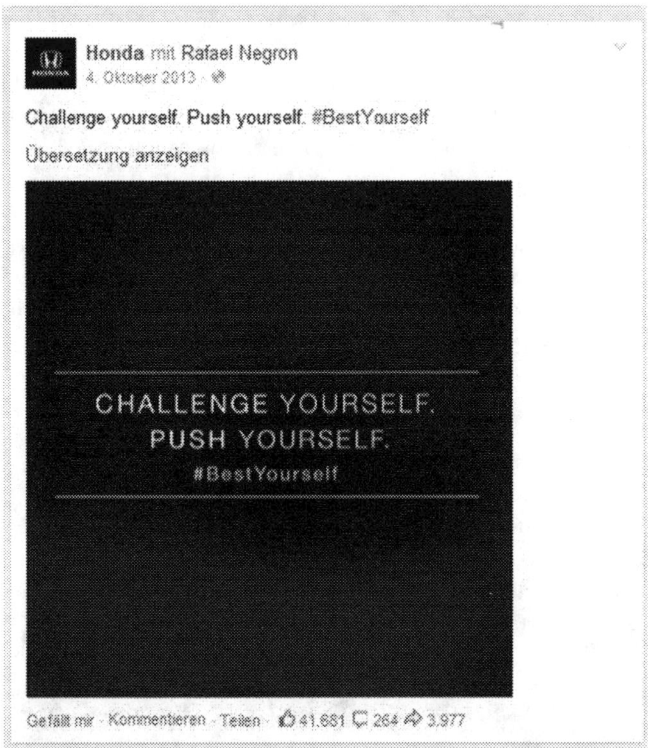

Fig. 8. Honda's "Best Yourself" campaign (Source: Honda, 2013 [50])

Overall, Honda utilized various media channels for their campaign, such as television and different social media channels (e.g., Facebook, Twitter, Instagram, and YouTube), and by integrating various media partners such as the HuffPost, Complex Media Network, and Ballislife.com. To further drive user engagement, Honda collaborated with Nick Cannon, a popular celebrity with millennials. In line with the co-branding pattern presented, integrating a celebrity helps to increase the potential audience and helps to leverage the large fan base of the celebrity. All in all, the post succeeded in communicating a strong and emotional message and successfully targeted the right audience through the right channels while utilizing the co-branding and campaign pattern.

4.3 Volkswagen's Post About the FIFA World Cup 2014

Volkswagen published by far the most shared post (see Fig. 9). The Volkswagen post shows a video of a soccer match between Volkswagen cars in the German national colors and cars in the Argentinian national colors. When the German team scores a goal, the text of the post states: "Now that's the power of German engineering. Congrats to #Germany, the new champions!" Additionally, the post provided a link to a longer version of the video on YouTube. In this post, the timing was crucial in order to make the post extremely successful. It was posted directly after the World Cup win of the German national soccer team on July 13th, 2014. In line with the previously mentioned timing pattern, this post tries to leverage the attention to this international sports event for the Volkswagen brand. Roughly three billion people worldwide watched at least one minute of one of the matches on TV, whereas one billion people watched the final game [51]. With this post, Volkswagen congratulated Germany for winning the soccer world championship. Volkswagen expected that users might share the post to celebrate the win with their team. Moreover, Volkswagen as a German brand, tried to link the success of the German national soccer team with their own brand in a creative way. All in all, this post is a perfect example of how marketing managers can use media coverage of specific events for their own purpose.

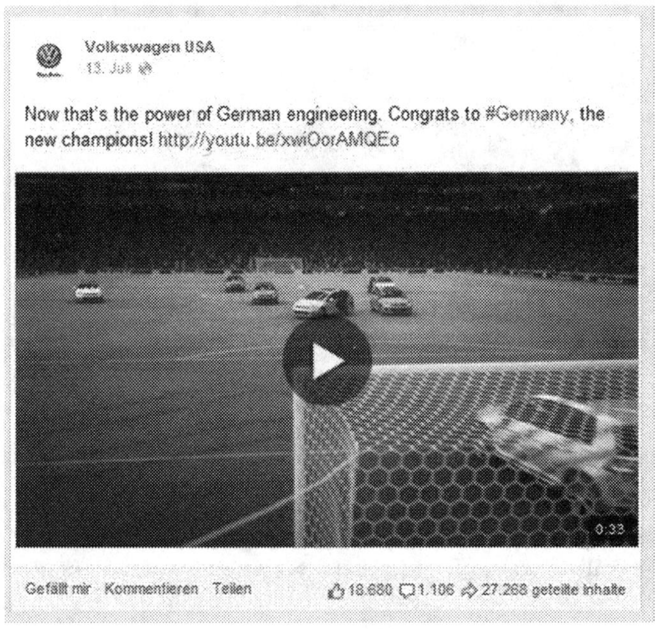

Fig. 9. VW's post about the FIFA World Cup 2014 (Source: Volkswagen, 2014 [52])

4.4 Toyota's Picture Puzzle

Toyota published the most successful post regarding the number of comments (see Fig. 10). The post showed a riddle with different sketches combined with letters and words that have to be decoded to reveal a piece of safe driving advice. In addition, the post made a clear call for action: "Put your detective skills to the test. Decode and find more safe driving tips at [link]". The reason for the high number of comments seems intuitive because answers to the quiz could only be entered in the comment below the post. The fact that a direct call for action leads to a higher level of user engagement is also supported by existing research [53]. Moreover, the post contains puzzling information that refers to the abovementioned cognitive task pattern. This makes users curious about the solution of the quiz. This curiosity may drive user engagement even more. Moreover, due to the nontrivial challenge, users might have different solutions, which led to a discussion, resulting in even more comments. Previous research shows that high variance in comments provokes other users to comment as well [53–55].

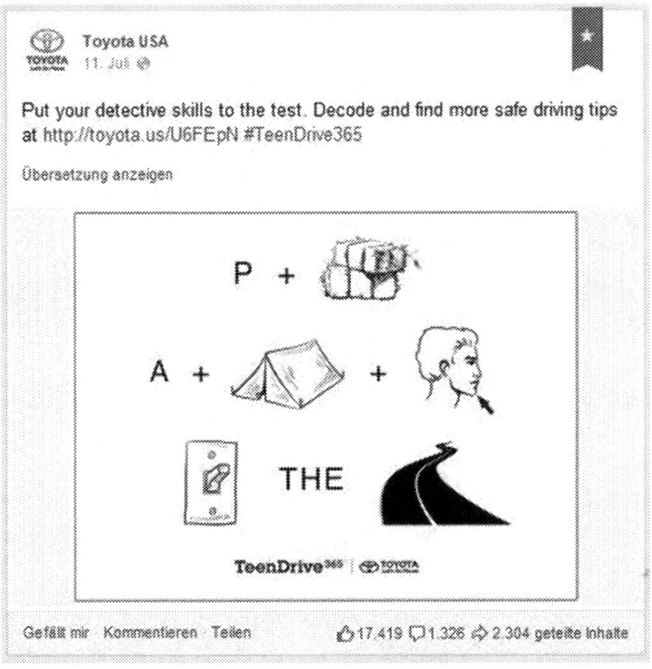

Fig. 10. Toyota's picture puzzle (Source: Toyota, 2014 [56])

5 Implications

Due to the growing importance of social media, and especially SNS, for marketing communication, scholars have put increasing emphasis on the antecedents of communication effectiveness. In this context, this article follows a unique research approach. Instead of exploring the antecedents of the success of a post in a large-scale sample, we focused particularly on SSP that have outperformed the majority of other posts, aiming to learn from these few success stories about the underlying mechanisms.

Our research offers several contributions. First of all, the findings are supposed to encourage practitioners to apply a broader view to social media posts. The results suggest looking beyond the typical features of a social media post and rather placing it in a broader media ecosystem or in much larger brand campaigns. Second, we revealed specific communication patterns that are regularly found in SSP. These patterns can help marketing managers to enhance their chances to replicate this success when planning future social media campaigns. Thus, the five derived patterns can inspire marketing managers in their daily work of designing social media posts. Third, this study adds to the body of literature on social media research through discussing various factors that make social media communication exceedingly successful regarding likes, shares, and comments. Our research approach also offers a new lens to look through when analyzing social media messages. Thereby, we hope to encourage future studies in this field.

The limitations of this article offer further research directions. First, the findings of this study are derived from case evidence from one specific industry. Thus, the results cannot simply be generalized across other industries. Further research could examine the applicability of our patterns in different industries. Second, this study focuses on the description of recurring patterns of SSP. Although we have relied on an empirical approach to generate our sample of SSP, the derived success patterns have to be validated further empirically. Future research could try to analyze our patterns in a broader sample or could explore underlying mechanisms in more detail to foster a sound understanding of the mode of action for each pattern.

References

1. Kietzmann, J.H., Hermkens, K., McCarthy, I.P., Silvestre, B.S.: Social media? Get serious! Understanding the functional building blocks of social media. Bus. Horiz. **54**, 241–251 (2011)
2. Mangold, W.G., Faulds, D.J.: Social media: the new hybrid element of the promotion mix. Bus. Horiz. **52**, 357–365 (2009)
3. SAS HBR. The New Conversation: Taking Social Media from Talk to Action. Harvard Business Review. Analytic Services, pp. 1–24 (2010)
4. Burton, S., Soboleva, A.: Interactive or reactive? Marketing with Twitter. J. Consum. Mark. **28**, 491–499 (2011)
5. Lin, K.Y., Lu, H.P.: Intention to continue using Facebook fan pages from the perspective of social capital theory. Cyberpsychology Behav. Soc. Netw. **14**, 565–570 (2011)
6. Lipsman, A., Mud, G., Rich, M., Bruich, S.: The power of "Like": how brands reach (and influence) fans through social-media marketing. J. Advert. Res. **52**, 40–52 (2012)
7. McCorkindale, T.: Can you see the writing on my wall? A content analysis of the Fortune 50's Facebook social networking sites. Public Relat. J. **4**, 1–14 (2010)
8. Trefzger, T.F., Dünfelder, D.: Unleash your brand! Using social media as a marketing tool in academia. In: Proceedings of the International Conference on Social Computing and Social Media, pp. 449–460 (2016)
9. Wagner, T.F.: Promoting technological innovations: towards an integration of traditional and social media communication channels. In: Meiselwitz, G. (ed.) SCSM 2017, Part I. LNCS, vol. 10282, pp. 256–273. Springer, Cham (2017). https://doi.org/10.1007/978-3-319-58559-8_22
10. Araujo, T., Neijens, P., Vliegenthart, R.: What motivates consumers to re-tweet brand content?: The impact of information, emotion, and traceability on pass-along behavior. J. Advert. Res. **55**(3), 284–295 (2015)
11. Ashley, C., Tuten, T.: Creative strategies in social media marketing: an exploratory study of branded social content and consumer engagement. Psychol. Mark. **32**, 15–27 (2015)
12. McAlexander, J.H., Schouten, J.W., Koenig, H.F.: Building brand community. J. Mark. **66**, 38–54 (2002)
13. Muniz, A.M., O'Guinn, T.C.: Brand community. J. Consum. Res. **27**, 412–432 (2001)
14. Kabadayi, S., Price, K.: Consumer–brand engagement on Facebook: liking and commenting behaviors. J. Res. Interact. Mark. **8**, 203–223 (2014)

15. Wallace, E., Buil, I., de Chernatony, L.: Consumer engagement with self-expressive brands: brand love and WOM outcomes. J. Prod. Brand Manag. **23**(1), 33–42 (2014)
16. Dholakia, U.M., Durham, E.: One café chain's Facebook experiment. Harv. Bus. Rev. **88**, 26 (2010)
17. Bagozzi, R.P., Dholakia, U.M.: Antecedents and purchase consequences of customer participation in small group brand communities. Int. J. Res. Mark. **23**, 45–61 (2006)
18. Williamson, D.A.: Worldwide Social Network Ad Spending 2011: A Rising Tide, eMarketer.com. http://www.emarketer.com/Report.aspx?code=emarketer_2000692. Accessed 5 Oct 2017
19. Hutton, G., Fosdick, M.: The globalization of social media: consumer relationships with brands evolve in the digital space. J. Advert. Res. **51**(4), 564–570 (2011)
20. Corstjens, M., Umblijs, A.: The power of evil: the damage of negative social media strongly outweigh positive contributions. J. Advert. Res. **52**, 433–450 (2012)
21. Naylor, R.W., Lamberton, C.P., West, P.M.: Beyond the "Like" button: the impact of mere virtual presence on brand evaluations and purchase intentions in social media settings. J. Mark. **76**, 105–120 (2012)
22. Wagner, T.F., Baccarella, C.V., Voigt, K.I.: Framing social media communication: investigating the effects of brand post appeals on user interaction. Eur. Manag. J. **35**, 606–616 (2017)
23. Sabate, F., Berbegal-Mirabent, J., Cañabate, A., Lebherz, P.R.: Factors influencing popularity of branded content in Facebook fan pages. Eur. Manag. J. **32**(6), 1001–1011 (2014)
24. Trefzger, T.F., Baccarella, C.V., Voigt, K.I.: Antecedents of brand post popularity in Facebook: the influence of images, videos, and text. In: Proceedings of the 15th International Marketing Trends Conference, pp. 1–8 (2016)
25. Trefzger, T.F., Baccarella, C.V., Scheiner, C.W., Voigt, K.-I.: Hold the line! The challenge of being a premium brand in the social media era. In: Meiselwitz, G. (ed.) SCSM 2016. LNCS, vol. 9742, pp. 461–471. Springer, Cham (2016). https://doi.org/10.1007/978-3-319-39910-2_43
26. Hettler, U.: Social media marketing: marketing mit Blogs, sozialen Netzwerken und weiteren Anwendungen des Web 2.0. Oldenbourg, München (2010)
27. Statista. Number of daily active Facebook users worldwide as of 3rd quarter 2017 (in millions). https://www.statista.com/statistics/346167/facebook-global-dau/. Accessed 5 Oct 2017
28. Baccarella, C.V., Scheiner, C.W., Trefzger, T.F., Voigt, K.-I.: Communicating high-tech products – a comparison between print advertisements of automotive premium and standard brands. Int. J. Technol. Mark. **11**, 24–38 (2016)
29. Millward Brown. Top 100 Most Valuable Global Brands 2014. https://www.millwardbrown.com/brandz/2014/Top100/Docs/2014_BrandZ_Top100_Chart.pdf. Accessed 5 Oct 2017
30. Baxter, P., Jack, S.: Qualitative case study methodology: study design and implementation for novice researchers. Qual. Rep. **13**, 544–559 (2008)
31. Washburn, J.H., Till, B.D., Priluck, R.: Co-branding: brand equity and trial effects. J. Consum. Mark. **17**, 591–604 (2000)
32. Ford Motor Company: For 50 years, Ford Mustang has unleashed the spirit of adventure and iconographic American culture. Today, I.C.O.N. 50 unites Ford Mustang with associates including Hard Rock Hotel, Mashable, Zagat, and ESPN The Magazine to showcase first movers – inspiring people and locations that are driving the future of pop culture, 1 October 2014. http://ford.to/1mS2GhN [Facebook post]. https://www.facebook.com/search/str/for+50+years+ford+mustang+has+unleash+the+spirit+of+adventure+and/keywords_search
33. Grossman, R.P.: Co-branding in advertising: developing effective associations. J. Prod. Brand Manag. **6**(3), 191–201 (1997)
34. Hillyer, C., Tikoo, S.: Effect of cobranding on consumer product evaluations. Adv. Consum. Res. **22**, 123–127 (1995)

35. Krishnan, H.S.: Characteristics of memory associations: a consumer-based brand equity perspective. Int. J. Res. Mark. **13**, 389–405 (1996)
36. Rao, A.R., Ruekert, R.W.: Brand alliances as signals of product quality. Sloan Manag. Rev. **36**, 87–88 (1994)
37. Smith, A.N., Fischer, E., Yongjian, C.: How does brand-related user-generated content differ across YouTube, Facebook, and Twitter? J. Interact. Mark. **26**, 102–113 (2012)
38. Moore, D.J., Harris, W.D.: Affect intensity and the consumer's attitude toward high impact emotional advertising appeals. J. Advert. **25**, 37–50 (1996)
39. Itti, L., Baldi, P.: Bayesian surprise attracts human attention. Vis. Res. **49**, 1295–1306 (2009)
40. Mercedes-Benz: From a vision to reality – the spectacular Mercedes-Benz Future Truck 2025 study will be providing a visually exciting and technically feasible take on the long-distance truck of tomorrow at the 2014 International Commercial Vehicle show (IAA). [Facebook post], 25 September 2014. https://www.facebook.com/search/str/From+a+vision+to+reality+-+the+spectactular+Mercedes-Benz+Future+Truck/keywords_search
41. Berlyne, D.E.: A theory of human curiosity. Br. J. Psychol. Gen. Sect. **45**, 189–191 (1954)
42. Volkswagen: Take your best guess: How many soccer balls are in this all-new Golf GTI? [Facebook post], 30 June 2014. https://www.facebook.com/search/str/Take+your+best+guess%3A+How+many+soccer+balls+are+in+this+all-new+Golf+GTi%3F/keywords_search
43. McCombs, M.E., Shaw, D.L.: The agenda-setting function of mass media. Public Opin. Q. **36**, 176–187 (1972)
44. Ford: At sixteen, failing your driver's test can seem like the end of the world. Especially the second time around. [Facebook post], 15 June 2014. https://www.facebook.com/ford/videos/10152462924095049/
45. Zajonc, R.B.: Attitudinal effects of "mere exposure". J. Pers. Soc. Psychol. **9**, 1–27 (1968)
46. BMW: Global launch campaign for BMW i8. https://www.press.bmwgroup.com/global/article/detail/T0179914EN/. Accessed 5 Oct 2017
47. Honda: Honda Taps into Millennial Mindset with 'Best Yourself' Civic Campaign Featuring Nick Cannon. http://news.honda.com/newsandviews/article.aspx?id=7313-en. Accessed 5 Oct 2017
48. BMW: A sports car from the outside. A pioneer from the inside. More dynamic. More efficient: the BMW i8, 9 June 2014. http://youtu.be/1mQhRSH5d9I [Facebook post]. https://www.facebook.com/BMW/photos/a.352379437268.193008.22893372268/10152512276447269/?type=3&theater
49. BMW: Take part in a journey from the impossible to the possible and discover the BMW story behind the BMW i8, 13 May 2014. http://youtu.be/n_5TwgtJLPU [Facebook post]. https://www.facebook.com/BMW/photos/a.352379437268.193008.22893372268/10152458489642269/?type=3&theater
50. Honda: Challenge yourself. Push yourself. #BestYourself [Facebook post], 4 October 2013. https://www.facebook.com/search/str/Challenge+yourself.+Push+yourself.+%23BestYourself/keywords_search
51. FIFA: FIFA Fussball-WM 2014™: 3,2 Milliarden Zuschauer, 1 Milliarde beim Finale. http://de.fifa.com/worldcup/news/y=2015/m=12/news=fifa-fussball-wm-2014tm-3-2-milliarden-zuschauer-1-milliarde-beim-fina-2745551.html. Accessed 5 Oct 2017
52. Volkswagen: Now that's the power of German engineering. Congrats to #Germany, the new champions! 13 July 2014. http://youtu.be/xwiOorAMQEo [Facebook post]. https://www.facebook.com/search/top/?q=now%20that%27s%20the%20power%20of%20german%20engineering.%20Congrats
53. de Vries, L., Gensler, S., Leeflang, P.S.: Popularity of brand posts on brand fan pages: an investigation of the effects of social media marketing. J. Interact. Mark. **26**, 83–91 (2012)

54. Moe, W.W., Trusov, M.: The value of social dynamics in online product ratings forums. J. Mark. Res. **48**(3), 444–456 (2011)
55. Schlosser, A.E.: Posting versus lurking: communicating in a multiple audience context. J. Consum. Res. **32**, 260–265 (2005)
56. Toyota: Put your detective skills to the test. Decode and find more safe driving tips, 11 July 2014. http://toyota.us/U6FEpN #TeenDrive365 [Facebook post]. https://www.facebook.com/search/str/Decode+and+find+more+safe+driving+tips+at/keywords_search

Dreaming of Stardom and Money: Micro-celebrities and Influencers on Live Streaming Services

Kaja J. Fietkiewicz(✉), Isabelle Dorsch, Katrin Scheibe, Franziska Zimmer, and Wolfgang G. Stock

Department of Information Science, Heinrich Heine University Düsseldorf, Düsseldorf, Germany
{kaja.fietkiewicz, isabelle.dorsch, katrin.scheibe, franziska.zimmer}@hhu.de, stock@phil.hhu.de

Abstract. Social live streaming services (SLSSs) are social media, which combine Live-TV with elements of Social Networking Services (SNSs). In social media and thus also in SLSSs, the so-called influencer and micro-celebrities play an important role, but to what extend are SLSSs' streamers motivated by fame or financial gain? We conducted a content analysis in order to investigate SLSSs' streamers (n = 7,667) on Periscope, Ustream and YouNow in respect to their general characteristics and streaming motivation being fame and financial gain. We have developed a research model referring to the platform used by the streamers, their gender, origin, age and streamed content (general characteristics), as well as the motivational aspects. Streamers of Ustream are mostly motivated by financial gain, whereas YouNow broadcasters seek to be famous. Considering the streamers age, older generations (Gen X, Silver Surfers) aspire after financial gain. With progressing age the motivation to become a star decreases. Mostly streamed content by streamers motivated by money is entertainment media. For streamers wanting to become a star chatting and making music are the preferred content categories.

Keywords: Social live streaming services · Micro-celebrities
Social media influencers

1 Introduction

Since the turn of the Millennium and increasing usage of the Internet and its applications, research on people becoming "celebrities" or "micro-celebrities" thanks to the new technology is gaining on popularity and importance [e.g., 17, 18, 30, 42]. Now, ordinary social media users can become important players of the so-called attention economy [31, 48] with the help of self-branding and presentation strategies [e.g., 23, 30, 37, 43]. We can find micro-celebrities and so-called social media influencers on YouTube, Instagram, or Snapchat. However, do users of a new kind of platforms like the social live streaming services also aspire "stardom and money"? This is an explorative study addressing this particular topic. First, we will shed light on the new form of social media – the social live streaming services as well as on the concepts of

© Springer International Publishing AG, part of Springer Nature 2018
G. Meiselwitz (Ed.): SCSM 2018, LNCS 10913, pp. 240–253, 2018.
https://doi.org/10.1007/978-3-319-91521-0_18

"micro-celebrity" and "influencer". Afterwards, we will elaborate on our applied methods and present results of our investigation based on observations of streamers on three different platforms. Finally, we will answer the question whether social live streaming users are indeed interested in fame and money.

1.1 Social Live Streaming Services

In recent years a new form of social media has established itself, the so-called Social Live Streaming Services (SLSSs). They combine Live-TV with elements of Social Networking Services (SNSs) as they include a backchannel between the viewers and the streamers as well as among the viewers. We can find such SLSSs as Periscope, Ustream, YouNow, YouTube Live, Facebook Live, Instagram Live, Snapchat Live Stories, niconico (in Japan), YiZhiBo, Xiandanjia, Yingke (all in China) or – for broadcasting e-sports or drawing – Twitch and Picarto, respectively. Such services allow their users to broadcast live anything they want and to everyone who is interested to watch.

The scientific research on SLSSs is gaining in importance as well as spectrum. In computer science, one can find studies on bandwidth [3], video quality [45] and the delay of comments' displays [39]. SLSSs find application in private contexts [41], but also in more serious environments, e.g. in teaching neurosurgery [35] or economics [6]. They can also be applied in marketing [22]. Furthermore, SLSSs are applied for live broadcasting sports events, however, this is also connected to some legal problems [1]. Despite broadcasting sports events, also other general legal and ethical implications may arise [2, 7, 15, 52]. There are studies on topic-specific SLSSs, e.g. in e-sports context on Twitch [e.g., 4, 12], and on general SLSSs (without any thematic limitation) [9–11, 41, 44, 46]. Studies found that general live streaming was appreciated for its authentic, uncurated, and interactive attributes [47] as well as for its role for sharing breaking news [46]. However, we miss studies, which systematically investigate the motivation of streamers to become micro-celebrities or influencers on the general SLSSs. We aim to close this research gap with the following investigation.

1.2 Micro-celebrities and Influencers on Social Media

Media like television have been instrumental in generating new "celebrities" parallel to the "film-celebrities", who enjoy slightly more popularity [21]. With time and creation of new TV genres, a new kind of celebrity like "reality TV stars" attracted attention of the crowds [21]. With increasing popularity of social media further types of celebrities emerged, for example, YouTube stars [49] or bloggers, usually reporting on both channels [21, 30]. This trending interest in uncensored (private) life of others and "Big Brother"-like shows is not unproblematic and became topic for many critical discourses, an example being the American movie "The Truman Show" [9].

Still, media change together with the concept of celebrity—from celebrity focused solely on mass and broadcast media, to the one active on a diversified media landscape, and then further to participatory media [20, 32]. More interestingly, this development enables not only famous people (from TV or films), but also non-famous people "to generate vast quantities of personal media, manipulate and distribute this content widely, and reach out to (real or imagines) audiences" [32]. Hence, increasingly "ordinary people" are being transformed into celebrities [20], or rather, thanks to social media and self-branding, they transform themselves into ones.

Marwick [32] points out two major changes in celebrity culture due to the shift towards participatory media. First, the "traditional" celebrities are using "social media to create direct, unmediated relationship with fans, or at least the illusion of such" [32]. This illusion of a real face-to-face friendships with celebrities created through watching TV shows or listening to music is the so-called "para-social interaction" [16, 32, 33], however, with use of social media this interaction can become more "social" and "increase the emotional ties between celebrity and fan" [32, 34, 36]. The second change is related to the phenomenon of "micro-celebrity", a form of celebrity that may have a small audience, but is still "able to inhabit the celebrity subject position through the use of technologies" [32]. As opposed to the "broadcast era" where "celebrity was something a person was; in the Internet era, micro-celebrity is something people do" [32]. The phenomenon of micro-celebrity is strongly linked to the notions of self-branding and strategic self-presentation, and requires "viewing oneself as a consumer product", and "image" that needs to be sold to the right target group [13, 27, 32]. Micro-celebrities view friends and followers on social media channels as their fanbase that needs to be managed by various affiliative techniques [34]. These trends have empowered many participants in the newly emerging "online reputation economy, where the reputation generated by social media platforms functions as a new form of currency, and more generally, value" [14, p. 203].

The emergence of online reputation economy has led to establishment of a new concept of the "micro-celebrity", namely the social media influencer (SMI). Such influencer "works to generate a form of 'celebrity' capital by cultivating as much attention as possible and crafting an authentic 'personal brand' via social networks, which can subsequently be used by companies and advertisers for consumer outreach" [14]. Businesses increasingly rely on social media influencers, on one hand "due to the sheer volume of advertising online, which drives down actual click-through rates and individual engagement levels", on the other hand, due to higher authenticity of claims made by "personal acquaintance" rather than by a rich celebrity [14, 29, 40]. Marketing strategists are looking for social media users with an extensive social network that is frequently used, as well as with "relevant or 'sticky' content about the product category, and whose personality 'resonates' with the tone and feel of the brand" [14]. This way ordinary social media users become social media influencers making money and their living by posting pictures, videos and blog posts—all the activities that other (non-influential) social media users do, but apparently not as good as the influencers.

Micro-celebrities and influencers will make money by advertising products or services. This also applies to social live streaming services. In addition to being paid by third parties, some of the services offer possibilities to make money by using the SLSS (of course, provided that the streamer attracts a considerable amount of viewers). Services like Facebook Live or Periscope allow pre-roll and mid-roll advertising as well as displaying overlay ads. Some of the gaming channels on YouTube also have access to sponsorships that are financed by the viewers who can purchase digital goods like badges and emojis and have access to "special perks" [51]. Very popular are also fan donations, for example, YouTube's Super Chat (viewers can get their chat message pinned to the top of the comments section by paying a small fee), or Bits on Twitch (viewers pay for affiliated streamers to receive a certain number of "Bits"). SLSSs as Twitch or Picarto offer monthly subscriptions [19]. On YouNow, streamers can earn money from tips and gifts. For this purpose, viewers can buy bars, with these they can buy gifts that they can give to a streamer who is a YouNow Partner (who in turn receives real money) [50].

To sum up, with new forms of media, new forms of celebrities and "influential" people emerged—the micro-celebrities and social media influencers. They earn money doing advertising for products and services (with product placement or reviews), or on some of the platforms, especially on social live streaming services, by subscriptions, donations and gifts from the viewers. They also gain recognition and approval of their fan-base, which for some of them is as attractive and important as financial gain for others. In this study we are going to investigate whether general SLSSs users indeed aspire to become micro-celebrities and/or to earn money with the help of these service. This is an explorative study that is supposed to shed light on the general characteristics of streamers dreaming of "stardom and money".

1.3 Research Questions and Research Model

In order to explore the general characteristics of SLSSs users (in particular, streamers or producers) motivated by fame or financial gain we formulated the following research questions:

- RQ1: Which channels (Periscope, Ustream, YouNow) are preferred by users motivated by fame or financial gain?
- RQ2: Are there gender-dependent differences regarding the streaming motivation being fame or financial gain?
- RQ3: Are there origin-dependent differences (Germany, Japan, USA) regarding the streaming motivation being fame or financial gain?
- RQ4: Are there age-dependent differences regarding the streaming motivation being fame or financial gain?
- RQ5: What are the contents streamed by streamers whose motivation is fame or financial gain?

According to our research model (Fig. 1), we focus on streamers that are either interested in financial gain or in becoming famous. These streamers will use a certain social live streaming platform. They will either stream by themselves (male or female streamer) or not (group of streamers). Furthermore, the streamers will have different origins (Germany, Japan or the USA). Moreover, there can be age-dependent differences between the streamers' motivations. And finally, they can stream different content types.

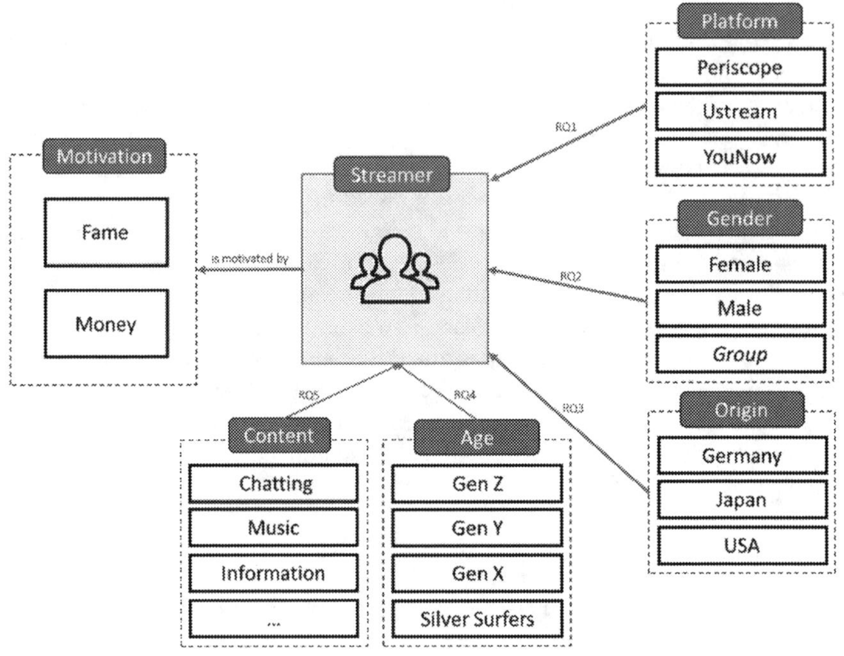

Fig. 1. Our research model.

2 Methods

2.1 Systematic Observation of Live Streams

In order to answer our research questions, we have conducted observations of the streams. We evaluated and compared SLSSs' users' streaming behavior as well as the content of a stream and motives of a streamer to produce a live stream [10, 52]. The empirical procedure of the content analysis included development of a codebook and a two-phased approach ensuring high reliability [24, 26, 38]. First, the directed approach was implemented with help of literature on social media, in order to get guidance for the research categories. Second, the conventional approach via observation of live streams was used to get a general idea of what people stream about. This way we were able to define the categories of content of a stream and motivation of the steamer.

The content categories include: to chat; to make music; to share information; news; fitness; sport event; gaming; animals; entertainment media; spirituality; draw/paint a picture; 24/7; science, technology, and medicine (STM); comedy; advertisement; nothing; slice of life; politics; nature; food; and business information. The motivation categories include: entertainment (boredom, fun, hobby); information (to reach a specific group, exchange of views), social interaction (socializing, loneliness, relationship management, need to communicate, need to belong), and self-presentation (self-improvement, self-expression, sense of mission, to become a celebrity, to make money, trolling). "No comment" was marked if the streamer did not state a motivation or no person could be reached via chat, for example if an animal was shown or a 24/7 stream (e.g. from a webcam) was broadcasted. However, for this investigation we focus on two subcategories of the self-presentation category, namely "to become a celebrity" and "to make money". Hence, for the investigation only observations were selected, where streamer confirmed to be motivated by one of these two factors.

Norm entries were used for the socio-demographic data like gender (male, female, group) and age of the streamer. The data about the streams from three general SLSSs (YouNow, Periscope, and Ustream) were collected from three different countries, namely Germany, Japan, and the USA. To ensure that the streams originated from those countries the declaration of the country for a broadcast on each platform was checked for every stream. Additionally, the data collectors had the required language skills for those countries. Twelve research teams (each consisting of two people) were evenly distributed between the three countries. Every coder received a spread sheet to code the observed data. Each stream was observed simultaneously but independently by two people for two to a maximum of ten minutes. Usually the streams were observed in two phases. First, the stream was watched and the data were collected. In phase two, if some aspects were not clear, for example the motivation of the streamer, the streamer was asked via the chat system of the service. In the end, a data set of 7,667 different streams in a time span of four weeks, from April 26 to May 24, 2016, was collected.

2.2 Data Preparation and Analysis

Our dataset consisted of mostly nominal data. There were three categories of the variable *platform* (YouNow, Periscope, Ustream) as well as three categories of the variable *origin* (Germany, Japan, USA). The variable *gender* was not binary-coded, but included categories male, female, group (for streams with more than one streamer, where specification of one gender was not possible), and n/a (not available, for streams where no streamer could be seen; these cases were subsequently defined as missing values). Finally, the *age* of the streamers was coded on a metric scale.

In order to investigate the possible influence of the age of the streamers on their motivation, we have aggregated the data into generational groups. For this purpose, we have followed the categorization applied in studies on generational cohorts of social media users [8, 28]. According to these studies, there is the Silent Generation (born between 1925 and 1945), the Baby Boomers (1946–1960), Generation X (1961–1980), Generation Y (1981–1998) and Generation Z (born after 1998). Due to low observation numbers of older steamers, we have merged the "Baby Boomers" and "Silent Generation" into one group called Silver Surfers (N = 33).

For the investigation we have applied descriptive statistics including frequencies and Pearson Chi-Square test for association, since almost all of our variables were nominal with more than 2 categories. The chi-square test determines whether there is an association between two nominal variables (in our case, association between the general characteristics and the motivation for using SLSSs being "fame" or "money"). Furthermore, we have measured the effect size using Cramer's V to investigate the strength of the respective association. The magnitude of effect size can be interpreted as small (0.1), medium or moderate (0.3) and large (0.5) [5, 25].

3 Results

In our study (observation of streams; N = 4,548 streams with single broadcasters; N = 1,082 of "groups"), we identified 61.2% male broadcasters and 38.8% females from Germany, Japan and the USA (Table 1). The results from Tang, Venial and Inkpen [47, p. 4774] confirm this gender distribution: about three fifths of SLSSs' users are male. The observed streams were almost evenly distributed among the three platforms, with the highest number of observations for Periscope (38.5%) and the lowest one for YouNow (26.4%). As for the distribution by the country of origin, the most streams were from the USA (41%) and the fewest from Japan (25%). Finally, we have aggregated the age of the streamers into generational cohorts. The most represented generation is the youngest one—Gen Z with 37.2% followed by Gen Y with 33.5%. The older generational groups, Generation X and Silver Surfers, are much smaller as they represent only 6.4% and 0.4% of the observed streamers, respectively. Since we could not estimate the age of all observed streamers or the ones streaming in groups, the number of observations within the *Generation* category is accordingly lower.

Table 1. Demographic data of observed streamers.

		Frequency	Valid percent
Platform [N = 7,667]	Periscope	2,960	38.6%
	Ustream	2,686	35.0%
	YouNow	2,021	26.4%
Gender [N = 5,630]	Female	1,766	31.4%
	Male	2,782	49.4%
	Group	1,082	19.2%
Origin [N = 7,667]	Germany	2,604	34.0%
	Japan	1,920	25.0%
	USA	3,134	41.0%
Generation [N = 4,937]	Gen Z	1,839	37.2%
	Gen Y	2,572	33.5%
	Gen X	493	6.4%
	Silver Surfers	33	0.4%

3.1 Platform-Dependent Differences

Platform-dependent differences regarding the motivational factor "making money" and "becoming a star" can be obtained from Table 2. With about 13%, the motivational factor money is highest for Ustream streamers, whereas becoming a star is of minor interest. Streamers of YouNow are mostly motivated by fame (9.65%). For Periscope streamers, neither factor plays a major role. A chi-square test for association was conducted for the platforms and the motivational factors. All expected cell frequencies were greater than five. There is a statistically significant association between the platforms and the motivational factors, however, the association is rather small (Cramer's V = 0.209 for money, and 0.179 for fame).

Table 2. Platform-dependent differences in motivation to *make money* and *become a star*.

Platform	Making money	Becoming a star
Periscope (N = 2.960)	1.32%	2.53%
Ustream (N = 2.686)	12.92%	1.15%
YouNow (N = 2.021)	4.60%	9.65%
Pearson Chi^2	$p < 0.001$	$p < 0.001$

3.2 Gender-Dependent Differences

Regarding the gender-dependent differences, males are slightly more motivated by financial gain than women (Table 3). Nevertheless, both motivational factors are highest for the group-streams. With regard to the factor fame, there are no major gender-specific differences ($p > 0.05$). The conducted chi-square test of independence between gender and the motivational factor money results in a very small (0.083) statistically significant association.

Table 3. Gender-dependent differences in motivation to *make money* and *become a star*.

Gender	Making money	Becoming a star
Female (N = 1,766)	2.49%	4.76%
Male (N = 2,782)	4.82%	5.14%
Group (N = 1,082)	7.49%	5.82%
Pearson Chi^2	$p < 0.001$	$p = 0.458$

3.3 Origin-Dependent Differences

Considering the streamers' origin and its influence on the motivations money and fame (Table 4), streamers located in USA are the ones most motivated by financial gain (7.45%), followed by Germans (5.91%) and Japanese (4.74%). In turn, fame is mostly aspired by German streamers (5.11%), followed by American (3.91%) and Japanese ones (2.34%). Even though there is a statistically significant association between origin and the motivational factors, the association is (similar to the gender-dependent differences) only small.

Table 4. Origin-dependent differences in motivation to *make money* and *become a star*.

Origin	Making money	Becoming a star
Germany (N = 2.604)	5.91%	5.11%
Japan (N = 1.920)	4.74%	2.34%
USA (N = 3.143)	7.45%	3.91%
Pearson Chi^2	$p < 0.001$	$p < 0.001$

3.4 Age-Dependent Differences

Differences in the streamers' motivation dependent on their age can be identified in Table 5. Unfortunately, we did not meet the assumption that all cells should have expected counts greater than five for one cell (12.5%), therefore, these result have to be interpreted with some caution. Apparently, with increasing age the intention to earn money rises, but simultaneously the goal to become a star decreases. 21.21% of the Silver Surfers seek financial gain, however, none of them wants to become a star. In contrast, 8.21% of the generation Z aim to become a star, whereas making money (2.39%) is a minor motivational factor. The chi-square test results in a statistically significant but small association (0.125 for money, and 0.092 for fame).

Table 5. Age-dependent differences in motivation to *make money* and *become a star*.

Generation	Making money	Becoming a star
Gen Z (N = 1.839)	2.39%	8.21%
Gen Y (N = 2.572)	4.12%	4.35%
Gen X (N = 493)	9.74%	2.64%
Silver Surfers (N = 33)	21.21%	0.00%
Pearson Chi^2	$p < 0.001$	$p < 0.001$

Table 6. Content of streamers motivated by *making money* (N = 479).

Content category		Pearson Chi^2
Entertainment media	40.29%	$p < 0.001$
Chatting	21.50%	$p < 0.001$
Share information	20.88%	$p < 0.05$
24/7	19.00%	$p < 0.05$
Make music	12.94%	$p < 0.05$
Advertising	11.90%	$p < 0.001$
News	11.27%	$p < 0.001$
Sport event	8.77%	$p < 0.001$
Slice of life	6.05%	$p < 0.001$
Business information	5.64%	$p < 0.001$
Comedy	4.18%	$p < 0.001$
Nothing	3.55%	$p < 0.001$
Food	3.55%	$p = 0.062$
Animals	3.34%	$p < 0.01$
Gaming	3.13%	$p < 0.05$
Politics	2.71%	$p < 0.05$
Nature	2.30%	$p < 0.01$
Draw/Paint picture	1.67%	$p < 0.05$
STM	1.46%	$p = 0.221$
Fitness	1.46%	$p = 0.96$
Spirituality	0.21%	$p < 0.001$

3.5 Differences in Streamed Content

Finally, we take a look at the potential differences in content streamed by broadcasters motivated by different aspirations. Streamers motivated by money (Table 6) provide mostly content evolving around entertainment media (40.29%), chatting (21.50%), sharing information (20.88%) and 24/7 (19.00%). Especially entertainment media is important for them. Likewise, such content is also in the Top 5 content categories for streamers motivated by fame (Table 7), but with only 13.29%. Even if chatting is the second most streamed content for streamers motivated by making money, it is more important for streamers motivated by becoming a star. Altogether, 67.77% of those fame-oriented streamers cover such content. This is followed by making music (42.86%), which is more popular among fame-oriented streamers than the ones motivated by money (12.94%). To share information is chosen equally often by both groups. Further noticeable differences, above 10%, exist for the categories 24/7 and news. Both were more frequently found in streams aiming for financial gain. Finally, there also exist statistically significant association between the motivational factor and most of the content categories ($p < 0.05$) as shown in Tables 6 and 7, however, all significant associations were only of small effect (<0.3).

Table 7. Content of streamers motivated by *becoming a star* (N = 301).

Content category		Pearson Chi^2
Chatting	67.77%	$p < 0.001$
Make music	42.86%	$p < 0.001$
Share information	19.93%	$p = 0.196$
Entertainment media	13.29%	$p = 0.394$
Slice of life	7.31%	$p < 0.001$
Advertising	6.31%	$p < 0.001$
Comedy	5.65%	$p < 0.001$
Nothing	4.98%	$p < 0.001$
Sport event	2.99%	$p = 0.394$
Fitness	2.33%	$p = 0.185$
Food	1.99%	$p = 0.710$
24/7	1.33%	$p < 0.001$
Business information	1.33%	$p = 0.935$
Gaming	1.33%	$p < 0.01$
Draw/Paint picture	1.00%	$p = 0.689$
Nature	0.66%	$p < 0.001$
Animals	0.66%	$p < 0.001$
News	0.33%	$p < 0.01$
Politics	0.33%	$p = 0.111$
STM	0.00%	$p = 0.085$
Spirituality	0.00%	$p < 0.01$

4 Discussion

This study investigated general characteristics of SLSSs streamers (on Periscope, Ustream and YouNow) motivated by fame or financial gain. For this purpose, we developed a research model and explored platform-specific characteristics (RQ1), gender-dependent differences (RQ2), origin-dependent differences (RQ3), age-dependent differences (RQ4) and contents streamed by broadcasters (RQ5) whose motivation is financial gain or fame.

From total 7,667 observed streams, 4,548 showed individual broadcasters (61.2% male and 38.8% females). The results indicate that Ustream is mostly applied by streamers motivated by financial gain, whereas YouNow by streamers aiming at becoming famous. This could also be related to generational aspects, since YouNow is a platform mostly applied by the younger generations [11]. Interestingly, the older generations (Gen X, Silver Surfers) are motivated by monetary aspects, whereas the younger ones (Gen Z) by fame. One reason for this could be the fact that nowadays the Social Media landscape and the associated attention economy is increasingly ruled by so-called micro-celebrities and influencers. These "career-paths" might often be associated with quick success, fame, appreciation, interesting offers (such as product samples, gifts), travel opportunities, the freedom to do what one likes or is interested in and also financial gain. Furthermore, such influencers and micro-celebrities often belong to the younger generations. These reasons may make it attractive for younger streamers to follow a similar path. More mature streamers may be more settled, grounded and mainly interested in the financial aspects.

According to our results, no strong association between gender and the motivation being fame or money exists. Females and males are equally interested in these aspects. There are, however, differences in content streamed by the broadcasters whose motivation is either money or fame. For streamers wanting to make money, "entertainment media" is the preferred content. We defined "entertainment media" as every action involving some form of media, e.g. displaying digital pictures, streaming a TV show or playing music. For the streamers seeking fame, the most important content categories are chatting and making music.

Since this study explores and is limited to general characteristics of SLSSs streamers and their motivation regarding fame and financial gain, further research could include qualitative interviews in order to explain our results in more depth. Besides, it would be interesting to conduct a long-term study to analyze if the streamed content (depending on the motivation) really leads the streamers to becoming a star or making money. Finally, investigation of established micro-celebrities and influencers is the next important step for our research. This study focused only on users aiming at becoming a star or influencer, however, this dream will come true only for the chosen ones.

References

1. Ainslie, A.: The burden of protecting live sports telecasts: The real time problem of live streaming and app-based technology. Southwestern Law School, Los Angeles (2015)
2. Alamiri, D., Blustein, J.: Privacy awareness and design for live video broadcasting apps. In: Stephanidis, C. (ed.) HCI 2016, Part I. CCIS, vol. 617, pp. 459–464. Springer, Cham (2016). https://doi.org/10.1007/978-3-319-40548-3_76

3. Bilal, K., Erbad, A., Hefeeda, M.: Crowdsourced multi-view live video streaming using cloud computing. IEEE Access **5**, 12635–12647 (2017)
4. Bründl, S., Hess, T.: Why do users broadcast? Examining individual motives and social capital on social live streaming platforms. In: Pacific Asia Conference on Information Systems Proceedings, PACIS, vol. 332 (2016)
5. Cohen, J.: Statistical Power Analysis for the Behavioral Sciences, 2nd edn. Psychology Press, New York (1989)
6. Dowell, C.T., Duncan, D.F.: Periscoping economics through someone else's eyes: a real world (Twitter) app. Int. Rev. Econ. Educ. **23**, 34–39 (2016)
7. Faklaris, C., Cafaro, F., Hook, S.A., Blevins, A., O'Haver, M., Singhal, N.: Legal and ethical implications of mobile live-streaming video apps. In: MobileHCI 2016, pp. 722–729. ACM, New York (2016)
8. Fietkiewicz, K.J., Lins, E., Baran, K.S., Stock, W.G.: Inter-generational comparison of social media use: Investigating the online behavior of different generational cohorts. In: Proceedings of the 49th Hawaii International Conference on System Sciences, pp. 3829–3838. IEEE, Washington, DC (2016)
9. Fietkiewicz, K.J., Scheibe, K.: Good morning…good afternoon, good evening and good night: adoption, usage and impact of the live streaming platform YouNow. In: 3rd International Conference on Library and Information Science, pp. 92–115. International Business Academics Consortium, Taipei (2017)
10. Friedländer, M.B.: And action! Live in front of the camera: an evaluation of the social live streaming service YouNow. Int. J. Inf. Commun. Technol. Hum. Dev. **9**(1), 15–33 (2017)
11. Friedländer, M.B.: Streamer motives and user-generated content on social live-streaming services. J. Inf. Sci. Theory Pract. **5**(1), 65–84 (2017)
12. Gros, D., Wanner, B., Hackenholt, A., Zawadzki, P., Knautz, K.: World of Streaming. Motivation and Gratification on Twitch. In: Meiselwitz, G. (ed.) SCSM 2017, Part I. LNCS, vol. 10282, pp. 44–57. Springer, Cham (2017). https://doi.org/10.1007/978-3-319-58559-8_5
13. Hearn, A.: 'Meat, mask, burden': probing the contours of the branded 'self'. J. Consum. Cult. **8**(2), 197–217 (2008)
14. Hearn, A., Schoenhoff, S.: From celebrity to influencer: tracing the diffusion of celebrity value across the data stream. In: Marshall, P.D., Redmond, S. (eds.) A Companion to Celebrity, pp. 194–212. John Wiley & Sons, Hoboken (2016)
15. Honka, A., Frommelius, N., Mehlem, A., Tolles, J.N., Fietkiewicz, K.J.: How safe is YouNow? An empirical study on possible law infringements in Germany and the United States. J. MacroTrends Soc. Sci. **1**(1), 1–17 (2015)
16. Horton, D., Wohl, R.R.: Mass communication and para-social interaction: observation an intimacy at a distance. Psychology **19**(2), 215–229 (1956)
17. Ingleton, P.: Celebrity seeking micro-celebrity: 'New candour' and the everyday in the sad conversation. Celebr. Stud. **5**(4), 525–527 (2014)
18. Kane, B.: Balancing anonymity, popularity, & micro-celebrity: the crossroads of social networking & privacy. Albany Law J. Sci. Technol. **20**(2), 327–363 (2010)
19. Kaser, R.: How to make money livestreaming. https://thenextweb.com. Accessed 26 Sep 2017
20. Kavka, M.: Reality TV. Edinburgh University Press, Edinburgh (2012)
21. Kavka, M.: Celevision: Mobilizations of the television screen. In: Marshall, P.D., Redmond, S. (eds.) A Companion to Celebrity, pp. 295–314 (2015). https://doi.org/10.1002/9781118475089.ch16
22. Keinänen, K.: The role of live streaming in marketing communications and corporate branding. Master's Thesis. Lappeenranta University of Technology, LUT School of Business and Management, Lappeenranta, Finland (2017)

23. Khamis, S., Ang, L., Welling, R.: Self-branding, 'micro-celebrity' and the rise of social media influencers. Celebr. Stud. **8**(2), 191–208 (2017)
24. Krippendorff, K.: Content Analysis: An Introduction to Its Methodology, 3rd edn. Sage, Thousand Oaks (2012)
25. Laerd Statistics, Chi-square test for independence using SPSS Statistics. https://statistics.laerd.com/. Accessed 8 Feb 2018
26. Lai, L.S.L., To, W.M.: Content analysis of social media: a grounded theory approach. J. Electron. Commer. Res. **16**(2), 138–152 (2015)
27. Lair, D.J., Sullivan, K., Cheney, G.: Marketization and the recasting of the professional self: the rhetoric and ethics of personal branding. Manag. Commun. Q. **18**(3), 307–343 (2005)
28. Leung, L.: Generational differences in content generation in social media: the roles of the gratifications sought and of narcissism. Comput. Hum. Behav. **29**, 997–1006 (2013)
29. Martin, R.: Cult of influence. Marketing: Advertising, media and PR in Canada. http://www.marketingmag.ca. Accessed 8 Feb 2018
30. Marwick, A.E.: Status Update: Celebrity, Publicity, and Branding in the Social Media Age. Yale University Press, New Haven (2013)
31. Marwick, A.E.: Instafame: luxury selfies in the attention economy. Public Cult. **27**(1), 137–350 (2015)
32. Marwick, A.E.: You may know me from YouTube: (Micro-)Celebrities in social media. In: Marshall, P.D., Redmond, S. (eds.) A Companion to Celebrity, pp. 194–212. John Wiley & Sons, Hoboken (2016)
33. Marwick, A.E., Boyd, D.: I tweet honestly, I tweet passionately: twitter users, context collapse, and the imagined audience. New Media Soc. **13**(12), 114–133 (2010)
34. Marwick, A.E., Boyd, D.: To see and be seen: celebrity practice on twitter. Converg. Int. J. Res. New Media **17**, 139–158 (2011)
35. Maugeri, R., Giammalva, R.G., Iacopino, D.G.: On the shoulders of giants, with a smartphone: periscope in neurosurgery. World Neurosurg. **92**, 569–570 (2016)
36. Muntean, N., Petesen, A.H.: Celebrity twitter: strategies of intrusion and disclosure in the age of technoculture. M/C J. **12**(5) (2009). http://journal.media-culture.org.au/index.php/mcjournal/article/view/194
37. Page, R.: The linguistics of self-branding and micro-celebrity in twitter: the role of hashtags. Discourse Commun. **6**(2), 181–201 (2012)
38. Recktenwald, D.: Toward a transcription and analysis of live streaming on Twitch. J. Pragmat. **115**, 68–81 (2017)
39. Rodríguez-Gil, L., García-Zubia, J., Orduña, P., López-Ipiña, D.: An open and scalable web-based interactive live-streaming architecture: the WILSP platform. IEEE Access **5**, 9842–9856 (2017)
40. Schaefer, M.: Return on Influence: The Revolutionary Power of Klout, Social Scoring, and Influence Marketing. McGraw Hill, New York (2012)
41. Scheibe, K., Fietkiewicz, K.J., Stock, W.G.: Information behavior on social live streaming services. J. Inf. Sci. Theory Pract. **4**(2), 6–20 (2016)
42. Senft, T.M.: Camgirls: Celebrity and Community in the Age of Social Networks. Peter Lang, New York (2008)
43. Senft, T.M.: Microcelebrity and the branded self. In: Burgess, J.E., Bruns, A. (eds.) A Companion to New Media Dynamics, pp. 346–354. Blackwell, Malden (2013)
44. Stohr, D., Li, T., Wilk, S., Santini, S., Effelsberg, W.: An analysis of the YouNow live streaming platform. In: 40th Local Computer Networks Conference Workshops, pp. 673–679. IEEE, Washington, DC (2015)

45. Stohr, D., Toteva, I., Wilk, S., Effelsberg, W., Steinmetz, R.: User-generated video composition based on device context measurements. Int. J. Semant. Comput. **11**(1), 65–84 (2017)
46. Tang, J.C., Kivran-Swaine, F., Inkpen, K., Van House, N.: Perspectives on live streaming: apps, users, and research. In: Conference on Computer-Supported Cooperative Work and Social Computing, CSCW 2017, pp. 123–126. ACM, New York (2017)
47. Tang, J.C., Veniola, G., Inkpen, K.M.: Meerkat and periscope: I stream, you stream, apps stream for live streams. In: Proceedings of the 2016 CHI Conference on Human Factors in Computing Systems, pp. 4770–4780. ACM, New York (2016)
48. Tufekci, Z.: 'Not this one' social movements, the attention economy, and microcelebrity networked activism. Am. Behav. Sci. **57**(7), 848–870 (2013)
49. West, L.: These YouTube stars you've never heard of have millions of teen fans [Blog Post]. http://jezebel.com/these-youtube-stars-youve-never-heard-of-have-millions-1493742066.s
50. YouNow. https://www.younow.com/partners. Accessed 8 Feb 2018
51. YouTube. https://support.google.com/youtubegaming/. Accessed 8 Feb 2018
52. Zimmer, F., Fietkiewicz, K.J., Stock, W.G.: Law infringements in social live streaming services. In: Tryfonas, T. (ed.) Human Aspects of Information Security, Privacy and Trust, pp. 567–585. Springer, Cham (2017). https://doi.org/10.1007/978-3-319-58460-7_40

Using Tiny Viral Messages on Social Networks to Spread Information About Science and Technology: Elements of a Theory of Nanovirals

Nick V. Flor(✉)

University of New Mexico, Albuquerque, NM 87131, USA
nickflor@unm.edu.com

Abstract. Viral messages reach a large number of people at almost no cost. However, the majority of viral messages are based on shocking or entertaining content. Is it possible to make other kinds of content go viral, such as science and technology news? I use conceptual blending analysis to analyze five representative, very small messages about solar technology that went viral (nanovirals). I identify four distinct viral strategies, that vary according to number of belief systems used, and whether the viral message confirmed or contradicted central beliefs. Finally, I use information systems modeling to depict a common viral mechanism underlying the strategies. I conclude with a practical heuristic to guide the design of nanoviral messages.

Keywords: Viral messages · Design science · Conceptual blending analysis Information systems modeling

1 Introduction

One of the challenges that technologists and scientists face is informing the general public about their innovations and discoveries. One solution is to conduct a national advertising campaign. However, for small businesses and most researchers, such campaigns are prohibitively expensive.

One promising, and low-cost alternative to a national advertising campaign is to use viral messages on social media to spread news about innovations and discoveries. Viral messages can reach a wide audience in a relatively short amount of time, with almost no cost except the time needed to develop the message. However, most messages that go viral contain shocking or humorous content.

Figure 1 is an example of a typical viral message with shocking content that was seen by over a hundred thousand individuals in a single day.

The research question I explore in this paper is: can you design a viral message around technology content rather than shocking or entertaining content?

While there are many popular books that discuss viral strategies [2] and some research analyzing the dynamic spread of viral messages [4], there is little formal research on designing viral messages. To answer the research question, I analyze tiny

© Springer International Publishing AG, part of Springer Nature 2018
G. Meiselwitz (Ed.): SCSM 2018, LNCS 10913, pp. 254–273, 2018.
https://doi.org/10.1007/978-3-319-91521-0_19

Jason Michael
@Jeggit
Follow
Believe it or not, this is a shark on the freeway in Houston, Texas. #HurricaneHarvy
12:00 AM - 28 Aug 2017
88,825 Retweets 149,459 Likes
7.2K
89K
149K

Fig. 1. An example of a typical viral message. https://twitter.com/Jeggit/status/902048241646280704

messages about solar technology that have gone viral. Before describing the method, I briefly clarify my distinction between viral messages and nanovirals.

2 Background: Viral Messages and Nanovirals

A viral message is information that spreads freely from person-to-person within a population often, but not necessarily, via social media. By "freely", I mean that people spread the information naturally—they do not have to be incentivized artificially to do so. A single viral message can reach hundreds of thousands to millions of people (see Fig. 1).

Viral messages differ in length. Viral news articles and viral videos are on the high end of the spectrum, and viral messages on micro-blogging, social media platforms like Twitter are on the low-end. My research focuses on very small viral messages, which I call nanoviral messages, or *nanovirals* for short. Figure 2 is an example of the smallest nanoviral—a single emoji depicting an expressionless face.

While "very small" is relative, generally, nanovirals are distinguished from longer viral messages in terms of length and operation—their length is typically less than

Fig. 2. The smallest viral message—a single emoji. "Nanovirals" are viral messages less than several hundred characters. https://twitter.com/KDTrey5/status/885318651728904192

several hundred characters, and they rely more on *retrieving* existing experiences to generate sudden insight, a process I call *apperception shift*, when compared to longer viral message which focus on *creating* an experience in a receiver via comprehension.

As suggested by Fig. 2, where the text is only a single emoji, the key to a message going viral is understanding the subtext of the message. A method is needed that helps discover the subtext from the text of a message.

3 Method

One method used in cognitive linguistics for analyzing the underlying meaning, or subtext, of a message is conceptual blending analysis [1]. It is based on the idea that people integrate elements of different beliefs mentally, to arrive at the meaning of a statement. The aim of conceptual blending analysis is to reconstruct how people mentally integrate elements of beliefs to arrive at meanings. Recently, it has been applied to analyzing meaning in advertisements [6]. Figure 3 depicts the process.

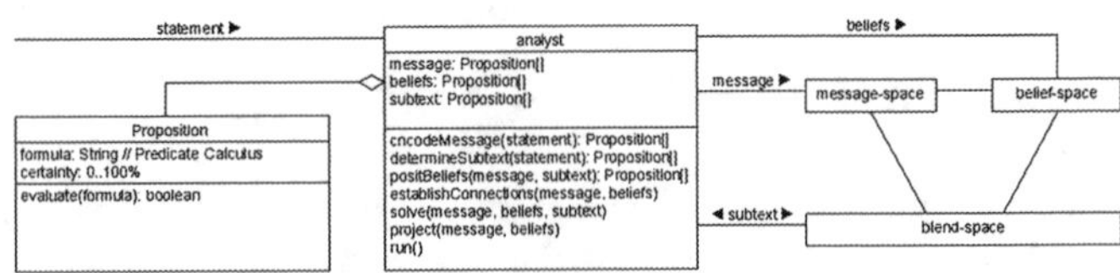

Fig. 3. Conceptual blending analysis on a nanoviral depicted as a hybrid class-communication model

Briefly, the unit of analysis is a statement. The analyst typically denotes the statement in propositional form, using the rules of predicate calculus. For example, the statement "Fred gave a rock to Wilma" in propositional form would be: Gave(fred, wilma, rock), where *Gave* is the predicate, and *fred*, *wilma*, and *rock* are terms.

Next, the analyst posits beliefs that the receiver of a message recalls in association with the message. These beliefs are also denoted in propositional form. The beliefs are put in a multi-column table, where each column denotes a *mental space*. Beliefs that have common (or synonymous) predicates or terms, are said to have a *pragmatic connection*, and are lined up row-wise in the table. Beliefs with pragmatic connections are special because their elements (predicates and terms) can substitute for one another in the blend.

Finally, the analyst selectively projects beliefs from the mental spaces into the blended space (or simply the "blend") to show the underlying meanings, the various subtexts, of the original statement. The blend is usually the last row in the table. An example should help clarify.

3.1 An Example of a Conceptual Blending Analysis

Table 1 depicts a conceptual blending analysis for the Kevin Durant tweet in Fig. 2. The timestamp of the tweet indicates that he posted it during ESPN's annual award show, the ESPY. During this show the emcee, Peyton Manning, made fun of Kevin Durant for switching teams in order to win a championship. When the camera panned to Kevin Durant he was not smiling, suggesting that he was mad, but one could not be certain based on the brief camera shot.

Table 1. Conceptual blending analysis for the Kevin Durant single-emoji tweet

Message (Tweet) Space	Belief (ESPY) Space
During(tweet, espy)	*peyton-manning* *kevin-durant*
	P1: MadeFun(peyton-manning, kevin-durant)
ExpressionLess(emoji)	*¬Smiling(kevin-durant)*
	R1: P1 & ¬Smiling(kevin-durant) → Mad(kevin-durant)
BLEND	

// Subtext: Kevin Durant is mad at Peyton Manning
// for making fun of him
//
// Derived by substituting ExpressionLess(emoji) for
// *¬Smiling(kevin-durant),* and projecting the substituted
// rule, R1, into the blend:
 	
P1 & Expressionless(emoji) → *Mad(kevin-durant)*	

Propositions representing the message are shown in the left column, the message (tweet) space. Possible beliefs retrieved by a reader as a consequence of the tweet occurring during the ESPY are shown in the right column, the belief (ESPY) space. This belief includes the rule that if Kevin Durant is not smiling he must be mad: … ¬Smiling(kevin-durant) → Mad(kevin-durant). There is a pragmatic connection between the expressionless emoji, ExpressionLess(emoji), and the proposition that Kevin Durant is not smiling, ¬Smiling(kevin-durant). In the blend, the expressionless emoji is substituted for this proposition, and readers of the tweet conclude that Kevin Durant is mad at Peyton Manning, which confirms their belief from watching the telecast.

3.2 Data and Apparatus

The data analyzed consisted of 330,827 tweets from the social media platform Twitter containing the hashtag #solar. I used SMEDA [5] as the social media scraping software. SMEDA is a custom module I wrote for Excel that scrapes tweets into an Excel worksheet. In addition to scraping it contains macros for organizing and sorting tweet content, and for building social network edges.

3.3 Procedure

SMEDA was run daily, over a two month period, from July 1, 2017—August 31, 2017. A total of 330,827 tweets were collected (N: 330,827; μ: 5335.92 tweets per day, σ: 1950.40). After the collection period, SMEDA was then used to sort tweets in descending order based on the number of retweets (shares). Tweets containing #solar, but unrelated to solar technology were thrown out. For example, there was a Korean music group who had a singer named Solar, and who would tag their tweets with #solar. All such tweets were deleted from the data set analyzed.

3.4 Procedure: Operationalizing Viral

Unlike viral messages containing entertaining or shocking content which receive thousands of retweets, messages with the #solar hashtag never received over a thousand retweets during the period scraped. Thus, rather than go with an absolute value to classify a tweet as viral, I used a relative measure. Specifically, given the author of a top-sorted tweet, I calculated the mean number of retweets over a week and the standard deviation. If the number of retweets was over one standard deviation I defined that as viral for that author, and the tweet was analyzed.

Figure 4 depicts the general hypothesis and theory building process. While conceptual blending analysis is a qualitative method, through iteration and triangulation, falsifiable theories can result.

It is beyond the scope of this paper to show every top tweet analyzed. Thus, in the results section I present just the analysis of five representatives of the top tweets.

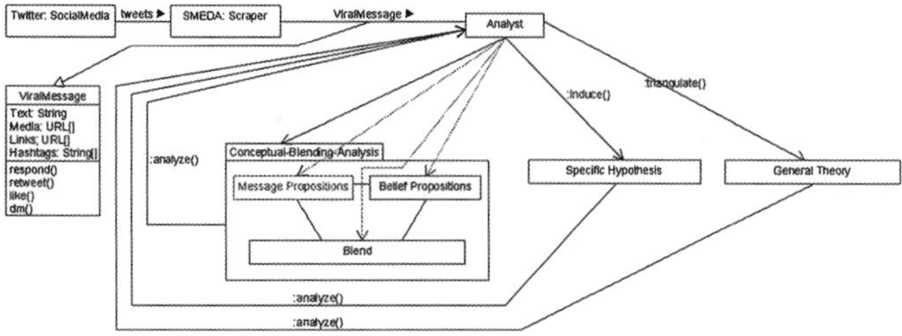

Fig. 4. The iterative procedure using conceptual blending analysis only on viral messages

4 Results

4.1 Representative 1: Fact Confirmation and Contradiction in Two Different Belief Systems—Progressive Version

The first tweet analyzed is from the DiCaprio Foundation (@dicapriodn) about the number of people employed in the solar industry versus the fossil fuel industry (see Fig. 5). The literal meaning of the text is clear: the solar industry hires more people involved in generating electricity than the fossil fuel industries combined.

The tweet contains: hashtags for Solar, Electricity, Oil, Coal, and Gas; a link to a Forbes news article for more information; and a user tag for @cleantechnica. The text of the tweet is taken from the title of the Forbes article that the tweet links to. Hashtags help spread the tweet to users searching on those tags, and a user tag displays the tweet on that user's mention timeline. Finally, there is a picture with a bar chart showing the number of people employed in the solar industry versus the fossil fuel industries. The picture sources the data to the Department of Energy.

As described in the method, I use conceptual blending analysis to discover possible subtext underlying the literal meaning of the text (refer to Table 2). The left-hand column contains propositions in predicate calculus form that correspond to the key content of the tweet. The right-hand column contains beliefs, both predicates and propositions (recall propositions are predicates filled-in with values), that the reader of the viral message could bring to mind as part in association with the text.

For example, the text mentions solar and fossil fuels. It is likely that readers will think of the beliefs of proponents of both solar and fossil fuels. If the reader is a renewable energy proponent, as many progressives are, a common belief is "solar is more important than fossil fuels", or in predicate calculus: Progressive(p) → MoreIm-portant(solar, fossil). The opposite is true if the reader is a fossil-fuel proponent, as many conservatives are: "fossil fuels are more important than solar", Conservative(c) → MoreImportant(fossil, solar). A reader may also recall general rules suggested by the text, in this case, "if some product x is more important than some other product y, more people will be employed making x than y"; in predicate calculus form: More-Important(x, y) → More(Employed(x), Employed(y)).

In the blend space the reader projects the "fact", or more precisely "a proposition

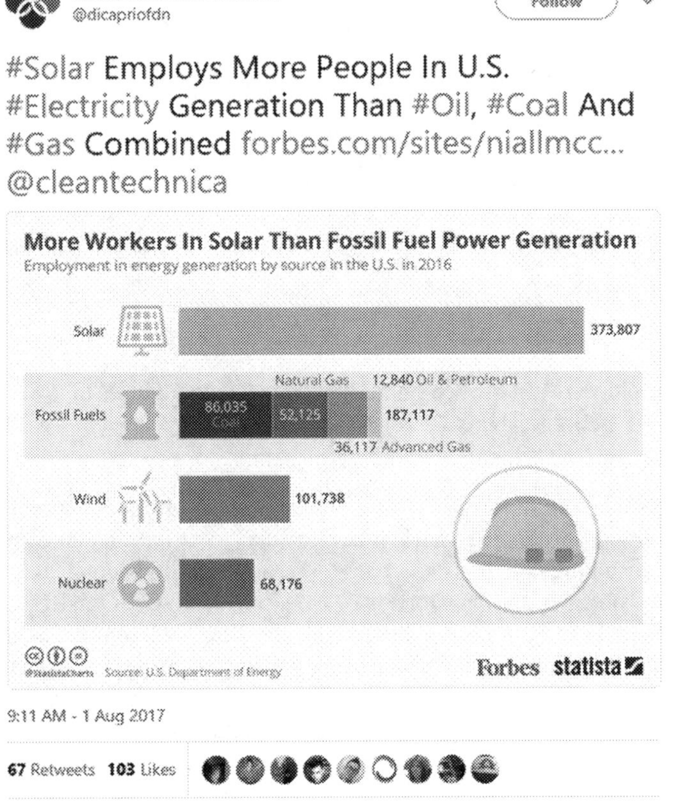

Fig. 5. Viral message from @DicaprioFdn, https://twitter.com/dicapriofdn/status/892417404428460032

with high certainty due to the source", that more people are employed in the solar than in the fossil fuel industry. The reader chains the propositions for both progressives and conservatives, with the general rule about product importance and employment, yielding a proposition that agrees with the facts in the case of the progressive belief, and disagrees with the facts in the case of the conservative belief.

Subtext confirming or discrediting widely-held, central beliefs is one of the most common occurrences in nanoviral messages, where I define "central belief" in terms of centrality in a network of propositions—a proposition that occurs in many of the propositional chains that constitute a belief system, c.f., node centrality in social networking theory. I call this the *confirm and contradict* strategy.

Table 2. Conceptual blending analysis for the @DicaprioFdn message

Message Space	Belief Space
More(Employed(solar), Employed(fossil)) Employed(solar, 373K) Employed(fossil, 187K)	*Progressive(p)* → *MoreImportant (*solar, fossil*)* *Conservative(c)* → *MoreImporant (*fossil, solar*)* *MoreImportant(x, y)* → *More (Employed(x), Employed(y))*
Blend Space: Agreement, Disagreement	

// Fact from message
More(Employed(solar), Employed(fossil)) // fact in message
 	
// Subtext 1: Progressives beliefs about solar energy are correct
// *via chaining to a proposition that agrees with fact*
Progressive(p) → *MoreImportant (solar, fossil-fuel)* → *More (Employed (solar), Employed(fossil))* // ***agrees*** *with fact in message*
 	
// Subtext 2: Conservatives beliefs about solar energy are incorrect
// *via chaining to a proposition that disagrees with fact*
Conservative(c) → *MoreImportant (solar, fossil-fuel)* → *More (Employed (fossil), Employed(solar))* // ***disagrees*** *with fact in message*	

4.2 Representative 2: Fact Confirmation and Contradiction in Two Different Belief Systems—Conservative Version

The next example shows a variation of the *confirmation and contradiction* strategy. @AndrewCFollet's viral message (see Fig. 6) is about old solar panels causing environment problems in China. Although lacking details about how the solar panels are causing problems, the literal meaning of the text is clear. The structure of the message is similar to the first one analyzed, namely, the user repeats the headline of an article in the text, includes hashtags, links to an article, and tags users. However, instead of creating hashtags from the title the user specified tcot (Top Conservatives On Twitter), tlot (Top Libertarians In Twitter), and AGW (Anthropogenic Global Warming). The picture caption elaborates on the meaning of "environmental crisis", stating that "Old Solar Panels … in two or three decades will wreck the environment".

As in the previous analysis, we can represent the key propositions from the message in the left column of our conceptual blending analysis table (refer to Table 3), and possible propositions in the right column. The bottom row blends elements from both columns. The key proposition in the message text is: ¬Helps(OldSolar(panel), Environment(china)). While the predicates Wrecks or Hurts could have been used instead of ¬Helps, it saves time in the analysis from writing synonym propositions.

Fig. 6. Viral message from @AndrewCFollett's, https://twitter.com/AndrewCFollett/status/892432713667547136

The possible beliefs include: old solar panels are solar panels; solar panels produce solar energy; progressives believe that solar energy helps the environment; conservatives believe the US should not focus on solar energy; we should not focus on energy technologies that harm the environment. These beliefs, stated as propositions in predicate form, are in the right column.

In the blend, the subtext includes: solar energy hurts the environment of the United States; progressives are wrong about solar energy benefiting the environment; and conservatives are right not to focus on solar energy. Unlike the previous example, this viral message contains a proposition that contradicts a widely-held progressive belief, while supporting a widely-held conservative one.

Although the details of the blending differ—both chaining propositions and substituting elements of propositions—the outcome of the blending is the same: a confirmation of a central belief in one belief system, and a contradiction of a central belief in another, opposing, belief system.

4.3 Representative 3: Confirmation and Counterfactual in a Single Belief System

Some users were particularly adept at creating viral messages. One user, @MikeHudema, often started off his tweets with the phrase "As Trump tweets" (see Fig. 7). In this case, the literal meaning of the text, masks complex subtext aimed at denigrating the current president via contrast with a former president. The structure of the tweet is:

Table 3. Conceptual blending analysis for the @AndrewCFollett message

<table>
<tr><th>Message Space</th><th>Belief Space</th></tr>
<tr><td>Environment(china)

¬Helps (OldSolar(panel), Environ-ment(china))</td><td>Environment(us)

Old(Solar(panel)) →Solar(panel)

Solar(panel) → Solar(energy)

Progressive(p) →
Helps(Solar(energy), Environ-ment(us))

Conservative(c) →
¬Focus(Solar(energy))

¬Helps(x, Environment(x)) →
¬Focus(x)</td></tr>
<tr><th colspan="2">Blend Space: Contradiction & Confirmation</th></tr>
<tr><td colspan="2">// Subtext 1: Old Solar Panels won't help the US environment either
// Derived via substitution from fact in Tweet space
¬Helps (OldSolar(panel), Environment(china)) →
¬Helps (OldSolar(panel), Environment(us)) // substitution

// Subtext 2: Solar Energy won't help the US environment
// Derived via substitution of Solar(Energy) for Solar(Panel)
Old(Solar(panel)) → Solar(panel) → Solar(energy)
¬Helps (Solar(energy), Environment(china)) // substitution
¬Helps (Solar(energy), Environment(us)) // substitution

// Subtext 3: Progressives are wrong about solar energy helping
// the environment
Progressive(p) →
Helps(Solar(energy), Environment(us)) // contradicts Subtext 1

// Subtext 4: Conservatives are right not to focus on solar energy
¬Helps(x, Environment(x)) → ¬Focus(x)
¬Helps (Solar(energy), Environment(us)) → ¬Focus(SolarEnergy)
Conservative(c) → ¬Focus(SolarEnergy) // confirms belief</td></tr>
</table>

message, hashtags, link to news article, and picture from news article. The hashtag #resist refers to a movement consisting of individuals against current-president Trump.

In the message space (refer to Table 4, left column), you have two actors, Trump and ex-president Jimmy Carter. There are also propositions that denote Trump tweets, that Jimmy Carter built a solar farm, and that the solar farm powers half the city. In the belief space (right column) you have the fact that Trump is president, and a progressive belief that Trump tweeting is a useless activity. There is a pragmatic connection between the solar farm powering half the city and the city using the solar farm.

Fig. 7. Viral message from @MikeHudema, https://twitter.com/MikeHudema/status/885139695377797121, 69.5K; source: futurism

Finally, you also have the general belief that if someone builds something used by others, then the builder is useful.

The blend contains three pieces of subtext. The first is that the current president is useless, which confirms a progressive belief. This is contrasted with the second subtext, which states that the former president is useful. The second subtext is important because it provides a kind of proof that progressives can cite if challenged on why they believe the current president is useless. Finally, we know that people constantly engage in counterfactual thought, and that it can result in negative emotions like anger and regret [3, 7]. The third subtext is the counterfactual: if current-president Trump had only built a solar farm, he would be useful.

Unlike the previous two examples—which employed two belief systems, confirmations, and a contradictions—this viral message employed a single belief system, confirmations, and a counterfactual. While one may argue that a counterfactual is a contradiction, I reserve the use of counterfactual for those contradictions involving the substitution of people and technologies in action propositions that have positive or negative consequences. I call this the *confirmation and counterfactual strategy*.

Table 4. Conceptual blending analysis for the @MikeHudema message

Message Space	Belief Space
trump Tweets(trump)	*President(*trump*)* *Progressive(p)* → Tweets(trump) → *Useless*(trump)
FormerPresident(carter) Build(carter, Solar(farm)) Powers(Solar(farm), Half(city))	*Build(b,y)* *Using(y, x)* *Build(b,y) & Using(y,x)* → *Useful(b)*

Blend Space: Confirmation & Counterfactual
// Subtext 1: Current President is Useless // *Derived by substitution into Progressive Tweets belief* *Progressive(p)* → Tweets(trump*)* → *Useless(*trump*)* → *Useless(President(*trump*))*
// Subtext 2: Former president jimmy carter is useful // *Derived by substituting predicate Power for Using* // *and parameter substitution* Build(carter, Solar(farm)) & *Using*(Solar(farm), Half(city)) → *Useful(carter)* → *Useful(*FormerPresident(carter))
// Subtext 3: if Trump built a solar farm he'd be useful // *Derived by parameter substitution of Trump for Carter* // *and parameter substitution* Build(trump, Solar(farm)) & *Using*(Solar(farm), Half(city)) → *Useful(trump)* ***// Counterfactual***

4.4 Representative 4: Wrong Economic Belief Indicating Technology Adoption

Not all viral messages about solar had political subtext. Another common type of viral message employed economic subtext (see Fig. 8). The literal meaning is straightforward: renewable energy will be cheaper than fossil fuels across the world in 3 years, according to the Morgan Stanley consulting firm. The structure of this message is like the previous examples, with the exception that the hashtags do not target specific political groups, and no other users are tagged.

The message space (refer to Table 5, left column) contains the proposition that the price of wind and solar will be less than the price of coal and gas in three years. The belief space (right column) includes the widely-held belief that renewable energies like wind and solar will always be more expensive than fossil-fuels, synonyms for renewables and fossil fuels, and the general belief that if the price of two equivalent items are similar, one should adopt the least expensive item.

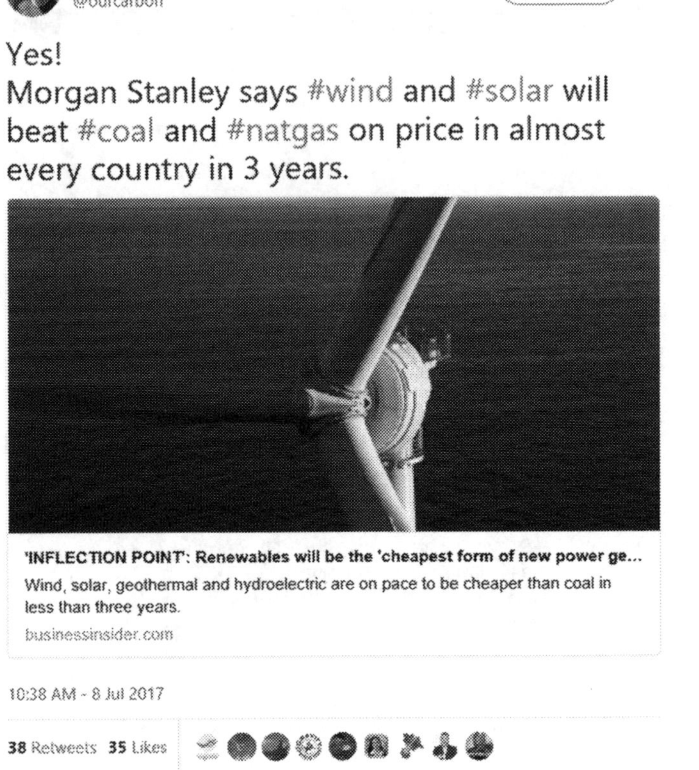

Fig. 8. Viral message from @OurCarbon, https://twitter.com/ourcarbon/status/883742030072922112

The blend contains two subtexts. First, that it is wrong to believe coal & gas will always be cheaper than wind & solar; second, that renewables should be adopted in three years.

In the case of this viral message, contradicting a widely-held belief leads to a conclusion to adopt a technology. I label this strategy *present economic case*.

4.5 Representative 5: Argument from Majority

The final kind of viral message that one finds about solar, are those that provide news about other groups of people creating, using, or adopting a technology. In this example it is Australian households adopting solar panels (see Fig. 9). The structure of the message is similar to the previous example: text, hashtags, a link to a news article, and a picture.

Table 5. Conceptual blending analysis for the @OurCarbon message

<table>
<tr><th>Message Space</th><th>Belief Space</th></tr>
<tr><td>□c ∈ countries, t ∈ date+3,
Price(wind, solar, c, t) < Price(coal, gas, c, t)</td><td>□c ∈ countries
Price(wind, solar, c, ∞) > Price(coal, gas, c, ∞)

wind, solar = renewables
coal, gas = fossil-fuels

Price(x)< Price(y) → adopt (x)</td></tr>
<tr><th colspan="2">Blend</th></tr>
<tr><td colspan="2">// Subtext: You are wrong to believe coal & gas are cheaper than
// wind and solar
// Derived by sample fact contradicting belief
□t ∈ date+3, Price(wind, solar, c, t) < Price(coal, gas, c, t) // fact
Price(wind, solar, c, ∞) > Price(coal, gas, c, ∞) // contradiction

// Subtext: You should adopt renewables
// Derived by equating wind & solar with renewables,
// projecting the time conditions, and
// substituting solar & renewables into general adopt rule
□c ∈ countries, t ∈ date+3, Price(x)< Price(y) → adopt (x)</td></tr>
</table>

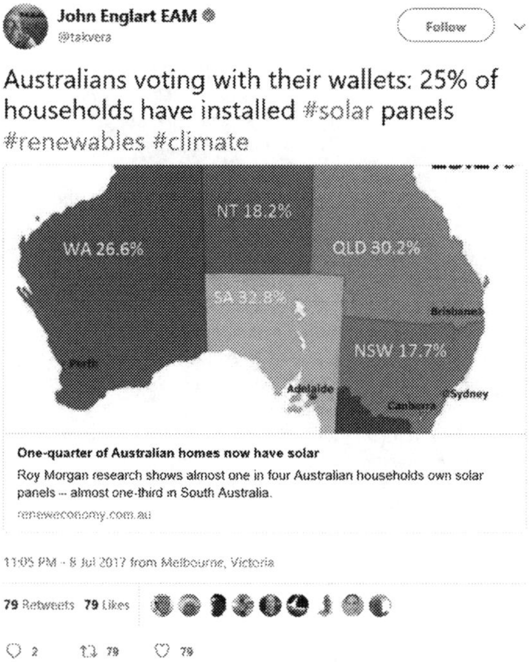

Fig. 9. Viral message from @Takvera, https://twitter.com/takvera/status/883929972812808193

In the message space (see Table 6, left column) are the propositions derived from the message, in particular that 25% Australian households have adopted solar panels. When a person reads such a message, it is natural to think of beliefs that compare or contrast the person's own group to the other group. In predicate calculus this is denoted by substituting predicates and parameters. Since the source group was Australian households, if the reader is American, the reader thinks of American households, and the fact that most American houses do not have solar panels installed. Whether or not this is bad depends on if these households are part of advanced nations, which in the case of Australia and America is true. Finally, there is the general belief that if an advanced nation is behind another advanced nation, it should catch up. Table 6, right column, summarizes potential propositions in the belief space.

In the blend, the subtext is that American households are behind Australians in terms of solar panel adoption and, being an advanced nation, should catch up. The viral message creates a new belief based on a propositions from the message combined with existing beliefs about progress. I label this strategy the *catchup strategy*.

Table 6. Conceptual blending analysis for the @takvera message (assumes reader is American)

Message Space	Belief Space
Australian(households) Have(Australian (households), Solar(panels), 25%)	*American*(households) Have(*American (households)*, Solar(panels), *LessThan*(25%)) *Have(x,y,z) & Have(a,y,LessThan(z) & AdvancedNation(x) & AdvancedNation(a) → Behind(a,x,y)* *Behind(a,x,y) → Catchup(a,y)*
Blend	

// Subtext: American households are behind Australian households
// in adopting solar panels and should catch up
// *Derived by substitution and chaining*
Have(Australian (households), Solar(panels), 25%) &
Have(*American* (households), Solar(panels), *LessThan*(25%)) &
*AdvancedNation(*Australian (households)*)* &
AdvancedNation(American (households)*)* → Behind(*American* (households), Australian (households) Solar(panels)) →
Catchup(American(households), Solar(Panels)*)*	

5 Discussion: Strategies and Common Mechanism

We have examined five different viral messages that appear to use four seemingly different strategies. Next we use systems modeling techniques to triangulate to a common underlying viral mechanism that will serve as the basis for a design theory of nanovirals.

5.1 Modeling: The Physical Dataflow

In systems analysis, physical dataflow diagrams depict a system as is, with the agents (both actors and technologies) exchanging data. Initially I assumed a model of viral messages with the following data flow (see Fig. 10):

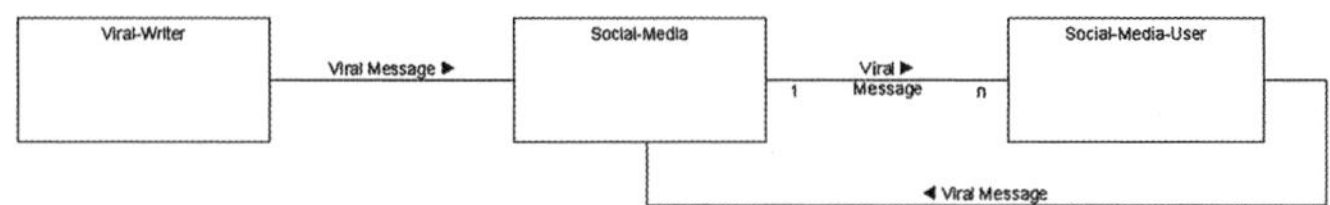

Fig. 10. Initial physical dataflow diagram

However, the analysis showed that news about events in the world was a central piece of every viral message. This news, created by some journalist and posted on a news website, can be understood as an input to the viral writer, as a key element of the viral creative process. Figure 11 depicts the revised diagram.

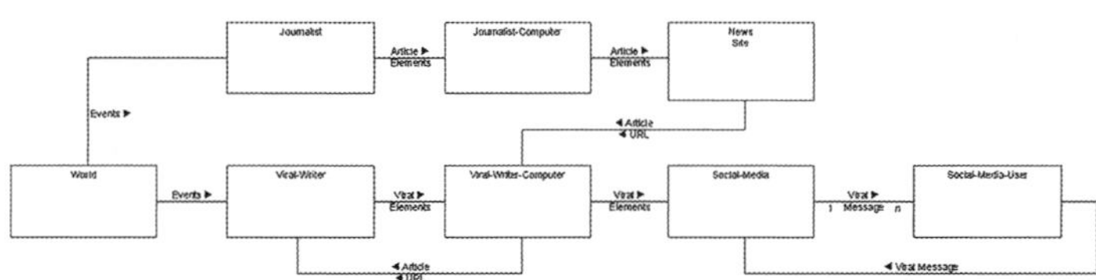

Fig. 11. Revised physical dataflow diagram based on the analysis

This revised diagram includes the viral writer's computer because it is a crucial tool used by the writer to search and organize news, as well as to compose the viral message. Note also that the diagram shows *viral elements* going from the writer to social media rather than a *viral message*. This is because the analysis made it apparent that social media formatted the final message viewed by users, which included the user's picture and information about date posted, retweets and likes.

From this diagram's inputs and outputs we can delineate four abstract processes to model: event, news, viral creation, and viral spreading (see Fig. 12). Finally, although not depicted explicitly in the process model, social media provides an input to the viral writer, serving as another source in the viral creation process.

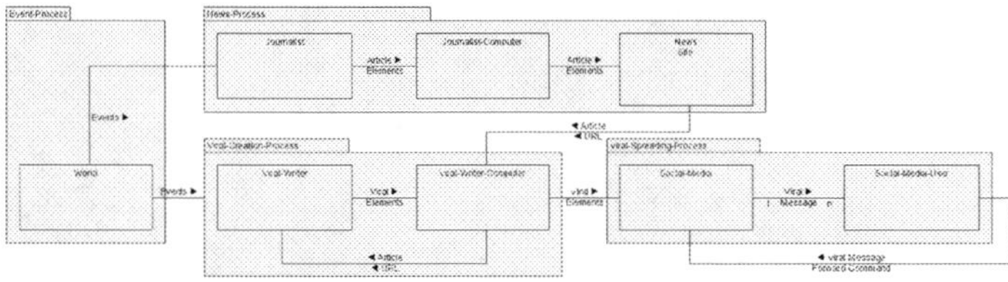

Fig. 12. The four processes to model. Messages from social media to the viral writer are implied but not shown.

To help construct a design theory of nanoviral messages that We will model two objects in two separate processes: the social media user in the viral spreading process, and the viral writer in the viral-creation process.

5.2 Modeling: The Social Media User in the Viral Spreading Process

Is there a common underlying mechanism in all the viral messages studied, which we can model? The analysis suggests, yes.

One can represent a viral message as a set of propositions. These propositions, through an associative mental process, retrieve beliefs from belief systems, which one can also represent as propositions.

Some of these beliefs are central to belief systems, e.g., "renewal energy is better than fossil fuel energy" in a progressive belief system, and vice-versa in a conservative belief system. I term such beliefs central beliefs, or central propositions. The intuition is that people use central beliefs to support explanations, predictions or actions. One can use centrality formulas from networking analysis to operationalize this term.

A social media user, given message propositions and central beliefs, will share a message if at least one of the propositions confirms or contradicts a central belief *and* the social media user determines that the confirmation or contradiction is not shared by his or her followers.

Figure 13 captures the main objects and the main information exchanged between objects.

Once shared, a message will continue to be shared if the belief systems of the followers (the message receivers) are consistent with the those of the sharer. This is likely why there are an abundance of political messages that go viral—progressive and conservative belief systems are consistent across followers, who in turn have followers with those belief systems.

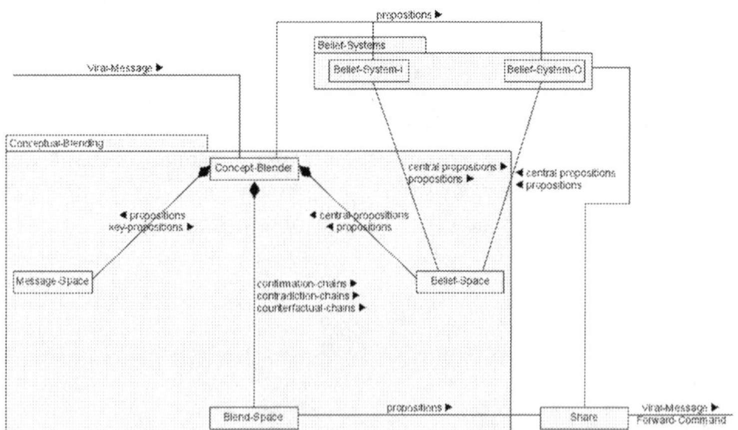

Fig. 13. Hybrid class-communication diagram for a social media user. The diagram depicts just two of many possible belief systems.

A message can both confirm and contradict central beliefs in separate belief systems, e.g., confirm a progressive belief while simultaneously contradicting a conservative one and vice versa, as the first two analyses showed. The decision rule for sharing is the same: if the sharer believes the confirmation and contradiction by followers, it will be shared.

5.3 Modeling: The Viral Writer in the Viral Spreading Process

The conceptual blending analysis analyzed the viral messages from the standpoint of a social media user reading them. While we did not analyze the viral creation process, it is possible that the same mechanism for comprehending a viral message, is used by a viral message writer to compose a viral message. Comprehension drives composition.

The primary difference is in input and output. An event happens in the world, which the viral writer either experiences directly or learns about via the news or social media. The viral writer represents the events, news, or social media messages as propositions, and if certain propositions confirm or contradict central beliefs, those propositions along with the central beliefs are the ingredients of a potential viral.

The decision to compose a viral message from those ingredients is similar to the sharing decision. A viral writer will create a viral message based on a contradiction or confirmation if the viral writer determines it is not shared by his or her followers.

The process of composing a viral message takes the confirmation or contradiction from the conceptual blending process and adds: supporting links, media (e.g., pictures and videos), user mentions, and hashtags. These viral elements are then sent to social media.

Figure 14 depicts how the viral creation process can leverage the viral spread process.

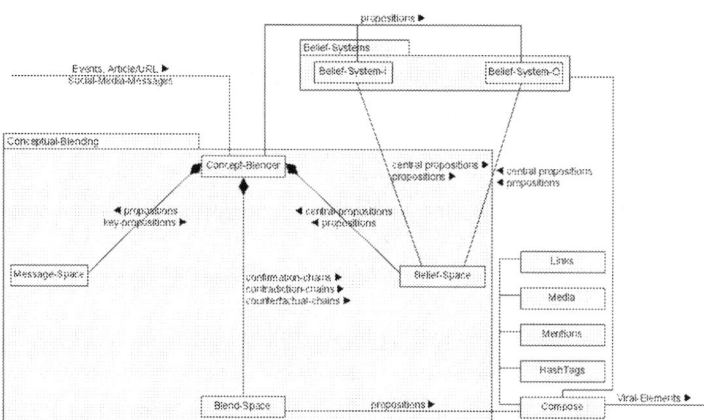

Fig. 14. Hybrid class-communication diagram for a viral writer. This composition process leverages the same conceptual blending & belief systems as the social media user.

6 Conclusion: A Heuristic and Future Research

In the old day of advertising, copywriters used formulas to help them write ads—Attention-Interest-Desire-Action (AIDA) was one, Picture-Promise-Prove-Push (PPPP) was another. There are analogous formulas for writing novels, screen plays, and video games as well. In the language of design science, these formulas are more properly thought of as "heuristics", because they don't guarantee success so much as they help focus one's effort in generating and in sequencing ideas for composition.

My analysis suggested the following heuristic—Check, Confirm | Contradict, Compose (CCCC):

- Check for news and other events, and based on that news
- Confirm central beliefs in shared belief systems, or
- Contradict central beliefs in shared belief systems
- Compose viral message around the confirmation or contradiction, adding in hash tags, mentions, media, and links.

The principle underlying the sharing of viral messages seems to be the *conservation of consistency in belief systems*. Messages are shared because they confirm beliefs that may be uncertain or, in the case of viral messages that show contradictions, they point out inconsistencies that must be repaired to maintain a consistent belief system.

The viral writer model in Fig. 14 suggests several areas for future research, which can help expand the theory. The first area is in terms of the source materials used by viral writers. For my #solar tweets, the source material was always a news article on some website. But the source could be a message read on social media from another user, or an event experienced first hand, or even a sudden realization of some confirmation or contradiction in one or more belief systems. How would the composition of the viral message change if the source was not a news article?

The second area of future research is belief systems. The viral messages I analyzed relied on no more than two belief systems. Is it possible to create a viral message that uses three or more belief systems? And it there are more than three belief systems, what are the "rules" for creating viral messages, beyond confirming and contradicting central beliefs? This is a question of process.

Process is the third area of future research. The viral rule in the data analyzed was the confirmation or contradiction of central beliefs. But the data also showed different ways for a news proposition to confirming a central belief, including early and late in a causal chain of propositions. The same was true of contradictions, especially in the use of counterfactual blends. More research is needed in specifying the details through which central beliefs get confirmed or contradicted. Finally, are there other rules beyond confirmation and contradiction, e.g., the connection of central beliefs from different belief systems.

The last area is composition. Future research is needed to clarify steps in this heuristic, especially the composition step. In particular, given a confirmation or a contradiction, or some other rule, what is the best way to state it and to support it with hashtags, mentions, media, and links. More research is needed, in particular, on the role of hashtags and mentions in making a message go viral, especially if the viral writer does not have a large follower base.

In conclusion, I focused my analysis on viral messages for the hashtag #solar, in an attempt to discover a way of spreading information virally about science and technologies. I discovered that it was not the existence of a new discovery, or a new innovation that made the news spread, nor was the spread due to a description of how it worked, or what it could do for the reader. Rather, information spread if it confirmed or contradiction widely-held, central beliefs. Furthermore, the belief systems may have very little to do with the discovery or innovation, as was the case with the progressive and conservative belief systems—political belief systems.

Scientists who want to spread discoveries may want to focus less on describing the details of their findings, or less on describing future benefits, and more on how the discovery confirms or contradicts existing widely-shared belief systems, which may not have much in common with the discovery.

Acknowledgments. This material is based partly upon work supported by the National Science Foundation (NSF) under CMMI –1635334. Any opinions, findings, and conclusions or recommendations expressed in this material are those of the author and do not necessarily reflect the views of the NSF.

References

1. Fauconnier, G., Turner, M.: The Way We Think: Conceptual Blending and the Mind's Hidden Complexities. Basic Books, New York (2008)
2. Gladwell, M.: The Tipping Point: How Little Things Can Make a Big Difference. Little, Brown, and Company, New York (2006)
3. Epstude, K., Roese, N.J.: The functional theory of counterfactual thinking. Pers. Soc. Psychol. Rev. **12**, 168–192 (2008)
4. Leskovec, J., Adamic, L.A., Huberman, B.A.: The dynamics of viral marketing. ACM Trans. Web **1**, 1–46 (2007)
5. Flor, N.: SMEDA — Social Media Exploratory Data Analytics Software. GitHub Repository (2017). https://github.com/professorf/smeda
6. Joy, A., Sherry, J., Deschenes, J.: Conceptual blending in advertising. J. Bus. Res. **62**, 39–49 (2009)
7. Roese, N.J.: Counterfactual thinking. Psychol. Bull. **121**, 133–148 (1997)

Product Sentiment Trend Prediction

Vatsal Gala[1], Varad Deshpande[1], Ibtihal Ferwana[2], and Mariofanna Milanova[3(✉)]

[1] Mumbai University, Mumbai, India
vatsalpgala@gmail.com, clustervarad@gmail.com
[2] Prince Sultan University, Riyahd, Saudi Arabia
ibtihalferwana@gmail.com
[3] University of Arkansas at Little Rock, Little Rock, USA
mgmilanova@ualr.edu

Abstract. The prospects of spectrum sentiment analysis are great and is a field that has been given very little research focus. We develop a system that can recognize human recognizable emotions and quantify them, the system can then predict the trend in the spectrum sentiments provided a chronological data. This paper discusses a lexicon-based approach for spectrum sentiment analysis. It further describes a quantification method to factor in the effects of time in trend prediction and a novel idea of using consecutive calculated values for current trend value calculation. The system is designed for e-commerce data but has flexibility to be used for other fields too. The system uses a simple neural network with image and text features as input and the trend values as output. This system can then be used to predict sentiment trend for newer or existing products. The system shows great prospects for multi-modal sentiment analysis of sentiments on spectrum range and can be advanced by using more complex approach.

Keywords: Trend prediction · Multi-modal approach
Spectrum sentiment analysis · Text mining

1 Introduction

The internet boom has been advantages to many sectors. One of them being commerce. E-commerce is one of the sector that has greatly prospered with the advent of internet (online shopping). With online shopping becoming the trend, the focus of sellers has shifted from profit margins and stock management to customer satisfaction and understanding the customer sentiment. E-commerce websites have contributed greatly to this shift of focus. With features such as descriptions of products and giving a space for the customer to give reviews for the products they are buying the websites have assisted a smoother exchange of information between the customer and the seller. This has profited both sides greatly. However now that the websites such as Amazon or E-bay have been growing at greater pace, the data generated through exchange of information between the customer and the seller has increased exponentially. Which in turn has given way for more deeper analysis of this exchange of information.

Before moving on to the study it is important to understand the exact process we are trying to study and enhance. The process being predicting the trend in sentiments for

© Springer International Publishing AG, part of Springer Nature 2018
G. Meiselwitz (Ed.): SCSM 2018, LNCS 10913, pp. 274–283, 2018.
https://doi.org/10.1007/978-3-319-91521-0_20

customer for a particular product. For a customer the first contact with the product on any e-commerce website is the image of the product. After which it starts to read its description and later browse through the reviews to check for any anomalies in the product reviews. Thus, we could say that the image and the product reviews act as key factors in influencing a customer sentiment towards a product. We give less importance to product description as it has mostly technical descriptions of the product and has little impact to the final customer sentiment[1]. We have now to process 2 kinds of data namely image and text. However merely analyzing text data cannot allow us to predict a trend. Trend can be predicted by bringing time as one of the factors for our analysis. The term trend in the context of our study means to find out how the sentiment of product changed overtime mainly from its date of launch till current time and using these data have estimation of how it will perform in the near future.

Images contain lots of information, however while considering the images to be the first impression for a product it becomes necessary to not analyze the detailed features of the image but stick to the features that give off a subtle information to the customer. Such an analysis partly falls into the domain of phycology and thus must be viewed so. Certain colors and certain levels of brightness act as pleasing feature for certain genre of products while sometimes a highly contrasted image and opposing colors can instigate the customer. Thus, the concept of what kind of image is necessary for influencing the customer becomes a topic of debate. However, it is quite clear that for a particular category the product images must have a similar appearance. Keeping this in mind we won't have to worry about the difference in appearance of different products as long as we operate on similar kind of products.

Text found on e-commerce websites in forms of reviews generally have a standard structure with minimal variations. Most of the reviews follow normal grammatical rules and don't have much spelling errors or use of slangs. However, it is quite important to understand that not all the reviews can be genuine and certain reviews can have a degree of deceit to it. There are methods to identify such reviews and segregate them from the dataset. The smaller the review the lesser importance it can have in the customer sentiment but if a review is longer than a few paragraphs, its authenticity must be questioned. Thus, though text plays a key role in sentiment analysis, it must be taken care that deceitful text can introduce errors in our prediction system.

The understanding of humans regarding emotions is equivocal however its quite clear that humans measure emotions on a scale and consider emotions to be of multiple types. Emotions range from anger to fear and from joy to sadness. Though we have slightly altered perception of how emotion is represented, we mainly use the same means of communicating the emotion i.e. words in a language. More often than not we use similar set of words to express similar emotions. The words "I'm furious." Can hardly mean anything but expression of anger. Likewise, a similar lexicon of words can be used to express similar emotions. We can greatly benefit from this aspect of language to develop a system that can recognize sentiments on a spectrum rather than the existing methods of

[1] We assume that the sentiment of customer towards a product is an indirect perception of the seller. Thus, the product description will certainly be same for various sellers, but the reviews and the images used are different from seller to seller.

recognizing bi-polar sentiments (positive and negative). Also, a system that can track multiple types of emotions can benefit the sellers more than a bi-polar one. Thus, our study will greatly focus on achieving spectrum sentiment analysis and analyzing a trend in it.

2 Literature Review

A lot of work has been done in the field of Sentiment analysis and its supporting subject such as NLP[2], text processing, semantic lexicon generation, image processing etc. Much of this research lays a foundation for our study and helps us in moving in the right direction. Much of such work that was inspirational for the study is discussed in this section.

2.1 Image

[1] shows how images can act as supplementary for sentiment analysis. The use of Flickr image dataset [2] for training a CNN[3] along with transfer learning proves to be quite effective. An approach of using domain specific dataset for feature extraction and machine training is rousing. A low-level feature extraction approach is synonymous to our approach as described in [3] with the difference of using a CNN to identify the features. We will plainly use the raw features available directly from the image. The reason for using raw features is to keep the system in check and not over-fit it with a particular feature. Features change form image to image, but the general genre shows a resemblance. A good reason to choose this is inspired by [4]. The paper describes how basic features of an image such as saturation and brightness can be used to derive Pleasure, Arousal and Dominance values. These values correspond to one of the 2 widely acclaimed models i.e. PAD model of emotion [5]. The model described in [4, p. 395] provides a spectrum value that our study aimed to achieve, thus acting as a stepping stone for our study.

2.2 Text

Much of the research has already been done in the field of sentiment analysis using text. However much of what is found focuses on the polar nature of words apportioning them into positive or negative. A semantic similarity approach in [6] using multiple sources has been motivating for using semantic similarity approach for spectrum emotion recognition.

2.3 Multi-modal Approach

Recently a lot of focus has shifted from performing sentiment analysis on a single type of data to having a multi-modal approach. A multi-modal approach ensures that the sentiment resonates with actual human sentiments as we tend to express emotions in different forms. Thus, an integration of text mining or opinion mining along with image analysis has proved to be of paramount importance for sentiment analysis. [7] describes

[2] NLP: Natural Language Processing

[3] CNN: Convolution Neural Network

a method for individually analyzing image and text for sentiments and then using a similarity-based classifier classifying the data for sentiments. [8] have too had early attempts of multi-modal sentiment analysis approach. Another yet innovative process is described in [9] which uses CNN and DNN[4] for analyzing the key features from text and images. This approach yields superior results but still lacks the spectrum approach nevertheless having greater accuracy.

The majority focus of research in sentiment analysis has been in the direct or multi-modal analysis but lesser attention has been given to spectrum approach and involving the time factor. Our study tries to tackle this while keeping the domain of e-commerce into mind.

3 Methodology

The following flowchart explains the process the system follows (Fig. 1).

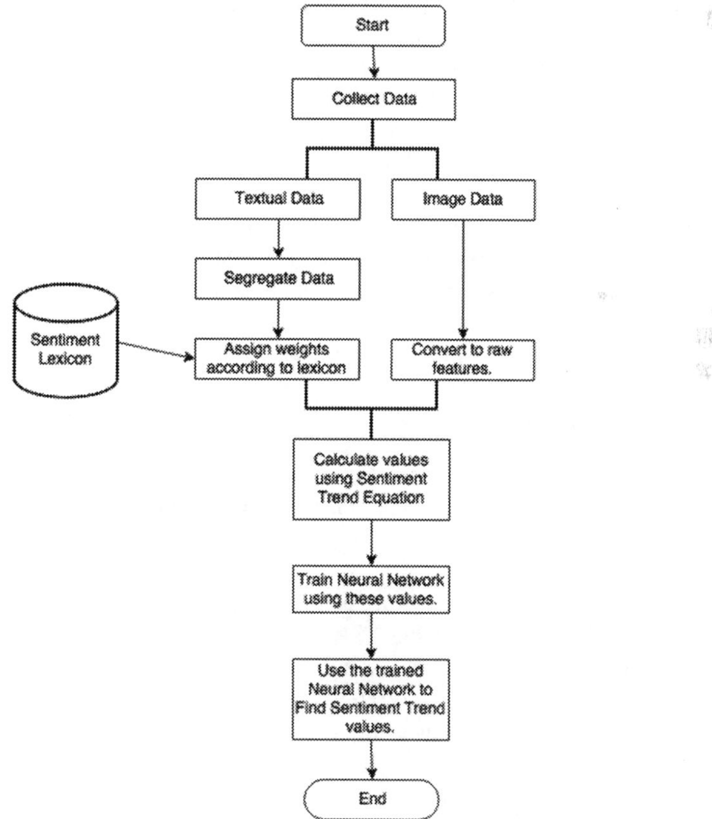

Fig. 1. The flowchart describing the process followed by the system for trend sentiment prediction.

[4] DNN: Deep Neural Network

3.1 Data

For accurate systems it is quite obligatory to have genuine data to work with. Since our field of study rests with the domain of e-commerce we require a dataset that is authentic in nature and quite descriptive. [10, 11] are a great source for such a data. [10] This dataset includes reviews (ratings, text, helpfulness votes), product metadata (descriptions, category information, price, brand, and image features), and links (also viewed/also bought graphs). Such data also contains the review timings to aid involving of time factor into sentiment trend analysis. This data is required to be processed in ways that make extracting features from it simpler. Unnecceary fields are removed and only the text (review content, review title), review timing, product id and the image (image link) are kept.

3.2 Image Processing

The raw features to be extracted from the images included: color features such as hue, saturation, brightness and structural features such as contrast, correlation and entropy. Hue, saturation and Brightness represent the general visual aspects of the image while the other three represents the structural aspects and play subtle role in determining the nature of the image. The basic pre-processing of the image includes resizing all of the image to a size of 160 px X 160 px. Images that have resolution below this size are discarded. Then using the individual pixel data calculate the most frequent hue, brightness level and saturation values. These values represent[5] the most probable values that the customer is to glance at. The rest three structural values of image viz. contrast, correlation and entropy were found out using the Mahotas python library [11]. It allows us to calculate the concerned values easily. All of these 6 values are stored separately with their product id. This concludes the basic pre-processing of image data.

3.3 Text Processing

Text processing has more steps as compared to image. Any sentence in any language has words that are important and words that are supplementary (stop-words). Firstly, the reviews and review titles were segregated by their length. If a review was too small say few words it was discarded, also reviews that had multiple paragraphs were discarded to reduce any error by introducing exaggerated or diminished valued text. Basic pre-processing of the text includes cleaning of these stop words. NLTK python library [12] is an excellent tool that was utilized for doing so. Later the words left were categorized by their POS[6]. Each set was associated with a product id. All of this data is stored along with their product id.

---

[5] A better representation of the hue, saturation and brightness could be taken by taking median or mean values of the image, however such a measure will not accurately represent the image. However, better methods to represent the image can be used at the expense of more complex computation.

[6] POS: Parts of Speech. Included verbs, noun, adjective and adverb

3.4 Trend Methodology

The data from both the image and text are combined into a single data file and the combination is done based on product id. The time data is stored for each review. The dataset contains multiple entries of image features with different reviews features. The words or parts of words extracted from the reviews are then utilized for calculating sentiment values. A process of identifying semantic similarity and assigning weights to the words is used. For this a simple lexicon is maintained for different sentiments classified based on POS. These act as core words for the semantic similarities. While checking for words against the lexicon a check for similar words is checked too. If the match is based on similar words a lesser weight is assigned to the word or phrase. A direct match yield higher weightage. Besides assigning weights based on individual words a n-gram of length 2 and 3 are used to have greater accuracy. Based on n-gram length, weights are assigned to semantic score found out between the n-gram and the available lexicon. At the end of the sentence all of these weighted values are summed up to calculate the final sentiment value for the particular sentence. This weighted sentence is then normalized based on length of important words in the sentence (Sentence without stopwords).

A heuristic approach is used to decide upon the weights for n-grams and similar words. All of the new sentiment values now found are stored alongside the image features. The next step involves factoring in the time constraint. The data now is grouped according to the product id and then chronologically. The data is then processed for individual product for factoring in time values.

For factoring in time, it is important to understand how time interacts with user sentiment. The 3 factors about time that are brought into the equation while factoring in time are:

1. Intensity: Intensity concerns with how often the value for a particular emotion for a product changes for a sentiment. A greater difference will indicate a greater change in the trend of customer sentiment. Also, consecutive increasing values will indicate an upsurge in the trend.
2. Direction: Direction involves the change in the direction of the sentiment. It involves the difference in sentiments from previous and current value.
3. Previous Value: For associating a trend it is apparent that the previous value should have certain impact on the next value to have a continuous trend.

The following equation explains the above factors:

$$y_n = f(t_{diff}, s_n, s_{n-1}, y_{n-1}) = g(t_{diff}) \times h(s_n, s_{n-1}) \times y_{n-1} + Sn \tag{1}$$

Where, y_n is the value for a trend to be calculated, t_{diff} is difference in time between 2 consecutive elements, s_n *and* s_{n-1} are current and previous sentiments respectively and y_{n-1} is the previous value of trend. Each of the individual functions are as follows:

$$g(t_{diff}) = \frac{\tanh(t_{diff})}{t_{unit}} \tag{2}$$

Here, t_{unit} is the idle time unit decided upon for measurement of time. The idle time unit is variable based on the frequency of the data available. Standard value can be assumed to be the minimum difference between any two consecutive values.

$$h(s_n, s_{n-1}) = \left|\frac{s_n - s_{n-1}}{r}\right|^r \tag{3}$$

Here, r is the range of possible values of sentiment.

The y_n values are individually calculated for all sentiments and the products are stored. These values i.e. hue, saturation, brightness, contrast, co-relation, entropy and the sentiment values (y_n) thus calculated will be used to train a neural network for recognizing the trend pattern for given values.

3.5 Training Neural Network

A neural network with 6 + n (n is the number of sentiments) input vectors is trained by taking the next successive y_n value to be the target value for $n - 1^{th}$ input values. Levenberg-Marquardt algorithm is used for training purpose and a simple 3-layer structure is used. Thus, a trained network will be able to generate the next value in trend for a particular product given current values for the product. A dataset of about 1900 reviews were used spanning more than 70 products from ‘pet supplies’ category. A separate dataset was maintained for testing purpose.

4 Observations

Figure 2 shows the result for one of the product that was tested on the neural network after training the network on dataset. The dataset showed great results as the values for the reviews corresponded with an increase in the popularity for the product. There is an increase in positive sentiment and a decrease in the anger sentiment as the product reviews span from the day 247 (the date for the first review) to the day 705 (the date for the last review).

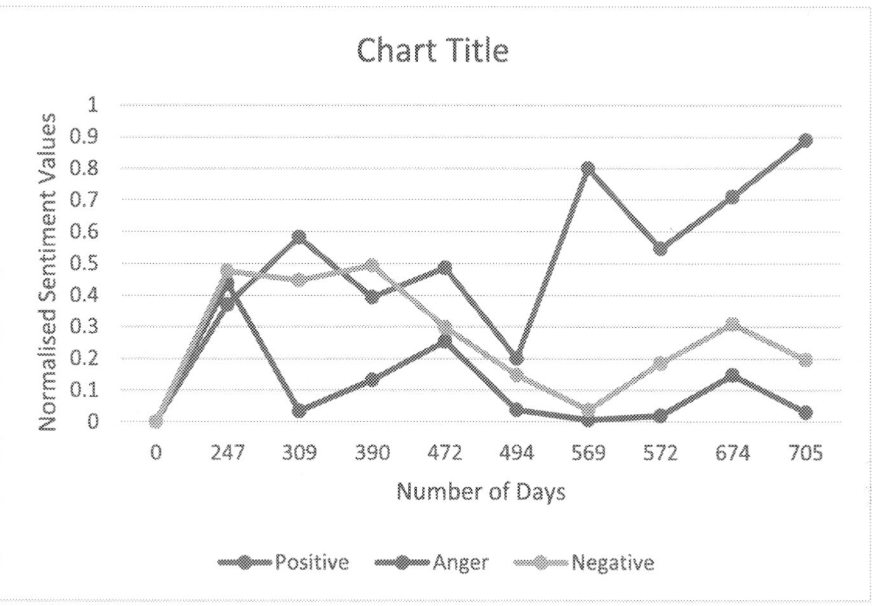

Fig. 2. Sentiment trend graph for product labelled 'natural balance'.

5 Conclusion

Sentiment Analysis is a tricky subject to deal with and the data that is to be used for the analysis usually contains lots of anomalies. E-commerce websites have been generating lots of relevant data that can be used for a spectrum sentiment analysis. A good volume of research has been done in multi-modal approach for sentiment analysis which inspired us for developing a spectrum sentiment trend prediction system. Authentic data from e-commerce website can be vital for developing an accurate system. Images play an effective role in generating genre-specific sentiment. The text plays a major role in sentiment analysis. The pre-processing methods used are primitive but show remarkable results.

The system we developed shows good results in recognizing sentimental values from review text and images. Sentimental Analysis on a spectrum range can be achieved at simpler level using the above-mentioned system. The results matched the human perspective of the sentiments from the reviews fairly and quantifiable measure allowed us to have a more certain view on the sentiments. (1) has parameters r and t_{unit} that control the depth and detail of the trend graph. This approach is a novel one (as per the knowledge of the researchers) and can prove to be fundamental in getting more accurate results. Whole of the system requires minimum data i.e. text, image and the time for the data and yields trend in the sentiments with simple computations. Such a system can easily be scaled to a much larger database to generate greater accurate results.

5.1 Benefit

The field of spectrum sentiment analysis has been given very little research focus given to it. Our study brings attention to the simplicity of such a field and the possibilities of much advanced methods to give more accurate results than just simply having a bi-polar sentiment analysis. A lexicon-based sentiment analysis approach has a benefit of simplicity, easy manipulation and scalability. The trend calculation has taken time factor into account by bringing difference between time for 2 consecutive entities for sentiment analysis. Moreover, since the system is developed with keeping e-commerce into account the idea of keeping time into the equation while calculating trend enriches the concept of trend prediction. Another one of the novel idea of the (1) is keeping the previous sentiment value for the same entity into the equation. This is done with an understanding that the trend should follow a smooth curve and each consecutive sentiment if in positive respect should increment the current trend value but if negative should only cancel out the previous sentiment trend value in proportion to the accumulative of the previous sentiments. That is to say a single negative sentiment or a single positive sentiment unless excessively positive or negative should not be able to cancel out the accumulated positive or negative values of the trend. The excessively positive or negative values are taken care of during pre-processing of text by eliminating text segments that are too long or seem fake.

5.2 Drawbacks

The system shows good similarities with human perspective and has considerable accuracy. However, the system is still in its primitive stages and requires much development. It has some drawbacks as identified by the researchers. The system is weak against smaller bits of text and cannot extensively recognize slangs or sarcastic remarks. A lexicon-based approach requires the system to have a substantial sized lexicon for all the respective sentiments. This can be a problem as the language is an ever-changing concept and does not necessarily mean the same every time. The lexicon must be category specific cause, different words can express different emotions in different domains. Though the difference is little and many of the words used for expressing an emotion are domain-neutral, lack of a dynamic lexicon will overtime bring in more error to the system.

5.3 Future Prospects

Since the system has few drawbacks, it opens up a new opportunity for perfecting the system and adding more features to the system. Some of them being using a CNN or DNN on the images for recognizing the domain and simultaneously training the neural network on text, using methods for identifying sarcastic remarks or use of slangs and appropriately assigning weights to them to tackle smaller text problems, using facial analysis systems for understanding the customer sentiments on social media sites and developing a sub-system to create a dynamic lexicon that can filter out words periodically and learn new words for each sentiment and dynamically add them while using the system. We strongly believe that the mentioned system has great prospects for further development and in itself is novel is a novel work of research.

Acknowledgement. We would like to thank University of Arkansas at Little Rock, AR - USA and Vishwaniketan (iMEET), INDIA for providing the opportunity to collaborate and work on this research project.

References

1. Jindal, S., Singh, S.: Image sentiment analysis using deep convolutional neural networks with domain specific fine tuning. In: International Conference on Infromation Processing (ICIP) (2015)
2. Jindal, S., Singh, S.: July 2015. http://dx.doi.org/10.6084/m9.figshare.1496534
3. Zhang, H., Yang, Z., Gönen, M., Koskela, M., Laaksonen, J., Honkela, T., Oja, E.: Affective abstract image classification and retrieval using multiple kernel learning. In: Lee, M., Hirose, A., Hou, Z.-G., Kil, R.M. (eds.) ICONIP 2013. LNCS, vol. 8228, pp. 166–175. Springer, Heidelberg (2013). https://doi.org/10.1007/978-3-642-42051-1_22
4. Valdez, P., Mehrabian, A.: Effects of color on emotions. J. Exp. Phycol. Gen. **123**(4), 394–409 (1994)
5. Wikipedia: https://en.wikipedia.org/wiki/PAD_emotional_state_model. Accessed July 2017
6. Li, Y., Bandar, Z.A., McLean, D.: An approach for measuring semantic similarity between words using multiple information sources. IEEE Trans. Knowl. Data Eng. **15**(4), 871–881 (2003)
7. Zhang, Y., Shang, L., Jia, X.: Sentiment analysis on microblogging by integrating text and image features. In: Cao, T., Lim, E.-P., Zhou, Z.-H., Ho, T.-B., Cheung, D., Motoda, H. (eds.) PAKDD 2015. LNCS (LNAI), vol. 9078, pp. 52–63. Springer, Cham (2015). https://doi.org/10.1007/978-3-319-18032-8_5
8. Maynard, D., Dupplaw, D., Hare, J.: Multimodal sentiment analysis of social media. In: BCS SGAI Workshop on Social Media Analysis (2013)
9. Yu, Y., Lin, H., Meng, J., Zhao, Z.: Visual and textual sentiment analysis of a microblog using deep convolutional neural networks. MDPI **9**(41), 4–5 (2016)
10. McAuley, J.: Amazon product data. http://jmcauley.ucsd.edu/data/amazon/. Accessed July 2017
11. Coelho L.P.: Mahotas: open source software for scriptable computer vision. J. Open Res. Softw. **1**(1), e3 (2013). http://dx.doi.org/10.5334/jors.ac
12. Loper, E., Bird, S.: NLTK (natural language toolkit). In: ETMTNLP 2002 Proceedings of the ACL-02 Workshop on Effective Tools and Methodologies for Teaching Natural Language Processing and Computational Linguistics, vol. I, pp. 63–70 (2002)

Paths Toward Social Construction of Knowledge: Examining Social Networks in Online Discussion Forums

David Raúl Gómez Jaimes[1(✉)] and María del Rosario Hernández Castañeda[2]

[1] University of New Mexico, Albuquerque, USA
dgomez25@unm.edu
[2] Universidad de Guadalajara, Zapopan, Mexico

Abstract. This mixed methods research project examined the relationship between social construction of knowledge and student centrality in three online discussion forums, which were part of a graduate online course on web conferencing in Spanish within the Mexican sociocultural context. The purpose of the study was to identify interaction patterns among twenty-one graduate students by analyzing discussion forum posts, measuring student centrality, and generating social network diagrams in order to explain the characteristics of posts and interaction dynamics that lead to social construction of knowledge. A sequential approach was used, starting with an interaction analysis model and social network analysis, followed by a combination of both analyses to shed light on interaction in online discussion forums carried out in Spanish.

Keywords: Online discussion forums · Social construction of knowledge Centrality · Interaction analysis · Social network analysis · Mixed methods

1 Introduction

Online courses lend themselves well to social constructivist instruction by providing students with opportunities to discuss ideas, work in teams to solve cases, problems, projects, and even assess themselves and their peers, which is part of the reason why online courses are as critical to the long-term strategy of higher education institutions around the world, as face-to-face courses. Furthermore, learning management systems and their user activity tracking and content archiving capabilities allow researchers to study online interaction among students in a relatively inexpensive way, technically speaking.

In this vein, what is the best way to orchestrate discussion forums that foster interaction in an online course? A current conundrum both in undergraduate and graduate online courses is interaction among students that leads to social construction of knowledge. In spite of a myriad of studies related to student-to-student interaction in online discussion forums, there is inadequate literature [1] about the orchestration of discussion forums that foster interaction aimed at generating social construction of knowledge.

© Springer International Publishing AG, part of Springer Nature 2018
G. Meiselwitz (Ed.): SCSM 2018, LNCS 10913, pp. 284–302, 2018.
https://doi.org/10.1007/978-3-319-91521-0_21

Social construction of knowledge is a phenomenon that can be defined as a function of interaction [2], which is understood as a reciprocal influence among individuals that engage in online interaction. Like a patchwork quilt, interaction is the collection of unique messages sewn together, resulting in socially constructed knowledge.

There are three themes in the literature about student-to-student interaction in online discussion forums, namely: (1) studies focused on the process of knowledge construction [3], (2) social networks [4], and (3) a combination of both [5]. However, most studies offer basic explanations of student-to-student interaction or do not provide practical solutions to the orchestration of discussion forums that promote interaction.

In U.S. Higher Education, around one in four students (28%) took at least one online course in 2015. Online students equaled a total of 5,828,826 students, which represented an annual increase of 3.9% compared to the 3.7% rate recorded in 2014. The total of 5.8 million online students included 2.85 million that took all of their courses online and 2.97 million that took some online courses. Public universities have the largest proportion of online students, with 72.7% of all undergraduate and 38.7% of all graduate-level students [6].

The 13th annual report of the state of online learning in U.S. Higher Education, reported even though the proportion of university leaders that say online courses are critical to their long-term strategy fell from 70.8% in 2014 to 63.3% in 2015, the proportion that rate the learning outcomes in online courses as the same or superior to those in face-to-face courses was at 71.4% in 2015. Furthermore, only 29.1% of university leaders reported that their faculty accept the value and legitimacy of online courses, as defined by Allen and Seaman's (2016) survey, and colleges with the largest online enrollments 60.1% reported faculty acceptance while only 11.6% of the colleges without online enrollments reported so.

In online courses "to have discussion for discussion's sake is not good instructional design. The discussions within an online distance education course must be well orchestrated to enable the learner to meet the learning outcomes, and build knowledge and insights" [7].

In the past scholars recognized online communication had the potential to represent a new generation of distance education [8] and paved the way for many studies on online and asynchronous group communication. For example, there was a study [9] about questions related to online interaction, particularly, related to the effects of the frequency of interaction, types of students, subject matter, alignment of interaction and learning objectives, and the effects of interaction on student satisfaction.

Later other scholars [10] published what would become a standard textbook on distance education in the USA, in which they devoted a chapter to technologies and media that included a section about learning management systems where they state online instructors "...have found the most valuable feature to be the asynchronous threaded discussion forum in text format. A discussion forum allows students and instructors to interact by posting and reading messages, while each has the flexibility regarding when they do it."

There is a scholarly reference on online distance education [11] that includes a chapter on interaction in the context of online courses, which presents a revamped version of a seminal idea of modes of interaction, namely: student-teacher interaction, student-to-student interaction, and student-content interaction. This reference states "although interaction among students has been studied most frequently, the various the [sic] forms and combinations of interaction discussed here would benefit from systematic and rigorous research using a variety of research tools and methodologies."

Thus, it is worth pointing out the following conundrum: what is the best way to orchestrate discussion forums that foster interaction in an online course? This is still a challenge both in undergraduate and graduate online courses as interaction among students may lead to social construction of knowledge.

1.1 Purpose of the Study

This mixed methods research project examined the relationship between social construction of knowledge and student centrality in three online discussion forums, which were part of a graduate online course on web conferencing. The purpose of the study was to identify student-to-student interaction patterns by analyzing discussion forum posts, measuring student centrality, and generating social network diagrams in order to explain characteristics of posts that lead or contribute to social construction of knowledge.

To approach said relationship, the Interaction Analysis Model [2]—commonly referred to as IAM—was utilized to determine if students constructed knowledge through interaction in discussion forums. In addition, SNA [12] was used to measure student centrality in order to account for the social aspect of knowledge construction. Graphing the structure of the social network that emerges from a discussion forum with social network diagrams is a way of "x-raying" interaction patterns with the ultimate purpose of identifying posts that provide potential paths to higher levels of knowledge construction.

This study was aimed at advancing the academic study of social construction of knowledge in online discussion forums previously reported [13–15], which demonstrated the adequacy of combining the Interaction Analysis Model and SNA. The relevance of supplementing the Interaction Analysis Model with measures of student centrality and social network diagrams that depict interaction patterns lies on the ability to advance previous studies not only by accounting for the social aspect of knowledge construction in social network terms, but by examining empirical data in Spanish within the Mexican sociocultural context.

Online instructors and instructional designers who develop online courses may find suggestions on the application of social constructivist principles to the design of discussion forums capable of fostering interaction. Also, this study may offer some clarification for university leaders on the alignment of discussion forums as a learning activity with the expected level of social construction of knowledge set by course and/or learning objectives as they relate to substantive and frequent interaction quality standards.

1.2 Research Question

How does social construction of knowledge relate to student centrality in online discussion forums?

2 Online Interaction and Social Network Analysis

There have been some research efforts to study interaction in online discussion forums, as it relates to both construction of knowledge and student centrality, from both a quantitative and qualitative perspective because textual data does not seem to be enough to explain the discussion process in a more visual manner and vice versa. For example, in studies where quantitative results were limited, several researchers [13–19] conducted mixed methods research to carry out supplemental analyses that explained social construction of knowledge and student centrality.

Taking into consideration the studies above, the main advantage of a mixed methods approach seems to be the ability for researchers to supplement their analysis with two or more perspectives, as opposed to being restricted to analysis techniques typically associated with qualitative research or quantitative research [20].

While the Interaction Analysis Model offers researchers a qualitative research technique that is subjective by nature to examine interaction in online environments (mediated by computer communication), SNA offers researchers different quantitative research techniques that are objective by nature to examine interaction in a variety of environments. Furthermore, the Interaction Analysis Model is an abstract way of outlining the process of social construction of knowledge and it is rooted in a theoretical framework based on social constructivism, on the other hand, SNA is a perspective rooted in sociology and social psychology, both of which focus on relationships or interactions among social entities and their patterns.

In addition, while researchers who use the Interaction Analysis Model argue for a complete post as the unit of analysis, social network analysts have developed unique techniques to analyze relation-based data, so they take a post as the unit of analysis in conjunction with the individual because though a post or an individual can fundamental units of analysis, in SNA they are not primary on their own because it is not theoretically sound to rely on separate units from this perspective, which requires researchers to operationalize concepts relationally.

To reiterate, the unit of analysis in SNA is also the post, but in connection to the student interaction, which occurs between members of the social network [21]. It is worth highlighting the fact interaction itself is the conceptual overlap between the Interaction Analysis Model and SNA that allows researchers to mix the two approaches because even though the attributes of these posts (e.g., the author, the message content) are primary to the first approach, they are secondary to SNA, but from a mixed methods perspective these attributes are key to the interpretation of the interaction patterns that are revealed by SNA.

The Interaction Analysis Model and SNA are similar in that both perspectives can be used to explain interaction and consider it equally relevant to analyze interaction patterns of independent relations as well as the totality of interconnected relations among social entities. Interaction is key in this study, because social construction of knowledge can be defined as a function of online interaction, which requires an information flow that coexists with a social relation among students.

2.1 The Interaction Analysis Model

The Interaction Analysis model [2] was created to examine knowledge construction in an online environment mediated by computer communication. The model's theoretical framework is based on social constructivist principles, so it considers knowledge construction as a function of interaction. The authors of this model put forward a definition of interaction that considers "the entire gestalt formed by the online communications among the participants" and presented an analogy between knowledge construction and a patchwork quilt as an organized whole with many unique messages sewn together. This definition of interaction is different than other definitions in that it does not focus only on individual relations, but on the totality of interconnected relations that emerge from online communication, so the authors argue for considering an entire message/post as the unit of analysis.

Due to the predominance of discussion forums as a fundamental ingredient for knowledge construction over other types of learning activities in online courses, it is worth listing the phases of the Interaction Analysis Model, which describes in detail five phases of knowledge co-construction, generally described as follows: Phase (I) sharing, comparing, Phase (II) dissonance, Phase (III) negotiation, co-construction, Phase (IV) testing tentative constructions, and Phase (V) agreement, application of new knowledge.

2.2 Social Network Analysis and Centrality

SNA [12] is a perspective that offers researchers both a set of algorithms and analysis techniques, which allows them to develop specific ways to measure phenomena and analyze relation-based data. Relation-based data is paramount in the operationalization of social networks because it is not sound to rely only on analytical techniques that consider separate individuals as primary. Studying phenomena from a network perspective requires that at least one theoretically significant concept be defined relationally e.g., social construction of knowledge—a function of interaction—involves an information flow that coexists with a social relation among students.

Researchers who study phenomena from a network perspective think about what kinds of networks are caused by different activities, such as interaction, which requires mapping sociological concepts onto particular network forms. Thus, when the effects of phenomena on networks is studied, the results are sociologically significant, in addition if something causes a network to be fractured so that there is a lack of relation or interaction between actors, the fracture matters because of the social effects it may have.

Social network analysts decide what kinds of networks and what kinds of relations they will study before collecting data [22]. There are two kinds of networks from which analysts must choose before starting to delimit the boundaries of their studies, namely: whole vs ego networks, and one-mode vs two-mode networks. Whole networks take a bird's-eye view of social structure, focusing on all actors rather than any particular one. These networks begin from a list of actors and include data on the presence or absence of relations between every pair of actors, for example, the network that emerges from students who interact in an online discussion forum. In contrast, ego networks focus on the network surrounding one actor, known as the ego.

Analysts use the whole networks approach to explain characteristics of social networks such as density, the average path length necessary to connect pairs of nodes, the average tie strength, the extent to which the network is dominated by one central actor (centralization) or the extent to which the network is composed of similar nodes (homogeneity) or of nodes with particular characteristics (composition), such as the proportion of network members who are women [23].

Most of the time, researchers who examine whole networks collect data on a single type of actor in networks where every actor could conceivably be connected to any other actor, therefore most of the networks they examine are one-mode networks. In contrast, two-mode networks, also referred to as affiliation networks, involve relations based on co-membership. In addition, researchers have to choose how to measure relations after selecting the kinds of networks they want to study and defining a theoretically significant concept relationally, and this choice is between directed or undirected and binary or valued relations [22]. Directed relations go from one actor to another and may be reciprocated, while undirected relations exist between actors in no particular direction. Both directed and undirected relations can be measured as binary relations that either exist or not within each pair of actors, or as valued relations that can be stronger or weaker.

Centrality. Only In SNA there is a group of metrics known as centrality measures, which quantify the relevance or influence of an individual in a social network based on her relations with other individuals. Central individuals or "actors are those that are extensively involved in relationships with other actors. This involvement makes them more visible to the others" [12], thus what is appealing for researchers studying interaction in online discussion forums is the relationship of students with higher centrality and social construction of knowledge.

With regards to social relations in online discussion forums, the question of who talks to who has important implications for information flow, so it is relevant to analyze interaction patterns of both independent relations and the totality of interconnected relations. Thus, student centrality is a concept that accounts for the social aspect of knowledge construction in that it serves as an indicator of student influence on other students. As I explained in my problem statement, the centrality of different individuals in a social network that emerges from a discussion forum can be analyzed with centrality measures and social network diagrams that depict interaction patterns.

From the SNA perspective, actors (also known as nodes) and their actions are viewed as interdependent rather than independent autonomous units, so the actors in this study will be students. Second, relational ties (linkages also known as arcs or edges) between students are interaction channels for transfer or "flow" of information through posts in discussion forums. Third, social network diagrams can represent patterns of interaction among students. Fourth, each student interacts with other students, each of whom interacts with a few, some, or many others, and so on. Therefore, the concept of social network refers to the finite set of students and their interactions in one discussion forum.

While centrality measures quantify the relevance or influence of an individual in a social network, there is a holistic measure of a social network that takes into consideration the totality of interactions named density [24], which defines "density, d, of a network" is the number of ties (interactions) in the network divided by the possible by number of ties (interactions). Thus, a well-connected social network—with high density—is one where everybody interacts with everybody else, enabling the flow of information in the presence of key students with high centrality (more influential), also known as "information brokers".

Social network diagrams provide visual representations of interaction in discussion forums that would otherwise be hidden to researchers, online instructors, instructional designers, and even students themselves, as demonstrated by some researchers [5, 25–27] who have used SNA to produce social network diagrams as a way of "x-raying" or mapping interaction patterns of online discussion forums to illustrate social construction of knowledge.

3 Methodology

3.1 Research Design

This was a sequential mixed methods study to examine interaction patterns of graduate students that participated in three discussion forums, which were part of an online course on web conferencing in a learning technologies master's degree at a Mexican university. The first stage of the analysis involved the application of Interaction Analysis Model to transcripts of discussion forums to find occurrences of social construction of knowledge by identifying qualitative characteristics of posts published by students. Next, centrality measures such as number of posts, in-degree, out-degree, and betweenness were taken to derive the degree of student centrality using SNA. Then, a comparison and contrasting of results from both methods was done, highlighting occurrences of social construction of knowledge of students with high centrality, and looking for the nature of the relationship between social construction of knowledge and student centrality.

3.2 Participants

Twenty-one graduate students between the age of 23 and 65 generated a dataset of discussion forum posts and gender was equally represented. This web-based secondary dataset of three online discussion forums contained de-identified authors, title, date, time, and posts extracted from a graduate course on web conferencing, which was part of a master's in learning technologies in a large public university in western Mexico. The discussion forums were deployed through the Moodle LMS, there was one discussion forum in the beginning, other in the middle, and another by the end of Spring 2015. These three forums were archived when the semester concluded in the university's LMS. The main inclusion criterion for this study was graduate students should have participated in discussion forums of the selected online course. There was not any sensitive information to be removed from any discussion transcript that could have compromised the identity of a student.

Due to the modular structure of the online course students were expected to study the content and participate in learning activities frequently as they had deadlines, so student-to-student interaction occurred primarily as required participation in discussion forums. At the beginning of the online course, students were studying factual information that introduced them to the subject, then as the course progressed gradually towards more analytical learning activities students were expected to engage in thought provoking discussions, and by the end of the course students worked in small groups preparing to host an educational web conference as a final project.

3.3 Unit of Analysis

The identification of the unit of analysis had to be reliable and encompass the phenomenon under study, so the post was chosen as the unit of analysis because it is objectively identifiable, meaning multiple coders can agree consistently on the total number of units; it produces a clearly delimited set of observations; and it has parameters determined by the author of the post. This choice addressed the lack of uniformity in the choice of the unit of analysis and inadequacies in reliability found in the literature. In addition, by concentrating on the post as the unit of analysis it was possible to report the intercoder reliability level in a straightforward fashion because coders did not need to argue about what a post is, as it is clearly defined by its author. Furthermore, the Interaction Analysis Model argues for a complete post as a unit of analysis.

In the application of the Interaction Analysis Model to examine transcripts of discussions, a post is taken as the unit of analysis and coded for as many occurrences or phases of social construction of knowledge as it contains, as opposed to mutually exclusive categories utilized in content analysis. When conducting SNA, a post can also be taken as the unit of analysis, but the post has to be taken in conjunction with the student because this perspective requires a relational concept such as the concept of interaction. Thus, the post in conjunction with the student become an actor (node) that may be connected to other students who interacted with each other in a discussion forum.

3.4 Data Collection and Analysis

The Interaction Analysis Model required discussion forum transcripts be extracted from a web-based secondary dataset archived in Moodle and exported both as PDF files and web archives, which offer great readability to human coders working with PDF readers or web browsers. Also, PDF files and web archives allow human coders to keep color highlights, annotations, and comments, keeping data safe in password protected computers with encrypted hard/flash drives. For example, posts were copied from said PDF files or web archives and pasted on a coding spreadsheet in order to have the text in the first column and then code with 1 or 0, as a way to improve precision.

SNA required network data, which had to be derived from the coding spreadsheets of the three discussion forums and processed using Microsoft Excel with NodeXL, a SNA plug-in [28], which facilitated entering posts as actors (nodes) with the actor labels being pseudonyms of students, and graph interaction(s) as edges or arcs. For example, if student A replies to student B, a directed edge (depicted as an arrow) was graphed from A to B. Directed edges were added with labels containing posting sequence number as well as the Interaction Analysis Phase of the post. NodeXL was also used to calculate the centrality measures and produce a social network diagram of interaction patterns. In the context of discussion forums, it is valuable to look at social network diagrams that show different interaction patterns and reveal student centrality. The preliminary step to generate these diagrams was to obtain the centrality measures of each student that published a post or replied to another student.

Interaction and social network analyses were conducted first to determine occurrences of social construction of knowledge and student centrality respectively. Then, results from both analyses were compared and contrasted, which involved the use of diagrams of interaction patterns in discussion forums that illustrate the centrality of different students. Finally, to explain the characteristics of the posts published by students with higher centrality, post excerpts were identified as textual evidence to complete the mixed methods design.

4 Results

4.1 Occurrence of Social Construction of Knowledge

Two coders who used the Interaction Analysis Model determined knowledge construction did occur through student-to-student interaction. To reiterate, the model's analysis procedure consists of reading every post from a discussion transcript and assigning them one or more codes for the purpose of identifying different phases of social construction of knowledge. When two coders use this type of model to code it becomes necessary to report the intercoder reliability level, which can be calculated using the percentage of agreement or Holsti's method [29]. In general, a Holsti's percent agreement higher than 90% or 0.90 is considered to be a high level of intercoder reliability and a percent agreement lower than 80% or 0.80 is considered doubtfully reliable [30].

As shown in Table 1 two coders concurred in determining a total of 24 occurrences of knowledge construction in forum 1 with a level of intercorder reliability of 87%.

Table 2 shows how the coders concurred in determining a total of 36 occurrences of knowledge construction in forum 2 with a level of intercorder reliability of 86%

Table 1. Occurrence of social construction of knowledge in forum 1.

Phase	Number of occurrences	Percentage
Phase I	18	75%
Phase II	0	0%
Phase III	6	25%
Phase IV	0	0%
Phase V	0	0%
Total	**24**	**100%**

Table 2. Occurrence of social construction of knowledge in forum 2.

Phase	Number of occurrences	Percentage
Phase I	13	36%
Phase II	11	31%
Phase III	11	31%
Phase IV	1	3%
Phase V	0	0%
Total	**36**	**100%**

Table 3 shows coders concurred in determining a total of 33 occurrences of knowledge construction in forum 3 with a level of intercorder reliability of 70%,

Table 3. Occurrence of social construction of knowledge in forum 3.

Phase	Number of occurrences	Percentage
Phase I	15	45%
Phase II	7	21%
Phase III	3	9%
Phase IV	8	24%
Phase V	0	0%
Total	**33**	**100%**

4.2 The Relationship Between Social Construction of Knowledge and Student Centrality

Student centrality accounts for interaction dynamics in the sense it can be a measure of the influence of a student in the social network that emerges from a forum due to an information flow that co-exists with a social relationship among students. In other words, students with high centrality have the quality of being at the core of the discussion as they are key to information flow.

Table 4 shows that the most prestigious student in forum 1 and the one with more potential access to information was S21, who reached a maximum phase of I by sharing/comparing information. The most influential student was S07, who reached a maximum phase of III by negotiating meaning/co-constructing knowledge. Although S02 did not have prestige or influence, she/he also reached a maximum phase of III. S03 and S17 did not have prestige or influence either, in addition to having the same potential access to information as everybody else, with the exception of S21, and they reached a maximum phase of I.

Table 4. Students with higher centrality in forum 1.

Student	Number of posts	In-degree	Out-degree	Betweenness centrality	Phases reached
S21	1	2	1	64	I
S07	2	1	2	34	I, III
S02	1	1	1	34	I, III
S03	1	1	1	34	I
S17	1	1	1	34	I

Table 5. Students with higher centrality in forum 2.

Student	Number of posts	In-degree	Out-degree	Betweenness centrality	Phases reached
S03	1	1	1	30	I, II, III
S16	1	1	1	22	I, II
S02	1	1	1	12	I, II, III
S12	1	1	1	12	I

Table 5 shows in forum 2 no student was more prestigious or influential than others. S03 had more potential access to information, followed by S16, the first student reached a maximum phase of III by negotiating meaning/co-constructing knowledge while the latter a maximum phase of II experiencing dissonance. S02 and S12 had the same potential access to information, which was less than that of S03 and S16, nevertheless S02 reached a maximum phase of III while S12 only shared/compared information reaching only phase I.

Table 6 shows in forum 3 no student was more prestigious or influential than others. S15 had more potential access to information and reached a maximum phase of IV by testing tentative constructions of knowledge. S04, S17 and S21 had the same potential access to information, which was less than S15, still S04 reached a maximum phase of IV, S17 a maximum phase of IV, but S21 a maximum phase of II by experiencing dissonance.

Table 6. Students with higher centrality in forum 3.

Student	Number of posts	In-degree	Out-degree	Betweenness centrality	Phases reached
S15	1	1	1	52	I, II, IV
S04	1	1	1	28	I, IV
S17	1	1	1	28	I, II, IV
S21	1	1	1	28	I, II

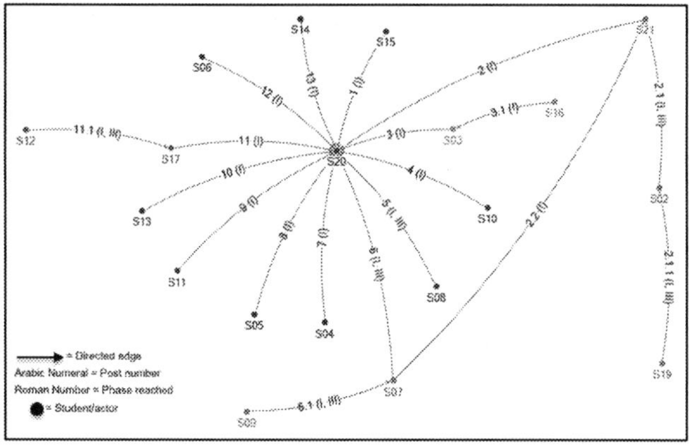

Fig. 1. Social network diagram of interaction patterns in forum 1

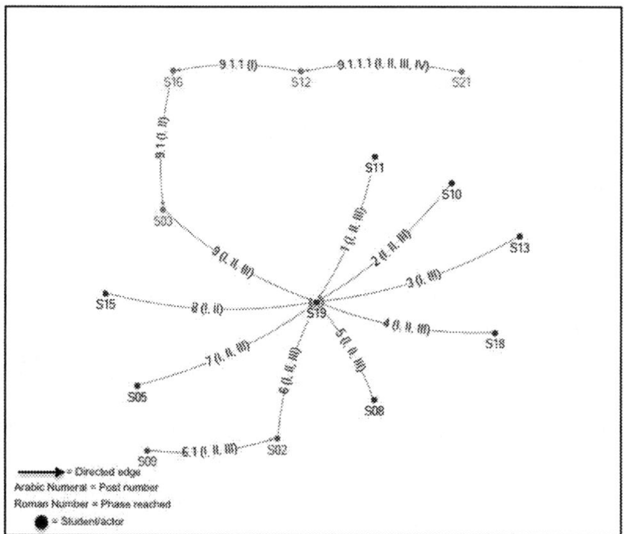

Fig. 2. Social network diagram of interaction patterns in forum 2

In the social network diagrams shown in Figs. 1, 2, and 3, edges are depicted with arrows showing the direction of information flow. Edges are labeled with an Arabic numeral indicating the sequence of the post and a roman number in parenthesis indicating the Interaction Analysis Model's phase reached by the student with that post.

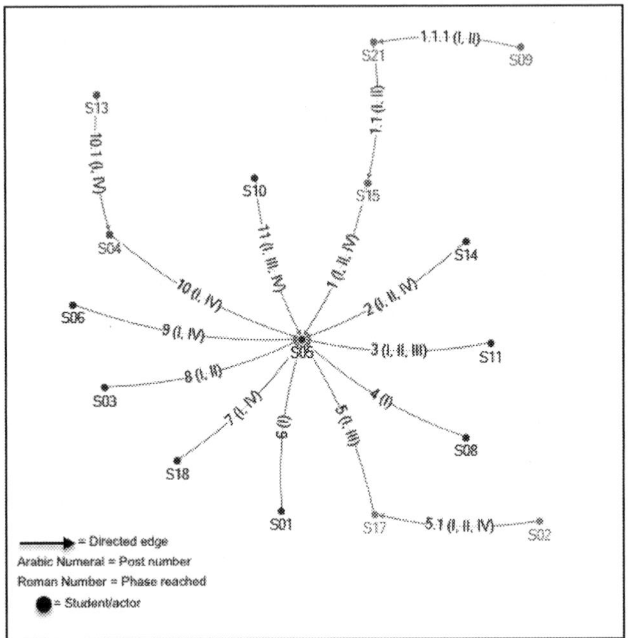

Fig. 3. Social network diagram of interaction patterns in forum 3

5 Findings

Interaction in online discussion forums was the main focus of this study as it is a construct that emerges from the conceptual overlap between the Interaction Analysis Model and SNA because interaction involves an information flow that coexists with a social relation among students. This conceptual overlap allowed the authors of this paper to mix both methods in the sense SNA supplemented the Interaction Analysis Model by accounting for the social aspect of knowledge construction with evidence of the basic generation of knowledge arising in and out of interaction within social networks of students that result from online discussion forums.

The model allowed two coders to measure occurrences of social construction of knowledge in three different discussion forums. The model's phases or coding categories were simple for coders to apply them to the discussion forum transcripts, which confirms the model's flexibility that has been appealing to many researchers who needed to identify the characteristics of posts that contribute or lead to higher levels of social construction of knowledge in discussion forums.

Knowledge construction occured through student-to-student interaction as evidenced by aggregate results of the three forums, which show a clear pattern of social construction of knowledge. There were a total of 93 occurrences in all forums spread out across forums as follows: 24 in forum 1, 36 in forum 2, and 33 in forum 3.

Almost half of occurrences in all forums, i.e., 46 out of 93 (49.46%) reached phase I, which is the lowest level where students share or compare information. Around one fifth of occurrences in all forums, 18 out of 93 (19.35%) reached phase II, which is a low level where students experience the discovery and exploration of dissonance or inconsistency among ideas, concepts or statements. Around one fifth of occurrences in all forums, 20 out of 93 (21.50%) reached phase III, which was arbitrarily set in this study as the standard to determine posts as higher level occurrences of social construction of knowledge because it is the one where students experience negotiation of meaning or co-construction of knowledge. Nine occurrences in all forums (9.67%) reached phase IV, where students experience the testing of tentative constructions of knowledge. There were not any occurrences that reached phase V, which is the highest level, where students experience agreement or application of new knowledge.

With regards to the trustworthiness of the Interaction Analysis Model's results from forum 3, an intercoder reliability level of PA = 70% can be considered doubtfully reliable based on a conservative interpretation of Holsti's percent agreement [30] to address the historical criticism to Holsti's method of being too liberal, statistically speaking. It is worth mentioning what other researchers who used the same model along with Holsti's percent agreement have reported, for instance some have reported a PA = 70% [31], while others [3] have reported a PA = 78% implicitly adhering to a more liberal interpretation of this statistic.

This insight into the results confirms a meta-analysis on literature about the Interaction Analysis Model [1], which reported its results were "quite similar to the results obtained in the original study [2]: there are low levels of complex thinking as the majority of operations coded remained in PhI. There is some evidence of operations in PhII and III, but they are almost non-existent in PhIV and V."

With regards to the sociocultural context, a previous study [32] attributed its finding of students making a leap from lower phases of social construction of knowledge to higher phases, without passing through intermediate phases to a lack of open disagreement, i.e., dissonance (phase II) was not evident in the data as open disagreement with ideas expressed by others might not to be appropriate or at least not a necessary element in the Mexican sociocultural context. The results of this study are inconclusive on this aspect because on the one hand previous findings [32] can be confirmed in forum 1, but not in the other two forums, where dissonance (phase II) accounted for a fifth (19.35%) of occurrences in all forums.

The centrality of different students in a social network that emerges from a discussion forum can be analyzed with centrality measures and social network diagrams that depict interaction patterns. Student centrality proved to be a concept that accounts for the social aspect of knowledge construction in that it serves as an indicator of student prestige and influence on other students as well as the degree of potential access to information as it flows through the discussion forum.

In-degree, out-degree, and betweenness account for student overall degree of centrality. The in-degree measure counts inbound posts with other students while out-degree counts outbound posts. These measures, when considered separately, are indicators of network "prestige" (in-degree) and influence (out-degree). "Prestige" results from the number of replies directed to a student's post and represents the degree to which other students seek out that student for interaction, thus students with high in-degree are notable because their information may be considered more important than others in the discussion forum. In contrast, students with high influence are in contact with many other students, as evidenced by the large number of discussion posts that they send to others, therefore students with low influence post fewer messages and do not contribute with information flow as much as other students. Betweenness reflects an individual's potential access to information as it flows through the network. The characteristics of a social network that emerges from an online discussion forum can be explained in terms of centrality measures obtained through SNA. Although the post is the most fundamental input required to take SNA measures, is not a centrality measure per se, but it is expected to be accounted for in centrality measures tables associated with online discussion forums as observed in the review of literature.

The characteristics of posts published by students with higher centrality in a given forum, can be explained in terms of social construction of knowledge by combining the Interaction Analysis Model and SNA. Social construction of knowledge involves phases such as sharing/comparing of information, dissonance, negotiation/co-construction of knowledge, testing tentative constructions of knowledge, and agreement/application of new knowledge, all of which require an information flow that coexists with a social relation among students. This information flow can in turn be explained with SNA in centrality measures terms, which reveal student centrality.

Whether higher student centrality contributes to a higher level of social construction of knowledge or not, is a question that can also be addressed by combining the Interaction Analysis Model and SNA as social network diagrams provide supplemental visuals related to the way information flows through the network that emerges from a discussion forum. One possible solution to establish what constitutes a higher level of social construction of knowledge is to use the Interaction Analysis Model to set the bar at phase III to determine posts at or above that phase as higher level occurrences of social construction of knowledge. In short centrality measures provide sound indicators of a student's ability to transfer information and exert influence over other students.

5.1 Limitations

One of the limitations of this study was the lack of access to other sources of data, which limited the scope of the study to the analysis of pre-existing and de-identified transcripts only. Even though this lack of access did not make it possible to ask follow up questions to any of the 21 participants or triangulate information, the research design did not require access to participants, the graduate program, or other type of documents, so in a way it would be appropriate to say it was a specialized analysis that required purposeful sampling, as opposed to a large number of participants, but the tradeoff was that as a specialized analysis it yielded a very specific answer to the research problem, like an "x-ray" that shows visual details, but still a visual from one angle and a given point in time.

Results should not be generalized because statistical tests were not conducted in this study due to the sample size of 21 participants. However, results should have a degree of transferability to similar contexts and settings. The sample size was the result of purposeful sampling, the rationale behind purposeful sampling was to select a set of participants that represented a typical case and it was not intended to make generalizations, which requires random sampling or selecting a large number of participants, as typically found in quantitative studies. In contrast, sample sizes are typically smaller in qualitative research, but sample sizes that are too small cannot adequately support claims of having achieved valid conclusions and sample sizes that are too large do not permit the deep, naturalistic, and inductive analysis that defines qualitative inquiry. Therefore, a sample size of 21 participants was a sound number the addressed the need of the researchers to reach middle ground through mixed methods.

Principles that guide SNA also limited the scope of the study in the sense the authors had to make certain assumptions to explain social construction of knowledge in social network terms. Again, the authors looked at relational data such as a social relation-information flow, not attributes of people such as age or income, in that, the social network approach was used to examine networks within a group of people not the group of people as a whole, which made it possible to make sense of people's centrality within networks, but not of people's centrality within the group. For example there were three social networks within the selected group of graduate students because there were three discussion forums in the dataset, therefore students may have had different centrality across the three forums and it is not appropriate to attribute an overall measure of centrality within the group.

Furthermore, relations were examined in a relational context, meaning the authors examined interaction patterns of a social network, not just relations between pairs or triads, which made it possible to account for the broader patterns of ties within the network to address the totality of interconnected relations that emerge from online interaction in a discussion forum. This strategy limited the study in that social networks had to be operationalized in a very specific way—carefully selected from a myriad of possibilities available to researchers—that addressed the phenomenon appropriately vis-à-vis the Interaction Analysis Model. Social networks were operationalized by focusing on whole networks, as opposed to ego networks, and on one-mode data, as opposed to two-mode data, and on directed ties. Other ways to operationalize social networks fall outside the boundaries of this study.

There was an abundance of literature on social construction of knowledge associated with the Interaction Analysis Model, SNA, and mixed methods about discussion forums carried out in English in undergraduate online courses from developed countries, but there was a scarcity of prior research reports on the same topic in connection to discussion forums in graduate courses from a different sociocultural context. This scarcity prompted the authors to limit the study to a graduate online course on web conferencing in Spanish within the Mexican sociocultural context.

The intercoder reliability level of forum 3, which can be considered doubtfully reliable with a PA = 70%, is not to be confused with negative results, but researchers are advised against adopting a more liberal position on the interpretation of this percent agreement statistic.

The aforementioned limitations matter in the sense that they point to the need for researchers to move forward aiming to address the relationship of social construction of knowledge and student centrality by taking into consideration the future research ideas.

5.2 Future Research

Future research should further investigate the association between social presence and the higher levels of knowledge construction according to the Interaction Analysis Model. Furthermore learning analytics can assist qualitative researchers in the application of techniques like data scraping, statistics, programming, and visualization to qualitative data, particularly when guided by models such as the Interaction Analysis Model to produce more robust findings.

5.3 Conclusion

"To have discussion for discussion's sake is not good instructional design. The discussions within an online distance education course must be well orchestrated to enable the learner to meet the learning outcomes, and build knowledge and insights" [7]. So, what is the best way to orchestrate discussion forums that foster interaction, which in turn can lead to social construction of knowledge. Researchers, online instructors, instructional designers and university leaders, need to gain insight into the orchestration of discussion forums to inform their decisions as they relate to online course offerings and substantive and frequent interaction quality standards.

The objective of this study was to identify student-to-student interaction patterns by analyzing discussion forum posts, measuring student centrality, and generating social network diagrams in order to explain characteristics of posts that lead or contribute to social construction of knowledge. The objective was met as different interaction patterns were identified and explained both in social construction of knowledge and social network terms. Furthermore, it was possible to explain the nature of the relationship between social construction of knowledge and student centrality and support this explanation with diagrams and information that put forward the idea of paths to higher levels of knowledge construction.

It is clear that interaction patterns in discussion forums have important implications for information flow, but the fact of the matter is, student-to-student interaction is important not because of the amount of posts, its frequency or timeliness, but because of its intent and form, which can be explained in terms of social construction of knowledge.

Having established a positive relationship between student centrality and the occurrence of higher levels of social construction of knowledge the authors put forward the notion that social interaction is as important as individual knowledge construction in a discussion forum, therefore there should not be trade-off between quantity of interaction and quality of information in student lead discussion forums, which suggests the balance of interaction lies on the proper alignment of student learning outcomes, specific learning objectives, materials, learning activities, but most important on providing students with explicit information such as grading rubrics, examples of posts, and other resources designed to set interaction expectations before students post as well as to make the social construction of knowledge explicit in a debate oriented forum, otherwise "student's won't know what they don't know," and knowledge construction will remain an hidden esoteric goal that only exists in abstract form in the online instructor's mind.

This study contributed to new knowledge about social construction of knowledge by explaining its relationship with student centrality and the same time it advanced the academic study of social construction of knowledge in online discussion forums previously reported [13–15] not only by accounting for the social aspect of knowledge construction in social network terms, but by examining data from a graduate level online course's discussion forums carried out in Spanish within the Mexican sociocultural context. In addition, this study supports the idea that the relationship between social construction of knowledge and student centrality helps researchers gain a better grasp of the characteristics of discussion postings and the degree of student centrality associated with potential paths to higher levels of knowledge construction. In sum, social network diagrams make the social dynamics of online learning tangible which extends the IAM analysis beyond its typical capacity of focusing on cognitive processes [33]).

References

1. Lucas, M., Gunawardena, C., Moreira, A.: Assessing social construction of knowledge online: a critique of the interaction analysis model. Comput. Hum. Behav. **30**, 574–582 (2014)
2. Gunawardena, C.N., Lowe, C.A., Anderson, T.: Analysis of a global online debate and the development of an interaction analysis model for examining social construction of knowledge in computer conferencing. J. Educ. Comput. Res. **17**(4), 397–431 (1997)
3. Chai, C.S., Tan, S.C.: Professional development of teachers for computer-supported collaborative learning: a knowledge-building approach. Teach. Coll. Rec. **111**(5), 1296–1327 (2009)
4. Gottardo, E., Noronha, R.V.: Social networks applied to distance education courses: analysis of interaction in discussion forums. In: Proceedings of the 18th Brazilian Symposium on Multimedia and the Web, pp. 355–358. ACM, October 2012
5. Toikkanen, T., Lipponen, L.: The applicability of social network analysis to the study of networked learning. Interact. Learn. Environ. **19**(4), 365–379 (2011)
6. Allen, E., Seaman, J.: Online Report Card: Tracking Online Education in the United States, February 2016. Accessed onlinelearningconsortium.org
7. Shearer, R.: Theory to practice in instructional design. In: Moore, M. (ed.) Handbook of Distance Education, pp. 251–267. Taylor and Francis, Hoboken (2013)
8. Lauzon, A., Moore, G.: A fourth generation distance education system: Integrating computer-assisted learning and computer conferencing. Am. J. Distance Educ. **3**(1), 38–49 (1989)
9. Kearsley, G.: The nature and value of interaction in distance learning. In: Third Distance Education Research Symposium, pp. 18–21, May 1995
10. Moore, M.G., Kearsley, G.: Distance education: a systems view of online learning, 3rd edn. Wadsworth-Cengage Learning, Belmont (2012)
11. Moore, M.G. (ed.): Handbook of Distance Education. Routledge, New York (2013)
12. Wasserman, S., Faust, K.: Social network analysis: methods and applications. Cambridge University Press, Cambridge (1994)
13. Aviv, R., Erlich, Z., Ravid, G., Geva, A.: Network analysis of knowledge construction in asynchronous learning networks. J. Asynchronous Learn. Netw. **7**(3), 1–23 (2003)
14. Li, Z.: Asynchronous discourse in a web-assisted mathematics education course. Thesis (Ph. D., Mathematics), University of Idaho (2009)

15. Buraphadeja, V.: An Assessment of Knowledge Construction in an Online Discussion Forum: The Relationship Between Content Analysis and Social Network Analysis. Thesis (Ph. D.), University of Florida (2010)
16. Dawson, S.: 'Seeing' the learning community: an exploration of the development of a resource for monitoring online student networking. Br. J. Educ. Technol. **41**(5), 736–752 (2010)
17. Mazza, R., Dimitrova, V.: Visualising student tracking data to support instructors in web-based distance education. In: Proceedings of the 13th International World Wide Web Conference on Alternate Track Papers & Posters, pp. 154–161. ACM, May 2004
18. Shea, P., Hayes, S., Vickers, J., Gozza-Cohen, M., Uzuner, S., Mehta, R., Rangan, P.: A re-examination of the community of inquiry framework: social network and content analysis. Internet High. Educ. **13**(1–2), 10–21 (2010)
19. Tirado, R., Hernando, A., Aguaded, J.I.: Aprendizaje cooperativo online a través de foros en un contexto universitario: Un análisis del discurso y de las redes. Estud. Sobre Educ. **20**, 49–71 (2011)
20. Creswell, J.W., Clark, V.L.P.: Designing and Conducting Mixed Methods Research. Sage Publications, Thousand Oaks (2007)
21. De Laat, M., Lally, V., Lipponen, L., Simons, R.J.: Investigating patterns of interaction in networked learning and computer-supported collaborative learning: a role for social network analysis. Int. J. Comput. Support. Collab. Learn. **2**(1), 87–103 (2007)
22. Carrington, P.J., Scott, J.: The SAGE Handbook of Social Network Analysis. SAGE, London (2011)
23. Freeman, L.: Centrality in social networks: conceptual clarification. Soc. Netw. **1**, 215–239 (1979)
24. Faust, K.: Comparing social networks: size, density, and local structure. Metodol. zv. **3**(2), 185–216 (2006)
25. Dawson, S., Bakharia, A., Lockyer, L., Heathcote, E.: "Seeing" networks: Visualising and evaluating student learning networks (Final Report). Australian Learning and Teaching Council Ltd, an initiative of the Australian Government, Canberra, Australia (2011). Accessed http://research.uow.edu.au
26. Haythornthwaite, C., De Laat, M.: Social networks and learning networks: using social network perspectives to understand social learning. Paper presented at the 7th international conference on networked learning, Aalborg, Denmark, May 2010. Accessed http://celstec.org.uk/
27. Firdausiah Mansur, A.B., Yusof, N.: Social learning network analysis model to identify learning patterns using ontology clustering techniques and meaningful learning. Comput. Educ. **63**, 73–86 (2013)
28. Hansen, D., Shneiderman, B., Smith, M.A.: Analyzing Social Media Networks with NodeXL: Insights from a Connected World. Morgan Kaufmann, Burlington (2010)
29. Holsti, O.R.: Content Analysis for the Social Sciences and Humanities. Addison-Wesley, Reading (1969)
30. Mao, Y.: Intercorder reliability techniques: Holsti method. In: Allen, M. (ed.) The Sage Encyclopedia of Communication Research Methods. SAGE Publications, Los Angeles (2017)
31. Tan, J., Ching, S.C., Hong, H.Y.: The analysis of small group knowledge building effort among teachers using an interaction analysis model. In: 16th Conference on Computers & Education, Taipei, Taiwan (2008)
32. Islas, L.: Collaborative learning at Monterrey Tech-Virtual university. In: Duffy, T.M., Kirkley, J.R. (eds.) Learner-centered theory and practice in distance education: cases from higher education, pp. 297–319. Lawrence Erlbaum Associates, Mahwah (2004)
33. Gunawardena, C.N., Flor, N.V., Gómez, D., Sánchez, D.: Analyzing social construction of knowledge online by employing interaction analysis, learning analytics, and social network analysis. Q. Rev. Distance Educ. **17**(3), 35 (2016)

Experimental Verification of Sightseeing Information as a Weak Trigger to Affect Tourist Behavior

Yuuki Hiraishi[1(✉)], Takayoshi Kitamura[1], Tomoko Izumi[2], and Yoshio Nakatani[1]

[1] Ritsumeikan University, Kusatsu Siga, Japan
is0230xp@ed.ritsumei.ac.jp
[2] Osaka Institute of Technology, Hirakata Osaka, Japan

Abstract. In this research, we verify information of sightseeing spots as a weak trigger which gives strolling tourists a chance to change their behaviors but does not specify the spot in a recommendation system. In a general recommendation system, the system provides complete piece of information about recommended spots. However, the provided information may deprive users of opportunities to discover interesting something by themselves. On the other hand, if no information is recommended to tourists, they may stroll in a restricted area because they have no hints of unfamiliar area. To reveal an appropriate information solving the above problems, we focus on the amount of information provided to users. Information about sightseeing spots is classified into the position and the feature information of a spot. For each information, we define the four categories of information according to the amount of information. We conducted the experiment with some subjects, and analyzed the impact on the information of these categories.

Keywords: User interface · Nudge · Suggestive methods
Sightseeing support system · Recommendation system

1 Introduction

1.1 Background and Motivation

Tourism trend is changing due to the development of information technology, especially the development of Social Networking Services (SNS) and mobile devices. In the previous tourism trends, tourists participated in a tour where all of destinations, routes to there, and time to spend there were predetermined by a tourism provider. In recent years, tourists plan their trips by their own will. That is, they decide their destinations and routes as they like. Such autonomous sightseeing has attracted attention and has been becoming a mainstream of tourism trend [1].

One of the reasons for this tendency is the expansion of SNS use. SNS users are able to share their own information about things they had and their experience of their sightseeing. Another user is easily able to obtain information about his/her interesting

© Springer International Publishing AG, part of Springer Nature 2018
G. Meiselwitz (Ed.): SCSM 2018, LNCS 10913, pp. 303–317, 2018.
https://doi.org/10.1007/978-3-319-91521-0_22

sightseeing spots in advance from the shared information on the SNS. That is, the services such as SNS make planning of sightseeing easy.

There many proposals of supporting sightseeing systems using sharing information on the SNS. One of the examples is a recommendation system. In [2], the system recommends sightseeing routes based on information sharing on photo-sharing sites. It is very convenient for tourists who visit an unfamiliar sightseeing area because they can know a suitable plan of their sightseeing in advance. However, these systems place importance on the efficiency of sightseeing, and then tourists who use one of the systems tend to follow the proposed plan. That is, it is possible that these systems limit free activities of tourists seeking autonomous sightseeing, and as a result, they may reduce opportunities which tourists encounter with accidental and interesting experience.

It is supposed that if tourists walk freely without any supporting system in a sightseeing area then they discover their favorite spots by themselves. Such experience may remain in tourists' memories more strongly than one which they visited recommended spots. On the other hand, if tourists have no information as a trigger to change their behavior, they may stroll in a restricted area. That is, no information proposed by a system may give free activity to tourists, but it may also restrict the various of their spontaneous actions.

1.2 Our Contribution

Accordingly, our goal in this research is to verify information of sightseeing spots as a weak trigger in a recommendation system. As a user, we consider tourists who enjoy strolling in a certain sightseeing area. The weak trigger means that it just gives tourists a chance to change their behaviors, and that it does not force them the changes. We assume that a detailed information about recommended spots may force tourists to visit there, and may deprive tourists of opportunities to discover interesting something by themselves. So, to give just a chance to discover interesting spots by themselves, a system does not show a recommended spot to them obviously. We consider the least amount of information about spots as a good trigger. For example, as for positions of spots, we set the four categories of information, a point, a direction, an area and no information. In our proposal, only a suggestive information about recommended spots according to one of the four categories are shown, e.g., an area in which the spots exist roughly. Such abstract information will trigger users to change their behaviors, but does not specify a spot.

A brief outline of this paper is as follows. In Sect. 2, we introduce other research related this study. Sections 3 describes classification of spot information. Section 4 describes the evaluation and consideration. Finally, we state our conclusions in Sect. 5.

2 Related Works

2.1 Sightseeing Support System Based on Inconvenient Benefit

In the research area about navigation systems, there are some studies that try to give tourists chances to change their behaviors by restricting provided information. This idea is based on the theory of the "FUrther BENEfit of a Kind of Inconvenience"

(FUBEN-EKI) proposed by Kawakami, which suggests that inconvenient things bring benefit in some cases [3, 4]. With advances in information technology, the notion of "anytime, anywhere" is taken for granted in modern society. However, there are benefits that has been overlooked because of too much emphasis on convenience and efficiency. What is important for supporting "inconvenient benefit" is not to create an inconvenient situation. It is to find an inconvenient mechanism in order to discover benefits that cannot be found by convenient tools.

Nakatani et al. [5] proposed a sightseeing navigation system based on handwritten routes. In this system, a user writes a sightseeing plan, such as destination and routes, by hand before his/her sightseeing, and then uses it as a reference during his/her sightseeing. Since the handwritten routes have many distortions, the user cannot know the exact routes on site. Tanaka et al. [6] proposed a navigation system which hides the map of area within a radius of 100 m around the user in accordance with the users' movement. Moreover, Takagi et al. [7] developed a system that navigates users only using information on direction and landmarks that are scattered throughout the tourist destination, without any detailed map information. These systems restrict map information given to tourists in order to promote actions of users such that they confirm their surrounding roads and buildings.

In these navigation systems, they focus on the information about map (i.e., route), not the information about recommended destination spots. For spots, these systems show detailed information, such as their locations, photos, or introductory sentences.

2.2 Recommendation System of Tourist Information

As for systems dealing with information about spots, recommendation systems have been studied actively. There are many studies about recommendation systems considering various conditions of spots or tourists. Oku et al. [8] proposed the methods to recommend spots based on posted information (e.g., tweets in Twitter, or photos taken in the spots) on the Internet. Misu et al. [9] evaluated the effect of the sightseeing application which provides spots in Kyoto based on the current feelings of tourists and the feature of sightseeing spots.

However, these studies focus on which spots should be recommended to users at the time, and they do not mention how to provide the information about the recommended spots. In the most of previous studies about recommendation systems, the detailed information about the recommended spots, such as their names, locations and photos are given to users.

2.3 Nudge and Suggestive Interface

Our goal is to provide suggestive information as a weak trigger to change tourists' behaviors, rather than the detailed information about the recommended spots. Such triggers or gimmicks are studied as a Shikakeology [10], which is the design method of suggestive triggers change people's behaviors or consciousness. Nudge is such the weakest triggers [11]. These triggers encourage them to a desired configuration, but do not prevent free behavior of users. Our proposal is the same concept of Nudge.

Kurata [12] proposed a sightseeing support system using a suggestive information. This system provides "Potential-of-Interest Maps", which have the similar characteristic to our purpose. The system visualizes the degree of attraction of spots which is calculated from the vast amounts of information that have been posted on photo-sharing sites. That is, the area where more photos are posted to the site are illustrated by deeper red. Users can know that the area has attracted attention of others, but cannot know what spots is in the area. However, in this research, the evaluation for only the one output design method was performed, and there is no comparative evaluation in terms of information provided to users.

3 Classification and Output Design of Information

3.1 Classification of Spot Information

As a user in this study, we consider tourists who enjoys strolling in a certain area and has no predefined destination. Our goal is to verify information of sightseeing spots as a weak trigger which gives strolling tourists a chance to change their behaviors but does not specify the spot in a recommendation system.

As information about sightseeing spots, there is research by Izumi et al. [13]. They categorized tourist information given to users into the location information of sightseeing spots and the characteristic information which introduce the summary of spots, and classified each information into four categories. However, in this previous work, some expression methods of these information is not able to properly express information on sightseeing spots. Furthermore, their evaluation experiments focused only on whether or not the subjects visited to the recommended spots, and no analysis about what category of information effect on the behaviors of the subjects. Therefore, in this research, we redefine the classification of spot information and show a suitable display design for the classification.

Information about sightseeing spots is divided to information about what and where the spot is. The first one is called the feature information and the other is called the position information of the spot. To verify information as a weak and a good trigger, we focus on the amount of information. We classify the position information and the feature information into the four categories.

First, we consider the feature information of spots. In a general guide book about sightseeing, information about the spots consists of category (e.g., restaurants, historical architectures), detailed introductory sentences, and photos of them. Among these information, photos have the largest amount of information and it gives tourists practical visual images of spots. Introductory sentences have the second largest amount of information, and categories have the least amount of information about spots. Therefore, we set the following four categories of the feature information about recommended spots:

- "None": There is no information about the feature information of a spot.
- "Category": A category of a spot is shown.
- "Text": Introductory sentences about a spot are shown.
- "Photo": A photograph of a spot is shown.

Regarding the position information of a spot, we also set the four categories depending on the dimension of the information expressing the position. That is, the exact position of a spot, called "pin" (i.e., a point), has the largest amount of information as its position. The one dimensional information corresponds to the "direction" to a spot, and the two dimensional information corresponds to the "area" in which a spot exists. The higher dimensional information has less information about location. Therefore, we set the following four categories of the position information about recommended spots:

- "None": There is no information about the position of a spot.
- "Area": An area in which a spot exists is shown roughly.
- "Direction": A direction to a spot from the current location is shown.
- "Pin": A pin is displayed at the exact location of a spot.

3.2 Output Design of Information

This section shows how to present the information described in the previous section on an actual system screens. The left figure in Fig. 1 shows the examples of the output for each pattern of information.

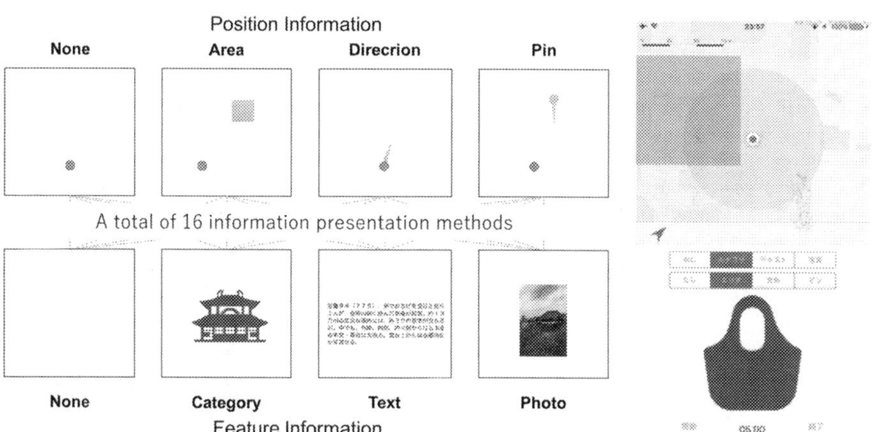

Fig. 1. Classification of provided information about spots (left) and an example of system screen (right). For the position information, the blue circle shows a current location of a user. The orange diagram shows the position information of a recommended spot. (Color figure online)

Each pattern of the feature information is shown in the following way (See Fig. 1):

- "None": The screen is blank.
- "Category": A category of a recommended spot is shown by using a corresponding pictogram. The categories of spots are determined based on the commercial guide books about a sightseeing area. Pictograms are generally used to present an intuitive image of a target. In some guide books, the pictograms are used to improve their

readability. In this study, we decide the design of the pictograms based on some guide books, and use the pictograms which are released free on the Web. Table 1 shows the pictograms for the categories we applied.

Table 1. The pictograms representing the category.

Pictograms	Categories
	Café
	Store or Souvenir shop
	Restaurant
	Temple or Shrine
	Strolling area
	Historical building

- "Text": The introductory sentences of a spot we got from the guide books are shown on the screen.
- "Photo": We took a photograph at each spot actually. The photograph is shown on a display.

The position information of a spot is provided on the electronic map with the current location of a user. On the map, the blue circle indicates the current position of a user, and the orange diagram (e.g., square, triangle) shows the position of a recommended spot (See Fig. 1). Each pattern of the position information is shown in the following way:

- "None": There is no information about position of a spot. Only the current position of a user is displayed on the map.
- "Area": A rough area in which a recommended spot exists is shown by a square. More precisely, a point in a square shape with the side 50 m whose center is a

location of a recommended spot is randomly selected. Then, a square shape with the side 100 m whose center is the selected point is shown on the map. By randomly setting the center of the displayed area, a user becomes difficult to estimate the position of a recommended spot.
- "Direction": An arrow represented by a triangle is displayed at the user's current position.
- "Pin": A pin is displayed at the exact location of a recommended spot.

On the screen, the position and the feature information are shown at a time. The right figure in Fig. 1 shows an example of the system screens. On the screen, the position information is shown in the upper half of the screen, and the feature information is shown in the lower half of the screen. As you see, we have the sixteen patterns of provided information about a recommended spot by the combination of the feature information and the position information.

4 Overview of Experiment

4.1 Recommendation Algorithm and Devices

For the experiment, we developed the prototype system on iOS terminal of Apple Inc. [14] and used it. The right figure in Fig. 1 shows the actual screen of the prototype system. In the figure, the screen shows the area of the position information and the category of the feature information of a spot.

To recommend a spot, we adopt the recommendation method by collaborative filtering [15] using NMF (Nonnegative Matrix Factorization). In the preliminary experiment, we required some subjects to answer their evaluation values for the target sightseeing spots, and prepared the base evaluation data for the recommendation. The results of the recommendation for a subject were outputted in a ranking order of spots. In the evaluation, the prototype system recommended a sightseeing spot with the first rank among spots which were not visited by the subject in the evaluation.

4.2 Experimental Procedure

We conducted this experiment with the cooperation of 24 college students (20 males and 4 females). In order to enjoy sightseeing having a conversation with a friend, we set up the pairs of the subjects so that the paired subjects are familiar with each other. So, there is no difficulty of their communication during their sightseeing. In the experiment, we got 12 pairs totally.

The experiment site was set to the sightseeing area in Shijo/Kawaramachi, Gion, and Kiyomizu Temple area in Kyoto city. In the area, there are many streets suitable for casual stroll on foot, and there are many sightseeing spots for every category, including historical architectures, shops for goods and souvenirs, restaurants, and strolling areas. The experiment site was divided to the four areas totally in order to have the pairs of the subjects used the four output patterns of information. The sizes of the areas and the numbers of sightseeing spots in the areas were set so that there were few difference between them.

Each pair of the subjects strolled freely by using the prototype system which outputted the information based on one of the sixteen patterns. Each pairs strolled four times in the four different areas totally, and used prototype systems with the different four output patterns with the same category in terms of the feature or the position information. Table 2 shows the patterns of the outputted information were applied to each pair. For example, the pair B saw the information about the area as the position and no feature information in the area 1. In the area 2, the pair B got the category information as the feature information, but the position information for the pair B was not changed, i.e., area. As shown in Table 2, each pattern was applied to three pairs totally. The day of the experiment was different for each pair, but each pair strolled four times on the same day.

Table 2. The pattern each pair of the subject used.

Pair	Area1	Area2	Area3	Area4
A, I	**None** None	**None** Category	**None** Text	**None** Photo
B, J	**Area** None	**Area** Category	**Area** Text	**Area** Photo
C, K	**Direction** None	**Direction** Category	**Direction** Text	**Direction** Photo
D, L	**Pin** None	**Pin** Category	**Pin** Text	**Pin** Photo
E	None **None**	Area **None**	Direction **None**	Pin **None**
F	None **Category**	Area **Category**	Direction **Category**	Pin **Category**
G	None **Text**	Area **Text**	Direction **Text**	Pin **Text**
F	None **Photo**	Area **Photo**	Direction **Photo**	Pin **Photo**

In the experiment, first, in order to recommend suitable sightseeing spots for each pair of the subjects, we asked them to answer their interests in a 5 scale for each spots of the 80 target spots before the experiment. On the experiment site, we explained how to use the prototype system and the purpose of the experiment. All of the pair strolled in the order of area 1, 2, 3, and 4. The subjects strolled in each area for 45 min. In each area, the system recommended three spots to the subjects. The three recommended spots were outputted from the prototype system at 5, 20, and 35 min after starting the experiment in each area. An observer walked together the subjects taking a distance. After the strolling in each area, we conducted a questionnaire about the applied pattern of the information in the area.

In this experiment, we focused on the following two points for the evaluation of each proposed information:

1. Does the provided information give the subjects chances to change their behaviors?
2. Does not the provided information make the subject notice the recommended spots?

To evaluate the points above, we recorded the positions of the subjects. In addition, the observer took the video data to get the situations of the subjects. In the questionnaire, we asked the questions about the degree of attentions of the subjects to the output from the system. The reason of this is that our goal is to verify the information as a weak trigger to change the behaviors of users, not to force the change of the behaviors. In the questionnaire, we set the four degrees of the attentions, such as, awareness of the output, watching the screen, seeing the contents, and decision based on the contents. Specifically, we set the following four questions in the questionnaire:

1. Were you aware of the output from the system?
2. Did you watch the screen of the system?
3. Did you see the contents provided by the system?
4. Did you decide your next actions based on the outputs from the system?

For these questions, the answer format were 5 scales, in which 1 corresponds to strongly disagree and 5 corresponds to strongly agree.

In the questionnaire, we also asked the following questions and the free opinions from the subjects for our interests:

5. Did you enjoy your strolls in this experiment?
6. Do you feel that your strolls using the system was more fun compared with your usual stroll?

4.3 Experimental Results

We explain the results using the logs of the subjects' positions and the voice or the video data of the subjects in this section. For each pattern of the provided information, while we got the data of the three pairs, we show one of them in the paper due to the limitation of space. Note that the shown one result has a similar characteristics or tendency as the other two results in every pattern.

In the most cases that no position information is provided to the subjects (i.e., "none"), there was few changes of the movements of the subjects. The reason of this is that the subjects had no idea which direction to go in order to visit the recommended spots even if the feature information was provided. In the paper, we show the results of all the cases excepting the case of no position information.

Table 3 shows the results in each case. Each map shows one of the results for the combination pattern of the corresponding row feature information and the corresponding column position information. That is, the top-left map is the result for the pattern of "none" and "area. On the maps, the voice data are mapped at the locations where the subjects acted of saying the words. The blue circled numbers indicate the

Table 3. The results of the movement history and the voice.

Feature Information	Position Information: Area	Position Information: Direction	Position Information: Pin
None	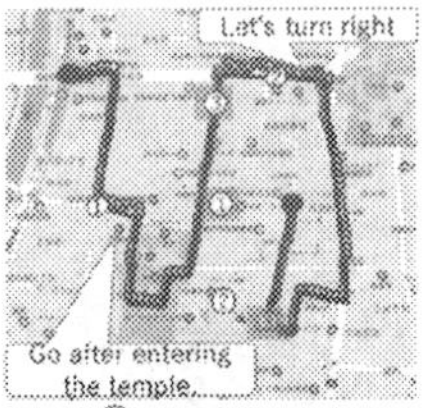 Let's turn right Go after entering the temple. Area2: ①Wabiyakorekido (Store or souvenir shop) ②Yasuikonpiragu(Shrine) ③Rakuraku(Café)	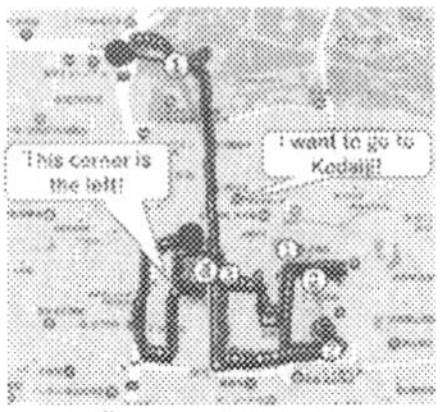 This corner is the left! I want to go to Kodaiji! Area3. ①Nenenomichi(Strolling area) ②SLOWJETCOFFEE(Café) ③Hisago(Restaurant)	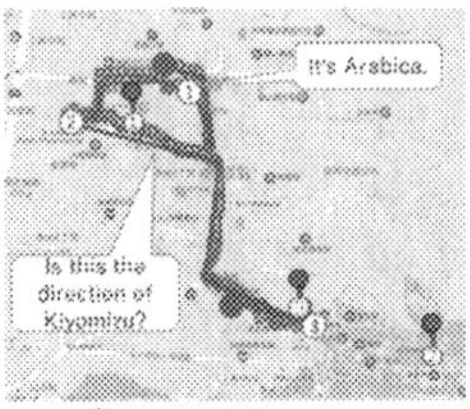 It's Arabica. Is this the direction of Kiyomizu? Area4. ①%Arabica(Café) ②Jisyu-jinja(Shrine) ③Ma-ruburansyu (Store or souvenir shop)
Category	 Just right, let's turn right here. Here left! Area2: ①Kenninji(Temple) ②Tsujiri(Café) ③Yasuikonpiragu(Shrine)	 The recommended spot is that direction! The recommended spot is this direction! Area2: ①Kenninzi(Temple) ②Tsujiri(Café) ③Yasuikonpiragu(Shrine)	 Turn right here! Here right? It is far to go from now. Area2: ①Kenninji(Temple) ②Yasuikonpiragu(Shrine) ③Tsujiri(Café)
Text	 Let's turn right here! Let's turn right next! Spot is here! Well then let's search for a little while ago! Area2: ①Kenninzi(Temple) ②Gyarusonkure-pu(Café) ③Yasuikonpiragu(Shrine)	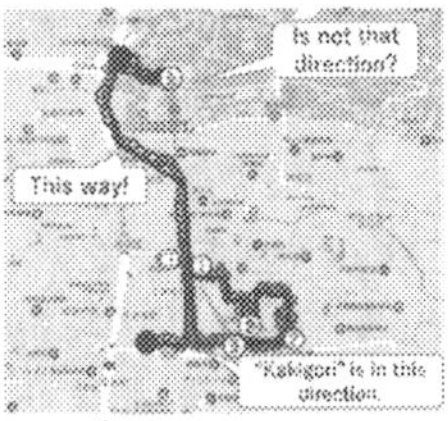 Is not that direction? This way! "Kakigori" is in this direction. Area3. ①Hisago(Restaurant) ②SLOWJETCOFFEE(Café) ③Pageone(Café)	 Let's go over there. Let's go around. Area3: ①SLOWJETCOFFEE(Café) ②Hisago(Restaurant) ③Jouvencelle(Café)
Photo	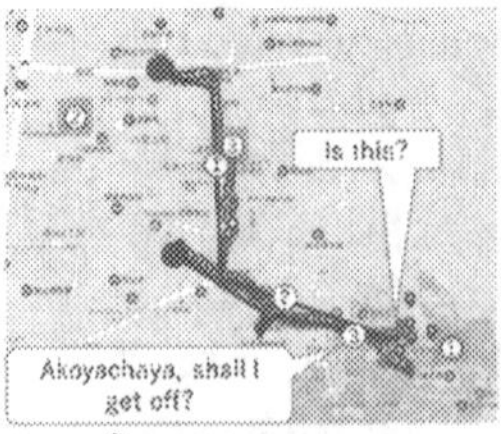 Is this? Akoyachaya, shall I get off? Area4: ①Jisyu-jinja(Shrine) ②%Arabica(Café) ③Akoyachaya(Restaurant)	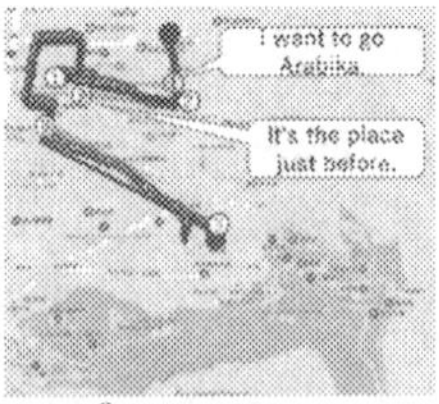 I want to go Arabika. It's the place just before. Area4: ①%Arabica(Café) ②Akoyachaya(Restaurant) ③Ma-ruburansyu (Store or souvanir shop)	 Let's get off the slope while tasting. This is the place of marriage in Kiyomizu. Area4: ①Akoyachaya(Restaurant) ②Jisyu-jinja(Shrine) ③Ma-ruburansyu (Store or souvenir shop)

locations of the subjects where the system outputted the information about the recommended spots. In the lower part of each map, the recommended spot at each location is shown. The red numbers indicate the locations of the recommended spots of the corresponding number.

The effect of the "area" as the position information: Regardless of the patterns of the feature information, the subjects moved to the direction to the recommended spots. The subjects decided their directions based on the outputted area on the corners. However, in the case that "none", "category", or "text" of the feature information, the subjects could not visit the recommended spots even though they went to the area near the recommendation. It seems that they could not detect the spots.

The effect of the "direction" as the position information: The results of the cases of the "direction" have the same tendency that of the "area" information. The subjects changed their direction of the movement based on the "direction" information of the recommended spots. In addition, they could not visit the recommended spots for the many cases of "none" and "category" information. However, in the case of "text" and "photo", the subjects detected the recommended spots, and moved there changing their direction exactly.

The effect of the "pin" as the position information: In all of the patterns, the subjects detected the recommended spots. That is, it is said that the exact position information of spots makes the guess of spots easy. Especially, by giving the exact position information on the map, the subjects could know the distance to the spots. Such knowledge has effect on the decision of changes the direction of the movement.

The effect of the "none" as the feature information: Even if no feature information was given to the subjects, they detected the recommended spots based on the position information. However, except the case of "pin" information, they did not notice the recommended spots.

The effect of the "category" as the feature information: By the effect of the position information, the movement to the direction to the recommended spots were caused. The observer saw the behavior of the subjects such that they looked for the spots based on their category information. Since they had the hints of the spots as their categories (i.e., temple, café), they could explore to find them, but they could not detect them.

The effect of the "text" as the feature information: In some texts, there were keywords for the recommended spots. See the result of the patterns of "text" and "direction". The subjects noticed the keywords "Kakigori" (it is Japanese shaved ice) of the recommended spot, and looked for the spots corresponding "Kakigori". In the experiments of the other pairs, there were many cases where the recommended spots were specified.

The effect of the "photo" as the feature information: The information of photo had a strong effect to the detection of the spots. The voice data from the subjects included the shop name or the keywords of the spots. The subjects saw their surrounding environment to take a matching to the given photo. Based on the visual image of the spots, some subjects understood them.

We will summarize the results for the position information. From the results, it is said that if no position information of the spots were given, few subjects moved toward them unless detailed feature information were given. Given the position information of "area", "direction", "pin", the subjects changed their direction of movement to the given spot in many cases. Especially in the case of "direction" or "pin", the subjects often grasped the position of the spots accurately. The exact position information also gave the relative position of the recommended spots.

For the feature information, even in the case of "none" and "category", we were able to confirm the actions of moving towards the spots based on the position information. However, in these cases, it was found that the recommended spot cannot be specified. When the feature information was presented by "text" or "photo", the recommended spots were sometimes specified. This is because the spots were easy to guess when the keywords in the texts or photos deeply related to the spots.

4.4 Questionnaire Results

Table 4 shows the results for the questionnaires about the degree of the attentions of the subjects to the output from the system. The results are shown in the average scores for the question in each combination of the information. The values of "Average" are the averages of the scores in the same row or column.

The results for the question about the awareness of the output shows that for the position information, the average values for "area", "direction", "pin" are over 3.00. Especially, the scores are high for "area" and "pin". For the feature information, the cases of "text" and "photo" have high scores of about 4.00. In particular, the combination of "area" and "text" has 4.50, and the combination of "pin" and "photo" has 4.67 score. Also, even if nothing is displayed about the position information, the subjects were aware the output from the system because they got the information of "text" or "photo" as the feature information.

The results for the question about the watching the output from the system show that in the cases that the position information is "direction" or "pin", the scores are high regardless of the feature information. There are no differences between the feature information. We can see this tendency in the results for the third question. The scores are high regardless the feature information when the position information is provided by "area", "direction", or "pin". From these results, it is said that the subjects would confirm the information whenever the system output. Especially, if the direction to a spot or the exact position of a spot is provided, this tendency becomes strong. For the feature information, if the information has some content, "category", text", or "photo", must of the subjects saw it.

The results for the last question show that for the position information, the scores are high values in the case of "area", "direction" or "pin". For the feature information, the case of "category" has a higher value than the others. In particular, the pattern of "direction" and "category" has the high average score 4.50.

From the results above, we summarize the results of the questionnaire. First, we consider the position information. In the case with no position information, the outputs from the system tended to be less noticed, and then, the subjects did not act based on the output. Furthermore, when the position information was given as "area",

Table 4. The results of the questionnaire.

Average scores for the questions						
Question: Were you aware of the output from the system?						
		Feature information				
		None	Category	Text	Photo	Average
Position information	None	1.83	2.50	3.67	3.33	2.83
	Area	3.33	3.17	4.50	3.83	3.71
	Direction	3.00	3.00	3.67	4.00	3.42
	Pin	3.33	3.00	4.17	4.67	3.79
	Average	2.88	2.92	4.00	3.96	
Question: Did you watch the screen of the system?						
		Feature information				
		None	Category	Text	Photo	Average
Position information	None	2.83	3.00	3.33	3.83	3.25
	Area	3.83	3.50	3.83	3.67	3.71
	Direction	4.50	4.33	4.33	4.33	4.38
	Pin	5.00	4.67	4.83	4.50	4.75
	Average	4.24	3.88	4.08	4.08	
Question: Did you see the contents provided by the system?						
		Feature information				
		None	Category	Text	Photo	Average
Position information	None	2.50	3.50	4.00	3.67	3.42
	Area	4.17	4.33	4.33	4.33	4.29
	Direction	4.00	4.67	4.33	4.33	4.33
	Pin	4.83	4.83	4.67	4.67	4.75
	Average	3.88	4.33	4.33	4.25	
Question: Did you decide your next actions based on the outputs from the system?						
		Feature information				
		None	Category	Text	Photo	Average
Position information	None	1.50	3.00	2.83	3.00	2.58
	Area	4.00	4.17	3.33	3.83	3.83
	Direction	3.83	4.50	4.44	3.67	4.00
	Pin	3.67	4.17	4.00	4.00	3.96
	Average	3.25	3.96	3.54	3.63	

(5 scales; 1 is strongly disagree, 5 is strongly agree)

"direction", or "pin", the subjects tended to be aware the information. In particular, the information of "direction" or "pin" attracted the attention of the subjects strongly.

For the feature information, if the output had some information then the subjects tended to be attracted their attention to the information regardless of the content of the feature information. The differences in the feature information are small. Even in the cases of “none” or “category”, the subjects were aware the information. The case of “category” has the higher score for the last question than the others. The observer confirmed that the subjects looked for somethings related to the given categories. That is, since the feature information of the recommended spots were not explicitly given, it is thought that the subjects freely guessed the recommended spots and walked looking for them. As a result, it is thought that scores for the questions became the high values.

4.5 Consideration

First, we consider the effect of the position information. From the results of the position logs of the subjects, it is said that there is no chance to change their movement if no position information is given to them. On the other hand, if the position information is given as “area”, “direction”, or “pin”, it was seen that the subject moved to the direction to the given position. Especially, some subjects detected the exact location of the spots based on the information of “direction” or “pin”. From these facts, it is said that the subjects tended to change their behaviors if some information about the position was given. However, if the position information has some exactness, the subjects detected the recommended spots. In the case of “area” and “direction”, the detections of the recommended spots were depending on the feature information.

Next, for the feature information, if no feature information was given to the subjects, they could not detect the recommended spots. The information of “text” and “photo” noticed the subjects the characteristics of the spots. Especially, because the “text” gives the precise keywords of the spots, it leaded the subjects to detect the spots. The information of “category” did not give the precise image of the spots to the subjects, but it promotes the actions of looking for the recommended spots.

From the above consideration, we consider the information as a weak trigger which gives strolling tourists a chance to change their behaviors but does not specify the spot in a recommendation system. Regarding the first conditions, it is necessary to give a position information. Moreover, the “direction” or “area” information of position suit to the second condition. Also, from the questionnaire results, it is said that the “direction” of information is attracted attention more strongly than “area”. The exact information of position (i.e., “pin”) make specifying the spots easy. For the feature information, the “text” or “photograph” of information gives a strong image of a spot, which does not satisfy the second condition. In the case of the “category” information, the subjects saw the output from the system, but could not find the recommended spots. In the experiment site, there were many spots corresponding to a category. So, the subjects could freely guess the recommended spots, and looked for them. From these results, it is said that the information of “area” and “category” of the spots is the desired weak trigger.

5 Conclusion

This paper considered the information of sightseeing spots as a weak trigger which gives strolling tourists a chance to change their behaviors but does not specify the spot in a recommendation system. We classified the information about spots into the position and the feature information, and then set the four categories of information for each. We conducted the experiment with some subjects. As a result of the experiment, it was indicated that the combination of area information as the position of a spot and category information as the feature of the spot is an appropriate trigger with a good balance. Our future works is to verify an interface to output the information, and moreover, to propose an interaction mechanism between the system and tourists during their sightseeing.

References

1. Ishimori, S.: The potentialities of autonomous tourism in the twenty-first century. Senri Ethnol. Rep. **23**, 5–14 (2001)
2. Lucchese, C., Perego, R., Silvestri, F., Vahabi, H., Venturini, R.: How random walks can help tourism. In: Baeza-Yates, R., de Vries, A.P., Zaragoza, H., Cambazoglu, B.B., Murdock, V., Lempel, R., Silvestri, F. (eds.) ECIR 2012. LNCS, vol. 7224, pp. 195–206. Springer, Heidelberg (2012). https://doi.org/10.1007/978-3-642-28997-2_17
3. Kawakami, H.: Towards a system design focusing on the utility of inconvenience. Hum. Interface Soc. Trans. **11**(1), 125–134 (2009)
4. Hasebe, Y., Kawakami, H., Hiraoka, T., Nozaki, K.: Guidelines of system design for embodying benefits of inconvenience. SICE J. Control Meas. Syst. Integr. (JCMSI) **8**(1), 2–6 (2015)
5. Nakatani, Y., Ichikawa, K.: Tourist navigation system that induces accidental encounter. Hum. Interface Soc. Trans. **12**(4), 439–449 (2010)
6. Tanaka, K., Nakatani, Y.: Sightseeing navigation system that promotes interaction with environment by restricting information. In: IEEE International Conference on System, Man, and Cybernetics (SMC), pp. 453–458 (2010)
7. Takagi, S., Izumi, T., Nakatani, Y.: Tour navigation system using landmarks that are customized by personal preference. In: The First International Symposium on Socially and Technically Symbiotic Systems (STSS), pp. 47-1–47-7 (2012)
8. Oku, K., Hattori, F., Kawagoe, K.: Tweet-mapping method for tourist spots based on now-tweets and spot-photos. Procedia Comput. Sci. **60**, 1318–1327 (2015)
9. Misu, T., Mizukami, E., Sugiura, K., Iwahashi, N.: Development of dialogue systems "'Kyo-no Hanna' and 'Kyo no Osusume'". J. Natl. Inst. Inf. Commun. Technol. **59**(314), 29–33 (2012)
10. Matsumura, N., Fruchter, R., Leifer, L.: Shikakeology: designing triggers for behavior change. AI Soc. **30**(4), 419–429 (2015)
11. Yamane, S.: Shikake as a nudge. J. Artif. Intell. Soc. **28**(4), 596–600 (2014)
12. Kurata, Y.: Potential-of-interest maps for mobile tourist information services. In: Fuchs, M., Ricci, F., Cantoni, L. (eds.) Information and Communication Technologies in Tourism, pp. 239–248. Springer, Vienna (2012). https://doi.org/10.1007/978-3-7091-1142-0_21
13. Izumi, T., Kitamura, T., Nakatani, Y.: A Suggestive recommendation method to make tourists "feel like going". In: The 13th IFAC/IFIP/ IFORS/IEA Symposium on Analysis, Design, and Evaluation of Human-Machine Systems, FriHTrack-31 (2016)
14. Apple - Official website: www.apple.com/. Accessed 17 July 2017
15. Segaran, T., Toyama, Y., Kamosawa, M.: Collective Knowledge Programming. O'Reilly, Sebastopol (1991)

A Middle-Aged Social Internet with a Millennial Exodus? Changes in Identifications with Online Communities Between 2009 and 2017 in Finland

Aki Koivula(✉), Teo Keipi, Ilkka Koiranen, and Pekka Räsänen

Economic Sociology, Department of Social Research, University of Turku, Turku, Finland
akjeko@utu.fi

Abstract. This study is focused on questions regarding online identifications. We intend to examine the extent to which different demographic groups and generations in Finland identify with online communities in 2009 and 2017. Our empirical data are derived from nationally representative surveys collected in Finland in 2009 (n = 1,202) and 2017 (n = 1,648). The findings indicated that identification with online communities has increased in Finland between 2009 and 2017. Notably, demographic differences have diminished over time as the popularity of online groups has increased among middle-aged citizens especially. Analysis showed an interesting interaction between age cohort and observed year. It seems that younger generations have experienced a communal backlash, in which identification to online communities has decreased, while identification with traditional social groups has not changed.

Keywords: Internet use · Online communities · Social media
Temporal change

1 Introduction

The emergence of new information and communication technologies (ICTs) has led to a Western society that is more connected than ever before. These days computers, mobile phones and the Internet serve as central means of communication, social interaction and entertainment. As a result, many existing social structures have had to re-formulate, raising questions over the possible impacts that these developments have on modern society.

The notion of online versus offline relationships has been an especially integral part of the debate concerning personal relationships, identification and impacts of digitalization [1, 2]. In the past, individuals based their identities on only a handful of social contexts, for example home, work, school and in the company of close friends. This is no longer the case, since the majority of individuals use the Internet and have hobbies through which they connect to many social networks online. Digitalization has indeed had a significant impact on society, serving as an influential factor in identity formation especially for younger generations. In this era, people have a multiplicity of general or

© Springer International Publishing AG, part of Springer Nature 2018
G. Meiselwitz (Ed.): SCSM 2018, LNCS 10913, pp. 318–332, 2018.
https://doi.org/10.1007/978-3-319-91521-0_23

specific groups to identify with through convenient access and effective communication tools online.

The tension between offline and online environments can be better understood by tracking where social ties are born in the first place. As Preece and Maloney-Krichmar have noted, "increasingly it is accepted that online communities rarely exist only online – either they start as face-to-face communities and then part or all of the community migrates on to digital media, or conversely, members of an online community seek to meet face-to-face" [3]. In this sense, if the members of these communities are feeling that they have formed social ties outside of online spaces in the first place, these platforms are more likely to be categorized as "real life" communities that are utilizing online platforms. On the other hand, if these social ties between community members are realized primarily on online platforms, these platforms are considered online communities. Taking this notion into account, it is crucial to locate where the social action and networks are discursively placed.

While the digitalization of society obviously continues, there are differing views in academic discourse about the relationship between offline and online networks [3, 4]. Central here is the extent to which offline social interaction has actually been replaced by online relationships. Is there a trade-off between online and offline communities or are online communities more likely to be extensions of "real life" communities? With this research we elaborate on this timely question and offer a new perspective for academic discussion on social identifications.

Earlier literature shows that age is one of the most important factors when considering online social action [5–7]. Younger generations have been utilizing new technology and new social media applications much earlier and at a much wider scale when compared to older population groups [8, 9]. On the other hand, however, a growing number of 'silver surfers' have emerged in recent years. Not only are older people becoming more frequent Internet users, but their activities online have become more versatile [10–12]. In this sense we may also ask whether the age differences in general online behaviours have diminished among age groups.

Our interest in this paper revolves around questions regarding online identifications. This is a profound question in the online setting, since social identification signals are not only how we define ourselves, but also show what we consider important in life. Therefore, in addition to examining the traditional quality and usage of ICT services, it is important to examine the social experiences involved in using different digital tools and services. We intend to examine the extent to which different age groups and generations in Finland identify with offline and online communities in 2009 and 2017. The empirical part of this paper focuses on identification with online communities and more traditional social groups in Finland. We are particularly interested in exploring the possible differences between age groups and generations. In addition, we are also interested in how other socio-demographic factors connect with the strength of identification. To summarize, we present the following three research questions:

RQ1: Were there temporal changes in identifying with online and offline communities between 2009 and 2017?

RQ2: Were there distinct socio-demographic profiles of identification with an online community in 2009 and 2017?
RQ3: Was there cohort variation in identifications with an online community between 2009 and 2017?

2 Identifications on the Internet, Platform Evolution and Online Communities

Social scientists have shown the importance of group memberships and social networks in past research, which remain a fundamental building block in society today as well despite fundamental changes due to ICT developments. All manner of social position are arranged through in-groups, including business deals, political influence, and positions of employment, among others. Here, advantages are afforded to individuals who are viewed as members, leaving outsiders who are structurally excluded at a disadvantage [13]. In addition to being a fundamental aspect of society, groups and communities act as a valuable psychological anchoring point, by bringing individuals of various groups a source of security, validation and self-esteem [14].

This bolstering of individuality and group belonging together meets deep needs in socialising individuals who find agreeable social connections. According to social identity theory [14], individuals make sense of their social environment by categorizing themselves and others into groups that can be compared with one another. Here, comparison of group behavioural norms and identifying characteristics help to determine the individual member's place in society. Notably, these social identities exist simultaneously and evolve over time; being a father, son, sports fan and member of a political party are all social identities of an individual, the strengths of which can change over time. Notably, certain social identities through experiences of identification are more likely for some population groups than others; age, gender, level of education and numerous additional factors can influence the likelihood of becoming attracted to identifying with a certain group [1]. As such, connection with others is central to this theme of membership and the social value ascribed to it.

The rapid transformation of Western societies through ICT developments has, at its core, been one of connectivity. Social networks and online communities have been key in this new level of social identification online where interacting partners are far more easily reachable regardless of geographic limitations. This feeds into a core need, namely for socialisation, which helps to explain the massive growth of social platforms online. This extension of the offline world has brokered countless new memberships and experiences of connection globally.

To understand what has happened to identification with online communities, it is important to understand changes over the years during which much online evolution took place. This study focuses on the years between 2009 and 2017, which can be described as the social media era, particularly in Western countries. When considering online communities' development, social media plays a crucial role, despite there already being online communities before the advent of social media. In the era of "Web 1.0", online communities were typically based on text-based platforms, such as various kinds of discussion forums, chat sites, and messaging services [15, 16]. These platforms exist today as well, but their popularity and importance is significantly diminished when comparing to modern social networking sites.

Social networking platforms began to emerge after in the early 2000s, enabling unparalleled methods of social interaction in the online environment. These new sites – such the first wave of platforms including Friendster, LinkedIn, and Myspace – had novel features of social interaction and featured new visual characteristics. After the boom of social media, several pioneering social networking sites were overridden by emerging social media giants, especially Facebook. Subsequently, Facebook grew to be by far the most popular social networking site and the most important media company in the world. Nowadays there are roughly two billion monthly active Facebook users worldwide [17]. In addition to Facebook, there are also many other popular social networking sites utilized today as platforms for online communities, such Youtube, WhatsApp, Facebook Messenger, Twitter, and Instagram.

When comparing "Web 1.0" online communities to modern day online communities, namely social networking sites, there are major differences between them. First, today's social networking sites are highly visual and packed with various features, far surpassing the user experience of older online community platforms. Secondly, during the first phase of online community platforms, people were gathered around various topics and issues, which they were interested in [16]. These "old school" online communities were more likely to be based on small niches of people sharing the same interest, than large-scale social networks extending users' offline experiences and social circles. Thirdly, today it is common, especially in platforms like Facebook, Twitter and LinkedIn, for users to interact and navigate using their own names and identities, while platforms such as discussion forums and chat rooms more prominent in the past allowed for use of a pseudonym or in some cases were fully anonymous [15, 18]. These differences in characteristics play a crucial role in the kinds of communities that these platforms are enabling.

Notably, the development of new emerging social networking sites can be seen as taking steps back to these first phase online community platforms. For example, platforms based on more private interactions and conversation, such as WhatsApp and Facebook Messenger, have gained popularity during this past decade. Also, platforms based on full anonymity, such as Jodel, have emerged. These new platforms mentioned are more concentrated in terms of features instead of being multi-media and multi-featured online platforms. When considering this reversion in social media services, it seems that demand remains for smaller groups and communities without the attention of the wider public and large-scale networks.

In the time before the spread of social media, online communities were typically made up of a group of people interacting based on a shared interests or purpose, who were guided by some form of protocols, norms, policies or rules, with the interaction supported or mediated by technology [3, 19]. Compared to the past, online environments are generally based on social ties from offline surroundings. In this sense, the line between online and offline space have become even blurred.

3 The Changing Social Media Landscape in Finland

Overall, meanings, settings and surroundings for online communities evolved over the past decade. Online social networks have become more important, commonly shared and more personal after the wide scale spread of social media. This transition of "web 1.0" platforms to social networking sites can be observed in how people are utilizing various platforms. For instance, at the starting point of our research in year the 2009, only 30% of the Finns have a registered profile on social media [6], while approximately the same amount of people were also participating in conversations on various discussion forums and news groups [20]. In the year 2017, the proportion of people who utilize discussion forums or news groups has diminished to less than 10% [21]. In turn, in the year 2017 roughly 65% of Finns were utilizing social networking sites. Figure 1 highlights these recent changes in the shares of those Finns who have registered for at least one SNS site.

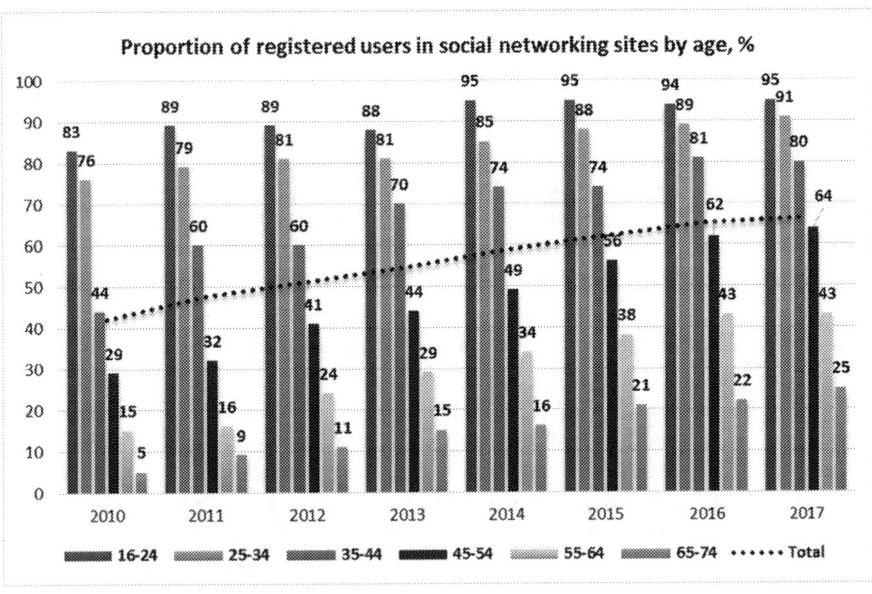

Fig. 1. Social media penetration rates in Finland, 2010–2017 (%)

As the figure shows, younger age groups are still more likely to be registered as SNS users than their older counterparts. In 2017, over 90% of under 35-year-olds were registered users, whereas less than 50% of over 55 year-olds belonged to this category. However, in 2010 the differences between the youngest and oldest age groups were even stronger. In addition, the middle-aged, namely 45- to 54-year-olds, have become clearly more active over this time period.

These observations indicate that there have been changes in the Finnish SNS landscape. Therefore, it would be necessary to address whether there have been certain qualitative differences in online identifications over time. If we assume that individuals' urge to define themselves through social media communities has generally increased, it would be especially important to know how people were identifying with different online communities earlier compared to now. Here, it is also important to compare online community identification with identification with different and more traditional communities formed primarily in offline surroundings.

4 Data and Methods

Our empirical data are derived from national-level surveys collected in Finland in 2009 (n = 1,202) and 2017 (n = 1,648). These nationally representative samples consist of respondents aged 18 to 74 years, thus providing an extensive look at the phenomenon in question. The surveys used simple random sampling as the respondents' home addresses were drawn from the Finnish population register database. Final samples are corrected with sample weights by balancing variation in sample sizes and response bias in terms of age and gender.

We use subjective measures of identifications as dependent variables. The variables were elicited with the question: "How strongly do you feel part of the following groups?" A total of eight items were displayed in the questionnaire for evaluation: city or town, church or religious community, a hobby group, an online community, residential community, colleagues at work or at school, friends, and family. Respondents gave their answers using a five-point Likert-type scale (ranging from 1 = "Not at all" to 5 = "Very much"). While no restrictions were given in the questionnaire, it is perhaps reasonable to assume that most of the respondents answered on the basis of their views relating to the circumstances of their daily life. In this way the interpretations of an online community or a hobby group, for instance, may vary between respondents. Nevertheless, possible ambiguities are taken into account in the interpretation of the results. In our analyses the dependent variables are treated as dichotomous measures as we estimate likelihood for identifying with different communities. In order to do this, we dichotomized original variable values (1 or 2 = "Not identifying"; 3, 4 or 5 = "Identifying").

Our main independent variable is age. In the first section of analysis, age was specified in the questionnaire as the year of birth, thus providing a continuous measure. In order to allow parallel comparisons with the other independent measures, age was categorized into seven groups: 18–24, 25–31, 32–38, 39–45, 46–52, 53–59 and 60–74.

In the second stage of analysis, we examine cohort effects while acknowledging problems related to the analysis of age, such as whether the phenomena discovered are related to certain life-cycle stage or whether they are actually typical of broader groupings, such as generations. However, we must bear in mind that our analysis is based on the utilization of cross-sectional data sets, which do not necessarily allow causal interpretations. In order to have observations from each age group in both survey periods, we had to re-categorize the data by establishing the following cohorts: 1943–49, 1950–56, 1975–63, 1964–1970, 1971–1977, 1978–84 and 1985–1991. As a result, the oldest age group, namely cohort 1935–1942, observed in 2009 and the youngest group, namely cohort 1992–1999, observed in 2017 were excluded from the cohort analysis.

The control variables include three demographic variables: gender, residential area and education. An earlier study has found that these variables have a strong association with identifying with an online community [5]. In the present study, residential area was measured simply by asking participants to choose their type of residential area, urban or non-urban. It can be argued that this variable reports unambiguously whether the respondent's residence is located in an urban or a non-urban setting. Education was measured in the data as vocational education. Here, the classification used consisted of four categories on the basis of ISCED classification [22]: (1) "Primary" (including all without at least secondary education), (2) "Secondary" (including lower, upper and post-secondary), (3) "Tertiary" (including tertiary and bachelor) and (4) "Master" (including master and doctoral degree).

The analytic techniques include logit models on the basis of which we post-estimate citizens' likelihood for identifying with offline and online communities. In the cohort analysis, we also equate effects of gender, education and residential area and show average year effects for each cohort. Before that we present unadjusted effects of applied background variables on identifying with online communities. All results are illustrated in figures by utilizing coefplots developed by Jann and figures are improved with schemes developed by Bischoff [23, 24]. Statistical tests are shown in the appendix in the Tables A1, A2 and A3.

5 Results

We begin our examination from the first research question (RQ1): Were there temporal changes in identifying with online and offline communities between 2009 and 2017?

Figure 2 shows estimations for identifications with various social communities. The values indicate probabilities of expressing identification experience on a scale from 0 to 1 for both years separately. The figure shows us that there have been changes over the time period examined. Social belongingness with online communities especially has grown stronger. In fact, the magnitude of increase is as much as 20 percentage points between the years 2009 and 2017 (from the likelihood of 0.29 to 0.49; logit coefficient = 0.84). This effect is statistically very significant (at the level of $p < 0.001$). Otherwise, the observed changes are relatively marginal. In general, identification experiences with most traditional communities have either increased or stayed at the same level. However, identification with a church or a religious community has weakened slightly. School and work colleagues as well as hobby groups have become more important sources of variation over time (the differences are not otherwise statistically significant). Statistical tests for temporal changes by item are given in Table A1.

The findings above indicate that Finns identified with an online community more intensively in 2017 when compared to 2009. This interpretation naturally applies only to the whole population on average. This is why we need to continue our investigation by looking at the possible differences between different population groups. This leads us to our second research question (RQ2): Were there distinct socio-demographic profiles of identification with an online community in 2009 and 2017?

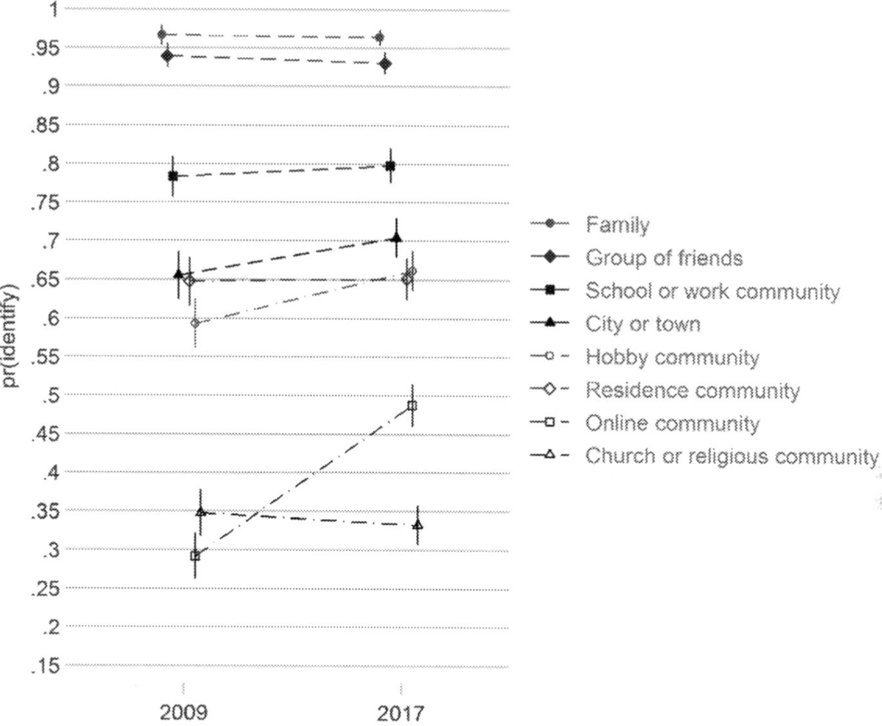

Fig. 2. Changes in the probabilities of identifying with different communities between 2009 and 2017

Figure 3 shows how different age and educational groups, men and women, and urban and non-urban dwellers identified with an online community in 2009 and 2017. The results are represented as unadjusted likelihoods (on a scale from 0 to 1), which make it possible to compare the effect of each independent variable between the years (the estimates are based on logit model shown in the Table A2). The most notable finding has to do with age. It appears that the disparities between the age groups have diminished. This is true especially between the youngest age group and others. In 2009, under 25-year-old Finns were a distinct category expressing a clearly stronger level of identification compared to all other age cohorts. In 2017 this is no longer the case. Moreover, what is noteworthy is that in the youngest age group the likelihood of identification has dropped by 13% points (from 0.80 to 0.67) between 2009 and 2017. All other age groups report stronger identifications in 2017 when compared to 2009.

The figure also shows that the differences between educational categories have become significant in 2017, while the differences were not statistically significant in 2009. Those with primary education identified less with an online community than others. The differences between males and females, as well as with urban and non-urban residents, were significant both years.

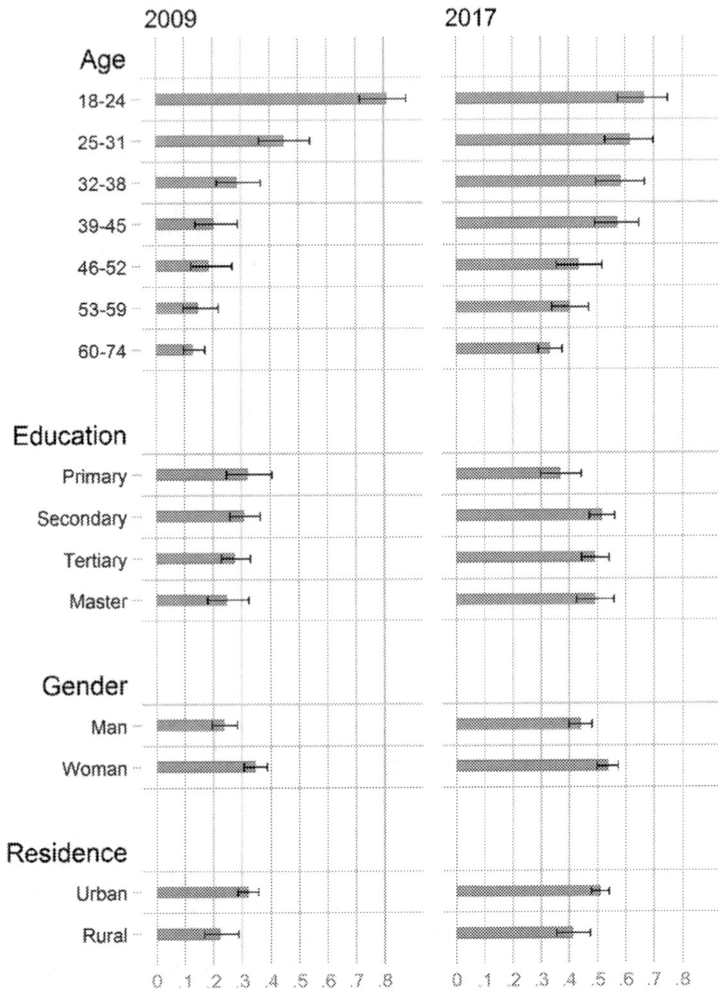

Fig. 3. Unadjusted probabilities of identifying with an online community by socio-demographic characteristics in 2009 and 2017

Together, these results indicate that certain temporal changes have taken place in online identifications between the years. The important changes relate to age. On one hand, the differences between all age groups have generally weakened. On the other hand, the youngest age group tend to be less attached to online communities than earlier. Given the fact that our data also allow us for comparing the responses within the age cohorts at different points in time, a further analysis was performed. This leads us to the final research question (RQ3): Was there cohort variation in identifications with an online community between 2009 and 2017?

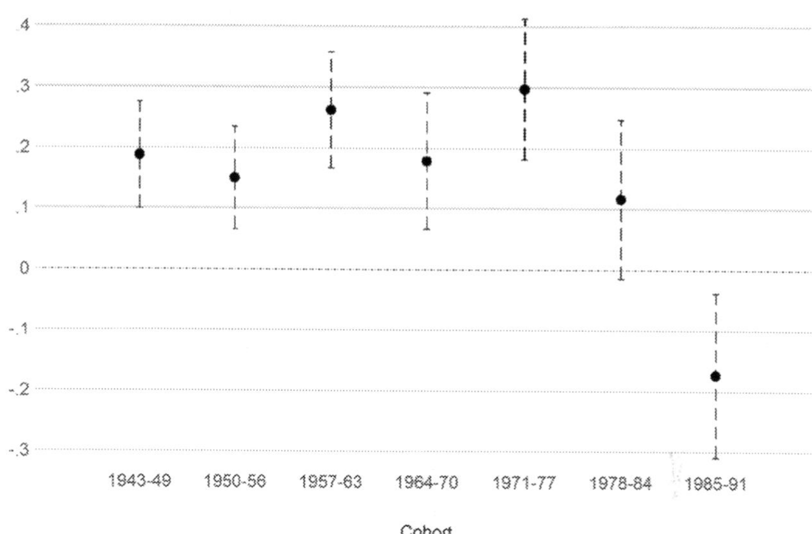

Fig. 4. Effects of year 2017 in predicting probability for identifying with an online community. Cohort analysis among 7-year age groups between 2009 and 2017 (average marginal effects adjusted for gender, education and residential area).

Figure 4 shows the average marginal effects for identification within each age cohort between 2009 and 2017. These estimates report the average change from 2009 to 2017 when taking into account respondents' level of education, gender and residential area (the estimates are based on logit model shown in the Table A3). The values are reported on a scale from −1 to 1, in which a positive value indicate an increase in the strength of identification and a negative one a decrease. The findings are clear: identifications have generally grown stronger within cohort groups over time. More importantly, however, the youngest age cohort (those born between 1985 and 1991) makes an exception here. Therefore, the findings are perfectly in line with our earlier interpretation regarding the youngest age group. It is indeed the youngest Finns who report weakened identification with an online community. What implications do these findings indicate?

6 Discussion and Conclusion

Our findings indicate that identification with online communities has become more prominent in Finland between 2009 and 2017. It is also noteworthy that demographic differences have also diminished between observed years especially in terms of age. Accordingly, the popularity of online groups has increased among middle-aged citizens especially. Also, women and people living in urban districts were more likely to be embedded in online communities. Interestingly, we did not find variation in terms of education level and online community identification in 2009, but in 2017 people with only primary level education were less likely to identify with online communities.

In the more detailed analysis, we found an interesting interaction between age cohort and observed year. Younger generations have apparently experienced a sort of communal backlash in which identification with online communities has decreased, whereas the probability of identifying with traditional social groups has not changed between the years 2009–2017.

According to the diffusion of innovations tradition, it is expected that the growth in the use and attachment of social media should begin to slow down within a period of time after a boom in popularity. In view of our findings, it seems that younger generations have tired of with the wide-scale online networks and multi-featured online platforms. This finding is in line with earlier statistics suggesting that younger users have started to abandon far matured and multi-featured social media platforms, such as Facebook [25]. This can be seen also in the adoption of nascent and recently growing social media platforms such as Snapchat and Instagram, which are more concentrated with a narrower set of features [26, 27].

However, these generational differences may also be explained by life stage transition. Even though the youngest generation still has the highest probability of identifying with online communities, online communities seem to be the first social groups they neglect after reaching the hurried thirties. This finding related to life transitions also underlines the order of importance between offline and online communities. As our results suggest, online communities are the first to be abandoned when prioritizing time consumption and when assessing what is essential and what is not.

On the basis of these findings, we argue that identifications with different communities can be seen as a good indicator of the impact that digitalization has had on society, especially on a generational level. Despite the fact that the growth rate of social media attachment has declined among the youngest groups, the growth is still strong among older population segments. However, it is also noteworthy here that traditional communities and social groups are still holding strong, even though the popularity of online groups has increased at the population level.

Notably, there are limitations in this study. Our findings represent the views of one European country alone, which do not allow us to make broader generalizations. It is also noteworthy to acknowledge that our data come from quantitative population surveys and not from qualitative in-depth interviews. However, the most difficult interpretations of the study deal with theoretical research implications.

Our findings partly support interpretations that online and offline communities can still be to be separated from each other. Namely, while online communities have become a much more essential part of people's lives, traditional offline communities, groups and networks have not been replaced by them. This suggests that the blurred line between online and offline in people's understanding is still possible to determine. However, it is evident that various offline social communities have embedded themselves in online environments, resulting in the blurring of past boundaries. As such, it has become even harder to make a distinction between online and offline communities. The question of what we are referring here is how users understand the term "online community."

For instance, is a WhatsApp group of close friends an online community or just an extension of a particular offline community? Do family members or colleagues at work cease to be offline community members on Facebook or ResearchGate?

Such questions lead us to wonder whether it is even possible to separate online and offline communities in the same way as before the era of social media. It is very likely that responses from different age cohorts would be contradictory with each other. Together, these notions indicate that more research is needed to assess this conceptualization of a possible boundary among social media users.

Acknowledgements. This research was supported by the Strategic Research Council of the Academy of Finland (decision numbers: 314171 [DDI] and 314250 [TITA]).

Appendix

Statistical tests for Figs. 2, 3 and 4

Table A1. Year effects in identifying with different communities. Logit coefficients with standard errors

	Family	Friends	School/Work	Neighbor	Online	City/Town	Church	Hobby
2009 (ref.)								
2017	−0.083	−0.158	0.085	0.015	0.837***	0.226*	−0.067	0.293**
	(0.250)	(0.187)	(0.105)	(0.092)	(0.093)	(0.094)	(0.089)	(0.090)
Obs.	2,548	2,549	2,427	2,526	2,521	2,512	2,538	2,520

Logit coefficients with standard errors in parentheses

*** $p < 0.001$, ** $p < 0.01$, * $p < 0.05$

Table A2. Unadjusted effects of background variables in identifying with online communities. Logit coefficients with standard errors.

	2009	2017
	Coef. (Std. Err.)	Coef. (Std. Err.)
18–24 (ref.)		
25–31	−1.655*** (0.329)	−0.218 (0.272)
32–38	−2.378*** (0.335)	−0.352 (0.269)
39–45	−2.828*** (0.359)	−0.406 (0.259)
46–52	−2.940*** (0.366)	−0.954*** (0.262)
53–59	−3.219*** (0.368)	−1.084*** (0.244)
60–74	−3.364*** (0.320)	−1.393*** (0.222)

(continued)

Table A2. (*continued*)

	2009	2017
	Coef. (Std. Err.)	Coef. (Std. Err.)
Primary education (ref.)		
Secondary	−0.054 (0.227)	0.605*** (0.183)
Tertiary	−0.213 (0.230)	0.510** (0.188)
Master	−0.371 (0.274)	0.508 (0.209)
Male (ref.)		
Female	0.530*** (0.155)	0.384*** (0.114)
Urban (ref.)		
Rural	−0.502* (0.197)	−0.386** (0.140)
Observations	1,132	1,389

*** p < 0.001, ** p < 0.01, * p < 0.05

Table A3. Predicting changing cohort effects on identifying with online communities. Logit coefficients with standard errors.

	Coef.	(Std. Err.)
2009 (ref.)		
2017	1.052***	(0.277)
1943–1949 (ref.)		
1950–1956	−0.707	(0.488)
1957–1963	0.098	(0.443)
1964–1970	0.073	(0.443)
1971–1977	0.654	(0.406)
1978–1984	1.525***	(0.410)
1985–1991	3.060***	(0.472)
Year#1943–1949 (ref.)		
Year#1950–1956	−0.071	(0.426)
Year#1957–1963	0.248	(0.395)
Year#1964–1970	−0.186	(0.406)
Year#1971–1977	0.198	(0.386)
Year#1978–1984	−0.582	(0.389)
Year#1985–1991	−1.911***	(0.459)
Observations	2,077	

*** p < 0.001, ** p < 0.01, * p < 0.05
Model control for effect of gender, education and residential area

References

1. Lehdonvirta, V., Räsänen, P.: How do young people identify with online and offline peer groups? A comparison between UK, Spain and Japan. J. Youth Stud. **14**(1), 91–108 (2011)
2. Eklund, L.: Bridging the online/offline divide: the example of digital gaming. Comput. Hum. Behav. **53**, 527–535 (2015)
3. Preece, J., Maloney-Krichmar, D.: Online communities: design, theory, and practice. J. Comput. Med. Commun., **10**(4), JCMC10410 (2005)
4. Fatimah, A., Gauntlett, D.: Young people's uses and understandings of online social networks in their everyday lives. Young **21**(2), 111–132 (2013)
5. Näsi, M., Räsänen, P., Lehdonvirta, V.: Identification with online and offline communities: understanding ICT disparities in Finland. Technol. Soc. **33**(1), 4–11 (2011)
6. Koiranen, I., Keipi, T., Koivula, A., Räsänen, P.: The different uses of social media – a population-level study in Finland. Working Papers in Economic Sociology (IX). University of Turku, Turku (2017)
7. Keipi, T., Koiranen, I., Koivula, A., Räsänen, P.: Assessing the social media landscape: Online relational use-purposes and life satisfaction among Finns. First Monday **23**(1) (2018). https://doi.org/10.5210/fm.v23i1.8128
8. Holtz, P., Appel, M.: Internet use and video gaming predict problem behavior in early adolescence. J. Adolesc. **34**(1), 49–58 (2011)
9. Merchant, G.: Unraveling the social network: theory and research. Learn. Media Technol. **37** (1), 4–19 (2012)
10. Näsi, M., Räsänen, P., Sarpila, O.: ICT activity in later life: internet use and leisure activities amongst senior citizens in Finland. Eur. J. Ageing **9**(2), 169–176 (2012)
11. Räsänen, P., Koiranen, I.: Changing patterns of ICT use in Finland – the senior citizens' perspective. In: Zhou, J., Salvendy, G. (eds.) ITAP 2016. LNCS, vol. 9754, pp. 226–237. Springer, Cham (2016). https://doi.org/10.1007/978-3-319-39943-0_22
12. Vošner, H.B., Bobek, S., Kokol, P., Krečič, M.J.: Attitudes of active older Internet users towards online social networking. Comput. Hum. Behav. **55**, 230–241 (2016)
13. Granovetter, M.: Economic action and social structure: The problem of embeddedness. Am. J. Sociol. **91**(3), 481–510 (1985)
14. Tajfel, H., Turner, J.C.: An integrative theory of intergroup conflict. Soc. Psychol. Intergr. Relat. **33**(47), 74 (1979)
15. Keipi, T.: Relatedness online: an analysis of youth narratives concerning the effects of internet anonymity. YoUng (2017). https://doi.org/10.1177/1103308817715142
16. Andrews, D.C.: Audience-specific online community design. Commun. ACM **45**(4), 64–68 (2002)
17. Statista: Number of monthly active Facebook users worldwide as of 3rd quarter 2017 (in millions) (2017). https://www.statista.com/statistics/264810/number-of-monthly-active-facebook-users-worldwide/
18. Keipi, T., Oksanen, A.: Self-exploration, anonymity and risks in the online setting: analysis of narratives by 14–18-year olds. J. Youth Stud. **17**(8), 1097–1113 (2014)
19. Porter, C.E.: A typology of virtual communities: a multi-disciplinary foundation for future research. J. Comput. Mediat. Commun. **10**(1) (2004)
20. Suomen virallinen tilasto (SVT 2009): Väestön tieto- ja viestintätekniikan käyttö [verkkojulkaisu]. Internetin käyttötarkoitukset 2009, prosenttia internetin käyttäjistä. Tilastokeskus, Helsinki (referred: 23.1.2018) (2009). ISSN 2341–8699. http://www.stat.fi/til/sutivi/2009/sutivi_2009_2009-09-08_tau_001.html

21. Suomen virallinen tilasto (SVT 2017): Väestön tieto- ja viestintätekniikan käyttö [verkkojulkaisu]. Liitetaulukko 19. Internetin käyttötarkoitusten yleisyys 2017, %-osuus väestöstä 1). Tilastokeskus, Helsinki (referred: 23.1.2018) (2017). ISSN 2341–8699. http://www.stat.fi/til/sutivi/2017/13/sutivi_2017_13_2017-11-22_tau_019_fi.htm
22. Schneider, S.L.: The international standard classification of education 2011. In: Birkelund, G.E. (ed.) Class and Stratification Analysis, pp. 365–379. Emerald Group Publishing Limited, Bingley (2013)
23. Jann, B.: Plotting regression coefficients and other estimates. Stata J. **14**(4), 708–737 (2014)
24. Bischof, D.: New graphic schemes for Stata: plotplain and plottig. Stata J. **17**(3), 748–759 (2017)
25. Kim, E.: This chart of teen sentiment shows why Facebook should be glad it bought Instagram. Business Insider (Referred 28.1.2018). (2015). http://www.businessinsider.com/most-important-social-networks-among-teens-2015-4?r=US&IR=T&IR=T
26. Phua, J., Jin, S., Kim, J.: Uses and gratifications of social networking sites for bridging and bonding social capital: a comparison of Facebook, Twitter, Instagram, and Snapchat. Comput. Hum. Behav. **72**, 115–122 (2017). https://doi.org/10.1016/j.chb.2017.02.041
27. Vaterlaus, J.M., Barnett, K., Roche, C., Young, J.A.: "Snapchat is more personal": an exploratory study on Snapchat behaviors and young adult interpersonal relationships. Comput. Hum. Behav. **62**, 594–601 (2016)

Effective Social Media Marketing Planning – How to Develop a Digital Marketing Plan

Marc Oliver Opresnik(✉)

Luebeck University of Applied Sciences, Public Corporation, Mönkhofer Weg 239, 23562 Lübeck, Germany
opresnik@fh-luebeck.de

Abstract. Marketing planning is also undergoing rapid development as the way of marketing communicating has changed forever. The increasing popularity of blogging, podcasting, and social networks enable customers to broadcast their views about a product or service to a potential audience of millions, and the proliferation of Internet access gives everyone who wants to the tools to address issues with products and companies. The traditional communications paradigm, which relied on the classic promotional mix to craft Integrated marketing communications (IMC) strategies, must give way to an effective digital marketing planning framework which includes all forms of social media as potential tools in designing and implementing IMC strategies. Consequently, there needs to be a sophisticated process to develop and implement a digital marketing plan in the social media environment.

Keywords: Social media marketing · Marketing planning
Marketing management · Web 2.0 · Marketing 4.0
Integrated marketing communication · Social computing · Social media

1 Introduction to Marketing Planning

Marketing is the organization function charged with defining customer tar-gets and the best way to satisfy their needs and wants competitively and profitably. Because consumers and business buyers face an abundance of suppliers seeking to satisfy their every need, companies and not-for-profit organizations cannot survive today by simply doing a good job. They must do an excellent job if they are to remain in the increasingly competitive global marketplace. Many studies have demonstrated that the key to profitable performance is knowing and satisfying target customers with competitively superior offers. This process takes place today in an increasingly global, technical, and competitive environment [1].

There are some key reasons why marketing planning has become so important. Recent years have witnessed an intensifying of competition in many markets. Many factors have contributed to this, but amongst some of the more significant are the following [2]:

A growth of global competition, as barriers to trade have been lowered and global communications improved significantly.

© Springer International Publishing AG, part of Springer Nature 2018
G. Meiselwitz (Ed.): SCSM 2018, LNCS 10913, pp. 333–341, 2018.
https://doi.org/10.1007/978-3-319-91521-0_24

The role of the multinational conglomerate has increased. This ignores geographical and other boundaries and looks for profit opportunities on a global scale.

In some economies, legislation and political ideologies have aimed at fostering entrepreneurial and 'free market' values.

Continual technological innovation, giving rise to new sources of competition for established products, services and markets.

The importance of competition and competitor analysis in contemporary strategic marketing cannot be overemphasized. Indeed, because of this we shall be looking at this aspect in more depth in later chapters. This importance is now widely accepted amongst both marketing academics and practitioners. Successful marketing in a competitive economy is about competitive success and that in addition to a customer focus a true marketing orientation also combines competitive positioning.

The marketing concept holds that the key to achieving organizational goals lies in determining the needs and wants of target markets, and delivering the desired 'satisfaction' more effectively and resourcefully than competitors [3].

Marketing planning is an approach adopted by many successful, market-focused companies. While it is by no means a new tool, the degree of objectivity and thoroughness with which it is applied varies significantly.

Marketing planning can be defined as the structured process of researching and analyzing the marketing situations, developing and documenting marketing objectives, strategies, and programs, and implementing, evaluating, and controlling activities to achieve the goals. This systematic process of marketing planning involves analyzing the environment and the company's capabilities, and deciding on courses of action and ways to implement those decisions. As the marketing environment is so changeable that paths to new opportunities can open in an instant, even as others become obscured or completely blocked, marketing planning must be approached as an adaptable, ongoing process rather than a rigid, static annual event [2].

The outcome of this structured process is the marketing plan, a document that summarizes what the marketer has learned about the marketplace and outlines how the firm plans to reach its marketing objectives. In addition, the marketing plan not only documents the organization's marketing strategies and displays the activities that employees will implement to reach the marketing objectives, but it entails the mechanisms that will measure progress toward the objectives and allows for adjustments if actual results take the organization off course [4].

Marketing plans generally cover a 1-year-period, although some may project activities and financial performance further into the future. Marketers must start the marketing planning process at least several months before the marketing plan is scheduled to go into operation; this allows sufficient time for thorough research and analysis, management review and revision, and coordination of resources among functions and business units.

Marketing planning inevitably involves change. It is a process that includes deciding currently what to do in the future with a full appreciation of the resource position; the need to set clear, communicable, measurable objectives; the development of alternative courses of action; and a means of assessing the best route towards the achievement of specified objectives. Marketing planning is designed to assist the process of marketing decision making under prevailing conditions of risk and uncertainty [2].

Above all the process of marketing planning has several benefits [3]:

Consistency
The individual marketing action plans must be consistent with the overall corporate plan and with the other depart-mental or functional plans.

Responsibility
Those who have responsibility for implementing the individual parts of the marketing plan will know what their responsibilities are and can have their performance assessed against these plans. Marketing planning requires management staff to make clear judgmental statements about assumptions, and it enables a control system to be designed and established whereby performance can be assessed against pre-defined criteria.

Communication
Those implementing the plans will also know that the overall objectives are and how they personally may contribute in this respect.

Commitment
If the plans are agreed upon by those involved in their implementation, as well as by those who will provide the re-sources, the plans do stimulate a group commitment to their implementation, and ultimately lead to better strategy-implementation.

Plans must be specific to the organization and its current situation. There is not one system of planning but many systems, and a planning process must be tailor-made for a particular firm in a specific set of conditions. Marketing planning as a functional activity has to be set in a corporate planning frame-work. There is an underlying obligation for any organization adopting marketing planning systems to set a clearly defined business mission as the basis from which the organizational direction can develop. Without marketing planning, it is more difficult to guide research and development (R&D) and new product development (NPD); set required standards for suppliers; guide the sales force in terms of what to emphasize, set realistic, achievable targets, avoid competitor actions or changes in the marketplace. Above all, businesses which fail to incorporate marketing planning into their marketing activities may therefore not be able to develop a sustainable competitive advantage in their markets [3].

2 The Main Stages in Developing a Marketing Plan

Marketing planning is a methodical process involving assessing marketing opportunities and resources, determining marketing objectives, and developing a plan for implementation and control. Marketing planning is an ongoing analysis/planning/control process or cycle. Many organizations update their marketing plans annually as new information becomes accessible. Once built-in, the key recommendations can then be presented to key stakeholders within the organization. The final task of marketing planning is to summarize the relevant findings from the marketing analysis, the strategic recommendations and the required marketing programs in a report: the written marketing plan. This document needs to be concise, yet complete in terms of presenting a summary of the marketplace and the business's position, explaining thoroughly the recommended strategy and containing the detail of marketing mix activities. The plan should be informative, to the point, while mapping out a clear set of marketing activities designed to satisfactorily implement the desired target market strategy [3].

Figure 1 illustrates the several stages that have to be gone through in order to arrive at a marketing plan. As illustrated, the development of a marketing plan is a process, and each step in the process has a structure that enables the marketing plan to evolve from abstract information and ideas into a tangible document that can easily be understood, evaluated, and implemented.

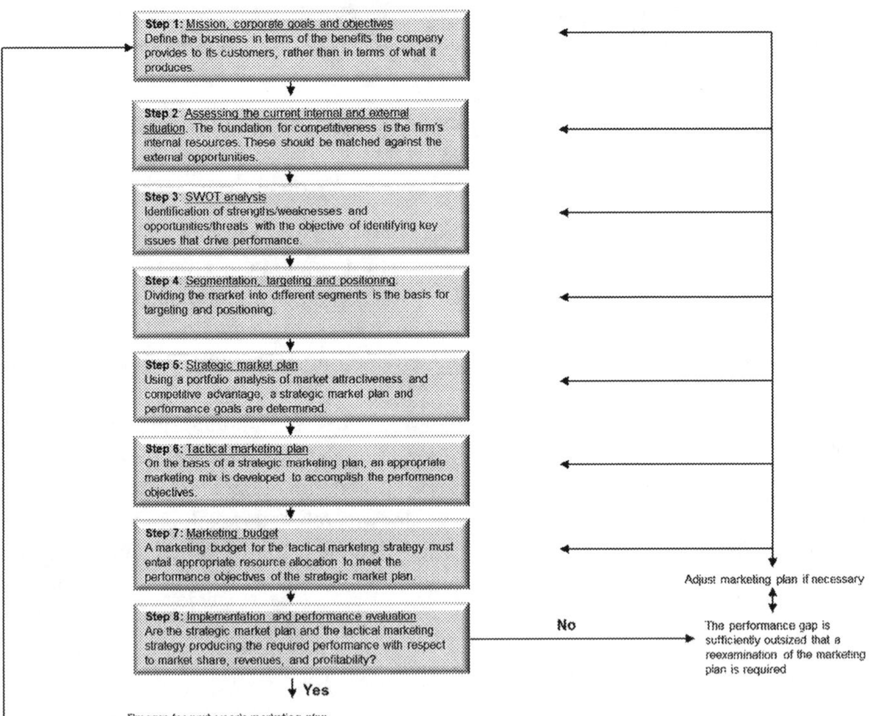

Fig. 1. The Stages of Building a Marketing Plan (Source: Hollensen and Opresnik, 2015)

3 The Main Stages in Developing an Effective Digital Marketing Plan

Against the background of the above-mentioned process of an overall marketing plan, a social media marketing plan is the summary of everything the company plan to do in social media marketing and hope to achieve for the business using social networks. This plan should comprise an audit of where the customers are today, goals for where you want them to be soon, and all the social media tools that the company wants to use to get there [2].

In general, the more specific the company can get with their plan, the more effective they will be in the plan's implementation. It is important to keep it concise. The plan will guide the company's actions, but it will also be a measure by which to determine whether the company is succeeding or failing. Figure 2. illustrates the several stages that should be gone through to arrive at a digital marketing plan [5, 6].

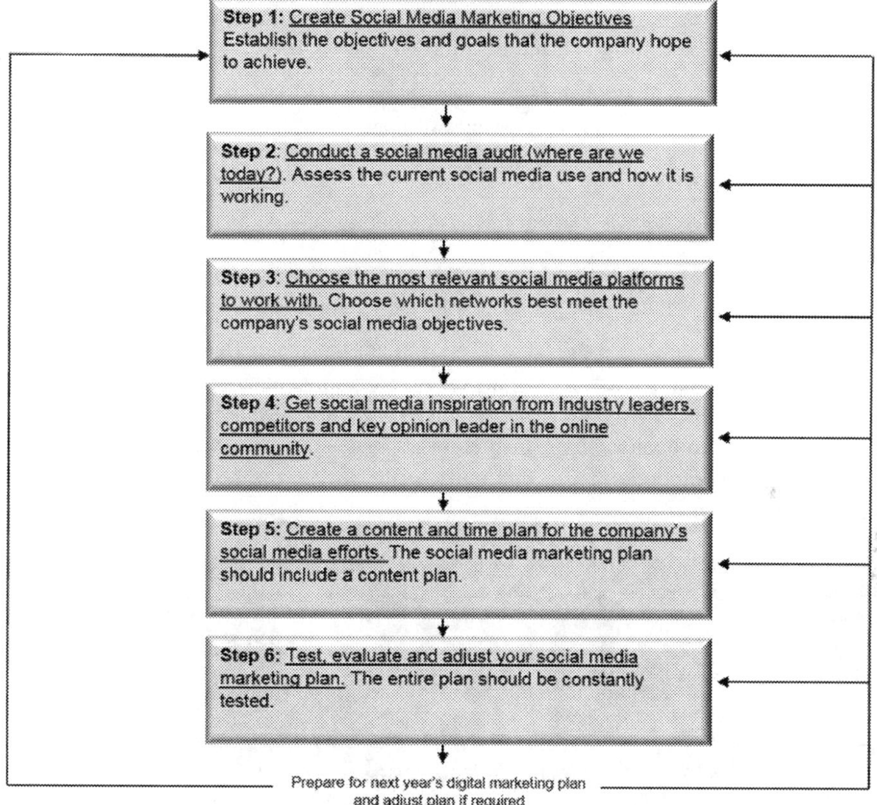

Fig. 2. The stages of building a digital marketing plan (Source: Hollensen and Opresnik, 2017)

Step 1: Create social media marketing objectives

The first step to any social media marketing strategy is to establish the objectives and goals that the company hope to achieve. Having these objectives also allows the company to quickly react when social marketing media campaigns are not meeting the company's expectations. Without objectives, the company has no means of evaluating success or proving their social media Return On Investment (ROI). These goals should be aligned with the broader marketing strategy, so that the social media efforts drive toward the business objectives. If the social media marketing plan is shown to support the overall business objective, the company is more likely to get executive and employee buy-in and investment. The company should try to go beyond popular metrics such as Retweets and Likes. Focus should be more on advanced metrics such as 'number of leads generated', web referrals, and con-version rate. The company should also use the SMART framework when setting their objectives:

Specific – target a specific area for improvement.
Measurable – quantify or at least suggest an indicator of progress.
Achievable – Agreed and aligned with corporate goals.
Realistic – state what results can realistically be achieved, given available resources.
Time-related – specify when the result(s) can be achieved.

Example: 'In Social Publishing we will share photos that communicate our company culture. We will do this by posting a total of ten photos a week on any of the photo sharing social media sites. The target for each week is at least in total 100 likes and 30 comments.'

A simple way to start the social media marketing plan is by writing down at least three social media objectives.

Step 2: Conduct a social media audit (where are we today?)

Prior to creating your social media marketing plan, the company needs to assess their current social media use and how it is working. This means figuring out who is currently connecting with the company and its brand via social media, which social media sites the company target market uses, and how the social media presence compares to the competitors. For this purpose, the following social media audit template can be used [2].

Presence on Social media platform	URL	Internal respon-sible for main-taining social media	Social media mission	Current number of followers	Main competitor's number of followers
etc.					

Fig. 3. Social Media Audit template (current situation) (Source: Kotler, Hollensen and Opresnik, 2017)

Once the audit is conducted the company should have a clear picture of every social media platform representing the business, who runs or controls them, and what purpose they serve. This audit should be maintained regularly, especially as the company scale up their business (Fig. 3).

It should also be evident which social media platforms (accounts) need to be updated and which need to be deleted altogether. If the audit uncovers for example a fake branded Twitter profile, it should be reported. Reporting fake accounts will help ensure that people searching for the company online only connect with the accounts that are managed by the company itself [1].

As part of the social media audit the company may also want to create mission statements for each social media platform (network). These one-sentence declarations will help to focus on a very specific objective for Instagram, Facebook, or any other social media network. They will guide the actions and help guiding back on track if the efforts begin to lag [2].

Example of Mission statement for a presence on the Snapchat platform: 'We will use Snapchat to share the CSR side of our company and connect with younger prospect customers among 15–40 years old.'

The company should be able to determine the purpose of every social media platform it has, for example Snapchat. If it cannot determine the mission for each social media platform, the platform & profile should probably be deleted.

Before it is possible to determine which social media platforms are right for the business, the company should find out who the audience is for each platform and what they want. The company should know which tools to use to gather demographic and behavioral data, and how to target the customers it wants.

Step 3: Choose the most relevant social media platforms to work with

Once you've finished with your social media audit, it is time to choose the online presence. Choose which networks best meet the company's social media missions and objectives. If there is not already a social media profile on each network/platform the company focuses on, it should build them from the ground up with the broader mission and audience in mind. Each social network has a unique audience and should be treated differently. If the company has some existing platforms, it is time to update and refine them to get the best possible results [2].

Optimizing profiles for SEO (Search Engine Optimization) can help to generate more web traffic to the company's online social media platforms. Cross-promoting social platforms can extend the reach of content. In general, social media profiles should be filled out completely, and images and text should be optimized for the social network in question [1].

Step 4: Get social media inspiration from Industry leaders, competitors and key opinion leader in the online community

If the company is not sure what kinds of content and information will get the most engagement, then the company, for inspiration, can look to what others in the industry are sharing. The company can also use social media listening to see how it can distinguish itself from competitors and appeal to an audience it might be missing [2].

Opinion leaders among consumers ('market mavens') can also offer social media inspiration, not only through the content that they share but in the way that they phrase their messages. The company can try and see how its target audience writes Tweets, and it could strive to write in a similar style. It can also learn their habits - when they share and why - and use that as a basis for the social media marketing plan [2–4].

A final source of social media inspiration is industry leaders. There are giants who do an incredible job of social media marketing, from Red Bull and Taco Bell to Turkish Airlines. Companies in every industry imaginable have managed to distinguish themselves through advanced social media strategies.

The company can follow industry leaders and see if they have shared any social media advice or insight elsewhere on the web.

Step 5: Create a content and time plan for the company's social media efforts

The social media marketing plan should include a content marketing plan, comprised of strategies for content creation, as well as an editorial calendar (time plan) for when the content should be shown online. Having great con-tent to share and the right timing will be essential to succeeding at social media marketing [2].

The content marketing plan should answer the following questions:

What types of content the company intends to post and promote on social media?
Who will create the content?
How often will the company post content?
What is the target audience for each type of content?
How you the company promote the content?

The editorial calendar lists the dates and times the company intends to publish blogs, Instagram and Facebook posts, Tweets, and other content that is planned to use during the social media campaigns.

The company can create the calendar and then schedule their messaging in advance rather than updating constantly throughout the day. This gives it the opportunity to work hard on the language and format of these messages rather than writing them on the fly whenever company employees have time. The company should make sure that the content reflects the mission statement that are assigned to each social media profile/platform. If the purpose of the LinkedIn account is to generate leads, the company should make sure that it is sharing enough lead generation content. The company can establish a content matrix that defines what share of the social media platform is allocated to different types of posts [2].

Step 6: Test, evaluate and adjust your social media marketing plan

To find out what adjustments need to be made to your social media marketing strategy, you should constantly be testing. Build testing capabilities into every action you take on social networks. For example, you could track the number of clicks your links get on a particular platform using URL shorteners. Furthermore, it is possible to measure track page visits driven by social media with Google Analytics.

Record and analyze your successes and failures, and then adjust your social media marketing plan in response [4].

Surveys are also a great way to gauge success - online and offline. The company can ask their social media followers, email list, and website visitors how they are doing on social media. This direct approach is often very effective. Then ask your offline customers if social media had a role in their purchasing. This insight might prove invaluable when you look for areas to improve [2].

The most important thing to understand about the social media marketing plan is that it should be constantly changing. As new networks emerge, the company may want to add them to their plan. As the company is attaining missions and objectives for each social media platform, it will need to set new targets. Unexpected challenges will arise that is needed to address. As the company is scaling up its business, it might need to add new roles or grow the social presence for different products or regions.

The company should rewrite its social media marketing plan to reflect its latest insights, and make sure that the team is aware of what has been updated [1].

4 Conclusion

As digital communication becomes an increasingly dominant way for people exchange and share information, a digital marketing plan becomes an essential tool for any company and organization. The process of creating a digital marketing plan will help organizations and companies clarify what they want to achieve, understand how to engage their target market online and outline the key activities they need to take to market your business digitally and measure the effectiveness of your actions.

References

1. Hollensen, S., Opresnik, M.: Marketing: Principles and Practice, 1st edn. Opresnik Management Consulting, Lübeck (2017)
2. Kotler, P., Hollensen, S., Opresnik, M.: Social Media Marketing – A Practitioner Guide. 2nd edn. Opresnik Management Consulting, Lübeck (2017)
3. Hollensen, S.: Marketing Planning: A Global Perspective. McGraw-Hill, Berkshire (2006)
4. Hollensen, S., Opresnik, M.: Marketing – A Relationship Perspective, 2nd edn. Vahlen, München (2015)
5. Gilmore, A., Carsons, D., Grant, K.: SME marketing in practice. Mark. Intell. Plan. **19**(1), 6–11 (2011)
6. Day, G.S.: Managing the market learning process. J. Bus. Ind. Mark. **17**(4), 240–252 (2002)

BrewFinder – An Interactive Flavor Map Informed by Users

Chandler Price(✉)

Georgia Tech Research Institute, Atlanta, GA, USA
chandler.price@gtri.gatech.edu

Abstract. Items that provide a complex experience to the user can be difficult to review and compare. Deciding what food, movies, or clothes to purchase can prove to be difficult on star ratings alone. While most digital storefronts allow the user to rate an item based on a qualitative score, the experience is more nuanced than a simple rating. Beer is one such beverage that can be difficult to describe on star rating alone and provides an experience that can vary greatly from one person to another. Craft beer in the U.S. is a fast-growing market, with volume growth totaling 18% in 2014 [4]. As the craft beer market grows, so do the number of options available to a customer at the store. Currently there are three main ways to describe a beer: its style (lager, ale, lambic), its alcohol percentage, and its bitterness rating. This project aimed to quantify user's beer experiences beyond these three characteristics, by performing a principal components analysis on user submitted reviews to RateBeer.com. Potential users were also involved in the development of an interactive information visualization, with the goal of allowing users to explore and navigate many beers at once by flavor. The research contained within this paper culminated in a live, fully functional informational visualization, which can be viewed online.

Keywords: Information visualization · Social media · Reviews
Products · Textual analysis · UI design · Consumers · Beer

1 Introduction

1.1 Background

Items that provide a complex experience to the user can be difficult to review and compare. Deciding what food, movies, or clothes to purchase can prove to be difficult on star ratings alone. While most digital storefronts allow the user to rate an item based on a qualitative score, the experience is more nuanced than a simple rating. Beer is one such beverage that can be difficult to describe on a star rating system and provides an experience that can vary greatly from one person to another.

Craft beer in the U.S. is a fast-growing market, with volume growth totaling 18% in 2014 [4]. As the craft beer market grows, so do the number of options available to a customer at the store. This leads to what is known as a long tail market place, where once a few key players dominated the market, several niche products instead become more popular [3]. As the number of beers available to a consumer increases, it can be hard for a consumer to find the beer they will enjoy the most. There are three main

© Springer International Publishing AG, part of Springer Nature 2018
G. Meiselwitz (Ed.): SCSM 2018, LNCS 10913, pp. 342–354, 2018.
https://doi.org/10.1007/978-3-319-91521-0_25

ways to describe a beer: its style (lager, ale, lambic), its alcohol percentage, and its bitterness rating. Unless you are a savvy beer consumer, understanding each beer style and what it means in terms of flavor can be difficult. Additionally, some consumers may find it hard to understand the bitterness rating of a beer, or even its alcohol percentage. Even when style is understood by a consumer, a beer may not follow the style to the letter and may instead opt to include several flavor additives that may change its overall flavor profile. While adding just a fruit flavor made be easy for a consumer to understand, some complicated beers such as a saison (a type of beer categorized by its herbal yeast) can be modified heavily. One such way to modify a beer is to add a biotic found in yogurt to produce a tart flavor.

Given the context of this problem, this project began by examining the current solutions available for users to browse and rate beers. One of the more popular apps used to rate and browse beers is Untappd. Untappd allows users to rate beers on a simple scale from 1 to 5, with 5 being the highest rating. Users can also add comments, although these are few and far between. Beers can generally only be browsed through a list, highlighting each beers rating.

In discussing Untappd's rating solution with local CEO of Creature Comforts Brewery Chris Herron, he stated; "I agree that I do not like the rating system of only 5 stars on untapped, but I think the ease of use is why so many people use it". Untappd's data alone wouldn't be very useful in creating a visualization, simply ranking beers by star or style rating wouldn't provide a novel or different beer browsing experience.

Many beer social networks while somewhat chaotic generally adhere to common themes. As outlined by Dwyer [5], several forces can manipulate or shape how a social network functions and discusses content. One principal of influence described by the paper is contextual imperative, in which groups aim to communicate using similar structure. An example of this can be seen at RateBeer.com. At RateBeer, each beer can be rated on the same 1–5 star rating system, however the community typically follows up their star rating with an additional textual review. In these reviews, the community generally discusses key flavors, and attempts to describe the beer using key adjectives, and tastes relating to other, more common foods. For richer, darker beers you may see the word "chocolate" used to describe the beer for example. The users are not required to fill this portion of the review out, however the community has itself self-imposed a general structure in describing beers using this format.

The goal of this research is to design an interactive visualization that could aid users in understanding beer beyond just its style or star rating. This tool would be informed directly by consumers, by directly translating their textual reviews and descriptions of a beer into a flavor scatterplot. This research aimed to follow the user-centered design process to aid in the creation of meaningful interactive elements [2].

2 Method

2.1 Beer Flavor Matrix Construction

The research began by collecting and examining user submitted textual beer reviews which were written on RateBeer.com. These long form textual re-views were copied and analyzed using several methods of textual visualization. First, word clouds were used to examine which words were most common. A master word list was created by ranking words by their commonality (with more common words appearing at the top). This was done to attempt to parse a taxonomy of flavor, removing words which were not adjectives or flavor descriptors such as great, nice, or mouthfeel. The flavor map created by Meilgaard et al. was used as inspiration for this taxonomy, by extending the 44 words used in the 1979 publication to describe beer into a list of 451 flavor descriptors based on commonality [8].

With a taxonomy in place, a matrix was of each beer and each of the 451 descriptors was populated with word scores. The max number of descriptors were used to help describe beers that may have unique flavors. While a good number of characteristics were not overall useful in calculating the PCA, they could still be helpful in describing the beer by its flavor. While beers that were described as tasting of coconut were rare, it was still important to record and display the flavor to the user to aid in their decision making process.

The word scores were computed by splitting each review sentence into individual words and counting how many times each filtered characteristic word appears across all beer reviews. A ratio was then computed, by taking the total number of times a specific word appeared and dividing it by the total number of words available in the corpus. This would then give a percentage score, of how much the specific characteristic appeared in relation to the total word count. The beers present in the matrix were filtered down to those which were made in Georgia (minus 11 macro beers) and which had at least 3 reviews.

With the matrix constructed, a principal components analysis (PCA) was created to aid in the future plotting of beers based on flavor [1]. In short, a PCA was used because each beer had a multitude of potentially inter-correlating variables. A PCA allows the creation of two uncorrelated variables which can be attributed to each beer. These variables allow for the creation of two major summary variables, that can be used to group similar beers together. The components are loaded onto either positively, negatively, or not at all by a characteristic. For example, a beer which had the word "chocolate" appear a lot in its reviews will load highly on the first component (X-Axis), while another beer with the word "golden" would load less for the same component. Thus the chocolate beer would be placed more to the right on the x axis, while the golden beer would be placed on the left side. Table 1 below shows the top characteristics for each axis and direction.

Table 1. Top loading characteristics for each axis and its direction.

Axis & Direction	1 Highest	2	3	4 Lower
Positive X	Tan	Chocolate	Dark	Black
Negative X	White	Golden	Yellow	Citrus
Positive Y	Hops	Pine	Bitter	Caramel
Negative Y	Tart	Yellow	Sour	Wheat

2.2 Information Visualization Technique Selection

With the data collected, and the matrix constructed, the user participated design process began by evaluating several key information visualization techniques. 8 Participants were asked to examine 3 separate visualization techniques in the context of beer. For each visualization, the participants were asked to rate how easily they were able to understand what the visualization was conveying.

The first visualization shown was a word cloud (Fig. 1) made up of the characteristics for one beer. The word cloud was generated using a website word cloud tool kit at voyant-tools.com. Word clouds can be helpful in demonstrating the general characteristics of one beer, but can take up a lot of space, making it difficult to compare multiple word clouds at once.

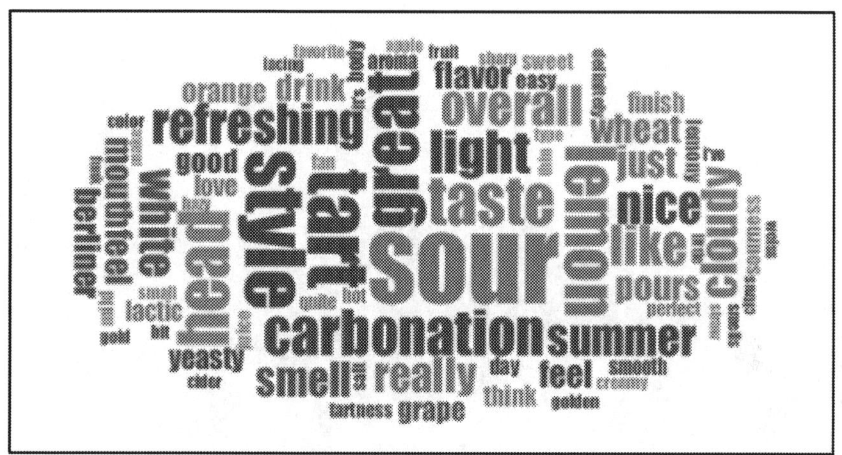

Fig. 1. A word cloud visualization taken from a beer's review texts.

The second visualization was a word tree (Fig. 2), which showed individual words emanating from a singular, key word across multiple reviews. The aim of this visualization is to give further context to individual words that might be of interest to the user. If a user desired to understand in what ways a beer is smooth for example, they could use a word tree to better understand how the word is used across several reviews. This visualization was made using the tool at jasondavies.com.

The final visualization shown to the participant, was inspired by a correlative textual data visualization created by Endert et al. [6]. This IN-SPIRE galaxy visualization technique (Fig. 3) is used to show how documents relate to one another based on concordance. In the context of this study, participants were asked to examine this visualization as if many beers were grouped together in "clouds" based on similarity. To put the design in further context, controls and beer-specific flavors were also included.

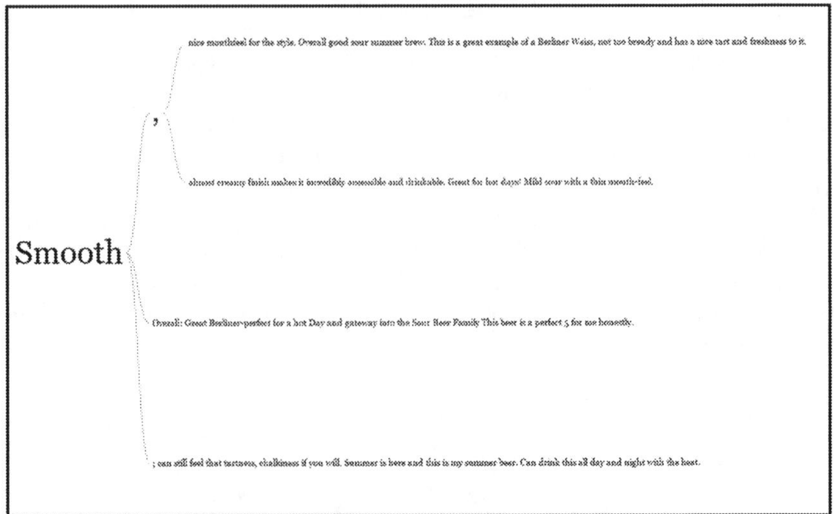

Fig. 2. A word tree with Smooth as the root word, following with individual reviews.

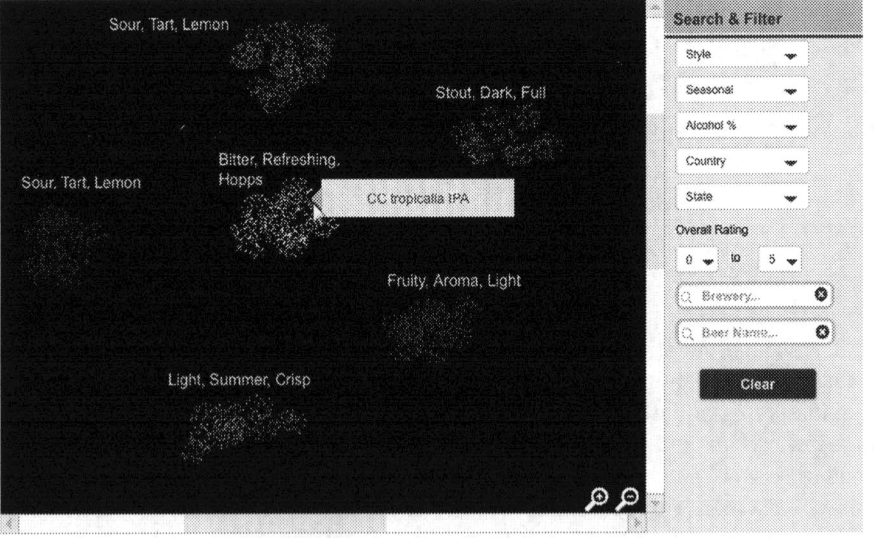

Fig. 3. Prototype interface using a galaxy view, with hover tool tip and search tools to the right.

2.3 Information Visualization Technique Feedback

These three designs were chosen based on their differing levels of granularity when displaying the data. The IN-SPIRE galaxy visualization could be used to review many beers at once. The word cloud view, could then be used to describe one beer selected within the IN-SPIRE galaxy view. Further, each word within the word cloud could then be examined in context with a word-tree. Each of these levels represented a drill down process in which a user could quickly explore many different beers and understand what makes each beer similar or different to others. This context was given to 8 participants before they were asked to review each visualization technique on its own, within the context of selecting a beer. Some comments which were influential in the prototype design are shown below.

Word Cloud Comments:

"A lot of turning and head twisting to read."
"Not sure if color means anything?"
"Show main contributors to flavor, such as summer and refreshing."
"Seems needlessly cluttered. Hard to pick out all of the traits."

Word Tree Comments:

"At first glance, unable to tell what it means. Not sure what it represents."
"With more interaction, I could see this being useful."
"Gives a limited idea of what the beer is like as compared to the cloud."

Galaxy View Comments

"Clustering of beer by characteristics, really want to interact with it."
"Trying to figure out the ordering, why are some clouds close to the top/bottom?"
"If it is a scatterplot, what is the x/y axis?"
"Would want to hover over each beer individually."

2.4 Functional Prototype Design

With my axes understood, and with user feedback, I moved onto plotting the data and designing the user interface using JavaScript, CSS, and HTML. The library D3.js was used to program most of the visualization, with the rest being handled in simple JavaScript or jQuery.

First, I attempted to remedy some of the initial questions users might have by creating a splash screen with help information that loads when the user arrives on the page (Fig. 4). This help screen describes the research, and general instructions. While most users may click away from this screen immediately upon arrival, a link in the bottom right-hand corner of the screen can be selected to revisit this menu.

Scatterplot Design

Once the scatterplot created from the PCA was rendered, an interesting pattern emerged. A triangle like structure was formed from the data-points, which could lend itself to general beer categories. These categories can be best described as high loading Y beers (hoppier flavor), low loading Y beers (fruitier flavor) and high loading X beers (richer flavor, such as chocolate or coffee). There is also a general fourth category of beers that emerged in the center of the triangle, which could be best described as blander beers, such as pilsners or lagers which did not contain these hoppy, fruity, or rich flavors. A view of the general structure can be seen in Fig. 5.

In designing the scatterplot visualization, inspiration was taken from the galaxy view, while attempting to emphasize the axes and position of the data points within it.

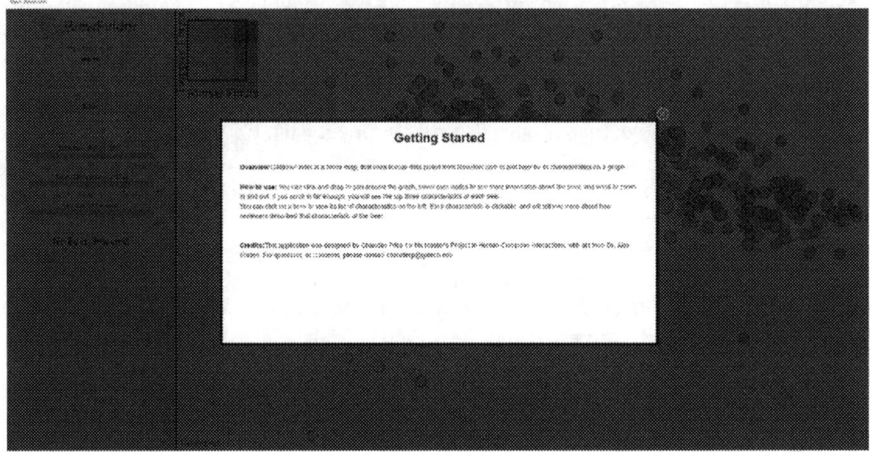

Fig. 4. BrewFinder help splash screen.

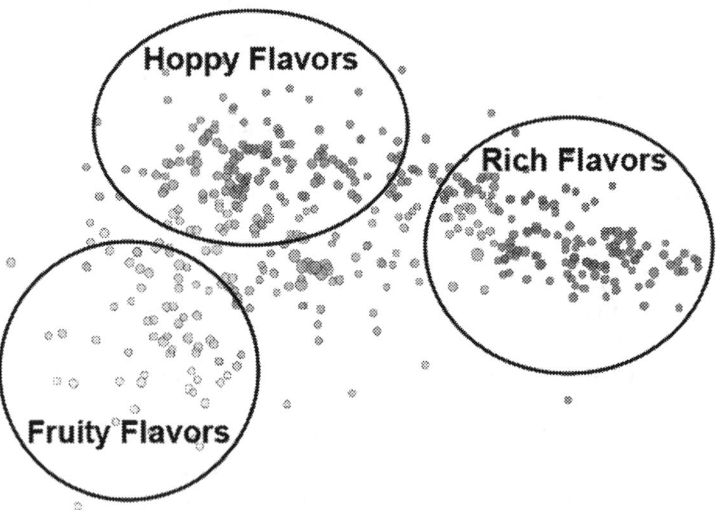

Fig. 5. PCA general structure in scatterplot with grouped labels.

To aid in user navigation of the scatter plot, a mini map was created with a black box to show the current user's view of the scatterplot, even when zoomed in. It would shrink if the user zoomed in and grow if the user zoomed out. The black box would also change its center position based on user pan movements. The mini-map was labeled with axes names and a direction. The x axis was labeled with "Richer Flavor" increasing to the right, and the y axis was labeled with "Hoppier Flavor" increasing in an upward direction.

A bivariate color map was also used to help differentiate nodal position [7]. The richer darker beers, were displayed using a darker blue shade in this ramp. Hoppier beers on the other hand, were displayed using a greener shade. Hops are typically shown on beer labels as green flower buds, so this connection may have some connection with other beer drinkers. Fruity/sour beers are generally colored in white. Colors were attributed to the points based on where that node fell on the x and y distribution. The size of the dot on the page corresponded to the number of reviews that have been written about the beer, or how “big” the beer is. Four different dot sizes were chosen based on specific review count ranges. If the cursor were to hover over a node, a tooltip would be displayed with an image of the beer, its style, its alcohol, and its bitterness rating (Fig. 6).

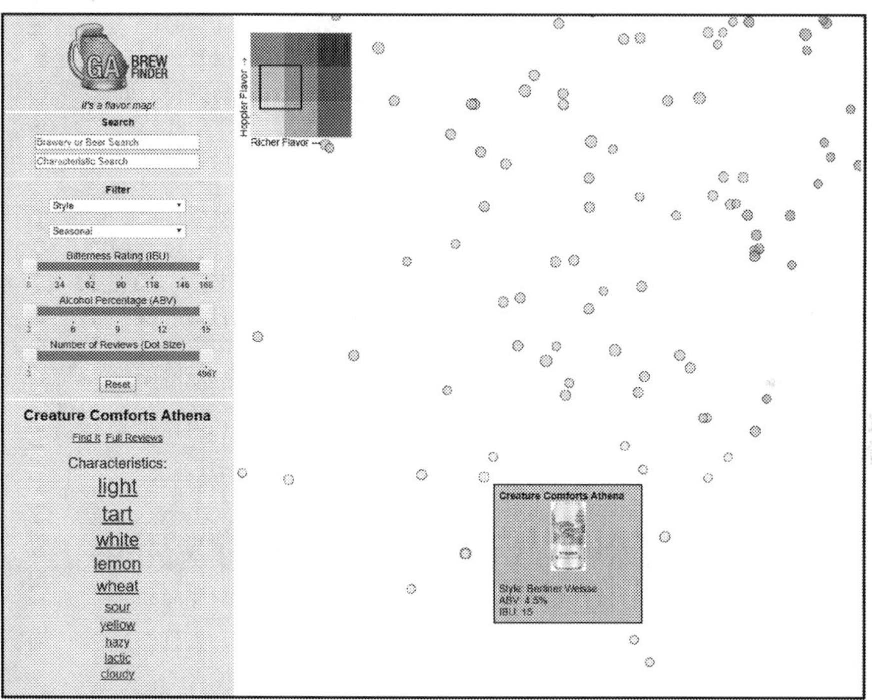

Fig. 6. BrewFinder overview, with a beer data point tooltip displayed, as well as selected with an ordered list of characteristics to the left of the screen.

Search and Filter Tools

Other tools were added to the toolbar to the left of the scatterplot view. At the top of the toolbar, are the search tools. The first search tool is the beer or brewery search, where a user can type in a specific beer or brewery name. When the user types in a name, or uses any filter control, beers that do not fit the criteria are hidden from the scatterplot. Below the brewery/beer search bar, is the characteristic search bar. This would allow someone to quickly filter for beers that contain a certain characteristic. For instance, a user could search for the term "fruity" and only beers described as fruity in their reviews would be displayed on the scatterplot. This could be very helpful to someone who is really looking for a specific flavor in their beer, such as chocolate, coffee, or caramel.

Below the search controls, are the general filter controls. Here the user can filter beers based on style, seasonal offering, bitterness rating, alcohol percentage, and the number of reviews. Each of these filters update automatically once selected and hide any beers that do not meet the given criteria. To clear both the search and filter controls, and show all beers on the scatterplot, a reset button was place just below the filter controls.

Selected Beer Name and Characteristic List

When a beer is selected, the beers name appears below the search and filter controls, followed by a couple external links and a character list. The first link takes the user to TapHunter.com, where the name of the beer is piped into the sites search field, so that a user could potentially find the beer at a bar or store near them. The second link to the right takes the user to the selected beer's full review page at the source, ratebeer.com.

Below the links, is the character list. Here a list of characteristics are displayed top to bottom, from most prevalent to least prevalent. Size is used here as well to convey the scale at which each word is used, similar to the size of a word in a word cloud. A list was used instead of a cloud here to improve readability of a selected beer. This results in fewer overall words that can be shown versus a word cloud, since only 10 can be displayed. If the user clicks on one of the words, they are able to see examples of the word in context through a popup display (Fig. 7). In this display, the user can understand what each reviewer meant in context when using a specific characteristic to describe a beer. This design is somewhat different from the word tree, however the general purpose of raising contextual awareness remains.

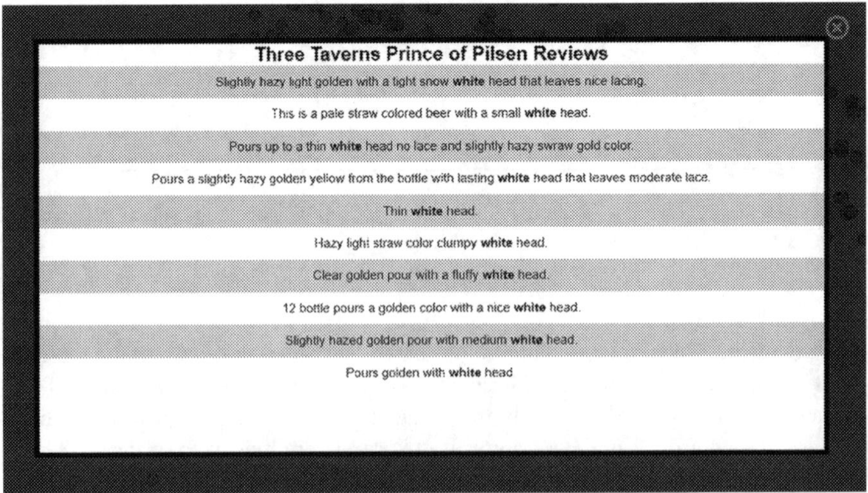

Fig. 7. Contextual characteristic view.

2.5 Functional Prototype Evaluation

The user evaluation was conducted with 8 participants, in a semi-structured interview format. The test was completed using the user's personal computer. This was done to obtain a variety of experiences based on the way they would use the website on their own. Users were first asked to rate their knowledge of both beer and interactive visualizations. Then, they were asked their initial reactions to the system once presented. The users were then asked to complete several tasks with the interface, and was encouraged to explore the data set on their own at the end of various tasks. In this way, emergent findings and discoveries could occur, where as a hard-scripted evaluation would be limited. One important task, was to ask the user to view the characteristics of a beer that they have had before, and answer whether the characteristics match their own past experiences. At the end of each task, several probe questions were provided, so that comments about the system could be recorded through note taking. Each user was also asked to rate the ease of use for each task on a five point Likert scale. Some post test questions were then asked at the end of the tasks, specifically concerning if they would use the tool for their own use, if they would recommend it to a friend, and any additional comments they may want to make about the system.

3 Results

3.1 Functional Prototype Summative Evaluation

For the demographics, the self-reported beer knowledge average across participants was reported at 4.25/10, with 10 meaning a large amount of knowledge, and 1 meaning very little. For the interactive visualization rating, participants averaged 6.5/10. Participants were also asked if they have had trouble purchasing beer in the past, and the answer was yes across every session. For each task, participants were asked to rate the ease of completing each task out of 5, with 5 being very easy, and 1 being not easy at all. The average ratings for each task were; help menu task (4.1), tooltip reading/hover task (4.8), understanding characteristics (4.7), zooming/panning/knowing where you are on the scale (4.4), using search and filter (4.4), and using reverse character search (5). The comments made after completing each task by participants are shown below.

Finding/understanding the help menu.
"Too many words."
"Expected the button to be somewhere at the top."
"(Button) should be placed in top right."
"Should be in bullets."
"Helps tell you what to do."

Reading tooltips, hovering over nodes.
"I can actually understand what the meaning of the color is."
"I like that the tooltip has a picture."
"Have a better explanation of why there are blank spaces on the scatterplot."
"Easy to read tooltip. You know it goes with that specific bubble."

Understanding characteristics.

“Matches the taste of the beer.”

“Characteristics totally match Guinness.”

“Characteristics accurately describe the beer.”

“Maybe not how I would personally describe it, but it describes it well.”

“Athena is kind of a love it or hate it thing, because it is tart and weird. If it has lemon, and you are not a big fan of that, you can click on it and see specifically what it is about.”

“I wouldn’t identify with the word pine for this beer, not a word I use.”

“Really like the sentence view by clicking on the word, helps get an understanding.”

Zooming/panning, knowing the scale.

“Can tell that the beers closest to the one I select are going to be similar.”

“Zooming in helps me understand beer at a glance.”

“Weird that the size of the dot doesn’t change on zoom, just the legend.”

“Easy to find what you are looking for once zoomed in to see the characteristics.”

“Zooming on touchpad is very difficult.”

“Text seems to overlap in some places when zoomed in.”

Search and filter.

“Really easy.”

“Had a hard time finding the brewery search at the top.”

“Name entry is confusing, uses a combination of brewery and beer name.”

“Sliders are intuitive.”

“Enter button doesn’t work to confirm search.”

Reverse characteristic search.

“The autocomplete list present while typing helps suggest things to you may not have thought of before.

“Would use this.”

“Seems to be case sensitive.”

“Would help me weed out the nasty stuff.”

“Super easy.”

4 Discussion

Feedback for the prototype was generally positive. Participant’s immediately found meaning within the characteristic list and could tell at a glance what it meant. Users also gravitated towards using the character search to find beers with specific qualities. One such user described beers they like as generally tasting “grassy”. They typed in the word and found several beers which had this trait listed as one of their top 10 traits. They then selected one of the beers and used the “Find It” link to locate the beer through TapHunter.com. They found the beer using the tool at a local grocery store and found the characteristics to be aligned with his own perception of the beer and stated that it worked as expected. Given the ease of use ratings for each task, the prototype overall was very usable, with a few key pain points that need to be addressed in a future revision.

One potential pain point across users who may not have deep knowledge about beer, may not identify with certain, potentially overly descriptive characteristics. Pine is one such characteristic a user was confused by and did not perceive it as a good word to describe flavor, while the word was quite popular among those who wrote reviews to describe certain beers. A difference in vocabulary can lead to poor communication between the interface, and those who may not understand potentially complicated descriptors.

Another potential pain point is the scatterplot itself, and its ability to display many beers at once. For simplicity's sake, this current revision focused on displaying only craft beers within the state of Georgia (along with several macro beers). To scale this up, would require either a filter for user geographic area (i.e. which beers are typically offered around them) and/or potentially a clustered, group view where the user is required to drill down to a specific area of the scatterplot (starting with a structure similar to the three-group design discussed earlier in the paper).

5 Conclusion

The BrewFinder interactive visualization is currently live at www.GABrewFinder.com. One general discussion had outside of the evaluation with those who have used the system, showed how much of a demand there is for a mobile version of the site. This makes sense, given the typical consumer's behavior of deciding which beer to purchase while they are out at a bar or at the grocery store, and may not have time or the time to browse the tool beforehand. I decided the desktop platform could provide the greatest amount of power to the user to explore the dataset at hand and could be used as a testing ground for which features specifically could be carried over into a mobile version. A new study could be done in the future which aims to take this application and bring it to mobile. This study could examine not the visualization techniques used here, and novel approaches that better suit the desired platform.

This research is also not just limited to the scope of beers and exploring their flavor. This same process and design, could be used to examine reviews about any type of product that may be hard to quantify, but it accompanied by length text reviews. Some potential candidates are movies, video games, or other media. With the growing appearance of the long-tail market in several areas, the ability to explore and draw the attention of consumers to their desired or intended product, is potentially very valuable.

References

1. Abdi, H., Williams, L.J.: Principal component analysis. Wiley Interdiscip. Rev. Comput. Stat. **2**(4), 433–459 (2010)
2. Abras, C., Maloney-Krichmar, D., Preece, J.: User-centered design. In: Bainbridge, W. (ed.) Encyclopedia of Human-Computer Interaction, vol. 37(4), pp. 445–456. Sage Publications, Thousand Oaks (2004)

3. Anderson, C.: The Long Tail: Why the Future of Business is Selling Less of More. Hyperion, New York (2006)
4. Berman, A.: Craft brewer volume share of U.S. beer market reaches double digits in 2014, 16 March 2015. https://www.brewersassociation.org/press-releases/craft-brewer-volume-share-of-u-s-beer-market-reaches-double-digits-in-2014/. Accessed Brewers Association
5. Dwyer, P.: An approach to measuring influence and cognitive similarity in computer-mediated communication. Comput. Hum. Behav. **28**, 540–551 (2012)
6. Endert, A., Fiaux, P., North, C.: Semantic interaction for visual text analytics. In: Proceedings of the SIGCHI conference on Human Factors in Computing Systems, pp. 473–482, May 2012
7. Stevens, J.: Bivariate Choropleth Maps: A How-to Guide, 18 February 2015. http://www.joshuastevens.net/cartography/make-a-bivariate-choropleth-map/. Accessed 3 May 2016
8. Meilgaard, M.C., Dalgliesh, C.E., Clapperton, J.F.: Beer flavour terminology. J. Inst. Brew. **85**(1), 38–42 (1979)

Co-designing for Co-listening: Conceptualizing Young People's Social and Music-Listening Practices

Michael Stewart[1(✉)], Javier Tibau[2], Deborah Tatar[2], and Steve Harrison[2]

[1] James Madison University, Harrisonburg, VA, USA
michael@hcientist.com
[2] Virginia Tech., Blacksburg, VA, USA

Abstract. Social networking applications have come to dominate the attention of technology-users of all ages, and are seen as the quintessential application of social media. They promise to connect us to our friends and family, but there are growing concerns over their ability to achieve this. We are interested in the potential of technology to connect people, but we question the approach of social networking apps and sites. Perhaps the only activity that competes with social networks for occupying so much of people's time is music-listening. Listening to music on personal devices is one of the most wide-spread forms of human-computer interaction. It also provides opportunities that could be characterized, positively, as privacy or, negatively, as isolation. To better understand the design space of people listening to music and their sociality, we examined the attitudes and practices of 26 semi-rural young people (9–15 years old) in the U.S. who are too young to drive and therefore cannot congregate at-will. Our study utilized semi-structured interviews, a design charrette, and user-testing of Colisten, our functional prototype. We found that the youth do not currently engage in widespread co-listening or even in the use of music recommendation systems. Indications are that the lack of co-listening is due to design gaps in sharing features rather than lack of interest. As one young person explained, co-listening would be "…more like a social thing, rather than 'I want to listen to music', more like, 'I want to hang out with my friend and listen to music…'". We present emergent design dimensions detailing how this population thinks about sociality and sharing media.

1 Introduction

Social networking sites and social media applications have come under increasing criticism. Most recently, the criticism is related to their ability to amplify disinformation or even hate-speech, and to conceal the true identities of the authors of such posts. While the potential to affect democratic processes such as public discourse, opinion, and elections is important, there are other important criticisms. Social media purports to connect us with our friends and family ("family"), but there are counter-indications (e.g. [1]). We look to other social contexts and forms of connectivity. We design for togetherness, rather than mere connectedness.

© Springer International Publishing AG, part of Springer Nature 2018
G. Meiselwitz (Ed.): SCSM 2018, LNCS 10913, pp. 355–374, 2018.
https://doi.org/10.1007/978-3-319-91521-0_26

In the past hundred years, the predominant situation of listening to music has moved successively from one of the public, shared consumption of live production, to one of the potentially shared consumption of recorded or broadcast production, to private consumption. "Many modern media consumption technologies provide us with a completely accessible and even personalized library. They also envision the consumption of music as an isolated act and indeed, the public consumption of music has overwhelmingly become an act that is at once isolated and often conducted in public" [2]. Personalization of music listening technology affords privacy, but has seemed to require that we sacrifice listening to music together [3].

People listening to music constitute an important population to the HCI community. Music-listening is one of the most wide-spread of human-computer interactions. Cell phones are ubiquitous and virtually all cell phones have music listening capabilities. People also listen to music on other devices in wide-spread use, such as iPods and MP3 players, tablets, and laptops computers. Although subscriptions are not necessary to use these devices to listen to music, there are over 140 million active Spotify users [4] and 30 million Apple Music Subscriptions [5].

Small, widely-available, high-capacity devices offer users high-quality, choice, and mobility. They also make music personal. The concept of a cell phone or an MP3 player is similarly individual. Furthermore, the ubiquity of headphones means that even when people listen to music in the presence of others, that experience may be quite private. Sometimes this is desirable, as when people use technology to "cocoon" in public, "escape from one's current environment through creating a kind of 'bubble' in which outside distractions are shut out" [6, p. 278].

Despite the prevalence of these devices and practices and the potential they present for design, they are infrequently investigated in design-oriented research. Other, related issues are explored, such as managing and sharing music libraries [7, 8] and engineering audio experiences [9, 10]; however, to our knowledge, Mainwaring, *et al.*'s work in large urban centers [6] is one of the few projects that explores sociality in the everyday experience of listening to music on these devices or their potential for sociality.

More recently, Kirk *et al.* explored the sociality of music sharing in public space using a low-fidelity prototype as a technology probe. They highlight some of their participants' speculations that remote co-listening may disappoint due to the mobile interface being a poor substitute for copresence [11].

While music-listening is popular in the general population, it is a critical part of identity-formation, relationship-building, and socialization for some young people [12–15], with ages 8–14 spending anywhere from 13 to 17 h per week [16, 17]. In urban areas, or settings with viable public transportation, or at least safe roads for cyclists, young people can congregate to socialize outside of school with relative ease. In some areas such as rural and semi-rural areas, or others with unsafe cycling conditions, young people are essentially as immobile as older adults in other communities [18] and cannot as easily congregate. Young people in these settings may turn to information and communications technologies (ICTs) to bridge the geographical gap.

However, these technologies were not designed for this specific purpose and may have drawbacks of their own, e.g. many teens feel obliged to respond with virtually no delay to text messages [1, 19, 20]. How might we facilitate isolated young people in "keeping in touch" with each other, while mitigating some of the common concerns raised by the usage of modern ICTs?

Following on our earlier work that explored young people's keeping in touch and music-listening practices, we endeavored to learn more about the design space of technology that might support relatively isolated young people in co-listening.

We define "co-listen," following common denotations of the "co-" prefix, as "with-" or "joint-" or "together-" listening. That is, a person listening "with" another. Historically "with" in relation to co-listening would have meant co-located listening or "collective listening" [21]. In our work, we define "co-listening" as consisting of people intentionally listening to the same thing, synchronously. This differs from the use of the term by Kirk, *et al.* who include in their definition asynchronous "co-listening" [11] (e.g. the "friend feed" feature supported by Spotify). In contrast, we do not require that participants be co-located but require that the activity be synchronous.

We conducted a lab study including semi-structured interviews, a design charrette, and user-testing to explore the relationship between listening to music and sociality for young people.

2 Related Work

People seek connection with others [22] in a variety of ways. Most technological ways of seeking connection do not involve music. Hassenzahl, *et al.* provide a review of the literature on connection under the label of "relatedness" [23]. He characterizes many technologies' strategies for supporting it. Many modern ICTs aim to be "social" or to "connect" us with our friends and family, but often those technologies approach this by supporting directed, explicit communication, which has long left a gap, wherein our more delicate and subtle sociality [24] is unsupported [3]. Recently various ostensibly personal technologies such as mobile devices and wearables have been exploited to support the social (e.g. [34]).

Media spaces have a long history of facilitating togetherness [9]. Media spaces can facilitated focused connection between people by supporting concurrent exercise such as in-home aerobics [30] and yoga [31], and even jogging [32]; however, because they create a persistent channel, they also facilitate sharing the quotidian [25–27], even those media spaces that are created ad hoc through user appropriation of ICTs [28, 29].

Young people are among those who seek connection [13, 15, 19, 33], and may utilize media spaces to hangout [28], but sometimes experience issues that we would characterize as related to access [19, 28].

Some approaches to connectedness do involve music, e.g. the Shakers (a Protestant religious sect) would "sing the same song at the same time of day as a way to feel connected across geography" [35, 1:50]. More recently, Leong and Wright studied social practices surrounding music, and found that sharing was a large part of people's modern music experiences [21]. Indeed for young people as well, music plays an important role in connecting with their peers, socializing, and identity-formation [13, 14, 36].

A first study in the current line of exploration [3] focused on young people's current practices and experiences with music-listening and ICTs via a diary-study. To summarize, in that study 19 children aged 9–15 responded to daily (SMS or email) diary prompts via following a link in the prompt and completing a short questionnaire that asked them a few multiple-choice questions about their practices, and asked them to elaborate as free text when appropriate. Participants were each prompted once per day for 14 consecutive days. In that work, we found that participants were interested in "keeping in touch" and listening to music, but rarely listened synchronously, and did so only while co-located.

Further work was necessary to determine young people's access to ICTs and entertainment technologies, interest in co-listening, technological support for co-listening, and social issues surrounding co-listening. The current study sought to replicate these findings, and learn more about some of the practices. Additionally, the current study advances the work by engaging the study participants as design informants [37] to design and test co-listening technologies [38].

3 Method

To find out more about semi-rural young people's thoughts, wishes and preferences, we recruited pairs of young people to come in together and engage in a semi-structured interview, a design charrette, and user-testing. Following Druin's framework for children's roles in research [38], we employed the child as informant method [37].

3.1 Interviews

The semi-structured interview guide started by replicating the questions in our previous 2-week long diary study [3] about young people's access to devices, music-listening, and communication. The entries in the diary study had been very short. We hoped that together with a friend in a face-to-face context in which we could pose follow-on questions, participants would elaborate more. The interview guide also contained further questions and topics to help deepen our understanding of participants' contexts. Additionally, as the sessions progressed, some exposed new topics that the interviewer appended to the interview guide for future sessions.

3.2 Design Charrette

The design charrette employed the participant as a design informant to design a co-listening technology. We asked participants to "design a technology that would let you listen to music with your friend, while you are each in your own homes." We gave each participant blank U.S. letter-size paper to begin with and put out a cup of markers and pens. The participants were told that they could use as much paper as they needed, and asked to inform us (we remained in the room) when they were done. Once all of the participants in the session finished their design, we asked them to explain their designs, and asked them questions about them.

3.3 Prototype Testing

Next, we asked participants to try out Colisten, our own prototype of such a technology. We handed an iPad to each participant (in the one session that involved four people (see Sect. 5), two participants shared each of the two iPads). We demonstrated how the app worked, and asked if they had any questions. Next the participants used Colisten to browse playlists and listen to music while sitting together. Part of this was listening to the same music on the two different devices at the same time. Finally, we asked participants for their feedback and questions. In this discussion, we often asked participants about their thoughts about the ways that Colisten was similar to or different from their design.

4 Colisten Prototype

We designed a streaming music player that utilized Spotify's API [39] to provide access to a Spotify user's playlists. The user could log in to and out of their Spotify account, browse their playlists (Fig. 1), play one of their playlists, start the playlist from a track of their choice, or change what they are listening to with by choosing another of these options. So, Colisten users (having authenticated with Spotify using their existing credentials, and authorized our Colisten app to access their Spotify account) were able to listen to full-length tracks and playlists. The novel component of the app was a Friends' Drawer (Fig. 2) that the user could open to reveal a feature to add a Colisten friend, accept or reject ("ignore") invitations to be a Colisten friend, and cancel a previously made invitation. In Colisten, a "friend" was a Colisten user (identified by their Spotify user name) with whom a person want to co-listen.

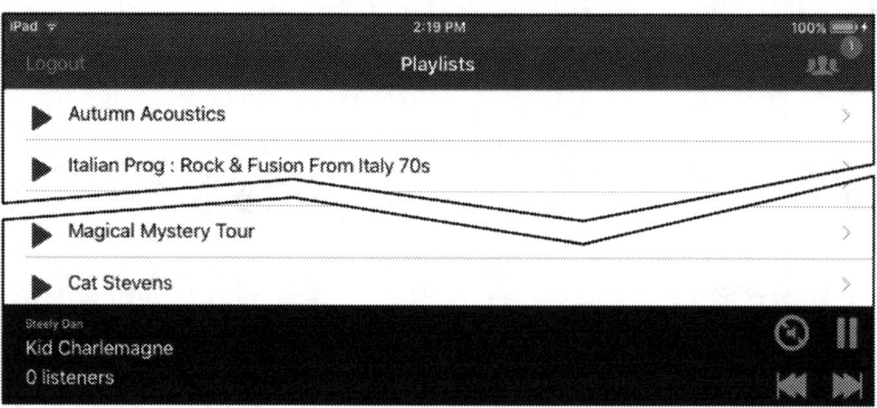

Fig. 1. User viewing playlists while listening to Steely Dan's "Kid Charlemagne" in Colisten. The top right "friends" button has an icon that indicates there is one friend online.

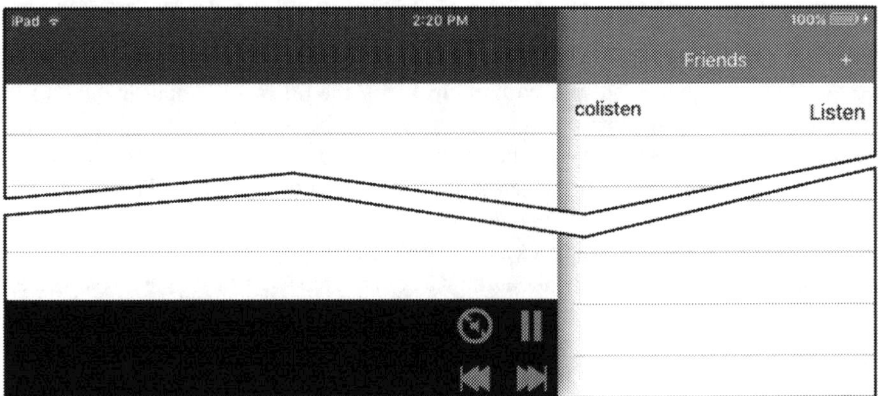

Fig. 2. Clicking on the friends button shows friends who are currently online and allows a request to Listen with them.

Additionally, a Colisten user can use the Friends' Drawer to (1) see if any of their Colisten friends are currently playing music and (2) begin listening to the same playlist, at the same track, at the same moment of that track. As the friend's music progresses so too does any of their friends who have chosen to co-listen with them. Usually, this means progressing through a playlist; however, if the friend pauses, skips a song or changes playlists so too will their currently listening co-listeners. When user A selects a friend (B) with whom to co-listen, B is notified that A is now listening. If B opens the app, the Now Playing bar will show the current number of colisteners (Fig. 3).

Importantly, users of the app cannot invite someone to co-listen with them (just to be their "friend") as they listen to some track or playlist. They can only be joined while listening, or join a friend who is listening. Once a user has accepted a friend, that friend can co-listen at any time the app is running. Only the originating user can pause/resume,

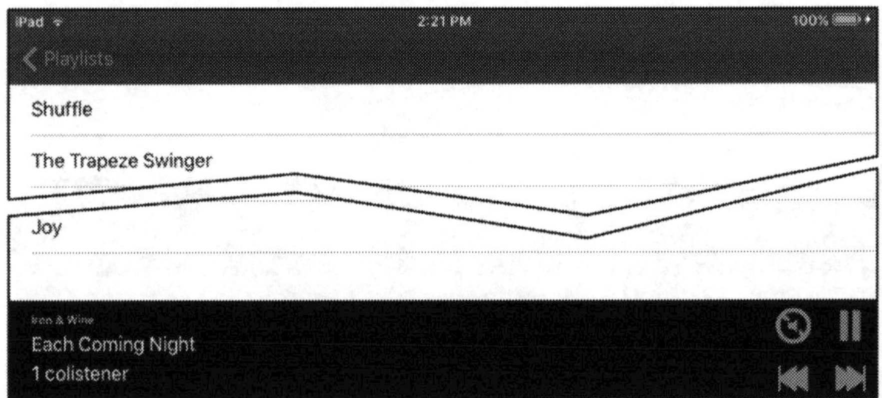

Fig. 3. Having clicked the Listen button, the user (not pictured) is now listening to the pictured user (who received a notification, and can now see that they have 1 colistener).

skip a track, or change playlists. If co-listeners wish to listen to something different, they can utilize other existing communications channels (e.g. SMS) to make an explicit request.

Listening to music, even with friends, is often not a foreground activity [40]. By making the design of co-listening light-weight and low-interruption and only ever notifying users when a friend invitation is received or that a friend has started/stopped listening with them, we can facilitate joining without interrupting.

4.1 System

Colisten was implemented using a client-server architecture. The server is an Amazon EC2 micro instance running (Amazon's Linux distribution and) the Colisten server application in the NodeJS environment. The client in this study is an iOS application written in Swift. To communicate between the client and server we use two different channels: Apple Push Notification Service (on the iOS client) and websockets.

4.2 Limitations

Colisten did not re-implement all features of a streaming music player. That is, since it required a Spotify account and utilized a Spotify API, we did not re-implement all of Spotify's features. For the prototype phase, Colisten users had to create their playlists via Spotify's web, mobile, or desktop applications.

The Spotify SDK (on which Colisten relied) required that users of third-party applications (like Colisten) sign in to a paid (premium or family) Spotify account. For our study (and development) we created several Spotify premium accounts.

5 Participants and Recruitment

Following a protocol approved by our Institutional Review Board (IRB), we recruited participants by sending emails to 6000+ graduate students at Virginia Tech and faculty in our department, word of mouth, and community outreach coordinators who used Facebook and email. Our recruitment text targeted 9–15 year-old children and advertised our interest about them, "feeling connected with each other." Further, we explained that participating children must bring a friend (also aged 9–15) and try out an app to listen to music together and talk about their schoolwork and social lives.

Parents contacted the researchers to express their child's interest and were asked to complete a recruitment questionnaire about the child's demographic information (age and gender) and the name of a friend with whom their child would participate. The recruitment questionnaire included informed consent information. Shortly after receiving the informed consent and recruitment information, the researcher contacted the consenting adults to arrange their respective children's session.

On day of their session, the participants arrived at our building and once both members of the pair had arrived, we obtained their informed assent (having already obtained their parents' consent). In one session, we had a quartet rather than a pair, as three of the participants were triplets and the fourth was friends with all three. Participants were compensated for participation with 20 USD.

Twenty-six child participants were included in the study. The children's gender was reported as female (12), male (14), or other (0). Their ages varied within our (9–15) recruitment range with an average of 11.5 years, and a standard deviation of 1.8 years (modal ages of 10 and 13, none were 14).

6 Analysis

Each part of the study collected different kinds of data, and was analyzed accordingly.

6.1 Semi-structured Interviews

To analyze the interviews, we numbered the topics to produce the beginning of our transcription guide. We then began transcribing the interviews and coding for those topics. As we encountered new topics, we added those to our transcription guide, and reviewed completed transcripts to ensure they considered the additional codes. Also, the transcriber (in most cases the interviewer) wrote memos into the transcripts to surface and record interesting observations. From these transcriptions we produced counts for several questions that elicited nominal data, and sought quotes that represent the data.

6.2 Design Charrette

We utilized a collaborative "data gallery" analysis approach. First, we taped all of the designs on the walls of our conference room (Fig. 4). Next, our research group got sticky-notes and a pen and walked through the designs, annotating them with sticky-notes. We each made a few passes until we agreed the data were saturated with annotations. Following this activity, we each discussed what annotations we made with the group. Some members reported they made similar annotations and coalesced their contributions, while others added nuance or generalization to the facets annotated by a colleague. Still others shared divergent observations.

6.3 Prototype Testing

Analysis of prototype testing was based on video transcriptions (see Sect. 6.1). In addition to coding the video for questions in the interview guide, we transcribed the prototype testing session and the discussion.

Fig. 4. Data Gallery – Participants designs are affixed to the walls. The researchers used sticky-notes to annotate interesting details, finding some commonalities and some distinguishing features.

7 Results

Study sessions ranged from 26–44 min (average of 35:10 with a standard deviation of 6:16) and occurred between October 2015 and April 2016.

7.1 Overview of Semi-structured Interview

Our interviews focused on questions about how participants kept in touch with their friends outside of school and on their music-listening practices.

As the interviews were semi-structured and conducted in pairs, not all participants were asked all questions (sometimes responses do not total 26, the number of our participants). Often interviews result in rich sources of primary source information that includes insightful and elucidating quotations. We value our participants and their contributions. However, perhaps due to their age, interest in the topics, or our interviewers' skills, they elaborated less than we had hoped (see Sect. 3.1), and so we rely less on their quotes to present the data than we would prefer.

Keeping in Touch. Like the prior work, we found that most of our participants kept in touch with their friends outside of school (16). Three participants said they do not listen to music, one clarified that they do "sometimes".

From the responses of twenty (20) of our participants, we know that they use multiple methods to keep in touch with each other: face-to-face, call, text (i.e. SMS), Instagram, Skype, voice chat in a (console) video game, Facetime, Snapchat, Facebook, Twitter, email, and Google Hangouts (video); with one participant using (the maximum we saw) 5 of these. Texting was the most popular medium (10 participants used texting, 5 of whom "mostly text"). The next most popular medium was voice calling (on a phone) with 5 participants. Three of the five who use voice calls indicated use only email as an alternative (1) or no other form of communication (2) outside of school.

Participants contact each other after school, in the evenings and on weekends, but not normally before school. The content of their talk was often characterized as "nothing" "anything", "what they were doing" or other communications that can be characterized as phatic, that is, language used for general purposes of communication rather than to convey information; however, as in the prior study, it also included plans [3].

The amount of "keeping in touch" varied widely. Some participants reported being either so busy with extracurricular after-school activities (e.g. gymnastics) and others not having their own devices that they reported keeping in touch less than once a month. Other participants reported keeping in touch with their friends once a month, one to two times per week, four to five times per day, and all the time, "Probably when I get home, I have it in my pocket, all the time, have my ringer turned all the way up, and usually I'm [also] on Xbox."

Listening to Music. Twenty-three of our participants reported listening to music, three reported that they did not. They report listening to a variety of music: "pop", "heavy metal", "not rap", "rap", "christian", "variety", "Sam Smith", "Kidz Bop", "country", "Hideaway", "Fetty Wop", "Drake", "Flo Rida", "Bieber". Our participants listen to music via several different sources (see Table 1). Perhaps surprisingly, the most popular music source for our participants was FM radio (24 participants listen to it). We also found it surprising that YouTube was such a popular music source (the second-most in our study with 14 participants using it for music), but apparently our participants fit a larger trend in this respect as during our study, YouTube released a new app, YouTube Music. 10 participants reported listening via Pandora.

Co-listening. More participants listened to music with friends or family than was found in our prior work [3]. 21 reported doing so, while one reported "not usually", and one said they did not. Our participants used a few different methods to listen with others, most popularly their mobile devices' speakers (12), automobile's stereo (11), or their home stereo (10). Only 2 participants reported sharing headphones (one earbud for each person), and only one participant reported listening to music with friends or family via their computer speakers. We asked the participants who co-listened with if they listen to the same music with others as when they listen alone. 8 said they listen to the same music, and 4 said they listen to different music (2 of these because their parents would be selecting the music whenever they are co-listening, i.e. in the car), but 1 listened alone to music his parents might object to.

Music Recommendation. Our participants reported engaging in music recommendation with their friends or family more often than found in our earlier study [3]. 14 indicated they participate in music recommendation, and 5 that they did not. None of these recommendations were made utilizing any explicit technological support or feature. The majority of these recommendations were made in a face-to-face setting (10), and only 3 reported sharing recommendations via text message.

Access. We asked some participants whether they had their own device to keep in touch, and learned that 10 do and 5 do not. Of those who do, five had iPhones, two have a cellphone, and one person each has an iPad, slide phone, tablet, or no such device.

Aside from those devices that are their own, we asked what technology was available in their homes for communication and entertainment. Televisions, computers, video game consoles and iPhones were the most frequent (see Table 2).

Table 1. Participants reported listening to music from a variety of sources.

Music source	Frequency
FM Radio	24
YouTube	14
Pandora	10
iTunes	6
Spotify	4
Google Music	2
SoundCloud	1

Table 2. Participants reported the availability of myriad technologies for communication and entertainment.

Technology	Frequency
Television	17
Video game console	17 (Xbox: 9, PlayStation: 8)
Computer	12
iPhone	8
iPad	5
FM radio	4
Tablet	3
Nintendo 3DS	2
Kindle	3 (Fire: 2, Paperwhite: 1)

Even when technology is available, access is gated by parental policies. Of 24 responses, nine indicated that they were required to finish homework and/or chores before they could use some of these devices for leisure, five said there were no rules, five said there was no specific rule, but if their parent came to feel that the young person had been using a device for a long time, they would tell them to get off. Other kinds of policies that our a few participants reported were: just to ask their parent before using such devices, only using them at a certain time of day (e.g. not in the morning when the family is trying to go to school/work), usage permission based on grades at school, or having a fixed amount of time that they were allowed to use the devices. One participant declined to tell us their parents' rules, telling us that the rule is, "kinda private."

We sought further understanding of the young people's music-listening practice by asking about how they discover new music and whether their family had any rules about what music they may listen to. FM radio was the most-used source of discovery (9), face to face recommendations (7) were the 2nd-most frequent. Less frequent music discovery options included Pandora, YouTube, CDs, Vine, and Facebook. Few of our participants indicated that their parents had rules about what they could listen to. Eleven said their parents had no rules, 3 had to ask first, 2 just have to make sure not to purchase the explicit version of any track, and one participant had unspecified rules where their parent may just not like a song and tell them not to listen to it.

7.2 Design Charrette

The participants' design charrettes varied in length from 3.5–15.2 min (avg. 6:13, with standard deviation of 3:15). The participants' designs, and the ways they were represented were fairly diverse. Some participants made a list of statements or features, many drew their design and added some text for explanation, and some tried to indicate successive steps in their design that were required to establish a co-listening session.

Five Emergent Design Dimensions. Variation along five dimensions emerged from the design gallery analysis: Initiation, Group Composition, Control, Hardware, and Activity Prioritization.

Initiation. One issue, how a co-listening session was initiated, was illustrated in several (13) of the designs. From these designs we would characterize the ways to initiate co-listening as a spectrum (Fig. 1). For simplicity, we have labeled the 2 participants of a co-listening session as the listener and the broadcaster to describe who is playing the music and who is listening. Note: in cases toward the center of the spectrum, these labels seem less appropriate.

For this dimension, we observed that in some designs, one user (in this case the "listener") could see what their friend is listening to, and just listen to them (in fact, this is how Colisten functions), in other cases this user had to request permission to listen (Fig. 11). In still other cases, a user could either "tune-in" to their friend or "invite" their friend. At the other extreme of the spectrum, a user could only invite a friend to listen with them (e.g. "An app that allows you to have friends and when you want to listen to music with somebody, you could send them a Request showing the song who its by and they will have the option to Accept or Decline."), or in other cases, could just play something that would also play for the friend (Fig. 6).

Group Composition. Participants' designs supported different co-listening group structures (Fig. 7). In the case of a Fixed Dyad (e.g. Figure 8), the participant's design illustrated that the dyad who could listen with each other was embedded in the system:

- "there's a specific product for this, it's not really like an iPhone, but it's like a music player… it's already connected to another one, so you can buy it like, it's like a… friendship thing, and they're like, automatically connected to each other".
- "if you have this thing, and if you have an app to connect with it, you can hear the same music that your friend is hearing"

In other cases, the group was still constrained to a dyad, but membership was configurable (e.g. Figure 5). At the other extreme found in our participants' designs, a group larger than a dyad could co-listen, and the membership of this group was configurable (again, this is how Colisten functions). One design employed 5 headphone "spaces" (Fig. 9).

Control. Some participants addressed issues of control. As part of initiation, some designs enabled an invited or a requested partner to deny, ignore, or block the requester (e.g. Fig. 11). Deciding what tracks would be played was implicit in some designs (e.g. a design that was [not suited to remote co-listening] just a smartphone which would have 5 headphone ports). While other designs more explicitly explored track selection (e.g. Fig. 9 indicates that the user is "not [the] song picker"). Some of these designs, implied the existence of a third-party arbiter that would be necessary to support their control features.

Hardware. Some participants seemed more constrained by, or at least more interested in existing technologies' hardware from older devices such as a boombox to touchscreen mobile devices (cf. Colisten). Some added a new hardware component to an otherwise existing/familiar technology, and some created entirely novel hardware (Fig. 8).

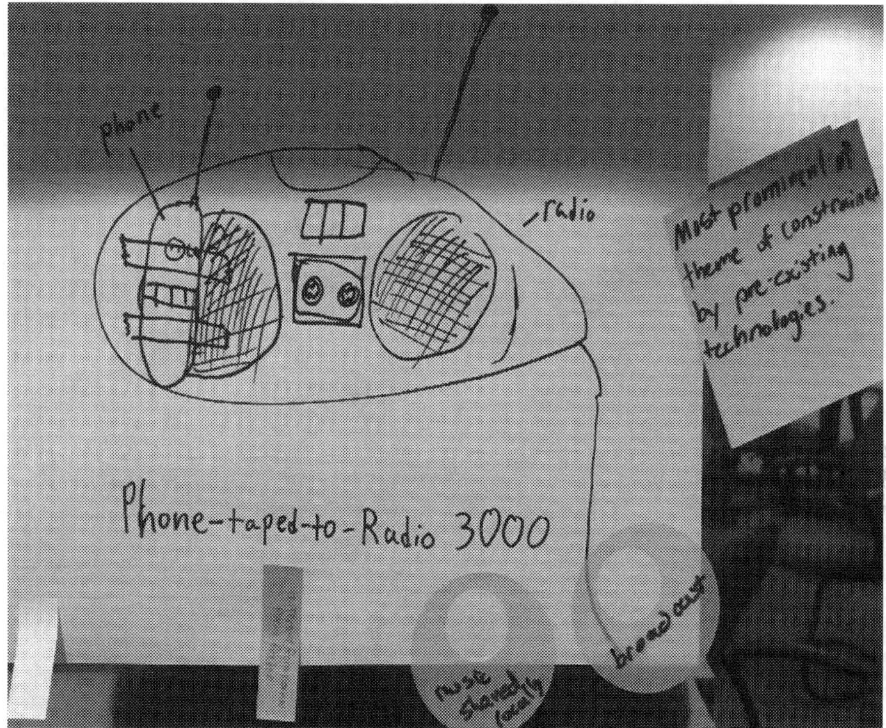

Fig. 5. Phone-taped-to-Radio 3000: a participant's design for a technology that would support remote co-listening.

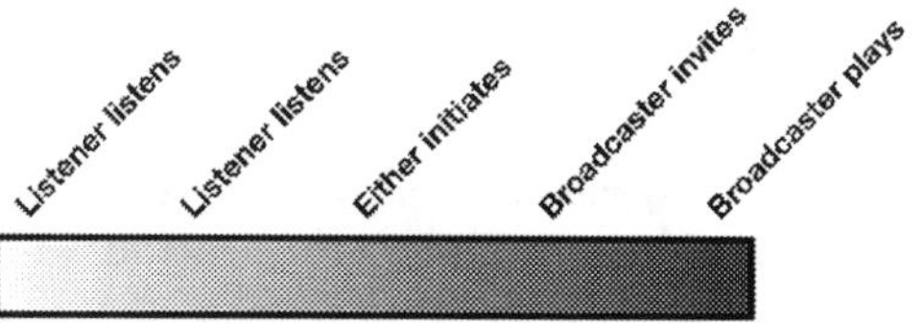

Fig. 6. Participants' designs depicted a range of co-listening initiation. At one extreme the listener can just listen to a friend, while at the other extreme, a friend can just start playing music that will be heard by another.

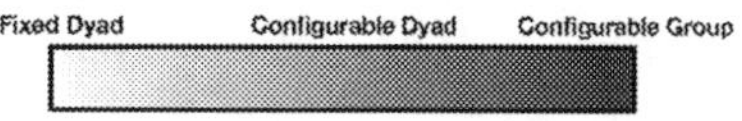

Fig. 7. Co-listening group compositions in participants' designs.

We observe that their designs reference a wide range of form factor from boom boxes to more modern touch-screen mobile devices. Some of their proposals were attempts to implement co-listening using existing technologies in combination "you tape a phone, to a speaker of a radio, and you turn on the radio, and you turn the music on low so it won't bust the person's eardrum" (Fig. 5).

Others imagined wholly new hardware (Fig. 8), perhaps as additions to existing hardware, such as a set of wireless earbuds that a user would have surgery to have implanted next to their eardrum, which would wireless communicate to an iPod as well as to another set of these owned by a friend. Many of the participants designed their co-listening technology as a new app that would run on an existing mobile device (Fig. 10), while others imagined adding only a new feature to an existing app (Fig. 13).

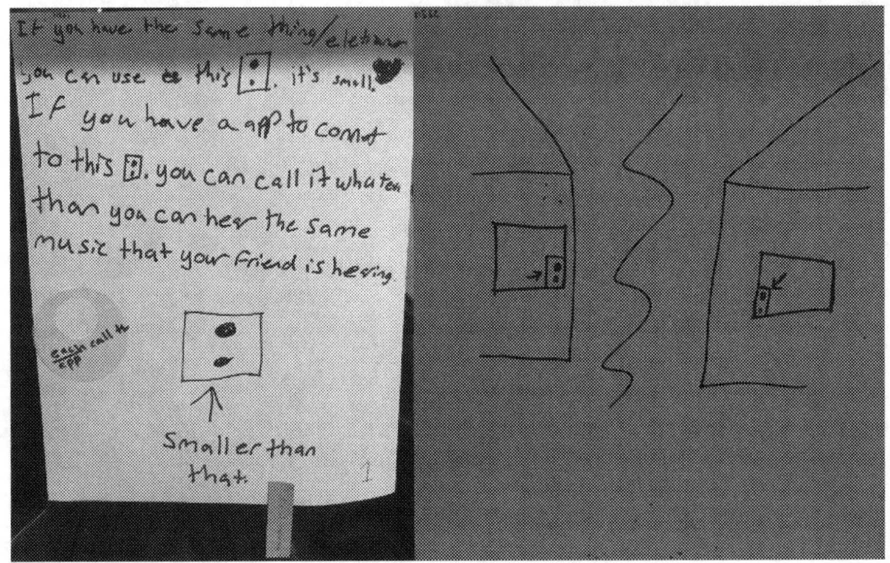

Fig. 8. A new piece of hardware could sit between a user's music player and their headphones to send music coming both to the headphone wire and, wirelessly, to another.

Activity Prioritization. It seems that all of our participants' designs were developed in one of two ways: (1) started from either a social technology (e.g. ICTs) and added a music-listening feature, or (2) that they began with a music-listening technology and added (social) support for co-listening (Fig. 12), which is more similar to Colisten's design.

Access. In some of our participants' designs we saw evidence of their concerns about device availability and connectivity that would affect their access to a co-listening technology. In some cases, (e.g. 5 headphones "spaces") there was an attempt to reduce the number of devices necessary below 1:1. In other cases the concern seems to focus more on connectivity (e.g. "…not have to use data or Wi-Fi…"). These concerns echo Grinter, et al.'s discussion of the home economics of media (in their case, SMS and IM) [19], but with a different larger of technologies.

Our participants as young as 9 years old were cognizant of the costs associated with their technology usage, especially when listening to music. Several of our participants who mentioned listening to music via YouTube indicated that it was because they

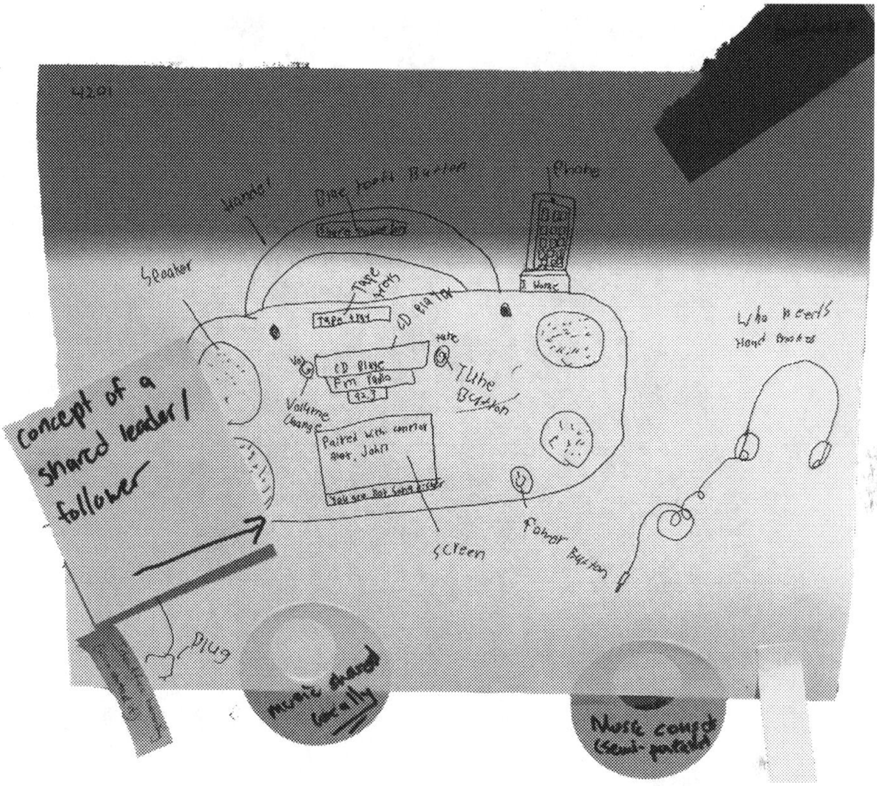

Fig. 9. Participant's design depicts a boombox that supports various music sources (cassette, cd, FM radio), and has a display to indicate with whom the user is co-listening, and that the current user is "not song picker".

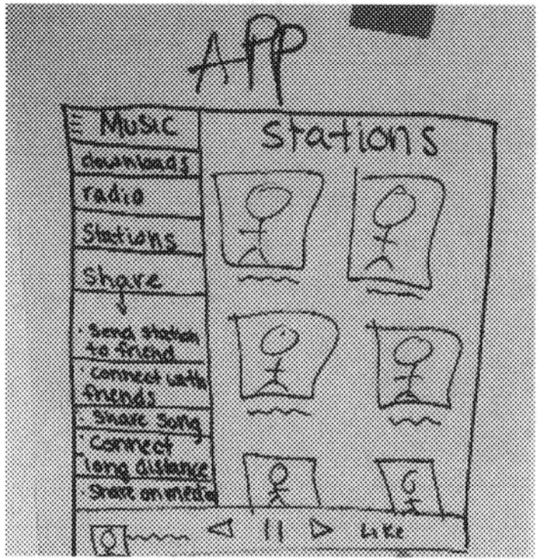

Fig. 10. Participant designed a new app for a tablet that would show each of their friends as a "station" that they could listen to.

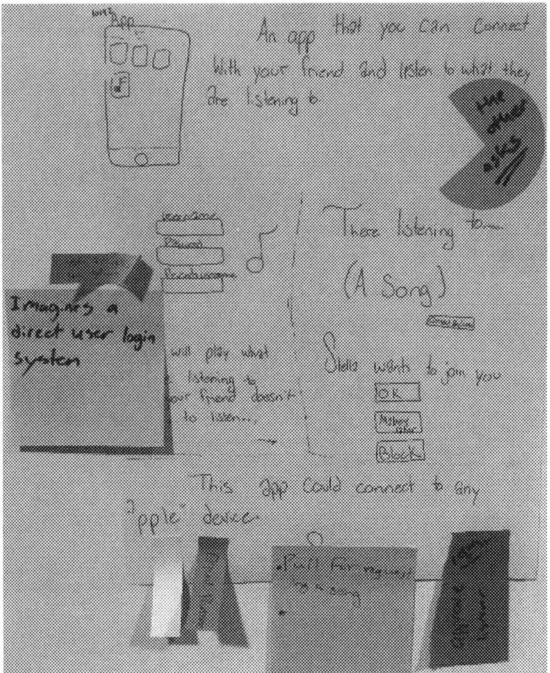

Fig. 11. Depicts a user interface for requesting to co-listen and for accepting or rejecting such a request.

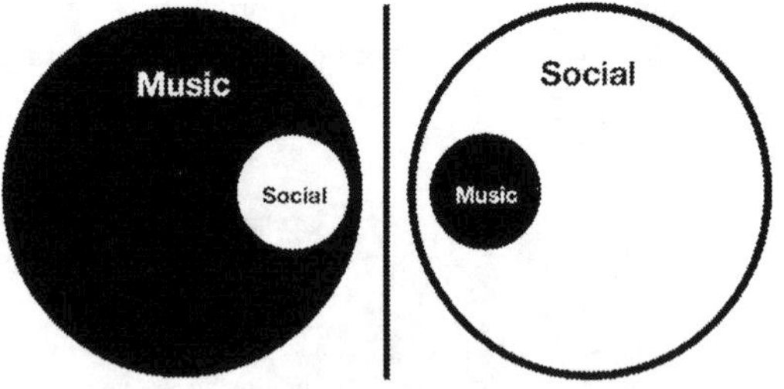

Fig. 12. Our participants' designs could be categorized as (left) adding sociality to music-listening, or (right) adding music-listening to social technology.

could do so for free. Others of our participants discussed the importance of songs being cached on their mobile device because they had little or no bandwidth available when away from Wi-Fi.

7.3 Colisten Testing

We briefly explained our prototype to the participants and then invited them to use it for a while. Participants used the app on average for 6 min (standard deviation of 3:30).

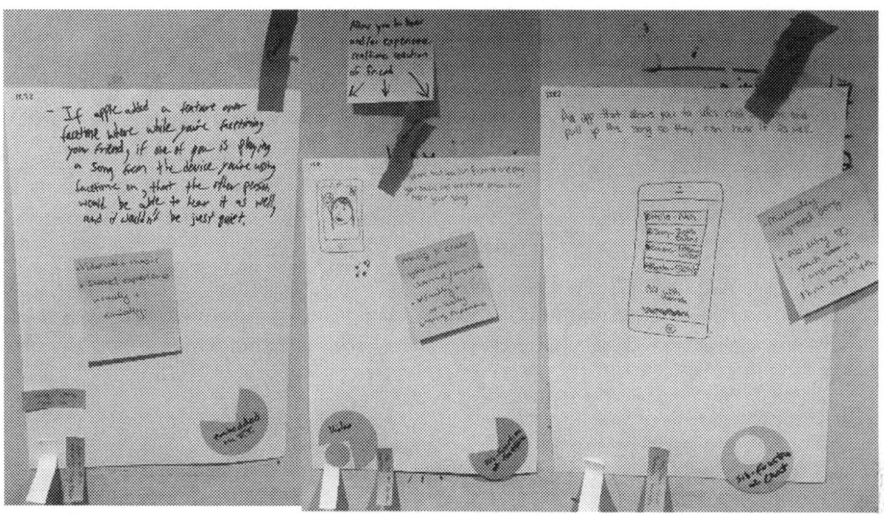

Fig. 13. Three participants all designed co-listening as integrated with voice/video chat on their mobile devices.

Many participants enjoyed the music they chose to play in Colisten, and began to bob their heads. When the participants clicked the "Listen" button on their co-participant, they often smiled when their device began playing the same track at the same time position.

Participants tested listening to different tracks and playlists and shuffling the playlists. They tried pausing/resuming and skipping tracks. They also tried our signature feature: listening with each other.

Nineteen participants said they would use Colisten. They reported their interest in using Colisten in a variety of circumstances: whenever not doing homework (4), while not in the car (i.e. at home where they have Wi-Fi, 4), weekends (1), and during carpooling (1). One participant who said that he doesn't usually listen to music said that he would like to use Colisten because with it, "you can see what mood your friend is in, sometimes you're in a mood for… see what your friend is doing".

From our discussions with participants about the differences between their designs and Colisten's implementation, we learned that 3 participants prefer to initiate co-listening as it is implemented in Colisten (listener just listens), 9 prefer to receive a request from their friends before that friend gets to listen. One participant said that (unlike the Colisten implementation) they, "would like to be able to listen to people who aren't my friends because they might have different taste and interesting music."

For some participants, we were able to think through, in more detail, various scenarios of Colisten use. In these cases, the young people did not seem to think it be would be weird for them if their friend were listening to them, and that friend had a friend of their own listening to them.

8 Discussion

Augmenting prior [3] and co-occurring work [11], we conducted a study utilizing semi-structured interviews, design charrettes, and user testing of a functional prototype to further understand the design space of young people's sociality, particularly in a semi-rural context, surface their attitudes about co-listening, and validate our design approach. We found that many participants are interested in listening to music, keeping in touch with their friends, and co-listening, but are concerned about data usage, about cost, and co-listening group composition. Five dimensions in particular emerged from our participants' designs: Initiation, Group Composition, Control, Hardware, and Activity Prioritization. These dimensions can help inform the design of co-listening technologies.

References

1. Turkle, S.: Alone Together: Why We Expect More from Technology and Less from Each Other. Basic Books, New York (2012)
2. Stewart, M., Tatar, D., Harrison, S.R.: Background, slow, and inattentive interaction: togetherness over connectedness. In: Designing Against the Status Quo Workshop of Designing Interactive Systems. ACM, Brisbane (2016)
3. Stewart, M., Tatar, D., Harrison, S.R.: Sharing, communication, and music listening: a diary study of technology use by pre-teens and adolescents. In: Proceedings of the 2016 International Conference on Collaboration Technologies and Systems (CTS 2016). IEEE, Orlando (2016)
4. Spotify Press: Spotify – About. https://press.spotify.com/us/about/
5. Karp, H.: Apple Music's Long Game: Why Jimmy Iovine Thinks They're "Not Even Close" to Success With Streaming. https://www.billboard.com/articles/news/magazine-feature/7980919/jimmy-iovine-zane-lowe-larry-jackson-interview-billboard-cover-story-2017
6. Mainwaring, Scott D., Anderson, Ken, Chang, Michele F.: Living for the global city: mobile kits, urban interfaces, and ubicomp. In: Beigl, M., Intille, S., Rekimoto, J., Tokuda, H. (eds.) UbiComp 2005. LNCS, vol. 3660, pp. 269–286. Springer, Heidelberg (2005). https://doi.org/10.1007/11551201_16
7. Brinegar, J., Capra, R.: Managing music across multiple devices and computers. In: Proceedings of the 2011 iConference, pp. 489–495. ACM, New York (2011)
8. Voida, A., Grinter, R.E., Ducheneaut, N., Edwards, W.K., Newman, M.W.: Listening in: practices surrounding iTunes music sharing. In: Proceedings of the SIGCHI Conference on Human Factors in Computing Systems, pp. 191–200. ACM, New York (2005)
9. Lenz, E., Diefenbach, S., Hassenzahl, M., Lienhard, S.: Mo. Shared music, shared moment. In: Proceedings of the 7th Nordic Conference on Human-Computer Interaction: Making Sense Through Design, pp. 736–741. ACM, New York (2012)
10. Seeburger, J., Foth, M., Tjondronegoro, D.: The sound of music: sharing song selections between collocated strangers in public urban places. In: Proceedings of the 11th International Conference on Mobile and Ubiquitous Multimedia, pp. 34:1–34:10. ACM, New York (2012)
11. Kirk, D.S., Durrant, A., Wood, G., Leong, T.W., Wright, P.: Understanding the sociality of experience in mobile music listening with pocketsong. In: Proceedings of the 2016 ACM Conference on Designing Interactive Systems, pp. 50–61. ACM, New York (2016)

12. Arnett, J.J.: Adolescents' uses of media for self-socialization. J. Youth Adolesc. **24**, 519–533 (1995)
13. Boer, D., Fischer, R., Strack, M., Bond, M.H., Lo, E., Lam, J.: How shared preferences in music create bonds between people: values as the missing link. Personal. Soc. Psychol. Bull. **37**, 1159–1171 (2011)
14. Schafer, T., Sedlmeier, P.: From the functions of music to music preference. Psychol. Music **37**, 279–300 (2009)
15. Selfhout, M.H.W., Branje, S.J.T., ter Bogt, T.F.M., Meeus, W.H.J.: The role of music preferences in early adolescents' friendship formation and stability. J. Adolesc. **32**, 95–107 (2009)
16. Lamont, A., Hargreaves, D.J., Marshall, N.A., Tarrant, M.: Young people's music in and out of school. Br. J. Music Educ. **20**, 229–241 (2003)
17. North, A.C., Hargreaves, D.J., O'Neill, S.A.: The importance of music to adolescents. Br. J. Educ. Psychol. **70**, 255 (2000)
18. Meurer, J., Stein, M., Randall, D., Rohde, M., Wulf, V.: Social dependency and mobile autonomy: supporting older adults' mobility with ridesharing ICT. In: Proceedings of the 32nd Annual ACM Conference on Human Factors in Computing Systems, pp. 1923–1932 (2014)
19. Grinter, R.E., Palen, L., Eldridge, M.: Chatting with teenagers: considering the place of chat technologies in teen life. ACM Trans. Comput. Interact. **13**, 423–447 (2006)
20. Ling, R.: The Mobile Connection: The Cell Phone's Impact on Society. Morgan Kaufmann, San Francisco (2004)
21. Leong, T.W., Wright, P.C.: Revisiting social practices surrounding music. In: Proceedings of the SIGCHI Conference on Human Factors in Computing Systems, pp. 951–960. ACM, New York (2013)
22. Ryan, R.M., Deci, E.L.: Self-determination theory and the facilitation of intrinsic motivation, social development, and well-being. Am. Psychol. **55**, 68 (2000)
23. Hassenzahl, M., Heidecker, S., Eckoldt, K., Diefenbach, S., Hillmann, U.: All you need is love: current strategies of mediating intimate relationships through technology. ACM Trans. Comput. Interact. **19**, 1–19 (2012)
24. Strong, R., Gaver, B.: Feather, scent and shaker: supporting simple intimacy. In: Proceedings of CSCW, pp. 29–30 (1996)
25. Judge, T.K., Neustaedter, C.: Sharing conversation and sharing life: video conferencing in the home. In: Proceedings of the SIGCHI Conference on Human Factors in Computing Systems, pp. 655–658. ACM, New York (2010)
26. Judge, T.K., Neustaedter, C., Harrison, S., Blose, A.: Family portals: connecting families through a multifamily media space. In: Proceedings of the SIGCHI Conference on Human Factors in Computing Systems, pp. 1205–1214 (2011)
27. Lottridge, D., Masson, N., Mackay, W.: Sharing empty moments: design for remote couples. In: Proceedings of the SIGCHI Conference on Human Factors in Computing Systems, pp. 2329–2338 (2009)
28. Buhler, T., Neustaedter, C., Hillman, S.: How and why teenagers use video chat. In: Proceedings of the 2013 Conference on Computer Supported Cooperative Work, pp. 759–768 (2013)
29. Greenberg, S., Neustaedter, C.: Shared living, experiences, and intimacy over video chat in long distance relationships. In: Neustaedter, C., Harrison, S., Sellen, A. (eds.) Connecting Families, pp. 37–53. Springer, London (2013). https://doi.org/10.1007/978-1-4471-4192-1_3
30. Judge, T.K., Neustaedter, C., Kurtz, A.F.: The family window: the design and evaluation of a domestic media space. In: Proceedings of the SIGCHI Conference on Human Factors in Computing Systems, pp. 2361–2370 (2010)

31. Muntean, R., Neustaedter, C., Hennessy, K.: Synchronous yoga and meditation over distance using video chat. In: Proceedings of the 41st Graphics Interface Conference, pp. 187–194. Canadian Information Processing Society, Toronto (2015)
32. O'Brien, S., Mueller, F.F.: Jogging the distance. In: Proceedings of the SIGCHI Conference on Human Factors in Computing Systems, pp. 523–526 (2007)
33. Grinter, R.E., Palen, L.: Instant messaging in teen life. In: Proceedings of the 2002 ACM Conference on Computer Supported Cooperative Work, pp. 21–30. ACM, New York (2002)
34. Woźniak, P., Knaving, K., Björk, S., Fjeld, M.: RUFUS: remote supporter feedback for long-distance runners. In: Proceedings of the 17th International Conference on Human-Computer Interaction with Mobile Devices and Services, pp. 115–124. ACM, New York (2015)
35. Sharon, S.: The Stewards of A Disappearing Faith — And 10,000 Songs (2016)
36. Tarrant, M., North, A.C., Hargreaves, D.J.: English and American adolescents' reasons for listening to music. Psychol. Music **28**, 166–173 (2000)
37. Scaife, M., Rogers, Y., Aldrich, F., Davies, M.: Designing for or designing with? Informant design for interactive learning environments. In: Proceedings of the ACM SIGCHI Conference on Human Factors in Computing Systems, pp. 343–350. ACM, New York (1997)
38. Druin, A.: The role of children in the design of new technology. Behav. Inf. Technol. **21**, 1–25 (2002)
39. Spotify Inc.,: Spotify Developer Resources. https://developer.spotify.com/
40. North, A.C., Hargreaves, D.J., Hargreaves, J.J.: Uses of music in everyday life. Music Percept. Interdiscip. J. **22**, 41–77 (2004)

A Study on the Differences in the Expressions of Emotional Cognition Between Bloggers and Users Based on the "Cloud Pet Keeping" Phenomenon

Chen Tang, Ke Zhong, and Liqun Zhang(✉)

Institute of Design Management, Shanghai Jiao Tong University, Shanghai, China
zhanglliqun@gmail.com

Abstract. "Cloud Pet Keeping" is a phenomenon rising in Chinese UGC background that social media users keep eyes on certain pets' growth by viewing the pictures and vocabularies released by pet bloggers. In the "cloud pet keeping" phenomenon, users can get emotional resonance through browsing the amusing pet photos shared by bloggers and are willing to contribute the consuming behavior. However, in the actual process of photos searching, users often fail to find the accurate content, as the users' tags are different from the tags input by bloggers, leading to the bad user experience. It shows that the differences may exist in the expressions of emotional cognition between bloggers and the users. This paper focuses on whether there are differences or not, and the detailed information may exist. For the analysis methods, first stimulus was chosen and associated emotional expression vocabularies were found through brainstorming. Then final emotional expression vocabularies were organized and extracted by SPSS cluster analysis and multidimensional scaling analysis. By calculating Euclidean distance with data from matching experiments of Emotional expression vocabularies and stimulus, the differences are obtained. The result indicated that there are differences existed between the bloggers and the users' expressions and it is apparent. The reason may be that bloggers do not systematically study the information contained in the photo when uploading. Also, users' individual differences and other factors impact the results of this study. Following research will focus on how to provide a tag selection mechanism on photo-sharing social media, to provide a better user experience.

Keywords: "Cloud Pet Keeping" · Expressions of emotional cognition
Stimulus experiment · User experience

1 Introduction

User-Generated Content (UGC) is an emerging mode for creating and organizing online information resources [1]. With the blooming of Web2.0, UGC provides endless possibilities for the content creation on the Internet. Users have become important part of the system [2]. "Cloud Pet Keeping" is a phenomenon rising in Chinese UGC

© Springer International Publishing AG, part of Springer Nature 2018
G. Meiselwitz (Ed.): SCSM 2018, LNCS 10913, pp. 375–387, 2018.
https://doi.org/10.1007/978-3-319-91521-0_27

background that social media users keep eyes on certain pets' growth by viewing the pictures and texts released by pet bloggers. The main reasons for the rise of "Cloud Pet Keeping" phenomenon are as follows.

Needs of Accompanying. As far away from hometown and relatives, urbanites are often living alone, lack of emotional sustenance and have no family life. Therefore, the accompanying needs of pets are getting more and more stronger [3].
Limited Conditions. Due to busy work and living in rented houses, urbanites do not have the time and space to keep pets. They also know little about the knowledge needed to keep pets.
Diverse Choices. In the social media, users can raise several different temperament pets according to their own multiple preferences at the same time.

In "Cloud Pet Keeping" phenomenon, users are willing to contribute the consuming behavior, which promote a series of industry chain. Users will often buy peripheral products, send cat food to the bloggers who is raising the cats they like. In additional, there are also pets' smart hardware, pets' household items, pets' foster hotels and adoption etc. The unlimited potential and huge profit of the pet industry highlights the value of improving the user experience [4]. Emotion means the subjective feelings or experiences of the individual [5]. Emotional experience is about the individual subjective experience of emotion [6]. In the "cloud pet keeping" phenomenon, users get emotional resonance from browsing these photos. However, in the actual process of photos searching the pets they would like to keep online, users often fail to find the accurate content, which lead to the bad user experience. This situation shows that the differences may exist in the expressions of emotional cognition between bloggers and the users. This paper focuses on whether there are differences or not. Research methods of emotional experience based on photos are feature analysis [7], user knowledge architecture [8, 11], quantitative study [9, 10]. While, there are few research to study the difference between the expressions of emotional cognition between bloggers and users.

The main contents conclude following parts:

- Collect vocabularies and photos which is about the object of "Cloud Pet Keeping" Phenomenon from bloggers and users as many as possible from social medias. Then professional researchers brainstormed, analyzed, summarized and correlated all the emotional expression vocabularies, and the correlation score matrix of these related vocabularies was obtained; The representative vocabularies were extracted through the multi-dimensional scale and cluster analysis by SPSS.
- The filtered representative vocabularies are used as the photo grouping category. The typical photos selected by the professional researchers, which match the representative vocabularies, were stimulus.
- By calculating Euclidean distance and statistical methods with data from multi matching experiments of emotional expression vocabularies and stimulus, the differences in the expressions of emotional cognition between bloggers and users are obtained.

2 Exploratory Research

In the "Cloud Pet Keeping" phenomenon, most common animals are cats, dogs, pandas and so on. Among them, cats get far more attention than the other. Several important reasons are as follows.

Quiet & Easy to Feed. For bloggers, cats are quieter than the other animals. They do not interfere too much the owner's life, and are easy to feed.

More Independent. Cats and dogs are the most popular pets in the family. Compared with dogs, cats are more independent or even cool. In daily life, urbanites lack specific channels to express their true emotions. Facing the cat on the other side of the screen, they can just imagine their emotions projected onto the cat and get spiritual support [12].

For these reasons, in this paper, cats' photos chosen from social networks are stimulus for experiments.

2.1 Emotional Expression Vocabularies Collection

The photos in this paper are mainly collected from Weibo and Lofter. The key word for searching is *cat*. According to the ranking from high to low, more than 700 photos were collected. Professional researchers brainstormed with these pictures to come up with as many Emotional expression vocabularies as possible. At the same time, the Emotional expression tags assigned by the blogger were obtained as supplementary reference.

The supplementary emotional expression vocabularies collected have several different sources (see Fig. 1).

- Tags input by the bloggers when they upload photos.
- Emotional expression vocabularies extracted from topic input by the bloggers when they upload photos.
- Emotional expression vocabularies extracted from text input by the bloggers when they upload photos.
- Emotional expression vocabularies extracted from sticker which made by the bloggers with the pet photos.

Combined with the results of brainstorming and various tags from bloggers, the final number of vocabularies to be analyzed is 46. They are *Serious, Sad, Tsundere, Adorkable, Puzzled, Angry, Dazed, Happy, Ignorant, Innocent, Horrified, Curious, Lazy, Adorable, Dignified, Shocked, Well-behaved, Overbearing, Tired, Annoyed, Sorrowful, Naughty, Dolorous, Mad, Afeared, Disgusted, Indifferent, Shy, Persistent, Despising, Leisurely, Cozy, Pitiful, Calm, Excited, Melancholic, Helpless, Chill, Astonished, Pleasant, Aggrieved, Focused, Depressive, Guilty, Fortunate, Anxious.*

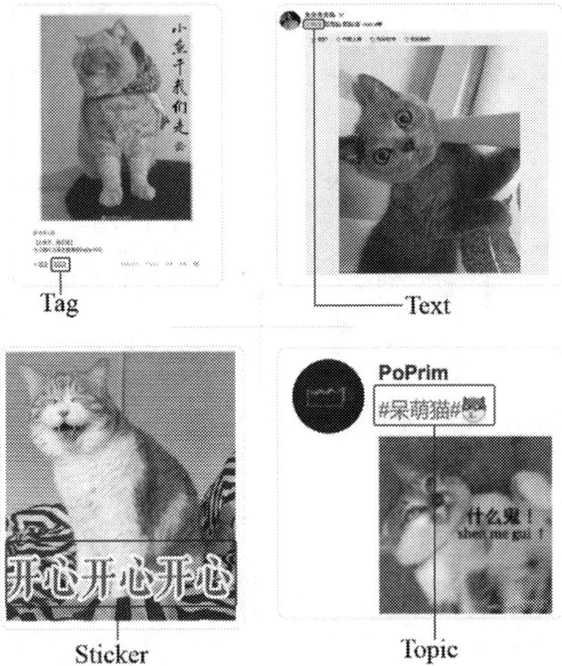

Fig. 1. Four different sources of the emotional expression vocabularies

2.2 Emotional Expression Vocabularies Correlation Analysis

For taking the 46 vocabularies as emotional expression vocabularies for the matching experiments will bring a huge experimental work, and the degree of relevance between these emotional expression vocabularies is quite different. Therefore, these 46 keywords need to be reprocessed. Professional researchers will grade vocabularies by their correlation to get the correlation matrix between them. Then results will be analyzed by SPSS tool. After SPSS analysis, more representative keywords will be extracted as the final emotional expression vocabularies.

Emotional Expression Vocabularies Correlation Score

Three professionals are invited to grade the correlation between any two of these 46 vocabularies. The professionals scored on a 9-point scale ranging from 'significant negative correlation' to 'significant positive correlation'. The point 1 means 'significant positive correlation' and the point 9 means 'significant negative correlation'. The relevant portion of the score result is shown in Fig. 2.

	Serious	Sad	Tsundere	Adorkable	Puzzled	Angry	Dazed	Happy
Serious	1	3	5	8	6	7	8	9
Sad	3	1	8	8	6	8	4	9
Tsundere	5	8	1	9	7	8	6	8
Adorkable	8	8	9	1	5	9	5	4
Puzzled	6	6	7	5	1	6	3	8
Angry	7	8	8	9	6	1	5	9
Dazed	8	4	6	5	3	5	1	9
Happy	9	9	8	4	8	9	9	1
Ignorant	8	7	8	3	4	7	2	6
Innocent	8	7	2	2	7	8	4	4
Horrified	9	3	7	7	3	5	5	8
Curious	8	7	7	4	4	8	4	7
Lazy	9	5	8	4	8	8	5	6

Fig. 2. Part of emotional expression vocabularies correlation score

Emotional Expression Vocabularies Correlation Matrix Analysis

Using the SPSS tool, the vocabularies correlation matrix is analyzed through multi-dimensional scale analysis and cluster analysis.

Of all three professionals' results, the highest reliability analysis results is as follows (see Fig. 3).

Reliability Statistics

Cronbach's Alpha	Cronbach's Alpha Based on Standardized Items	N of Items
.706	.749	46

Fig. 3. Reliability statistics result

Cluster Analysis

Cluster analysis calculates the distance between variables in a multidimensional space. The method classifies by following the closest distance principle (see Fig. 4) [12].

Multidimensional Scaling Analysis

Multidimensional scaling analysis is also referred to as "Similarity structure analysis". The method reduces research objects (samples or variables) in multi-dimensional space to the low-dimensional space to locate, analyze and classify. At the same time, original relationship between objects is preserved (see Fig. 5).

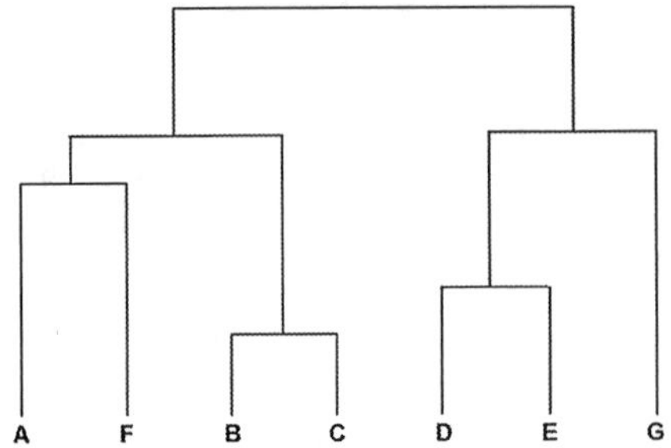

Fig. 4. Hierarchical clustering

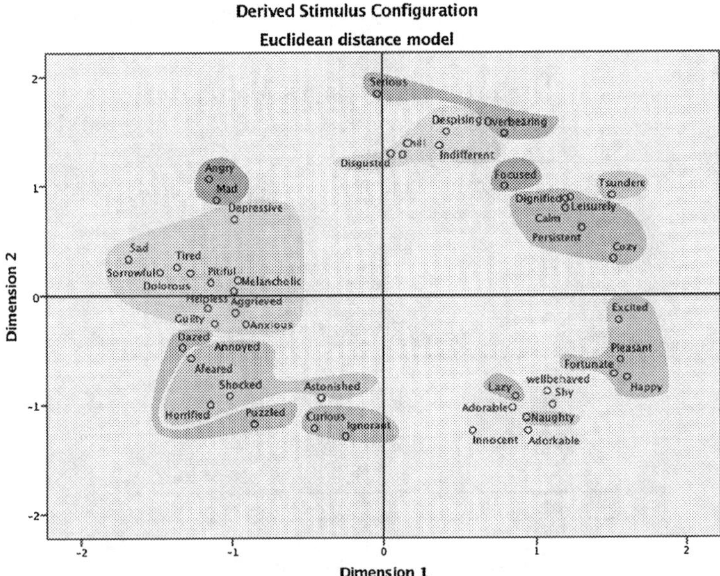

Fig. 5. Multidimensional scaling analysis

Combining these two kinds of analysis results, researchers conclude and extract emotional expression vocabularies in a more scientific way (see Fig. 6).

Finally, 14 vocabularies are extracted from 46 vocabularies as indicators for later experiment. The vocabularies are *Happy, Naughty, Adorable, Disgusted, Serious, Tsundere, Calm, Focused, Lazy, Curious, Shocked, Angry, Dazed, Sad.*

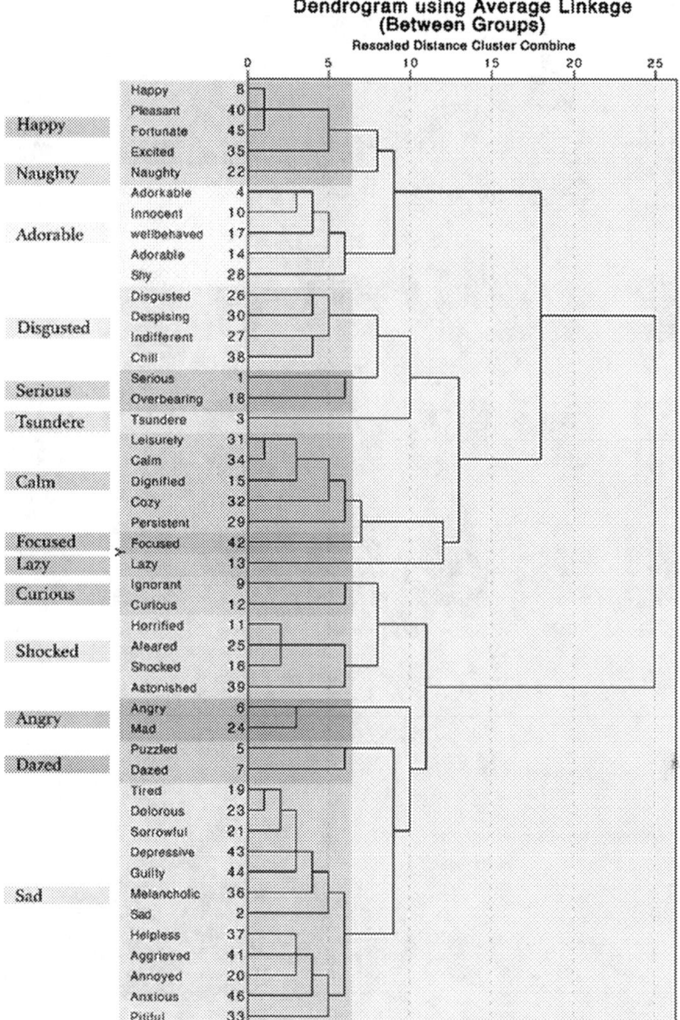

Fig. 6. Dendrogram and word extraction

3 Stimulus Experiment

3.1 Stimulus Experimental Design

Stimulus Images Preparation

Based on the 14 emotion expression vocabularies obtained previously and the tags assigned by bloggers, the researchers categorized more than 700 images collected from social platforms. The photos that have similar tags are divided into the same one category. Then, the researchers carefully selected 5 typical pictures in each category as

Fig. 7. Part of photos for experiment

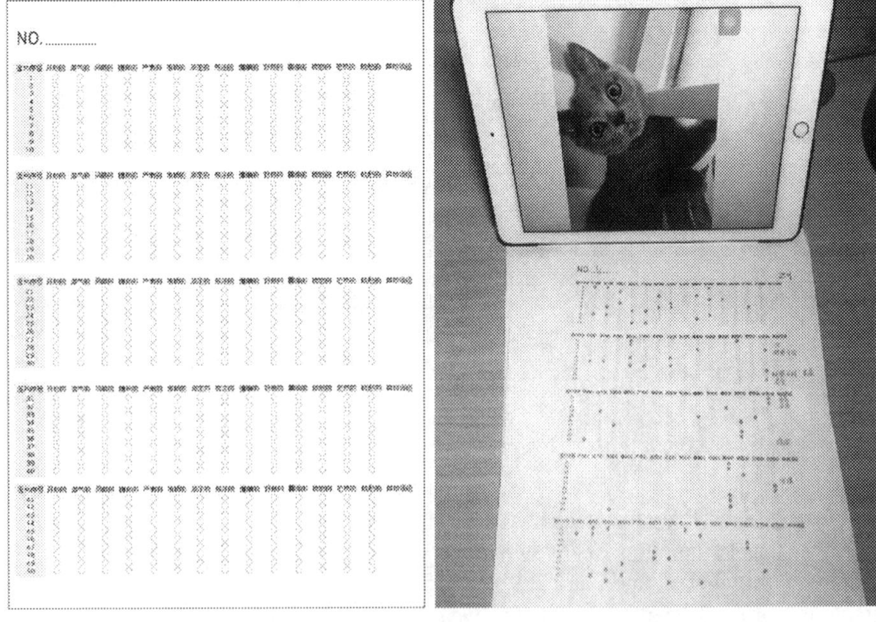

Fig. 8. Card designed for this experiment and the actual use

stimuli. Finally, 70 stimuli photos are obtained. Selected experimental photos do not contain any text, the decoration such as the expression effect given by apps, and the cat itself does not have any dress. The 70 photos are randomly disrupted and labeled with number from 1 to 70. Experiment photos are shown in Fig. 7.

Experimental Method

Each time, the expert user browses one photo, and decided whether the emotional vocabulary can be the photo's tag, describing the photo. If it can, record 1. If not, record 0. If expert users think that there are other vocabularies that differ greatly from those provided, they can also write the vocabulary down separately. In order to let the expert user record more conveniently, card designed for this experiment and the actual use is shown in Fig. 8. Data recording is shown in Fig. 9.

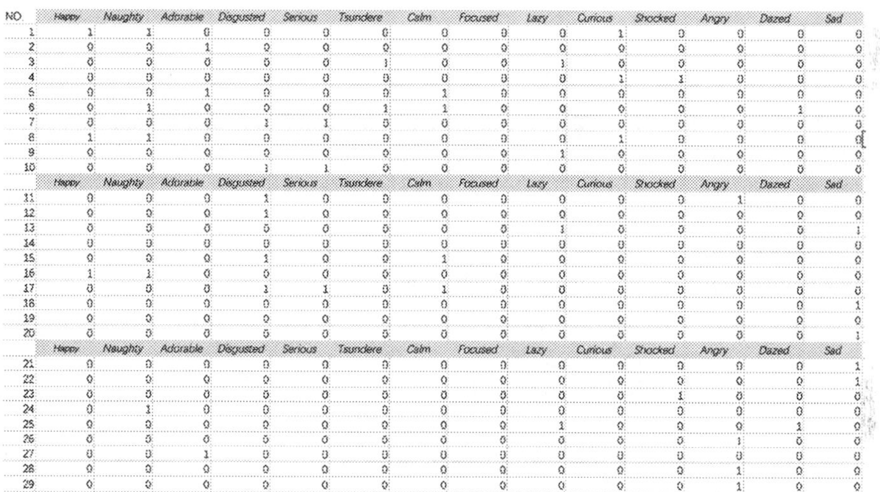

NO.	Happy	Naughty	Adorable	Disgusted	Serious	Tsundere	Calm	Focused	Lazy	Curious	Shocked	Angry	Dazed	Sad
1	1	1	0	0	0	0	0	0	0	1	0	0	0	0
2	0	0	1	0	0	0	0	0	0	0	0	0	0	0
3	0	0	0	0	0	1	0	0	1	0	0	0	0	0
4	0	0	0	0	0	0	0	0	0	1	1	0	0	0
5	0	0	1	0	0	0	1	0	0	0	0	0	0	0
6	0	1	0	0	0	1	1	0	0	0	0	0	1	0
7	0	0	0	1	1	0	0	0	0	0	0	0	0	0
8	1	1	0	0	0	0	0	0	0	1	0	0	0	0
9	0	0	0	0	0	0	0	0	1	0	0	0	0	0
10	0	0	0	1	1	0	0	0	0	0	0	0	0	0
	Happy	Naughty	Adorable	Disgusted	Serious	Tsundere	Calm	Focused	Lazy	Curious	Shocked	Angry	Dazed	Sad
11	0	0	0	1	0	0	0	0	0	0	0	1	0	0
12	0	0	0	1	0	0	0	0	0	0	0	0	0	0
13	0	0	0	0	0	0	0	0	1	0	0	0	0	1
14	0	0	0	0	0	0	0	0	0	0	0	0	0	0
15	0	0	0	1	0	0	1	0	0	0	0	0	0	0
16	1	1	0	0	0	0	0	0	0	0	0	0	0	0
17	0	0	0	1	1	0	1	0	0	0	0	0	0	0
18	0	0	0	0	0	0	0	0	0	0	0	0	0	1
19	0	0	0	0	0	0	0	0	0	0	0	0	0	0
20	0	0	0	0	0	0	0	0	0	0	0	0	0	1
	Happy	Naughty	Adorable	Disgusted	Serious	Tsundere	Calm	Focused	Lazy	Curious	Shocked	Angry	Dazed	Sad
21	0	0	0	0	0	0	0	0	0	0	0	0	0	1
22	0	0	0	0	0	0	0	0	0	0	0	0	0	1
23	0	0	0	0	0	0	0	0	0	0	1	0	0	0
24	0	1	0	0	0	0	0	0	0	0	0	0	0	0
25	0	0	0	0	0	0	0	0	1	0	0	0	1	0
26	0	0	0	0	0	0	0	0	0	0	0	1	0	0
27	0	0	1	0	0	0	0	0	0	0	0	0	0	0
28	0	0	0	0	0	0	0	0	0	0	0	1	0	0
29	0	0	0	0	0	0	0	0	0	0	0	1	0	0

Fig. 9. Data recording of cards

3.2 Stimulus Experiment of Expert Users

This experiment invites 35 participants on behalf of the experts. There are 16 males and 19 females. They all using weibo over years, browsing cats' photo every day and some of them keep several different cats online. Before the experiment, researchers explain in detail about the experiment process and rules.

4 Data Analysis

The schematic diagram of data of each photo is shown in Fig. 10.

For each experiment photo, the blogger's tag was obtained when downloading it. Therefore, this dimension was counted as 1. In the remaining dimensions, all counted as 0 points.

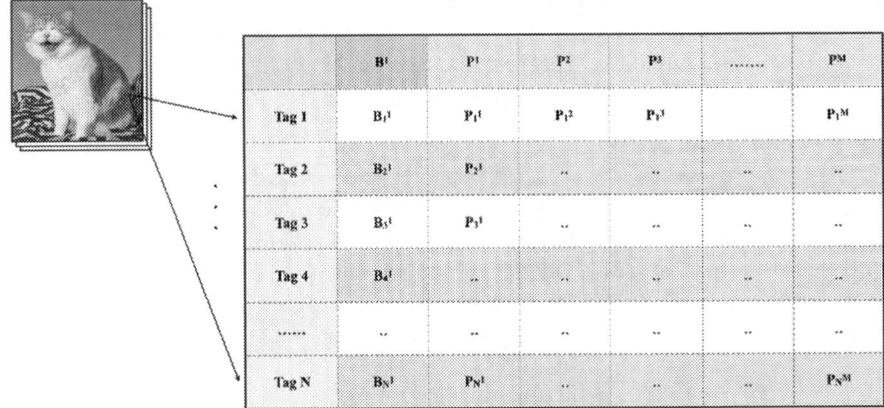

Fig. 10. Schematic diagram of data of each photo

The 14 emotional expression vocabularies represented 14 separate dimensions. If the expert user writes down a new word, the researcher would consider whether it was appropriate and if it was, then the vocabulary will be added as a new dimension to calculate.

For each picture, Euclidean distance *Dis* is calculated to obtain the distance between the blogger and each expert user in all dimensions.

- B_i^x *: the blogger's cognition* P_i^x*: the expert user's cognition*
  Supermark: the serial number of the participants
  Subscript: the serial number of TAG
- Dis_i^2
  Subscript: the serial number of photo

For the convenience of calculation, using Dis_i^2 represent the Dis_i for they have the same value. It is calculated by(taking photo1 as an example):

$$Dis_1^2 = \left(B_1^1 - P_1^1\right)^2 + \left(B_2^1 - P_2^1\right)^2 + \left(B_3^1 - P_3^1\right)^2 + \ldots + \left(B_N^1 - P_N^1\right)^2 \quad (1)$$

Divide the sum of the square of *Dis* by the number of participants N to get the average difference between bloggers and users. By calculating the experimental results of 70 pictures, the picture of the emotional cognitive distribution between users and bloggers of the 70 photos *AverageDis* is finally obtained.

$$AverageDis = \sum \left(Dis_1^2 + Dis_2^2 + Dis_3^2 + \ldots + Dis_{70}^2\right)/N \quad (2)$$

The analysis tools used is EXCEL.

The Experimental Results

After obtaining all 35 expert users' data, the analysis is calculated. The Fig. 11 shows the distribution of emotional cognitive differences between bloggers and users. The abscissa

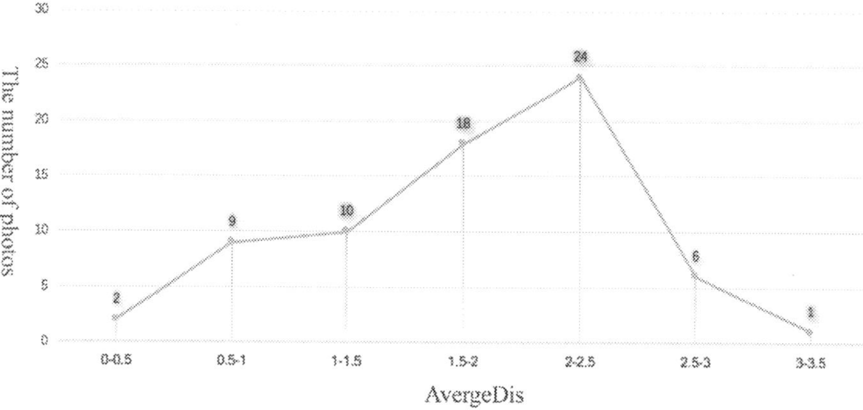

Fig. 11. Distribution of emotional cognitive differences between bloggers and users.

of the graph Fig. 11 is the average difference between the blogger and the users on each photo. According to the calculation results, *AvergeDis* are divided into 0–0.5, 0.5–1, 1–1.5, 1.5–2, 2–2.5, 2.5–3, total of 7 groups. The vertical axis represents the number of photos. In 0–0.5 range, there are 2 photos; in 0.5–1 range, there are 9 photos, in 1–1.5 range, there are 10 photos; in 1.5–2 range, there are 18 photos; in 2–2.5 range, there are 24 photos; in 2.5–3 range, there are 6 photos; in 3–3.5 range, there are 1 photos. The differences more than 1 have a total of 10 + 18 + 24 + 6+1 = 59 photos (see Fig. 11).

For the correlation between each two vocabularies is far and the selected images are very typical, the difference between the selected vocabularies is still very large, there are 70% of the pictures, bloggers and users of the difference vocabulary reached 1.5 and above. It shows that there are actually differences existed between the bloggers and the users' in the expressions of emotional cognition and it is significant.

For each emotional expression vocabulary category, the consistency between bloggers and users' choosing *Con* are calculated by: (X is the serial number of the tag chosen by the blogger)

$$Con = \left(\sum \left(P_X^1 + P_X^2 + P_X^3 + \ldots . + P_X^N\right)\right)/N \tag{3}$$

Vocabularies with high difference existed between the bloggers and the users' in the expressions of emotional cognition are *Dazed, Tsundere, Adorable, Calm, Focused and Disgusted* (see Fig. 12). Their Con's value is lower than 0.3.

Photos having a quite high level of consistent with bloggers emotional cognition vocabulary: *Shocked, Angry, Lazy*. It can be speculated that users have a clearer cognition and expression of these emotions of cats.

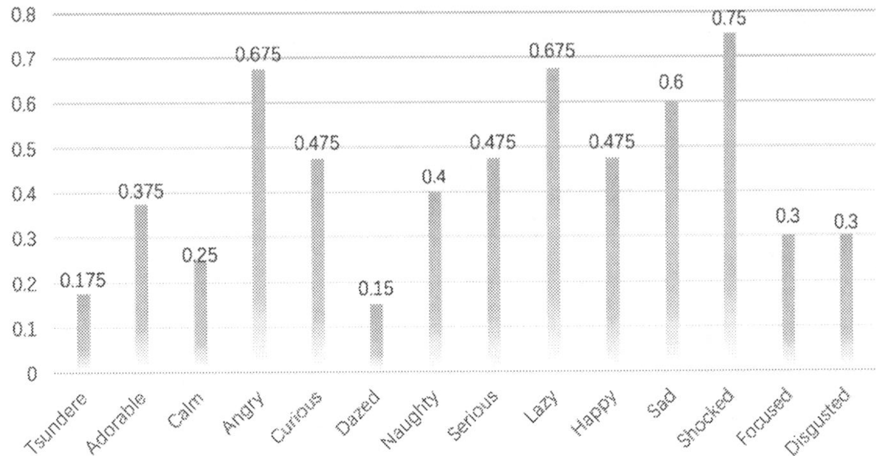

Fig. 12. Consistency between bloggers and users' scoring of each emotional kind

5 Conclusion

This paper explores the differences in the expressions of emotional cognition between bloggers and users based on the "Cloud Pet Keeping" phenomenon. The main study method includes three parts: Firstly, collecting vocabularies and photos about the object of "Cloud Pet Keeping" Phenomenon as many as possible. Then professional researchers brainstormed, summarized and correlated all the emotional expression vocabularies. The representative vocabularies were extracted through the multi-dimensional scale and cluster analysis. Secondly, the filtered representative vocabularies are used as the photo grouping category. The typical pictures selected by the professional researchers, which match the representative vocabularies, were stimulus. Thirdly, by calculating Euclidean distance and statistical methods with data from multi matching experiments of Emotional expression vocabularies and stimulus, the differences are obtained.

The result indicated that there are actually differences existed between the bloggers and the users' expressions and it is apparent. Vocabularies such as *Dazed, Tsundere, Adorable* etc. are hard for users to unify their expression of recognition. On this basis, researchers can find more differences between the bloggers and users and the reasons.

The reasons for the difference in this experiment may be that the bloggers do not systematically study the information contained in the photo when uploading. The bloggers may use the photo in a funny way on purpose to attract users. Users may not know the story happened between the blogger and the cat. Also, the expert users' individual differences and other factors have the impact on the results of this study. Following research will focus on how to provide a tag selection mechanism on photo-sharing social media, to provide a better user experience for the users.

References

1. Yuxiang, Z., Zhe, F., Qinghua, Z.: Conceptualization and research progress on user-generated content. J. Lib. Sci. China **5**, 008 (2012)
2. Zhe, F., Zhu, Q., Zhao, Y.: Department of Information Management, Nanjing University, Nanjing 2100932 School of Law and Politics and Management, Nantong University, Nantong 226019. A Review of the Research on UGC in Web2. 0 Environment. Lib. Inf. Serv. 22 (2009)
3. 张连成.社交媒体网络"云养猫"热潮浅析 [J/OL]. 新媒体研究. **02**, 55–56 (2018). https://doi.org/10.16604/j.cnki.issn2096-0360.2018.02.023. Accessed 09 Feb 2018
4. Liu, X.: Pet industry marks a robust increase. China's Foreign Trade **02**, 50–51 (2016)
5. Fox, E.: Emotion Science Cognitive and Neuroscientific Approaches to Understanding Human Emotions. Palgrave Macmillan, New York (2008)
6. Carstensen, L.L., Pasupathi, M., Mayr, U., Nesselroade, J.R.: Emotional experience in everyday life across the adult life span. J. Pers. Soc. Psychol. **79**, 644 (2000)
7. Wang, D., Liang, N., Zhong, J., Zhang, L.: Mining and construction of user experience content: an approach of feature analysis based on image. In: Marcus, A. (ed.) DUXU 2016. LNCS, vol. 9748, pp. 223–234. Springer, Cham (2016). https://doi.org/10.1007/978-3-319-40406-6_21
8. Liang, N., Zhong, J., Wang, D., Zhang, L.: The exploration of user knowledge architecture based on mining user generated contents – an application case of photo-sharing website. In: Marcus, A. (ed.) DUXU 2016. LNCS, vol. 9748, pp. 180–192. Springer, Cham (2016). https://doi.org/10.1007/978-3-319-40406-6_17
9. Xie, M., Zhang, L., Liang, T.: A quantitative study of emotional experience of *Daqi* based on cognitive integration. In: Marcus, A., Wang, W. (eds.) DUXU 2017. LNCS, vol. 10288, pp. 306–323. Springer, Cham (2017). https://doi.org/10.1007/978-3-319-58634-2_24
10. Liang, T., Zhang, L., Xie, M.: Research on image emotional semantic retrieval mechanism based on cognitive quantification model. In: Marcus, A., Wang, W. (eds.) DUXU 2017. LNCS, vol. 10290, pp. 115–128. Springer, Cham (2017). https://doi.org/10.1007/978-3-319-58640-3_10
11. Zhong, J., Wang, D., Liang, N., Zhang, L.: Research on user experience driven product architecture of smart device. In: Marcus, A. (ed.) DUXU 2016. LNCS, vol. 9748, pp. 425–434. Springer, Cham (2016). https://doi.org/10.1007/978-3-319-40406-6_41
12. **张盖伦. 云养猫：一场低成本疗愈仪式** [N]. **科技日报** (008), 27 October 2017
13. **邓铸, 朱晓红. 心理统计学与** [M] SPSS **应用. 华东师范大学出版社** (2009)

Consumer Behavior of Foreign Residents in Japan for Service Industry

Zhen Wang[1(✉)] and Noriyuki Suyama[2]

[1] Doctoral Program in Environmental Clothing Studies, Graduate School of Bunka Gakuen University, Tokyo, Japan
wu17da0001@bunka-wu.ac.jp

[2] Department of Fashion Sociology and Science, Bunka Gakuen University, Tokyo, Japan
bun161034@bunka.ac.jp

Abstract. Globalization refers not only economic globalization, which is the increasing economic interdependence of national economies across the world through a rapid increase in cross-border movement of goods, services, technology, and capital (Joshi (2014) International Business, Oxford University Press, New Delhi and New York.) but an essential aspect of globalization is also movement of people. International movement of labor is often seen as important to economic development ("Mainstreaming of Migration in Development Policy and Integrating Migration in the Post-2015 UN Development Agenda"). Like the European Union, Japan has started to accept people from overseas, who are eligible to live, study and work in this country (Nikkei Newspaper (2017/1/28)). Since Japan has been a mono-ethnic nation because 99% are Japanese, every company in Japan needs to build its business model to meet immigrants' expectation. This research paper examines what types of the segments of foreign residents exist and how effectively to respond to their needs, ending up increasing customer satisfaction. The focus is on the service industry whose share among the GDP of Japan exceeds 70%. The data, which collected from questionnaires regarding hair salon usage conducted towards foreign residents in Japan, are analyzed for the sake of clarifying distinctive consumer behavior.

Keywords: Segmentation · Foreign residents in Japan · Service/tertiary sector

1 Introduction

The Globalization Working Group, which was established under the Cabinet Office of Japan, has been taking action toward globalization since 2003. According to the group, the progress of globalization has resulted in fierce competition among corporations all over the world and then for Japan, such a world represents an opportunity, but can also constitute a threat, due to the onset of competition not only among corporations but also in the global market for human resources. They urge Japan to aim to take advantage of globalization and become a country where there is a high degree of

© Springer International Publishing AG, part of Springer Nature 2018
G. Meiselwitz (Ed.): SCSM 2018, LNCS 10913, pp. 388–399, 2018.
https://doi.org/10.1007/978-3-319-91521-0_28

satisfaction in terms of income, work, security, the environment, and people's daily lives. Accordingly, the Ministry of Justice started promoting to accept foreign workers proactively. This means foreigners are not only workers or laborers but also consumers in the Japan market.

Service or tertiary sector[1] service is the non-material equivalent of a good. Service provision is defined as an economic activity that does not result in ownership, and this is what differentiates it from providing physical goods. Services (also known as "intangible goods") include attention, advice, access, experience, discussion, and affective labor and may involve in the area of the entertainment, government, telecommunication, hospitality industry/tourism, mass media, healthcare/hospitals, public health, information technology, waste disposal, consulting, gambling, retail sales fast-moving consumer goods (FMCG), franchising, real estate, education, financial services and professional services. The service industry comprises 71.4% of the nominal gross domestic product in 2012[2] of Japan, which is a huge and attractive market.

The purpose of this research paper is to understand the current situation of foreign residents' market in Japan, explore any issue Japan is facing if any for foreign residents to use the service and obtain insights into their consumer behavior, so as to meet or exceed their expectation from the viewpoint of the service industry. Beauty salon business is used as a representative of the service industry to verify the research.

According to e-Sta[3], the number of beauty salons in Japan accounts for 243,360, among which fierce competitions occur. The number of establishment is huge compared to that of convenience stores in Japan (56,222)[4], that of ramen shops (31,988)[5] and that of fast food shops (32,958)[6]. The Metropolitan Tokyo area accounts for 22,064 in 2016 in Japan, which means 9.1% of the total salons are located in Tokyo area. The total market size of the industry is estimated by JPY1,578 billion[7] in 2013 and the market is so fragmented that no player can perform with its market share more than 1% (Table 1).

2 Precedence Studies

In the past studies, one of major criteria a hair salon needs to set is to decide a store location (Craig et al. 1984). Furthermore, the definition of "good" location is measured by return on investment (Krause-Traudes et al. 2008), assumingly being regarded sales as a representative criteria of ROI.

---

[1] The word was developed by Allan Fisher, Colin Clark and Jean Fourastié.

[2] GDP (nominal): International Monetary Fund, World Economic Outlook Database, April 2012.

[3] Statistics of Japan, e-Stat is a portal site for Japanese Government Statistics.

[4] http://todo-ran.com/ (2016).

[5] Town Page (2017).

[6] Japan Food Association (Jan 2019).

[7] Yano Research (2013) Beauty Marketing.

Table 1. Number of hair salons in Japan (Top 5, 2016) (The Portal Site of Official Statistics of Japan, e-Stat, Statistic Bureau, Ministry of Internal Affairs and Communications)

Prefecture	# of hair salons	Share
Tokyo	22,064	9.1%
Osaka	15,985	6.6%
Aichi	11,971	4.9%
Kanagawa	11,281	4.6%
Saitama	10,739	4.4%
Total	243,360	100.0%

Masaoka and Ninomiya (2015) conducted the survey, which examined users' tendency to patronize a hair salon, influential factors of fostering loyalty to a hair salon like service, prices, hairdressing techniques or handsomeness of stylists, etc. However, the surveyees of this research are only Japanese. Since each nation has its own history, culture and religion (including atheism), perception of or a sense of hairdressing is assumed different from each country.

A study, which looked into the hotel industry, analyzes its structure of customer satisfaction, and evaluates it on the basis of both a price and a quality of service. According to the result, its correspondents do not necessarily seek for an "affordable" price but a "value" for money like convenience, food and shopping environments.

As a result of my searching preceding research, there are many academic papers, in which conducted experimental research of influential factors of customer satisfaction or segmentation. However, no research was carried out, which deals with the service industry or more specifically, a hair salon business and reveals influential factors of customer satisfaction or shop loyalty of foreign residents in a domestic market.

3 Objectives

The objectives of this study are to comprehend what sorts of foreign residents are in Japan through segmentation, develop strategic insight into their satisfaction and suggest a comprehensive solution to address issues the service sector is facing. In fact, there exist several papers, which describe or analyze customers' satisfaction in the service or tertiary industry toward, but no one refers to the industry in Japan from the viewpoint of foreign customers.

Plenty of companies in the tertiary sector of Japan are recently provided with POS systems that can file up tons of transaction data related with sales activities. But only a few firms, especially those in the hair salon industry make better use of the valuable data toward their marketing strategies and tactics, even though the hair salon industry in Japan has been in the condition of cutthroat competition since a few decades ago.

4 Data Description

4.1 Profiles of Data

The period of a survey conducted is from June 17, 2016 to August 17, 2016. The method of a survey is a written questionnaire around Tokyo. The total number of the survey results accounts for 716 records, which were collected in a manner that the data would be unbiased as much as possible. The survey implemented in the places of shopping areas, studying institutes like a university and a college, railway stations, parks, bus terminals and so on. 460 surveyees answered she/he had experience in using a hair salon in Japan among those 716 results, which were applied for the study.

The profile of data includes his/her age, gender, occupation, address, nationality, duration staying in Japan, purpose of staying, etc., as for surveyee's profile.

In addition, the criteria for selecting a hair salon, a preferred stylist, frequency of hair service, number of a hair salon used, usage of service menus, average payment for hair service, degree of satisfaction for using a hair salon (service, price, technique, atmosphere, time, location and communication), way of communication with a stylist, preference a hair salon between Japan and attractiveness between Japan's and his/her own country's, etc., as for the usage of a hair salon and its service.

4.2 Data Selection

The number of foreign residents participated in the survey, who have viable visa, excluding short stay visitors like tourists and businesspersons. 460 surveyees have experience in using a hair salon in Japan. Among those 460 respondents, 44.6% of the hair salon users are Chinese, followed by Korean 8.0% and American 6.7%. The proportion of gender is female 62% to male 38%. The distribution of the occupation is students by 68.9%, workers by 24.4%. The age groups comprise 67.4% of totals are 20's, followed by 30's 16.7% and 10's 10.1% (Figs. 1, 2, 3 and 4).

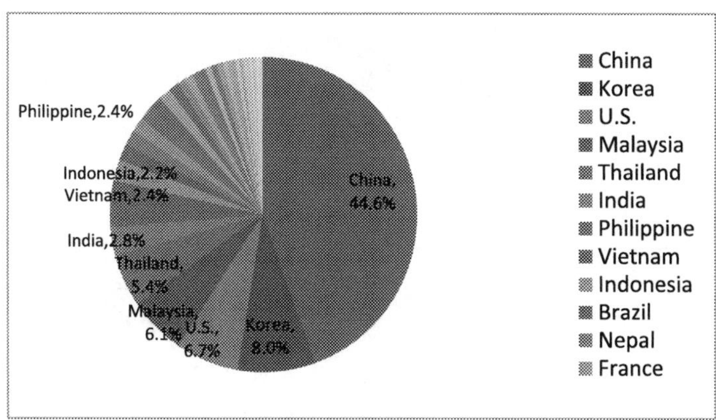

Fig. 1. Nationality

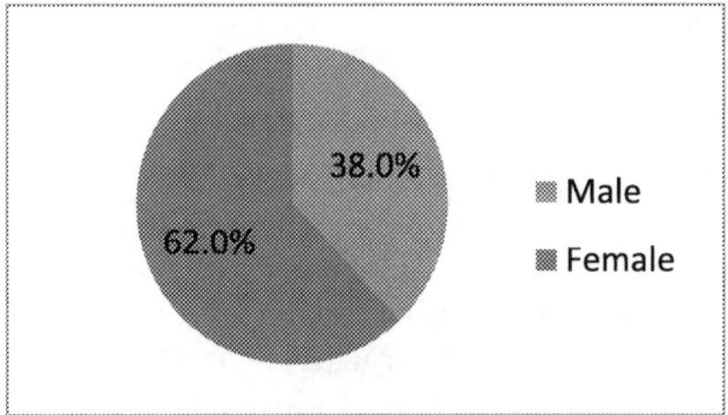

Fig. 2. Genders

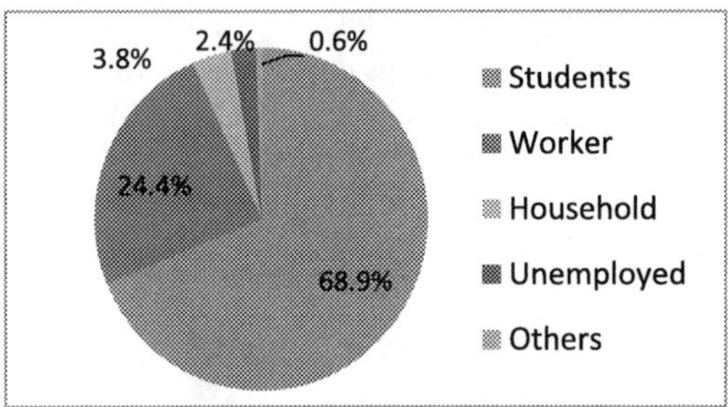

Fig. 3. Occupation

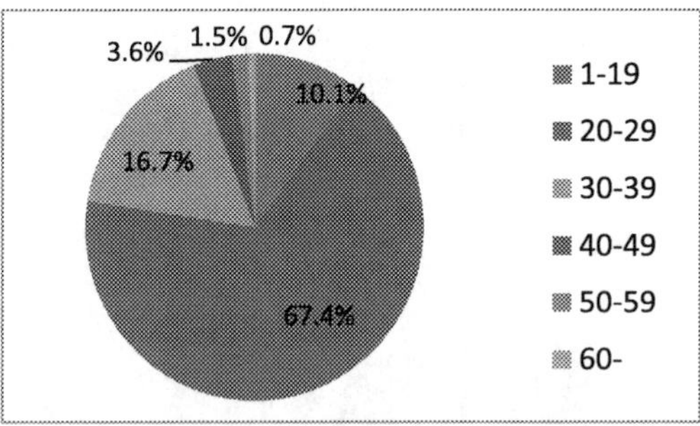

Fig. 4. Age

Figure 5 indicates how long they stay in Japan. Each period looks evenly distributed. Living in Japan from 2 to 3 years are more than any other groups (22.5%). One out of three foreign residents have just come to Japan within 6 months or 12 months.

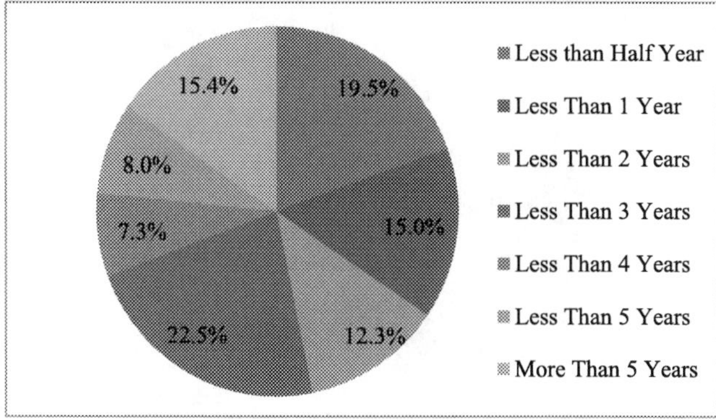

Fig. 5. Duration of staying in Japan

Figure 6 visualizes the current usage of hair service in Japan. The residents who use only one hair salon occupy by 50.2% (loyalty customers), while 17.0% do not decide a specific salon. Furthermore, the frequency of using hair service (Fig. 7) indicates that 43% of the people patronize a hair salon in four times a year and 23.5% are twice a year. Additionally, Fig. 8 shows how much the customers spend in hair service each time in Japan.

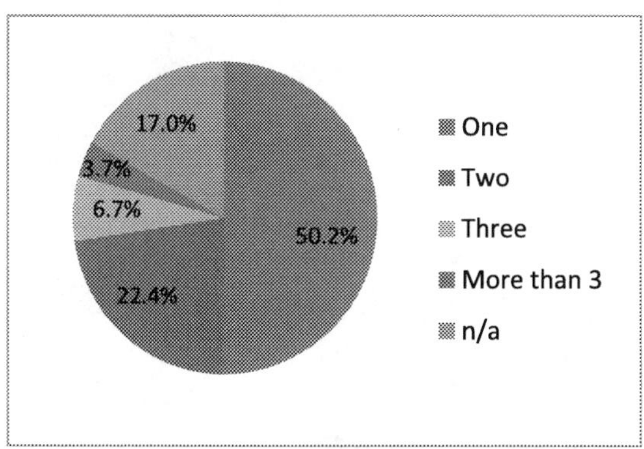

Fig. 6. # of hair salon usage

Figure 9 indicates 35.7% foreign residents regard Price as the criteria of choosing the hair salon, followed by technique 31.1%. At last, 460 respondents gave a degree of satisfaction rating on the hair service, which they experienced in Japan.

Table 2 is the survey result of customer satisfaction toward hair salons in Japan. Among 460 respondents, approximately 85% are satisfied a hair salon in Japan based on their usage. On the other hands, 15.2% of them are not satisfied. Furthermore, seven attributes are examined to get an insight of their ratings in Table 3.

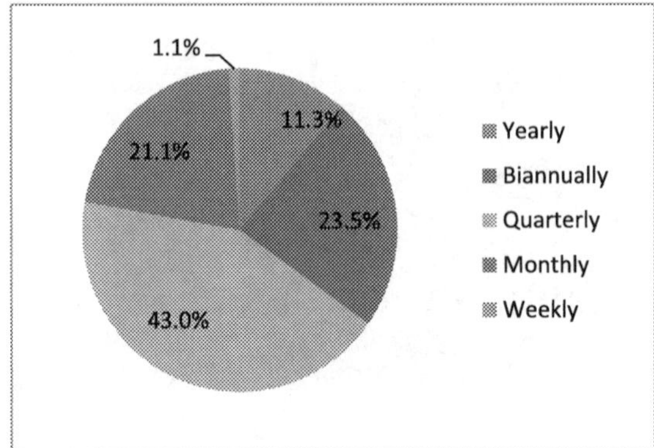

Fig. 7. Frequency

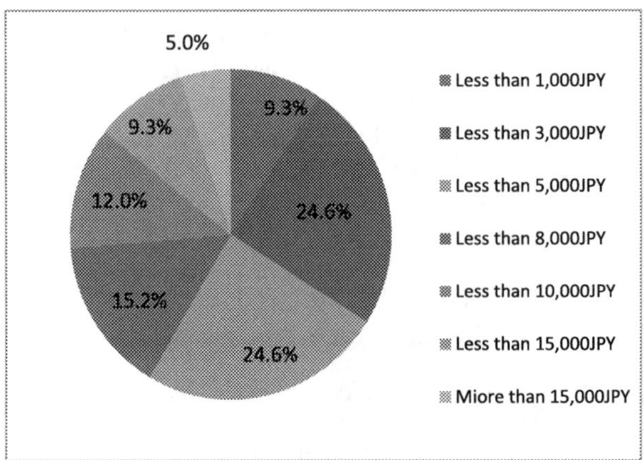

Fig. 8. Payment amount in Japan

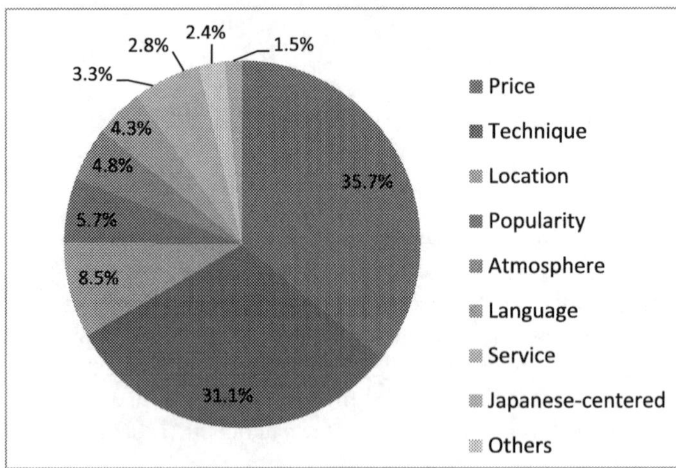

Fig. 9. Criteria to select hair salon

Table 2. Degree of satisfaction among 460 foreign residents

Hair salon	Extremely satisfied	Satisfied	Unsatisfied	Extremely unsatisfied	Total
Total	28.5%	56.3%	12.8%	2.4%	100.0%
	84.8%		15.2%		

The surveyees are comfortable with Japanese Service, Atmosphere, and Location, and, are at the same time, relatively satisfied with Technique, and Time. However, their ratings of Price and Communication are not highly evaluated, which are proven as one of hypotheses addressed in the section of research objectives.

5 Clustering and Fitting Regression Models

The first step is to categorize data, so that each parameter can be estimated and accordingly each segment can show specific aspects regarding satisfaction factors. Although hierarchical clustering has the distinct advantage that any valid measure of distance can be used, a non-hierarchical model is said to be suitable for big data analysis, and then k-means clustering, one of a popular for cluster analysis in data mining and a non- hierarchical methods of vector quantization is adopted (k = 4).

Table 4 indicates the result of K-means analysis with parameters and its visualization in the Fig. 10. Customers of cluster 1 are satisfied with service and conversation but not really with prices or atmosphere, a hospitality-oriented group. The cluster 2 emphasizes prices, techniques and time, who have tendency to face reality. The customers of cluster 3 are average consumers, to whom every factor looks important for their satisfaction. The cluster 4 is meticulous about time, maybe who are busy people in his/her business or life.

Based on the analysis result shown above, in the next step, two types of multiple regression models are used to predict model parameters more precisely: linear regression (1) and binary logistic regression models (2). A linear multi regression model described in (1) is adopted in this study.

$$y = \alpha + \beta_1 x1 + \beta_2 x2 + \beta_3 x3 + \beta_4 x4 + \beta_5 x5 + \beta_6 x6 + \beta_7 x7 \quad (1)$$

$$y = \frac{1}{1 + e^{-(\alpha + \beta 1x1 + \beta 2x2 + \beta 3x3 + \beta 4x4 + \beta 5x5 + \beta 6x6 + \beta 7x7)}} \quad (2)$$

The reason for an estimation used by a linear regression model can perform better than binary logistic regression model to select a model, which estimates more statistically significant. Both models applied “whether to prefer the hair salon service currently used” as a dependent variable and seven independent variables: a degree of each customer satisfaction (1) service, (2) price, (3) technique, (4) atmosphere, (5) time, (6) location and (7) communication. Table 5 provides the linear regression coefficients for each of classes. Each parameter’s adequacy is examined with two criteria applied: an R-square and a p-value.

Table 3. Breakdown of degree of satisfaction among 460 foreign residents based on business functions

Service	Total
Extremely satisfied	42.8%
Satisfied	51.3%
unsatisfied	4.3%
Extremely unsatisfied	1.5%
Total	100.0%

Time	Total
Extremely satisfied	34.8%
Satisfied	52.0%
unsatisfied	12.0%
Extremely unsatisfied	1.3%
Total	100.0%

Price	Total
Extremely satisfied	30.0%
Satisfied	51.5%
unsatisfied	16.7%
Extremely unsatisfied	1.7%
Total	100.0%

Location	Total
Extremely satisfied	49.1%
Satisfied	44.3%
unsatisfied	5.0%
Extremely unsatisfied	1.5%
Total	100.0%

Technique	Total
Extremely satisfied	35.9%
Satisfied	52.0%
unsatisfied	10.9%
Extremely unsatisfied	1.3%
Total	100.0%

Communication	Total
Extremely satisfied	34.3%
Satisfied	43.9%
unsatisfied	18.3%
Extremely unsatisfied	3.5%
Total	100.0%

Atmosphere	Total
Extremely satisfied	40.2%
Satisfied	51.5%
unsatisfied	7.6%
Extremely unsatisfied	0.7%
Total	100.0%

Table 4. Result of K-means clustering

Satisfaction	Cluster 1	Cluster 2	Cluster 3	Cluster 4
Service	1.87	0.67	0.59	-0.59
Price	-2.58	-0.32	0.19	0.05
Technique	0.35	-0.40	0.02	-0.44
Atmosphere	-2.41	-0.99	0.82	0.36
Time	-0.17	0.92	0.64	0.74
Location	1.45	-0.58	-0.76	-0.71
Conversation	2.87	-0.80	0.35	0.12

High Satisfaction	Low Satisfaction

Even though the profiles of each four segment do not have so much difference, the amount of spending in hairdressing service indicates distinctive characteristics (Table 6). The customers of Cluster 1 would pay less than any other cluster in both Japan and their own countries. The size of this price-sensitive group is not so large but further investigation of the cluster is supposed to be necessary for obtaining deep insight into the specific group. Cluster 2, which emphasizes time and atmosphere,

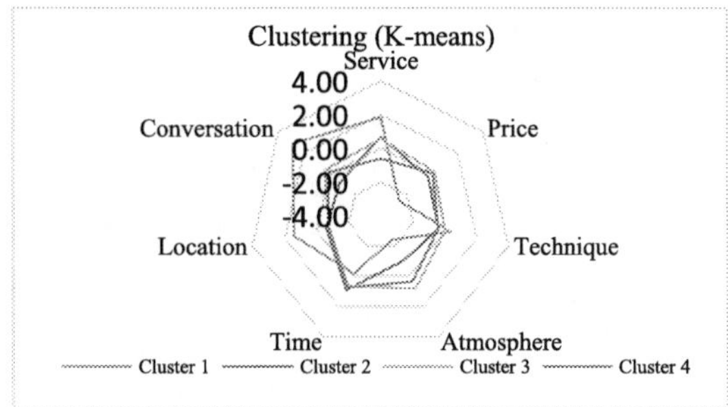

Fig. 10. Result of K-means clustering

Table 5. Result of linear regression analysis

Cluster 1	Partial regression coefficient	R2	P-value	*: P < 0.05 **: P < 0.01	Cluster 3	Partial Regression Coefficient	R2	P-value	*: P < 0.05 **: P < 0.01
Service	2.263	0.2084	0.018	*	Service	-1.322	0.1259	0.132	
Price	–0.160		0.798		Price	–0.398		0.356	
Technique	–1.260		0.060		Technique	0.019		0.979	
Atmosphere	–1.368		0.060		Atmosphere	1.095		0.170	
Time	–0.216		0.680		Time	0.228		0.738	
Location	–0.625		0.357		Location	–0.860		0.192	
Conversation	0.082		0.890		Conversation	–0.012		0.984	
Cluster 2	Partial regression coefficient	R2	P-value	*: P < 0.05 **: P < 0.01	Cluster 4	Partial regression coefficient	R2	P-value	*: P < 0.05 **: P < 0.01
Service	1.127	0.4174	0.363		Service	0.617	0.1973	0.172	
Price	–2.521		0.009	**	Price	–0.427		0.146	
Technique	–2.062		0.019	*	Technique	–0.178		0.657	
Atmosphere	–1.900		0.080		Atmosphere	–1.560		0.000	**
Time	–1.198		0.211		Time	0.519		0.155	
Location	2.008		0.006	**	Location	–0.727		0.032	*
Conversation	–0.274		0.649		Conversation	–0.244		0.363	

relatively spend more in Japan although they do less in their mother nations. This group may have tendency to be fond of Japanese-styled atmosphere and then do not meticulous about payment so much. The customers of Cluster 3 look average people, whose lifestyle is family-centered, diligent in working and not aureate. This group is the largest among four (200 customers). The last segment, Cluster 4 is the most payable and efficiency-centered in both Japan and their own countries. Since the amount of payment to hairdressing service account for 8,118 yen per time and there exist 153 people in the survey, this segment sound suitable to a target market from the viewpoint of marketing.

Table 6. Result of linear regression analysis

	# of sample	Strong factors	Average age	Duration of stay in Japan (Yr)	Average spending (Japan)	Average spending (Mother country)
Cluster 1	32	Conversation, price	27.5	3.8	¥5,406	¥2,750
Cluster 2	75	Time, atmosphere	27.0	3.8	¥7,173	¥3,453
Cluster 3	200	Atmosphere, location	26.4	3.9	¥6,045	¥4,115
Cluster 4	153	Time, location	27.1	3.9	¥8,118	¥5,327

6 Results and Discussions

This research shows how to segment customers and in fact, to apply the analysis result to real business world with multivariate analysis, which is more effective than normal multi-tabulation. Cluster 4 seems the best target market for hair salon business, both of whose spending and size are large enough to invest marketing costs. In addition, as each segment has each strength and weakness, the segments need to be managed as a portfolio. Based on those analysis outcomes, other marketing frameworks like internal analysis and a value chain analysis may be worked well together, and then a deliberate marketing strategy could be planned.

7 Conclusion

This research shows how to estimate more effective parameters, which are related to competitive advantages. These findings contribute to the development of studies for customer satisfaction in the hair salon industry. On top of that, the result can be applicable to other service section business like retail business and fast food business.

The current study, however, has several limitations. Overall, more statistically significant results are required so that the model can work more effectively. Moreover, in addition to two regression models, other methods of multivariate analysis should have been adopted.

References

Craig, C.S., Ghosh, A., McLafferty, S.: Models of the retail location process: a review. J. Retail. **60**(1), 5–36 (1984)

Joshi, R.M.: International Business, 2nd edn. Oxford University Press (2014)

Krause-Traudes, M., Scheider, S., Reping, S., Mebner, H.: Spatial data mining for retail sales forecasting. In: 11th AGILE International Conference on Geographic Information Science 2008, University of Girona, Spain (2008)

Masaoka, M., Ninomiya, S.: A research on the marketing strategy of beauty salon. Depart. Bull. Pap. Osaka Univ. Econ. **66**, 185–212 (2015). (in Japanese)

Merriam-Webster. https://www.merriam-webster.com/dictionary/immigration

Ministry of Justice: The Foreign Residents in Japan as of The End of 2016, Press Release, 17 March 2017. http://www.moj.go.jp/nyuukokukanri/kouhou/nyuukokukanri04_00065.html. (in Japanese)

Oxford Living Dictionaries. Oxford University Press. https://en.oxforddictionaries.com/definition/us/immigration

Suyama, N., Namatame, T.: Competitive advantages of hair salon business influencing on consumer behavior. In: Proceedings of the 17th Asia Pacific Industrial Engineering & Management Systems Conferences, 6 p. (2016)

Tanaka, K.: Policy for promoting inbound tourism to Japan. Transp. Policy Stud. **10**, 11–21 (2007). (in Japanese)

Yano Research Institute Ltd.: Hairdressing Market 2013, Market Report, March 2013. (in Japanese)

Yano Research Institute Ltd.: Research for Hairdressing market in Japan, Implementation of the survey for Hairdressing Industry, Press Release, 21 April 2017. (in Japanese)

A Content Analysis of Social Live Streaming Services

Franziska Zimmer(✉)

Department of Information Science, Heinrich Heine University Düsseldorf, Düsseldorf, Germany
Franziska.Zimmer@hhu.de

Abstract. Social Live Streaming Services (SLSSs) are a new and exciting area of Social Networking Services (SNSs), with Periscope, Ustream, and YouNow representing some of the most commonly used international services. SLSSs offer the opportunity to examine the human-computer interaction between the streamers and their medium, the live stream, as well as the streamers' information behavior. To get a better understanding on the information behavior on SLSSs and who produces which kind of live streams we conducted a systematic observation of live streams (N = 7,667) in a time-span of four weeks. We implemented a content analysis and investigated if differences between gender and the produced content, as well as the motivation of a person can be observed. Furthermore, the content was analyzed by country (U.S., Germany, and Japan) as well as by the service (Periscope, Ustream, and YouNow) to gain insight into the question if the streamed content depends on the applied services or the cultural background.

Keywords: Social Live Streaming Services (SLSSs) · Content analysis Social Networking Services (SNSs) · Users · Information behavior YouNow · Periscope · Ustream

1 Introduction: HCI Research on SLSSs

Social Networking Services (SNSs) like Instagram and Facebook are an important element in people's everyday life. Now, Social Live Streaming Services (SLSSs) are a new emerging field of SNSs, which could possibly even gain the same status in peoples' lives. They combine elements of social networks, for example the interaction with the viewer and vice versa, and are a form of live-TV, with the streamer being, most of the time, boss, producer, and manager of his or her channel, deciding when he goes online and what content he wants to stream. The most popular examples of streaming platforms are Periscope[1], Ustream[2], YouNow[3], YouTube Live[4], or Facebook Live[5]. Some services are focused on a specific interest or business, for example the

[1] https://www.pscp.tv/.
[2] http://www.ustream.tv/.
[3] https://www.younow.com/.
[4] https://www.youtube.com/channel/UC4R8DWoMoI7CAwX8_LjQHig.
[5] https://live.fb.com/.

© Springer International Publishing AG, part of Springer Nature 2018
G. Meiselwitz (Ed.): SCSM 2018, LNCS 10913, pp. 400–414, 2018.
https://doi.org/10.1007/978-3-319-91521-0_29

service Picarto[6] which is mostly used to broadcast drawings and art, or Twitch[7], the main representative of the e-Sports industry. Some services are mainly used in a particular country, niconico[8] in Japan or YY[9] in China, for example. To stream oneself, or something else like a live concert, a camera of a mobile phone or webcam is used that can be connected worldwide and sometimes even the whole day. An example would be a webcam on the ISS which streams videos of the earth from space every day. The streams can be found by searching for hashtags which were attached to the stream before broadcasting, or, like in the case of Periscope, via a world map on which current live streams are marked. Since many platforms offer multi-channel options, which means the linking of other SNSs like Instagram or Facebook to the streaming channel, potential viewers get a notification when the streamer is broadcasting.

Since SLSSs are a relatively new field of social media, they are not thoroughly studied yet. A few general studies on SLSSs and information behavior were conducted. Information behavior is classified as human behavior in relation to information and knowledge, for example HII: Human Information Interaction, and information and communication technologies, e.g. HCI: Human Computer Interaction. We found human information interaction studies on Twitch [10, 18] and a few studies on the general SLSSs YouNow, Periscope, and Ustream [5, 6, 8, 9, 20]. General SLSSs are appreciated for the interaction between the streamer and the viewer, and the fact that SLSSs are very authentic [24], since everything happens in real life and cannot be staged. Furthermore, an SLSS can be used to broadcast breaking news and is adopted by citizen and professional journalists [7, 19]. A study on Periscope determined its role in the context of the economics classroom, since its grants the students to gain insight into other people's lives in distant cultural areas [3]. Some studies were focused on legal problems concerning SLSSs [4, 11, 28]. Results showed that music and video copyright violations could be observed, furthermore, personality rights are another important factor in context with SLSSs. Gamification is an important aspect of SLSSs, since it engages the viewer or streamer while one is streaming or watching a broadcast [21]. Sometimes, the user even experiences flow, which is a state of total immersion into whatever one is doing, making him potentially forget about the time spend with the service [9]. A study on Twitch focused on the user-interaction with the live-stream and if the content can be influenced by the viewers. A "Twitch Plays Pokémon"-like setting was developed in which the audience shares the control of the main character of a game, thus leaving the control of the content mainly by the audience [14].

In accordance with the proposed model for information behavior research on SLSSs [29], this study investigates the aspect of content production.

Since some SLSSs found their expert community, it will be interesting to investigate what kind of content is produced on general SLSSs and if it differs among diverse services. Furthermore, distant cultural areas could be interested in contrasting

[6] https://picarto.tv/.

[7] https://www.twitch.tv/.

[8] http://www.nicovideo.jp/.

[9] http://www.yy.com/.

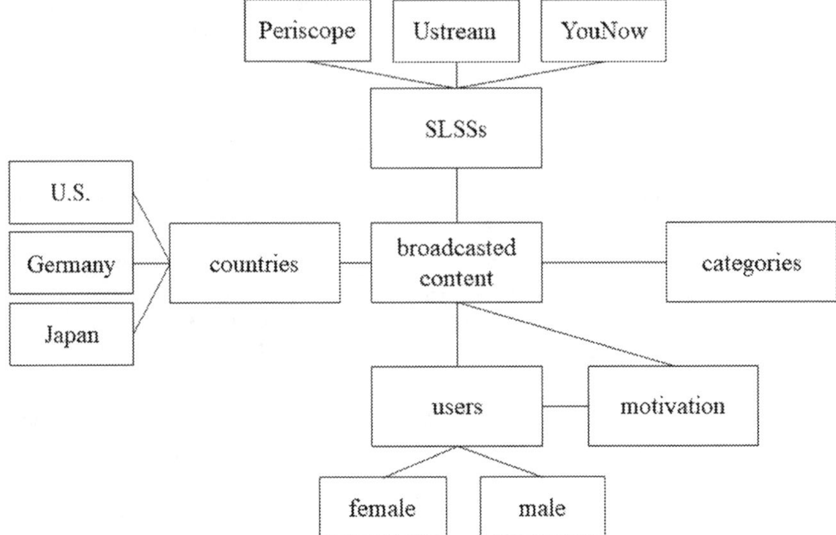

Fig. 1. Research model for this content analysis of SLSSs.

kind of topics, making it another compelling aspect of this content analysis. Another aspect is the gender of the streamer that will be investigated in context with the content.

A study on information behavior in connection with the content of SLSSs and how it is related to the users, countries or the service, is still missing. To this end, a content analysis was conducted after a research model (Fig. 1) and the following research questions (RQs) formulated:

- RQ1: What kind of content is broadcasted on SLSSs?
- RQ2: Are there differences regarding the streamed content between genders?
- RQ3: Does the country of the streamer have an impact on the kind of content that is produced, in this case Japan, the U.S., and Germany?
- RQ4: Is the content divergent for the three SLSSs Periscope, Ustream, and YouNow?
- RQ5: Does the motivation of the streamer influence the content?

2 Method

To get a better understanding on the possible kinds of topics and content that people may talk about, perform or display while streaming, a content analysis [13, 15, 16] was implemented. To this end, a codebook based on literature regarding the use of social networking services was developed to get standardized data sets. A team of researches assessed, evaluated, and compared SLSSs' users' streaming behavior and the produced content. To guarantee a qualitative content analysis and a high reliability two different approaches were applied. First, the directed approach was used by selecting literature to get guidance on what kind of content gets produced on social network services. Second, the conventional approach was implemented via the observation of live streams to get an idea on what people stream about [12]. Several steps were taken in our content analysis. According to McMillan [16], our steps were: the drafting of the research questions and hypotheses. Then, a sample was selected by watching streams to get a general overview on them. The time span of the collection was set to four weeks, which is the third step. As a result, a spread sheet with the different categories and formalities was generated.

The *content categories* were marked in a tally chat and are the following: to chat; make music (m. music); share information (share infor.); news; fitness; sport event (sports); gaming; animals; entertainment media (ent. media); spirituality; draw/paint a picture; 24/7; science, technology, and medicine (STM); comedy; advertisement; nothing; slice of life; politics; nature; food; business information (busi. infor.).

The *motivation categories* were modeled after the uses and gratifications theory [1, 17]: entertainment (boredom, fun, hobby); information (to reach a specific group, exchange of views); social interaction (socializing, loneliness, relationship management, need to communicate, need to belong); self-presentation (self-improvement, self-expression, sense of mission, to become a celebrity, to make money, trolling).

Furthermore, the gender (male, female) of the streamer was listed as well.

The data was collected from three countries, namely Germany, Japan, and the United States of America, to inspect if differences between cultural areas are present. Furthermore, the research team had the required language skills for the three countries. To ensure that the streams originated from these countries, the declaration of the country was checked for each broadcast and service. Twelve research teams á two persons (advanced students of information science in Düsseldorf) were formed. Every coder received a spread sheet and marked everything in it that was applicable to the stream. While watching the stream the 'four eyes principle' was used. Every stream was watched by the two coders simultaneously, but independently for two to ten minutes. To reach a 100% intercoder reliability the entries were compared and if consensus could not be reached, the item was discussed. The streams were observed in two phases. First, the data was collected by observing the stream. If questions remained, for example about the age or the motivation of the streamer, he or she was asked via the chat system of the service.

The streams were not recorded, since it would require the consent of the streamer, but not every streamer communicates with the viewer or agrees with the recording. In the end, 7,667 streams in a time span of four weeks, from April 26 to May 24, 2016, were observed.

Descriptive statistics were calculated for the distribution of the content categories among the motivations, gender, countries, and services. Furthermore, for the inductive statistical analysis, the phi coefficient (mean square contingency coefficient) or Cramér's-V and p-value were determined were applicable. The phi coefficient was used for two binary variables, meaning the content and motivations were matched, as well as the gender and content categories. To determine the correlation between the services and the content, as well as the countries and content, the Cramér's-V measure was used. The value for the phi coefficient as well as the Cramér's-V ranges from 0 to 1, with 0 representing no relationship and 1 equal values. Each resulting correlation was checked with their respective *p*-value for statistical significance.

3 Results

3.1 Content Categories

The produced content on SLSSs will be discussed in the following section.

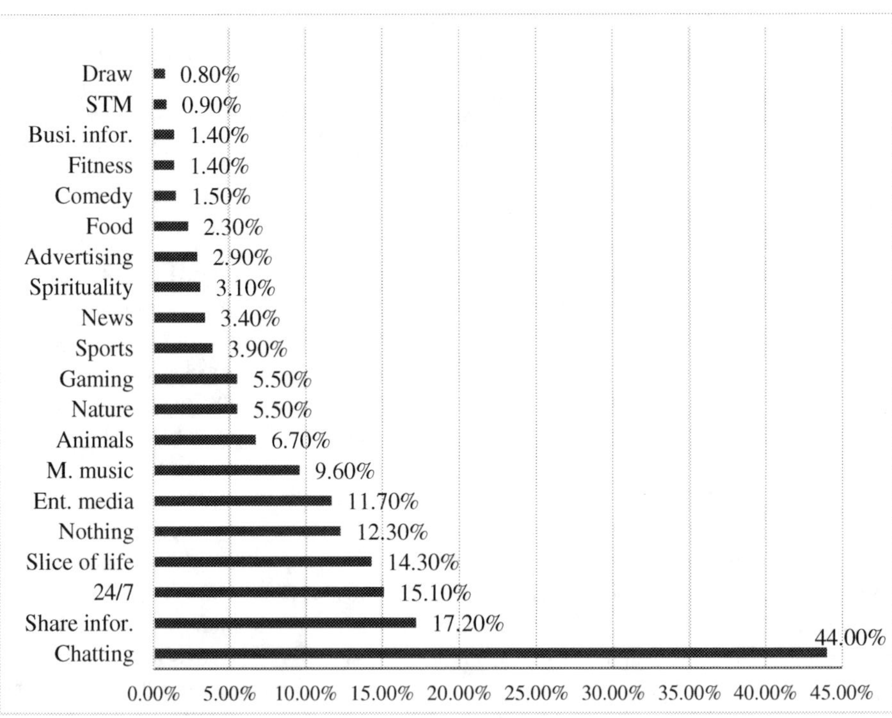

Fig. 2. Distribution of the content categories on SLSSs (N = 7,667).

Like Tang et al. [24, p. 4773] concluded, chatting (44.0%) is the category that can be found the most frequent on SLSSs (Fig. 2). Other content categories like sharing information on various topics (17.2%), 24/7 streams (like the ISS webcam) or slice of life (14.3%) are often represented. Entertainment media (11.7%) is another favored kind of content. Interestingly, a big category is "nothing", no streamer was present, or any other activity could be observed, just an empty room. All these categories do not need any kind of preparation or a high amount of cognitive effort [25]. Presumably, this means that the lower the relative frequencies of the content categories are, the more effort is needed. If one wants to talk about politics (1.4%), business information (1.4%)

or science, technology and medicine (STM) information (0.9%), one needs knowledge about these areas.

The data also suggests that an SLSS that specializes in some form of content has the monopole for it. The content categories drawing (0.8%) and gaming (5.5%) are not well represented on the general SLSSs YouNow, Periscope, and Ustream since Picarto (for drawing) and Twitch (for gaming) are the specialized SLSSs for those areas.

3.2 Content and Gender

In the following section, the differences of the content categories distributed among the genders will be determined. Overall, of the 7,667 streams, 4,548 streams (59.32%) were broadcasted by streamers who identified with either male or female. This means the streamer stated his or her gender, it was displayed in their channel description or the streamers assigned themselves the corresponding tag, e.g. #boy or #girl. For the other streams, the streamers either did not state their gender, or no person could be seen, if an animal was shown, for example. 2,782 (61.17%) streamers were male and 1,766 (38.83%) female. This distribution of genders among streamers is confirmed by the research of Tang, Veniola, and Inkpen [24, p. 4774] as well.

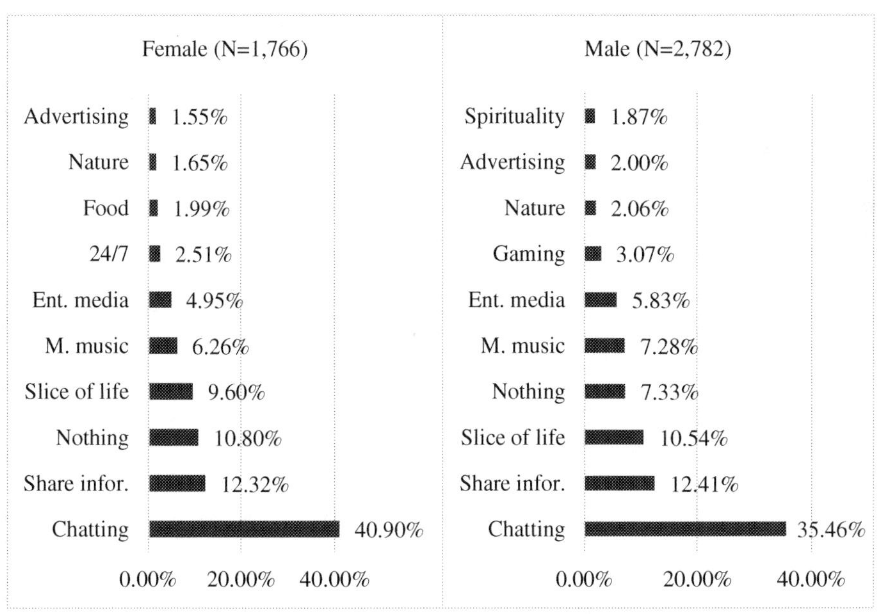

Fig. 3. Distribution of the content categories among the gender (N = 4,548).

For the distribution of the content categories among the gender, only a few differences were observed (Fig. 3). Men seem to prefer topics related to spirituality (1.87%) and gaming (3.07%), whereas women are interested in streaming food related videos (1.99%) or 24/7 streams (2.51%). It can be assumed that overall, men are generally more drawn to gaming related streams than women, which statistics for the e-sports SLSS Twitch also determine [23]. The distribution of the chatting categories among the gender suggests that women are more likely to talk to their viewers (40.90%) than men (35.46%), but the phi coefficient does not support this assumption (.087), which is highly significant (.000).

Table 1. The phi coefficient and p-value for the top ten content categories in relation to gender.

Content	phi coefficient	p-value
Chatting	.087	.000
Share information	–.003	.873
24/7	.030	.040
Slice of life	–.021	.151
Nothing	.079	.000
Entertainment media	–.025	.090
Make Music	–.027	.072
Animals	.020	.181
Nature	–.019	.189
Gaming	–.093	.000

Looking at the correlations, more specifically, the phi coefficient of the content distribution among the gender, none of the categories seem to be related to either one gender (Table 1). Even for the category gaming (phi coefficient –.093; p-value .000) we only find a very small or non-existent correlation. Since the data is not normal distributed, it is only possible to determine some trends, but because all correlations do not imply any relationship between gender and content at all, there seem to be no trends.

In conclusion, there are only a few differences in information production behavior between the genders; this result seems to differ in comparison to other SNSs. Seymour [22] states that striking differences exist between men and women when it comes to sharing personal information on sites such as Facebook, and YouTube. Furthermore, only men with greater degrees of emotional instability were more regular users of SNSs [2].

3.3 Content and Services

Following in this section, the distribution of the content categories among the countries for each service, namely Periscope, Ustream, and YouNow, will be analyzed. Overall, we watched 2,960 streams on Periscope, 2,686 streams on Ustream, and 2,020 (2,021) streams on YouNow. Since YouNow is unknown in Japan, we have only one stream originating from there (excluded in this part of our analysis).

Overall, Periscope and YouNow seem to share their most popular content categories (Fig. 4). Here, just chatting with the audience and sharing information are favored. But a few differences can be marked. The content of YouNow seems to be more self-produced, for example performing a comedy sketch (2.04%), making fitness related videos (1.42%), e.g. showing or explaining exercises, or playing a video game (1.92%). Whereas on Periscope, the content is more passive, for example being in nature (3.18%), or broadcasting a sport event (1.88%). Those differences and the more active engagement of YouNow's streamers with their content can possibly be explained with some users' motivation to become a celebrity [5], which can be compared to success stories of some YouTube stars like Justin Bieber or Lindsay Stirling.

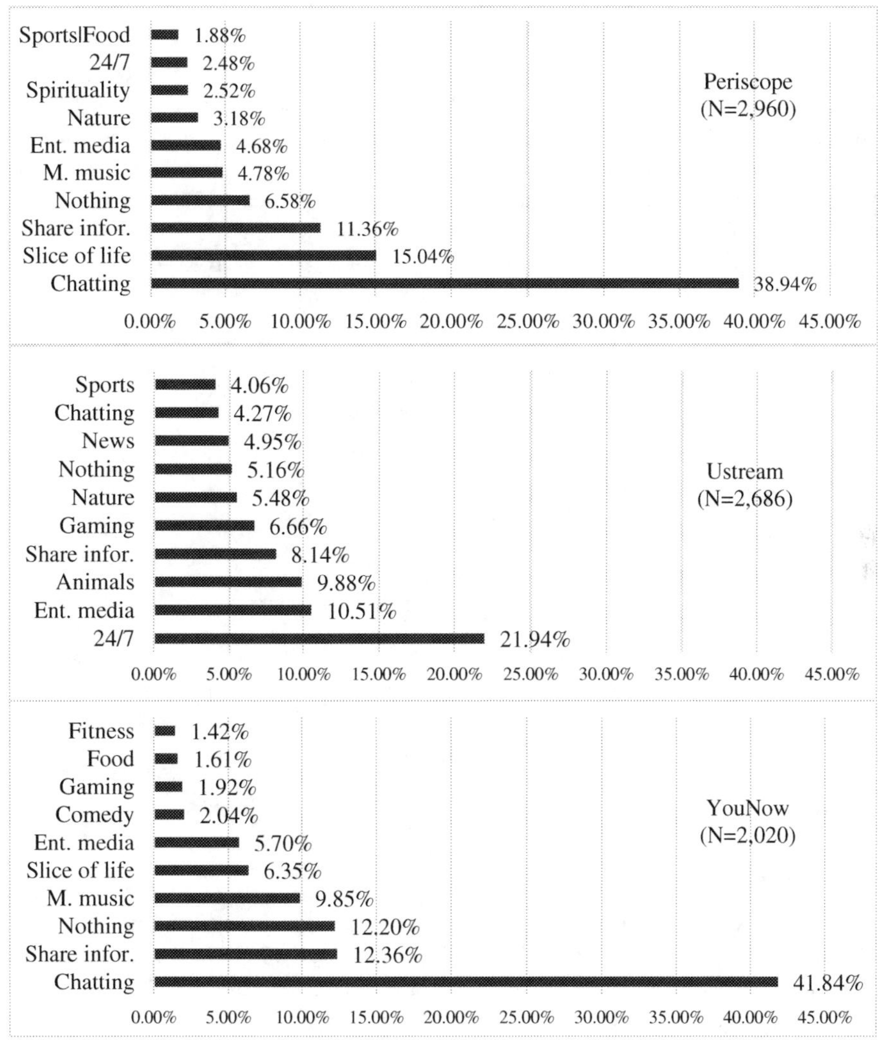

Fig. 4. Distribution of content categories among Periscope, Ustream, and YouNow; sometimes multiple assignments.

In contrast, Ustream offers another spectrum of content, which is more focused on different kinds of media. Entertainment media (10.51%), gaming (6.66%), news (4.95%), or sport events (4.06%) are broadcasted. Chatting (4.27%) or interacting with the viewer is not a big part of the streaming behavior on Ustream. Furthermore, streaming videos of animals (9.88%) and/or nature (5.48%) for 24 h a day can be observed often as well. The findings are in line with Ustreams agenda of being a SLSS which aims to educate its viewers, as NASA is an official customer of Ustream, as well as offering a platform to companies giving them the opportunity to broadcast live events, like concerts or sport events, for example [26, 27].

The calculations of the Cramér's-V between the three services for 24/7 (.458; p-value .000) and chatting (.542; p-value .000) streams also provide clues that especially two of these categories are strongly dependent on the service (Table 2). On Ustream, only 4.27% of the streamer chat with their audience, whereas on YouNow, it is 41.84% and on Periscope 38.94%. For all other categories, there are at least low correlations; however, all correlations are statistically significant (.000).

The results could be explained with the three services being used for various kinds of topics and are not focused on one expert area yet, like Picarto for drawing related content, for example.

Table 2. The Cramér's-V and *p*-value for the top ten content categories in relation to the services.

Content	Cramér's-V	*p*-value
Chatting	.542	.000
Share information	.059	.000
24/7	.485	.000
Slice of life	.218	.000
Nothing	.131	.000
Entertainment media	.156	.000
Make Music	.126	.000
Animals	.312	.000
Nature	.154	.000
Gaming	.204	.000

3.4 Content and Countries

All three countries share chatting as the main content category (Fig. 5). But the number of relative frequencies differ among them. Nearly 40% of all German streams are just people chatting with the audience, in Japan, they are nearly 30% of the overall distribution, however, in the U.S. only about 18%.

The most 24/7 streams (14.78%) originate from Japan; here other big foci on animals (6.60%) and nature (6.50%) exist as well. In contrast, the streams' content that is popular in Germany is more related to entertainment, like entertainment media (10.18%), making music (5.34%), gaming (4.89%), or sport events (1.36%).

In the U.S., a lot of 24/7 streams (7.67%) can be found as well, but also, the highest number of streams in which nothing (9.04%) is happening. The U.S. is also the only country of the three which has advertisements (2.85%) as one of its top ten categories (Fig. 5), suggesting that for now, only streamer from the U.S. see live streaming as an opportunity to generate financial gains.

But taking a closer look at the Cramér's-V correlations, all the categories are highly significant weakly correlated (Table 3).

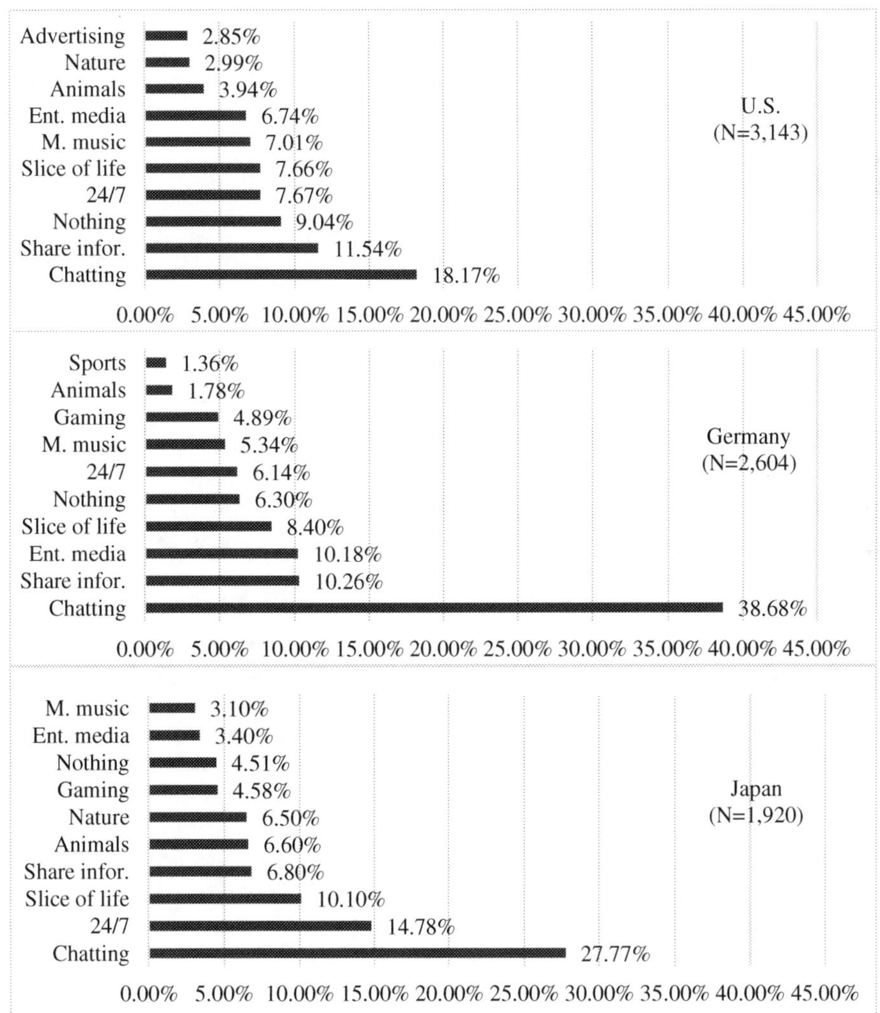

Fig. 5. Distribution of content categories among the U.S., Germany, and Japan; sometimes multiple assignments.

Table 3. The Cramér's-V and *p*-value for the top ten content categories in relation to the countries.

Content	Cramér's-V	*p*-value
Chatting	.130	.000
Share information	.153	.000
24/7	.155	.000
Slice of life	.059	.000
Nothing	.163	.000
Entertainment media	.116	.000
Make music	.142	.000
Animals	.127	.000
Nature	.149	.000
Gaming	.072	.000

Overall, the streamers of the countries seem to have slightly different tastes in their favored content, but the correlations suggest that even people from distant cultural areas just want to fight their boredom [9] and talk to other people while sharing their thoughts and daily lives.

3.5 Content and Motives

Are there any differences among the distribution of the content categories and the motives, as well as any correlations?

For chatting, the percentage values shrink with the decreasing ranking of the four main motives (Table 4). But overall, the percentages of the content categories do not differ greatly among the four motives.

Some highly significant phi correlations can be observed. Streamers seem to broadcast themselves and just want to chat if they are bored (.349, *p*-value .000) or want to socialize (.365, *p*-value .000). Furthermore, the motive of the need to communicate leads the streamer to share information (.318, *p*-value .000). The content of the information that is being shared can range among various topics. But this also implicates legal issues, since the streamer could handle sensitive data carelessly [28]. This suggests that streamers who are bored or need some form of human interaction just want to chat with their audience to idle away time.

The highest correlation (.401, *p*-value .000) can be found between streams that showcase spiritual aspects, for example a Holy Mass or citing quotes from the Quran, and if the streamer has a sense of mission. Furthermore, the streamer has a need for self-presentation (3.65%) and information (4.08%). In this case, the streamer feels a need to broadcast his believes or moral values to his or her audience, therefore using the service for an explicit goal. Since Periscope has the highest percentage of spiritual related content, it can be assumed that especially this service is used to this end.

Table 4. Distribution of motives and content categories; sometimes multiple assignments; N = 7,667.

Entertainment (32.26%)		Social interaction (30.07%)		Self-presentation (21.79%)		Information (15.87%)	
Content	Rel. freq.	Content	Rel. freq.	Content	Rel. freq.	Content	Rel. freq.
Chatting	37.93%	Chatting	34.84%	Chatting	25.89%	Chatting	18.73%
Share infor.	11.23%	Share infor.	17.31%	Share infor.	14.61%	Share infor.	16.76%
Slice of life	9.60%	Slice of life	8.61%	M. music	11.41%	Ent. media	7.20%
Nothing	7.20%	Nothing	5.79%	Ent. media	9.93%	24/7	7.17%
M. music	6.66%	Ent. media	5.31%	Slice of life	5.42%	Slice of life	6.20%
Ent. media	6.63%	M. music	4.76%	24/7	4.48%	M. music	5.51%
Gaming	4.50%	24/7	3.95%	Nothing	3.88%	Advertising	4.28%
24/7	2.35%	Advertising	2.63%	Advertising	3.79%	Nothing	4.17%
Nature	1.91%	Nature	1.95%	Spirituality	3.65%	Spirituality	4.08%
Animals	1.86%	Food	1.89%	Sports	2.31%	Animals	3.83%

4 Discussion

The content production on three general SLSSs (Periscope, Ustream, and YouNow) from three different countries (U.S., Germany, and Japan) was analyzed. The distribution of the content categories was determined among gender, countries, SLSSs, and motivations. Furthermore, the phi correlation/Cramér's-V as well as p-values were calculated for the content and the genders, countries, SLSSs, and motivations.

The most produced content overall is chatting, followed by sharing information, 24/7 streams, slice of life, and also, nothing, e.g. streams in which only an empty room is being broadcasted. These kinds of topics imply that not much cognitive effort is needed for preparing the stream, in contrast to broadcasts that display the streamer performing fitness routines or talking about politics.

For the distribution of the gender, there are more male than female streamers active on the analyzed SLSSs. All in all, the calculations of the correlations show that the produced kind of content does not differ between the genders.

If the service is concerned, highly significant strong correlations can be observed for the categories chatting, and 24/7 streams, implying that 24/7 streams are more broadcasted on Ustream, and streams in which the streamer talks to his or her audience on Periscope and YouNow.

The country from which the streamer broadcasts is not correlated to the content, which is supported by highly significant p-values. Even though the countries are culturally different, the streamers only want to chat with other people.

Further correlations can be observed for the motivation of the streamer and the content he or she produces. If people are bored or want to share information, they usually just chat with their audience. Another finding was that streamers who have a strong sense of mission broadcast spiritually related content and vice versa. They also seem to like to present themselves and distribute information.

The study shows some limitations. It would be interesting to analyze other SLSSs and compare the produced content, for example Instagram Live, Facebook Live, or YouTube Live. Another aspect that should be studied is the origin of the stream. Even though we did not find contrasts for the categories and the three countries we choose, there could exist differences; for example, from countries in the middle east, especially Saudi Arabia, where YouNow is popular, or from China, where SLSSs are heavily used. Furthermore, it should be analyzed if the content of the streams differs per service in relation to the streams' origin. This means that the content that was broadcasted on Periscope could be different for Germany or the U.S.

Another aspect that should be analyzed is the age of the streamer, since generational differences can be observed for SNSs, so it is likely that there are divergent results for SLSSs as well. Moreover, a closer look should be taken at the motivation of the streamer. Since we found statistical significant results for the content and the streamers' motives, the age and gender could play an important role in this context as well.

All in all, the content analysis on SLSSs is still in its early beginning, but some valuable results could be observed and show a promising groundwork for future research.

Acknowledgement. The author wants to thank Wolfgang G. Stock for his valuable and much needed insights and help with this paper.

References

1. Blumler, J.G., Katz, E.: The Uses of Mass Communications: Current Perspectives on Gratifications Research. Sage, Newsbury Park (1974)
2. Correa, T., Hinsley, A.W., de Zúñiga, H.G.: Who interacts on the web? The intersection of users' personality and social media use. Comput. Hum. Behav. **26**(2), 247–253 (2009). https://doi.org/10.1016/j.chb.2009.09.003
3. Dowell, C.T., Duncan, D.F.: Periscoping economics through someone else's eyes: a real world (Twitter) app. IREE **23**, 34–39 (2016). https://doi.org/10.1016/j.iree.2016.07.003
4. Faklaris, C., Cafaro, F., Hook, S.A., Blevins, A., O'Haver, M., Singhal, N.: Legal and ethical implications of mobile live-streaming video apps. In: Proceedings of the 18th International Conference on Human-Computer Interaction with Mobile Devices and Services Adjunct, pp. 722–729. ACM, New York (2016). https://doi.org/10.1145/2957265.2961845
5. Fietkiewicz, K.J., Dorsch, B., Scheibe, K., Zimmer, F., Stock, W.G.: Dreaming of stardom and money: micro-celebrities and influencers on live streaming services. In: Meiselwitz, G. (ed.) SCSM 2018. LNCS, xx–yy, vol. 10913. Springer, Cham (2018)
6. Fietkiewicz, K.J., Scheibe, K.: Good morning… good afternoon, good evening and good night: adoption, usage and impact of the live streaming platform YouNow. In: Proceedings of the 3rd International Conference on Library and Information Science, pp. 92–115. International Business Academics Consortium, Taipeh (2017)
7. Fichet, E., Robinson, J., Dailey, D., Starbird, K.: Eyes on the ground: emerging practices in Periscope use during crisis events. In: Tapia, A., Antunes, P., Bañuls, V.A., Moore, K., Porto, J. (eds.) ISCRAM 2016, Conference Proceedings – 13th International Conference on Information Systems for Crisis Response and Management, pp. 1–10. Federal University of Rio de Janeiro, Rio de Janeiro (2016)

8. Friedländer, M.B.: Streamer motives and user generated content on social live-streaming services. J. Inf. Sci. Theor. Pract. **5**(1), 65–84 (2017). https://doi.org/10.1633/JISTaP.2017.5.1.5
9. Friedländer, M.B.: And action! Live in front of the camera: an evaluation of the social live streaming service YouNow. Int. J. Inf. Commun. Technol. Human Dev. **9**(1), 15–33 (2017). https://doi.org/10.4018/IJICTHD.2017010102
10. Gros, D., Wanner, B., Hackenholt, A., Zawadzki, P., Knautz, K.: World of streaming. motivation and gratification on Twitch. In: Meiselwitz, G. (ed.) SCSM 2017. LNCS, vol. 10282, pp. 44–57. Springer, Cham (2017). https://doi.org/10.1007/978-3-319-58559-8_5
11. Honka, A., Frommelius, N., Mehlem, A., Tolles, J.-N., Fietkiewicz, K.J.: How safe is YouNow? An empirical study on possible law infringements in Germany and the United States. J MacroTrends Soc. Sci. **1**(1), 1–17 (2015)
12. Hsieh, H.F., Shannon, S.E.: Three approaches to qualitative content analysis. QHR **15**(9), 1277–1288 (2005). https://doi.org/10.1177/1049732305276687
13. Krippendorff, K.: Content Analysis: An Introduction to Its Methodology, 2nd edn. Sage, Thousand Oaks (2004)
14. Lessel, P., Mauderer, M., Wolff, C., Krüger, A.: Let's play my way: investigating audience influence in user-generated gaming live-streams. In: Proceedings of the 2017 ACM International Conference on Interactive Experiences for TV and Online Video, pp. 51–63. ACM, New York (2017). https://doi.org/10.1145/3077548.3077556
15. Lai, L.S.L., To, W.M.: Content analysis of social media: a grounded theory approach. J. Electron. Commer. Res. **16**(2), 138–152 (2015)
16. McMillan, S.J.: The microscope and the moving target: the challenge of applying content analysis to the World Wide Web. JMCQ **77**(1), 80–98 (2000). https://doi.org/10.1177/107769900. 007700107
17. McQuail, D.: Mass Communication Theory. Sage, London (1983)
18. Recktenwald, D.: Toward a transcription and analysis of live streaming on Twitch. J. Pragmat. **115**, 68–81 (2017). https://doi.org/10.1016/j.pragma.2017.01.013
19. Rugg, A., Burroughs, B.: Periscope, live-streaming and mobile video culture. In: Geo-blocking and Global Video Culture (Theory on Demand; 18), pp. 64–73. Institute of Network Culture, Amsterdam (2016)
20. Scheibe, K., Fietkiewicz, K.J., Stock, W.G.: Information behavior on social live streaming services. J. Inf. Sci. Theor. Pract. **4**(2), 6–20 (2016). https://doi.org/10.1633/JISTaP.2016.4.21
21. Scheibe, K., Göretz, J., Meschede, C., Stock, W.G.: Giving and taking gratifications in a gamified social live streaming service. In: Proceedings of the 5th European Conference on Social Media. Academic Conferences and Publishing International, Reading (2018)
22. Seymour, C.: Social media and the gender gap: what do you share? EContent **35**(3), 8–10 (2012)
23. Statista: Distribution of Twitch Users Worldwide as of 3rd Quarter 2016, by Gender (2016). https://statista.com
24. Tang, J.C., Veniola, G., Inkpen, K.M.: Meerkat and Periscope: i stream, you stream, apps stream for live streams. In: Proceedings of the 2016 CHI Conference on Human Factors in Computing Systems, pp. 4770–4780. ACM, New York (2016). https://doi.org/10.1145/2858036.2858374
25. Tyler, S.W., Hertel, P.T., McCallum, M.C., Ellis, H.C.: Cognitive effort and memory. J. Exp. Psychol. Hum. Learn. Mem. **5**(6), 607–617 (1979). https://doi.org/10.1037/0278.5.6.607
26. Ustream: Our Company (2018). https://www.ustream.tv
27. Ustream: Live Event Streaming Services & Solutions (2018). https://www.ustream.tv

28. Zimmer, F., Fietkiewicz, K.J., Stock, W.G.: Law infringements in social live streaming services. In: Tryfonas, T. (ed.) HAS 2017. LNCS, vol. 10292, pp. 567–585. Springer, Cham (2017). https://doi.org/10.1007/978-3-319-58460-7_40
29. Zimmer, F., Scheibe, K., Stock, W.G.: A model for information behavior research on social live streaming services (SLSSs). In: Meiselwitz, G. (ed.) SCSM 2018. LNCS. Springer, Cham (2018, in Press)

Privacy and Ethical Issues in Social Media

Moral Disengagement in Social Media Generated Big Data

Markus Beckmann[1], Christian W. Scheiner[2(✉)], and Anica Zeyen[3]

[1] Chair of Corporate Sustainability Management, Friedrich-Alexander-Universität Erlangen-Nürnberg, Nuremberg, Germany
markus.beckmann@fau.de

[2] Institute of Entrepreneurship and Business Development, Universität zu Lübeck, Lübeck, Germany
christian.scheiner@uni-luebeck.de

[3] Centre for Research into Sustainability, Royal Holloway University of London, Egham, UK
anica.zeyen@royalholloway.ac.uk

Abstract. Big data raises manifold ethical questions. While there is a certain consensus on general principles for addressing these issues, little is known about when and why decision-makers display such ethical conduct or opt for unethical behavior with regard to collecting, storing, analyzing, or using big data. To address this research gap, we draw on the concept of moral disengagement. Moral disengagement describes psychological mechanisms by which individuals rationalize and thus disengage themselves from unethical conduct. We develop a theoretical model in which the motivation for monetary benefits as well as the motivation for hedonic benefits is set into relation to moral disengagement and the tendency to make unethical decisions in the context of social media generated big data. Our model spells out four sets of testable propositions that invite further research.

Keywords: Moral disengagement · Big data · Intrinsic motives
Extrinsic motives · Unethical behavior

1 Introduction

The past years witnessed an increasingly rapid digitization of not only business processes but of basically all fields of society and human life. This development goes hand in hand with the exponential growth of digital data. In fact, "big data" has emerged as a phenomenon characterized as a multifold shift in how data becomes available and potentially relevant in our society [1]. First, in terms of volume, big data refers to data sets that include huge amounts of data thanks to both digital storing technologies and the diffusion of data-creating devices such as smart phones. Second, in terms of variety, big data reflects that the type and nature of data is changing thanks to new sensors and the ability to store text, sound, images, etc. Third, in terms of velocity, big data is linked to the potential real-time availability of data. Due to the volume and complexity of such data sets, big data challenges conventional methods of capturing, storing,

© Springer International Publishing AG, part of Springer Nature 2018
G. Meiselwitz (Ed.): SCSM 2018, LNCS 10913, pp. 417–430, 2018.
https://doi.org/10.1007/978-3-319-91521-0_30

analyzing, and using data. At the same time, it opens up new possibilities for data analysis as well as ethical issues such as data privacy, data security, and data property rights [1].

Within the past years, practitioners and researchers have focused their attention especially on areas where ethical issues occur and have suggested possibilities to avoid the unethical use of big data from the beginning [e.g. 1–4]. Simultaneously, misuse of big data can be observed frequently. In order to understand why big data is used in an unethical way, it is important to examine the psychological and cognitive processes of decision-makers with respect to moral reasoning and ethical decision making.

For that reason, we develop a conceptual framework, linking extrinsic and intrinsic motives with moral disengagement, and the tendency to make unethical decisions in the use of big data.

2 Theoretical Background

2.1 Big Data and Unethical Behavior

Discussions about challenges and ethical issues in big data become more and more pronounced both through the increasing number of principles and guidelines as well as increasing research. The following section aims to highlight the most prominent concern and challenges as they relate to an ethical use of big data. To be precise, this section does not refer to any technical issues within big data but focusses on normative challenges in regard to its use. In particular, it emphasizes issues that can arise from intended as well as unintended use of big data.

Discrimination: Big Data analysis can lead to positive and negative discrimination of certain individuals or groups of individuals [2, 5, 6]. Such discrimination can range from customized pricing strategies based on previous purchases, personal likes and dislikes as well as socioeconomic status [cf. 7] to decisions as to which kind of healthcare receives investment in low and middle income countries [8]. To overcome this, most ethics codes call for considering the benefits and harms of each analysis [6, 9].

Privacy: Privacy is defined as the "state of being free from public attention" [10]. While the extent to which privacy is considered important differs across cultures [11], there is nonetheless a call for strict privacy guidelines [5, 12]. The collection and storage of big data increases the possibility of breaching an individual's privacy. Many public debates on this often reference the idea of the 'right to be forgotten'. Some core ethical issue here is the question of how long data can be stored and what kind of data should be stored and who should control the data [2] or own the data [13, 14]. To further protect individuals, anonymizing data is a called for practice [6, 12, 14–16] to avoid compromising personal identities [3, 17].

Surveillance: Surveillance or dataveillance is another ethical challenge for big data users [2, 18]. Take the example of smart cities [1]. While big data can help optimize traffic flow or the general flow of movement during peak hours to avoid severe traffic jams or overcrowded public transport, it can also be used to track the movements of individuals and survey their movements throughout the day. Similarly, the mass surveillance of social media activities can lead to suppressed speech [7].

Limited knowledge of users: This issue is particularly tricky. Here, the question is less what analysist or owners of big data are using it for, but rather if and to what extent users are aware of how their data is used. This aspect is challenging as (1) most people do not read any terms and conditions supplied by companies before they provide their data and (2) often do not understand potential harm that could come to them. To least overcome parts of this problem, guidelines call for a transparent communication about how the data is used [3, 5, 6, 16] so as to provide data providers with the necessary information to make an informed decision.

Data use outside of context: On top of new laws in some countries which prohibit companies to use any data collected from individuals outside the explicit use these individuals have agreed to, it is a commonly agreed upon ethical rule that data should not be used for any purpose except for which they were provided initially [5, 12, 13, 16, 19, 20]. Moreover, it is important to understand the data in its wider context so as not to misinterpret findings [14].

On top of the publications on ethical behavior in regards to big data, there is a discussion on knowing when to break rules [12] – e.g., in situations of natural disaster, emergencies or potential threats to security. This discussion substantiates that the issues above are inherently ambivalent issues. That is to say that they are neither black nor white. The potential of negative consequences very much depends on conscious and unconscious decisions by those in charge of collecting, processing, analyzing and using big data. However, these decisions are not made by an individual but often by many different individuals who might not always be aware of potentially negative consequences of their respective decisions [4, 21] – which in turn further complicates the ethical use of big data.

2.2 Moral Disengagement

Based on social cognitive theory [22], Bandura [23] developed the notion of moral disengagement. Social cognitive theory takes an agentic perspective where people exercise control over their own thoughts and actions [22, 24]. This regulatory system operates through the three self-monitoring, judgmental, and self-reactive functions [25]. Hence people monitor constantly their behavior which is then evaluated against their own (moral) standards and situation-related characteristics [25]. Depending on the moral judgement, positive self-reactions or negative self-sanctioning anticipate behavior and motivate individuals to behave in accordance to their moral standards [23]. This self-regulatory system is, however, not immutable as self-influences operate solely if they are activated. There are, however, numerous psychological mechanisms by which individuals can disengage themselves from unethical conduct and therewith from self-sanctioning [25]. Attribution of blame, dehumanization, disregarding or distorting the consequences, diffusion of responsibility, displacement of responsibility, advantageous comparison, euphemistic language, and moral justification illustrate key mechanisms through which individuals can disengage themselves from harmful behavior and do not activate self-influences [22]. Attribution of blame and dehumanization can enable individuals to morally disengage from detrimental actions by making the victim herself/himself personally responsible for such behavior. In case of attribution of blame, it is argued that the victim has provoked harmful outcomes on herself or himself by their own doings [23]. When victims are dehumanized, individuals feel no longer obliged to evaluate their actions against their moral values as their victim does not belong to the same group [23]. Disregarding or distorting consequences, diffusion of responsibility, as well as displacement of responsibility enable individuals to neglect or ignore own harmful actions. With disregarding or distorting the consequences, harm for others is ignored. This is especially given, when consequences for others are not visible to the individual or occur with a temporally delay [23]. When individuals question or deny personal accountability, diffusion of responsibility is given. Personal accountability can be reduced in cases where group decisions are taken or when division of labor is given and such collective behavior causes harm [23]. Displacement of responsibility diffuses personal accountability as individuals reject their personal role for causing harm. Displacement of responsibility especially occurs where individuals feel obliged to follow orders from legitimate, authorized people. Advantageous comparison, euphemistic language, and moral justification help individuals to misinterpret harmful behavior as morally acceptable or even as completely benevolent. Advantageous comparison allows to downplay own wrongdoing, by comparing it with even more harmful actions. The more malign the contrasting behaviors, the easier it gets to see one's own conduct as acceptable [23]. With euphemistic language, individuals reduce or neglect detrimental conduct by using neutral language or by verbally sanitizing these kinds of actions [23]. Moral justification describes the mechanism by which individuals excuse harmful conduct with a moral imperative. Detrimental conduct is therefore serving moral purposes or is at least personally and socially justifiable from a moral standpoint [23].

3 Propositions and Framework Development

This paper develops a theoretical framework where the motivation for monetary benefits as well as the motivation for hedonic benefits is set into relation to moral disengagement and the tendency to make unethical decisions in the context of big data generated by social media. In limiting the underlying motivational basis on two contrasting types of motivation, we follow previous research [26, 27].

Following Amabile [28], individuals are "extrinsically motivated when they engage in the work in order to obtain some goal that is apart from the work itself" (p. 188). Motivation for monetary benefits illustrates therewith a generic expression for extrinsic motivation.

In the work place, monetary benefits occur manifold from regular wages to variable forms of compensations, financial rewards or pecuniary advantages and have been found to potentially evoke in general unethical behavior [e.g. 29].

The linkage of monetary benefits and moral disengagement has also attracted the attention of researchers and has been object of scientific research [e.g. 26, 30, 31]. Baron et al. [26] found for instance that financial gains and moral disengagement are positively related among entrepreneurs. Moore [30] connected organizational corruption with moral disengagement and argued that moral disengagement can be a crucial factor for organizational corruption as it affects the initiation, facilitation, and perpetuation of corruption in the workplace. Monetary benefits can especially be found in the perpetuation of organizational corruption, as individuals who are more likely to make unethical decisions in the interest of the organization have a higher probability of organizational advancement and in turn higher monetary benefits. Shepherd and Baron [31] examined the assessment of business founders with respect to the attractiveness of business opportunities which cause harm to the natural environment. They found that moral disengagement enabled entrepreneurs to perceive opportunities as highly attractive even if they would harm the environment.

Given these previous findings, it can be assumed that motivation for monetary benefits can cause deviant behavior in all work-related facets. For that reason, we propose that also in the context of big data motivation for monetary benefits is positively related to moral disengagement.

Proposition 1a: Employees' motivation for monetary benefits is positively related to moral disengagement.

In contrast to motivation for monetary benefits, motivation for hedonic benefits illustrates a generic intrinsic motivation as behavior is not triggered by an externally offered incentive but is conducted out of interest for the activity itself [32]. Intrinsically motivated individuals "seek [subsequently] enjoyment, interest, satisfaction of curiosity, self-expression, or personal challenge in the work" [32, p. 188].

Intrinsic motivation has been found to generally impact positively different work-related activities [e.g. 33, 34], the selection of specific career paths [35], and has proven to affect performance on some tasks more positively than conditions related to extrinsic motivation [32].

Despite the general notion that intrinsic motivation can influence individuals to morally disengage, recent research examined the relationship of moral disengagement and intrinsic motivation and came to contradicting conclusions [26, 27]. In their study on entrepreneurs, Baron et al. [26] found a negative relationship between intrinsic motivation for self-realization and moral disengagement. Scheiner et al. [27] examined the motivation for hedonic benefits and moral disengagement in the context of an idea competition and found also partial support for the negative relationship.

In light of previous findings, individuals with a high intrinsic motivation seem to be less likely to morally disengage. For that reason, we propose that motivation for hedonic benefits is negatively related moral disengagement in the context of big data.

Proposition 1b: Employees' motivation for hedonic benefits is negatively related to moral disengagement.

One key aspect of big data in the context of social media is that there are novel ways to collect data, both with regard to new data sources (such as browsers, smartphones, health trackers etc.) and with regard to different types of data (such as text, sound, pictures, and other metrics generated in online search behavior, login personal or financial information, or motion and health data). This new volume and variety of data that can be collected certainly creates opportunities for innovations that benefit not only companies but also consumers, citizens, and society at large. As already reviewed above, these novel options for data collection also give rise to ethical questions such as privacy concerns and with regard to the property rights of the data collected from individuals.

Against this background, numerous industry and policy guidelines have formulated standards for the ethical conduct of big data collection [36–38]. Two principles are particularly important in this regard. First, the principle of *voluntary consent* highlights that personal data should only be collected from people with their explicit and voluntary agreement [36]. Second, the principle of *transparency* requires that the people whose data is collected are informed about how, what kind of data is actually gathered (and treated later on) [36]. Given these two principles, unethical conduct in the data collection phase can fall into several categories: Data could be collected without the consent (or even against the will) of individuals. Data collection could occur without individuals having full knowledge of what kind of data is actually collected. Finally, data collectors could fail to display transparency or could legally live up to the transparency principle in ways that themselves fail to be transparent, e.g. when the terms of agreement or the data privacy statement are hard to find, in extra small print or difficult to read/understand because of its technical wording, or the sheer length of the text.

From a business perspective, it is tempting to have as few restraints in the data collection as possible and therefore to violate the aforementioned ethical principles. Moral disengagement could increase the tendency towards such unethical conduct through several of its underlying mechanisms. Attribution of blame [22] would occur if data collectors shifted the blame onto individuals who do not protect or even freely share their data, e.g. by claiming that people can and should decide for themselves how to protect their data or that individuals are responsible in the first place [23] if they download a social media app that collects motion data via a smartphone. Another moral disengagement mechanism that could favor unethical behavior in big data would be advantageous comparison. As there are drastic examples of how personal and sensitive data was collected against the will of individuals in other areas of digital life – e.g. the alleged spying through web-cams –, decision-makers could always euphemistically downplay their own wrongdoing [23].

In short, as the collection of big data in the context of social media creates various options for unethical behavior and as several moral disengagement mechanisms could rationalize such actions, we propose:

Proposition 2a: Employees' moral disengagement is positively related to their tendency for unethical conduct with regard to the collection of big data.

In addition to the issues of data collection, the volume and velocity of big data also raise questions of data storage in the context of social media. Ever bigger amounts of data need to be stored at reasonable cost, should often be available in real-time and accessible irrespective of where the data was collected or is needed. As a consequence, new data storage architectures, often cloud-based, emerge.

As new storage solutions create opportunities, they also create risks that call for a responsible data storage management to address potential concerns of data privacy, data sovereignty, and data security. Similarly to the data collection, various ethical principles have emerged to govern these issues. With regard to data privacy, respecting the privacy of individuals requires that personal information that reveal someone's identity should either be blinded or only be stored if absolutely necessary, with the respective individuals giving their consent to the storage of personalized data [38]. With regard to data sovereignty, individuals should know what kind of data is stored about them, should be able to check this data record and have the ability to call for correction if the data is faulty [39]. In fact, if faulty data is stored and people cannot check and correct it, they might be unjustly blocked, for example, from attaining credit or health insurance [40]. Finally, with regard to data security, sensitive data – ranging from passwords to private conversations and health data – needs to be protected not only against being lost but also against being stolen or manipulated by third parties. Otherwise, issues of identity theft, credit card fraud, privacy infringements etc. could significantly harm the individuals who cannot protect themselves against such risks once their data is stored.

While guidelines for the ethical conduct of storing big data thus exist, keeping such standards can be costly, require effort, or limit a company's options, thus creating the temptation to violate them. Unethical conduct with regard to data storage then spans various practices: Decision-makers could store personal, sensitive information of individuals without their knowledge or even against their will; they could leave opaque which information is stored and difficult to check and correct it; and they could fail to invest in necessary IT security, thus tolerating poor IT architectures with known security weaknesses.

Given the nature of these issues, moral disengagement mechanisms could enhance the likelihood for unethical conduct with regard to data storage in several ways. To start with, attribution of blame [22] could mean that individuals whose data is stored are attributed responsibility because their own behavior allowed the data collection and storage in the first place. Diffusion of responsibility could occur when decision-makers such as managers in big data enterprises refer to technological system constraints that allegedly make a different conduct unfeasible, with the responsibility diffused to ICT engineers, software developers etc. Advantageous comparison could, again, refer to bigger scandals, e.g. to Yahoo's infamous 2016 data breach [41] where the sensitive information of 500 million users was hacked– thus effectively downplaying one's own wrongdoing [23] if data security does not live up to the desired standards.

In short, as the storage of big data creates specific options for unethical behavior and as moral disengagement mechanisms can be argued to rationalize such actions, we propose:

Proposition 2b: Employees' moral disengagement is positively related to their tendency for unethical conduct with regard to the storage of big data.

The sheer variety and volume of big data leads to various challenges when analyzing big data. To this end, new tools have been developed and are still being developed to navigate the volume of data.

These new tools can be highly effective in analyzing patterns and supporting the identification of idea solutions. At the same time, big data analysis entails many potential ethical challenges. Some of those challenges relate to the actual tools used in the analysis and tackle challenges known from statistical analysis such the outlier problem. Moreover, while guarding anonymity is a principle already readily used in statistical analysis, this challenge's magnitude increases significantly in the context of big data in social media. This is because by pulling data from various social media sources, it would be possible to reconstruct an individual's life quite accurately. Therefore, to safeguard the identity of individuals, many principles in big data analysis call for anonymization of the data prior to running any analysis [6, 12, 14, 15] and to implement measures that disallow re-identification of individuals [12]. Indeed, recently guidelines for research and analysis of data from specific social media platforms have started to emerge [e.g. 16].

Furthermore, the volume, variety and velocity of big data makes its analysis very complex. As such, there is a danger that analytics used fully or partly ignore the context in which the data was collected. This effect is made more complicated by the variety of not only data types but by also of data sources. These complexities notwithstanding, many principles in big data analysis very clearly point to the importance of the context of data [15, 16, 42] in order to fully understand its meaning.

The process of data anonymization and especially the avoidance of re-identification can be very complex and therefore costly. Moreover, companies might have a vested interest in being able to identify individuals in order to target them with specific products or service offerings. Similarly, implementing mechanisms that robustly ensure that the context of the data is respected increases the complexity of big data analysis and might even impede certain types of analyses.

Against the backdrop of these challenges, moral disengagement mechanisms could increase the likelihood for unethical conduct during data analysis. First, disregarding or distorting consequences [22] might lead individuals who are in charge of big data analysis to ignore the context of the data analyzed. Here, individuals might simply choose to ignore potential consequence of not respecting data context in order to simplify their work or to be able to use a greater volume or variety of data in their analysis. Similar to data storage, big data analysis might not follow anonymization and re-identification avoidance principles by displacing responsibility to individuals who provided the information in the first place. Moreover, big data analysis is rarely done by one individual [4]. Indeed, most companies use pre-build software to analyze their data. Thus, the programmer of the software and the user might have no link to each other. Therefore, both sites – software programmer and software user – might make use of diffusion of responsibility due to the potentially large number of people involved in a single analysis.

In sum, as the analysis of big data creates specific options for unethical behavior and as moral disengagement mechanisms can be used to rationalize such actions, we propose:

Proposition 2c: Employees' moral disengagement is positively related to their tendency for unethical conduct with regard to the analysis of big data.

The use case for big data is enormous. Big data use can range from applications for public safety [43] to smart cities by optimizing traffic flows based on movement profiles of commuters and targeted advertisement and investment decisions. Especially in the area of development [44], big data has led to reduced costs in decision-making and has been applied in areas such as underwater animal tracking [45] or providing information on where best to build schools to protect them from droughts [46].

To ensure that big data is used ethically, various principles have emerged. For instance, The Ten Commandments of Computer Ethics by the Computer Ethics Institute call for using big data in way that is respectful to people [20]. Similar guidance can also be found in publications by Accenture [5], Zook and colleagues [12], the ICO [6], Davies and Patterson [13] and Narayanan and colleagues [9]. The discussion of potential negative consequences of big data uses is very prominent in the field of big data research [e.g. 16, 47].

Despite these guidelines and calls for a respectful use of big data that keeps in mind potential negative effects to both data providers and the general public, there are many reasons why such ethical approaches might not be fully implemented. One such use is surveillance of citizens or customers. Such techniques allow big data users to profile individuals [2, 14] and use it to, for example, predict behaviors and movement patterns. Furthermore, big data can be used to positively or negatively discriminate individuals or groups of people [2, 5, 6, 42]. Potential consequences of discrimination include customized pricing strategies based on previous purchases, personal likes and dislikes and socioeconomic status [48] as well as decisions which impact healthcare investment in low and middle income countries [8]. In higher education, big data is used more and more frequently to develop performance prediction tools for individual students [49]. While such information can help an education institution to better support students, as in the case of Arizona State University, it can also be used to predict students who intend to transfer to another university [50].

As the decision on how to use big data clearly lies with individual decision-makers, misuse might be rooted in mechanisms of moral disengagement. For instance, in the case of higher education institutions using big data to preempt student transfer and its consequent loss of income, the decision-makers might engage in advantageous comparison by pointing to other institutions that engage in similar activities or who might use data to preemptively expel them. Dehumanization may occur where consequences of big data use impact many people or people who are far away and therefore might seem less "real" to decision-makers. If we consider a scenario where a smaller group of individuals that are identified as being more likely to have a costly disease are excluded from healthcare services, moral justification could be used to argue that it is in the interest of everyone else to keep their healthcare costs down.

In sum, we thus propose:

Proposition 2d: Employees' moral disengagement is positively related to their tendency for unethical conduct with regard to the usage of big data.

Consistent with our line of argumentation, we propose that the relationship between motivation for monetary benefits and the tendency to make unethical decisions in the context of data collection, data storage, data analysis, and data usage can be explained, in part, through moral disengagement. In cases where individuals are motivated by monetary benefits, they are less likely to evaluate their doing from a moral standpoint. For that reason, decision-makers are more likely to actively morally disengage from self-regulation and self-sanctioning, which could lead to a higher tendency to make unethical decisions. We thus posit:

Propositions 3a: Moral disengagement mediates the positive relationship between employees' motivation for monetary benefits and the tendency for unethical conduct with regard to the collection of big data.

Propositions 3b: Moral disengagement mediates the positive relationship between employees' motivation for monetary benefits and the tendency for unethical conduct with regard to the storage of big data.

Propositions 3c: Moral disengagement mediates the positive relationship between employees' motivation for monetary benefits and the tendency for unethical conduct with regard to the analysis of big data.

Propositions 3d: Moral disengagement mediates the positive relationship between employees' motivation for monetary benefits and the tendency for unethical conduct with regard to the usage of big data.

Given our previous propositions, we suggest that the relationship between motivation for hedonic benefits and the tendency to make unethical decisions in the contexts of data collection, data storage, data analysis, and data usage can be explained, in part, through moral disengagement processes. Individuals motivated by hedonic benefits are more likely to evaluate their behavior from a moral perspective. Thus, they are less likely to disengage from self-regulation and self-sanctioning. This should result in a lower likelihood to make unethical decisions. Consequently, we propose:

Propositions 4a: Moral disengagement mediates the negative relationship between employees' motivation for hedonic benefits and the tendency for unethical conduct with regard to the collection of big data.

Propositions 4b: Moral disengagement mediates the negative relationship between employees' motivation for hedonic benefits and the tendency for unethical conduct with regard to the storage of big data.

Propositions 4c: Moral disengagement mediates the negative relationship between employees' motivation for hedonic benefits and the tendency for unethical conduct with regard to the analysis of big data.

Propositions 4d: Moral disengagement mediates the negative relationship between employees' motivation for hedonic benefits and the tendency for unethical conduct with regard to the usage of big data.

Based on these propositions, our overarching theoretical framework is represented graphically in Fig. 1. While our model starts with the assumption that different types of motivation are relevant for ethical (mis)conduct in the context of big data, our framework puts the concept of moral disengagement at its core. We posit that moral disengagement is not only related to employees' tendency for ethical misconduct with regard to the collection, storage, analysis, and usage of big data. We also propose that moral disengagement mediates the relationship between extrinsic/intrinsic motives and ethical conduct.

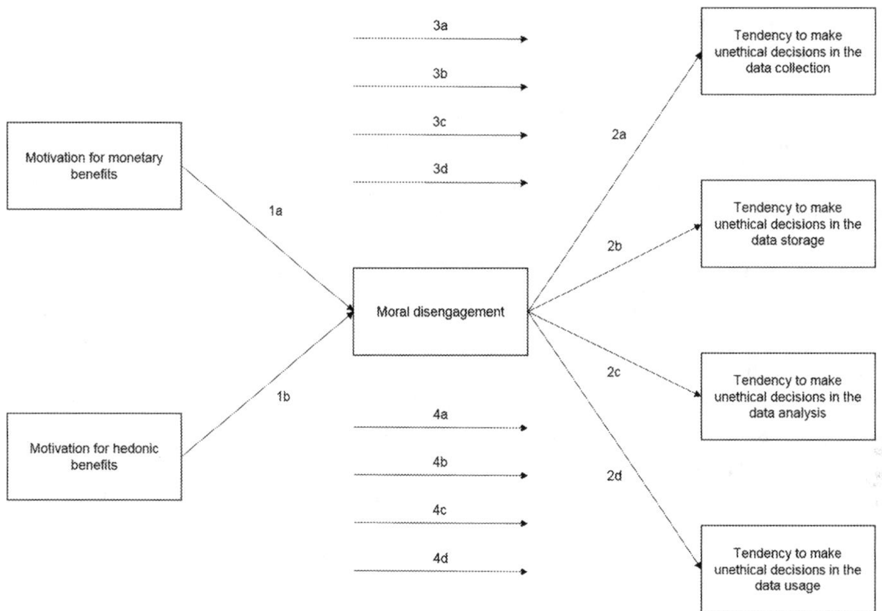

Fig. 1. Theoretical framework

4 Conclusions

As one facet of the mega trend of digitization, "big data" has received increasing attention by practitioners and academics alike. Early on, this debate acknowledged that big data does not only raise technological issues and questions about business use cases. Big data also invokes ethical questions. In fact, numerous guidelines, principles, and standards have emerged that seek to canonize an emerging consensus on how to ethically deal with big data.

While there is thus ample research on the normative implications of big data and on rules for ethical conduct, so far little is known about when and why decision-makers abide by these rules or opt for unethical behavior instead. The purpose of this paper was to address this research gap. To this end, we identified and discussed relevant factors that influence decision-makers' tendency for unethical conduct in the context of big data generated by social media. At the center of our theoretical framework stands the concept of moral disengagement. Moral disengagement occurs when decision-makers who perceive a certain behavior as unethical find ways to rationalize such behavior, thus disengaging themselves from unethical conduct and therewith from processes of self-sanctioning that would otherwise inhibit the unethical behavior.

To elaborate the role of moral disengagement, our framework derived four groups of propositions. First, we theorized that different types of motivation relate differently to moral disengagement. While extrinsic motives tend to be positively related to moral disengagement, we proposed a negative relationship for intrinsic motives. Second, we differentiated decision-making in the context of big data to fall into the four domains of big data collection, big data storage, big data analysis, and big data usage. We then proposed that moral disengagement is positively related to unethical conduct in each of these domains. For our third and fourth proposition sets, we propose that moral disengagement works as a mediator for the relationship between motives and ethical (mis)conduct.

Needless to say, our study is not without limitations. While we hold that motivations and moral disengagement play an essential critical role for ethical (mis)conduct), there are certainly other situational and personality factors that we have not explored despite their potential relevance. Further research is thus needed to expand our conceptual framework. We hope that our contribution may serve as a useful starting point in this regard. In terms of future empirical research, our framework builds upon testable propositions that can be used in further studies. As big data continues to play an ever bigger role in our lives, so will the question of when and why decision-makers choose to respect or violate principles for its ethical use. Moral disengagement research can help to illuminate this question.

References

1. Kitchin, R.: The real-time city? Big data and smart urbanism. GeoJournal **79**(1), 1–14 (2014)
2. IBE: Business Ethics Brief: Business Ethics and Big Data. Issue 52 (2016). https://www.ibe.org.uk/userassets/briefings/b52_bigdata.pdf. Accessed 8 Feb 2018
3. Richards, N.M., King, J.H.: Big data ethics. Wake For. Law Rev. **49**(2013), 393–432 (2014)
4. Zwitter, A.: Big data ethics. Big Data Soc. **1**(2), 1–6 (2014)
5. Accenture: Universal Principles of Data Ethics: 12 Guidelines for Developing Ethics Codes (2014). https://www.accenture.com/t20160629T012639Z__w__/us-en/_acnmedia/PDF-24/Accenture-Universal-Principles-Data-Ethics.pdf. Accessed 8 Feb 2018
6. ICO: Big Data, Artificial Intelligence, Machine Learning and Data Protection (2017). 20170904. https://ico.org.uk/media/for-organisations/documents/2013559/big-data-ai-ml-and-data-protection.pdf. Accessed 8 Feb 2018
7. Boyd, D., Crawford, K.: Critical questions for big data: provocations for a cultural, technological, and scholarly phenomenon. Inf. Commun. Soc. **15**(5), 662–679 (2012)
8. Dereli, T., Coşkun, Y., Kolker, E., Güner, O., Ağirbaşli, M., Özdemir, V.: Big data and ethics review for health systems research in LMICs: understanding risk, uncertainty and ignorance-and catching the black swans? Am. J. Bioeth. **14**(2), 48–50 (2014)
9. Office of the Australian Information Commissioner: Privacy Management Framework: Enabling Compliance and Encouraging Good Practice. Office of the Australian Information Commissioner Privacy (2018). https://www.oaic.gov.au/resources/agencies-and-organisations/guides/privacy-management-framework.pdf. Accessed 8 Feb 2018
10. Google Dictionary: Privacy (2018). https://www.google.co.uk/search?safe=active&client=firefox-b-ab&dcr=0&q=Dictionary. Accessed 8 Feb 2018
11. Bellman, S., Johnson, E.J., Kobrin, S.J., Lohse, G.L.: International differences in information privacy concerns: a global survey of consumers. Inf. Soc. **20**(5), 313–324 (2004)
12. Zook, M., Barocas, S., Boyd, D., Crawford, K., Keller, E., Gangadharan, S.P., Goodman, A., Hollander, R., Koenig, B., Metcalf, J., Narayanan, A., Nelson, A., Pasquale, F.: Ten simple rules for responsible big data research. PLoS Comput. Biol. **13**(3), 1–10 (2017)
13. Davis, K., Patterson, D.: Ethics in Big Data. O'Reilly Media Inc., Sebastopol (2012)

14. Mittelstadt, B.D., Floridi, L.: The ethics of big data: current and foreseeable issues in biomedical contexts. Sci. Eng. Ethics **22**(2), 303–341 (2016)
15. George, G., Haas, M.R., Pentland, A.: From the editors: big data and management. Acad. Manag. J. **57**(2), 321–326 (2014)
16. Rivers, C.M., Lewis, B.L.: Ethical research standards in a world of big data. F1000Research **3**, 38 (2014). Online First
17. Crawford, K., Schultz, J.: Big data and due process - toward a framework to redress predictive privacy harms. BCL Rev. **55**(1), 93–128 (2014)
18. Lyon, D.: Surveillance, snowden, and big data: capacities, consequences, critique. Big Data Soc. **1**(2), 1–13 (2014)
19. Chessell, M.: Ethics for Big Data and Analytics. IBM Corporation (2014). http://www.ibmbigdatahub.com/sites/default/files/whitepapers_reports_file/TCG%20Study%20Report%20-%20Ethics%20for%20BD%26A.pdf. Accessed 8 Feb 2018
20. Metcalf, J.: Ethics Codes: History, Context, and Challenges (2014). https://bdes.datasociety.net/council-output/ethics-codes-history-context-and-challenges/. Accessed 30 Oct 2017
21. Data Science Association: Data Science Code of Conduct (n.d.). http://www.datascienceassn.org/code-of-conduct.html. Accessed 15 Oct 2017
22. Bandura, A.: Social Foundations of Thought and Action. Prentice-Hall, Englewood Cliffs (1986)
23. Bandura, A.: Moral disengagement in the perpetration of inhumanities. Pers. Soc. Psychol. Rev. **3**(3), 193–209 (1999)
24. Detert, J.R., Treviño, L.K., Sweitzer, V.L.: Moral disengagement in ethical decision making: a study of antecedents and outcomes. J. Appl. Psychol. **93**(2), 374–391 (2008)
25. Bandura, A., Barbaranelli, C., Caprara, G.V., Pastorelli, C.: Mechanisms of moral disengagement in the exercise of moral agency. J. Pers. Soc. Psychol. **71**(2), 364–374 (1996)
26. Baron, R., Zhao, H., Miao, Q.: Personal motives, moral disengagement, and unethical decisions by entrepreneurs: cognitive mechanisms on the "slippery slope". J. Bus. Ethics **128**(1), 1–12 (2014)
27. Scheiner, C.W., Baccarella, C., Bessant, K., Voigt, K.-I.: Participation motives, moral disengagement, and unethical behaviour in idea competitions. Int. J. Innov. Manag. **22**(4), 1850043-1–1850043-24 (2018)
28. Amabile, T.M.: Motivational synergy: toward new conceptualizations of intrinsic and extrinsic motivation in the workplace. Hum. Resour. Manag. Rev. **3**(3), 185–201 (1993)
29. Litzky, B.E., Eddleston, K.A., Kidder, D.L.: How managers inadvertently encourage deviant behaviors. Acad. Manag. Perspect. **20**(1), 91–1036 (2006)
30. Moore, C.: Moral disengagement in processes of organizational corruption. J. Bus. Ethics **80**(1), 129–139 (2008)
31. Shepherd, D.A., Baron, R.A.: "I Care About Nature, But…": disengaging values in assessing opportunities that cause harm. Acad. Manag. J. **56**(5), 1251–1273 (2013)
32. Utmann, C.: Performance effects of motivational state: a meta-analysis. Pers. Soc. Psychol. Rev. **1**(2), 170–182 (1997)
33. Zhang, X., Barthol, K.: Linking empowering leadership and employee creativity: the influence of psychological empowerment, intrinsic motivation, and creative process engagement. Acad. Manag. J. **53**(1), 107–128 (2010)
34. Dewett, T.: Linking intrinsic motivation, risk taking, and employee creativity in an R&D environment. R&D Manag. **37**(3), 197–208 (2007)
35. Carsrud, A., Brännback, M.: Entrepreneurial motivations: what do we still need to know. J. Small Bus. Manag. **49**(1), 9–26 (2011)

36. Council of Europe: Guidelines on the Protection of Individuals with Regard to the Processing of Personal Data in a World of Big Data (2017). https://rm.coe.int/16806ebe7a. Accessed 13 Feb 2018
37. Information and Privacy Commissioner of Ontario: Big Data - Guidelines. Hg. v. Information and Privacy Commissioner of Ontario (2017). https://www.ipc.on.ca/wp-content/uploads/2017/05/bigdata-guidelines.pdf. Accessed 8 Feb 2018
38. van Rijmenam, M.: Big Data Ethics: 4 Guidelines to Follow by Organisations. Datafloq B.V. o.O. (2017). https://datafloq.com/read/big-data-ethics-4-principles-follow-organisations/221. Accessed 17 May 2017
39. U.S. Department of Health, Education and Welfare, Secretary's Advisory Committee on Automated Personal Data Systems, Records, computers, and the Rights of Citizens viii: The Code of Fair Information Practices, cited from Electronic Privacy Information Center (EPIC): The Code of Fair Information Practices. Hg. v. epic.org (1973).https://epic.org/privacy/consumer/code_fair_info.html. Accessed 8 Feb 2018
40. King, S.: Big Data: Potential und Barrieren der Nutzung im Unternehmenskontext. Springer, Wiesbaden (2014). https://doi.org/10.1007/978-3-658-06586-7
41. Trautman, L.J., Ormerod, P.: corporate directors' and officers' cybersecurity standard of care: the Yahoo data breach. 66 Am. Univ. Law Rev. **1231** (2017). https://ssrn.com/abstract=2883607 or https://doi.org/10.2139/ssrn.2883607. Accessed 13 Feb 2018
42. IAF: Big Data Assessment Framework and Worksheet. IAF Big Data Ethics Initiative (2015)
43. Toole, J.L., Eagle, N., Plotkin, J.B.: Spatiotemporal correlations in criminal offense records. ACM Trans. Intell. Syst. Technol. **2**(4), 38:1–38:18 (2011)
44. Hilbert, M.: Big data for development: a review of promises and challenges. Dev. Policy Rev. **34**(1), 135–174 (2016)
45. IBM News: IBM Ushers In Era Of Stream Computing. Press release, 13 May. IBM, Armonk (2009). https://www.ibm.com/blogs/research/2009/05/ibm-ushers-in-era-of-stream-computing/. Accessed 18 Jan 2018
46. GFDRR (Global Facility for Disaster Reduction and Recovery): Open Data for Resilience Initiative (OpenDRI). Global Facility for Disaster Reduction and Recovery. World Bank, Washington, DC (2012). https://www.gfdrr.org/sites/default/files/documents/3%20OpenDRI.pdf. Accessed 8 Feb 2018
47. Metcalf, J., Crawford, K.: Where are human subjects in big data research? The emerging ethics divide. Big Data Soc. **3**(1), 1–14 (2016)
48. Danna, A., Gandy, O.H.: All that glitters is not gold: digging beneath the surface of data mining. J. Bus. Ethics **40**(4), 373–386 (2012)
49. Johnson, J.A.: The ethics of big data in higher education. IRIE Int. Rev. Inf. Ethics **4**, 3–10 (2014)
50. Parry, M.: College Degrees, Designed by the Numbers, The Chronicles of Higher Education. https://chronicle.com/article/College-Degrees-Designed-by/132945/. Accessed 8 Feb 2018

Privacy Protecting Fitness Trackers: An Oxymoron or Soon to Be Reality?

Kaja J. Fietkiewicz(✉) and Maria Henkel

Department of Information Science, Heinrich Heine University Düsseldorf, Düsseldorf, Germany
{kaja.fietkiewicz,maria.henkel}@hhu.de

Abstract. The rapid technological advancements are supposed to simplify our everyday life. They are also increasingly utilized to support an active lifestyle with diverse tracking devices, like fitness trackers or smart watches. However, they do not seem to make the life of legislators and data privacy advocates easier. In contrary, with better and faster technology our (health-related) private data faces more and more threats. To better understand the current status of the intersecting domains of devices like fitness trackers and the data privacy, we have analyzed the development of general data privacy regulations in the EU as well as the data transfer modalities between EU and USA. Afterwards, we reviewed scientific publications on fitness trackers (or smart watches) and data privacy, in order to identify, whether there is interest in this topic among scholars and if so, which aspects do they investigate in particular.

Keywords: Fitness trackers · Data privacy · GDPR · Privacy Shield

1 Introduction

New technological advancements like smart and wearable devices or the Internet of Things (IoT) simplify not only our everyday life, but also “the tracking and logging of data” in order to support an active lifestyle [17]. Such fitness trackers are getting smaller and more affordable [17], while offering more and more options to track our health and activity. This is possible due to the economies of scale that drastically reduced the costs of production, whereas “concurrent advances in technology have expanded their physiological recording capabilities” [22]. In turn, they are also increasingly employed in medical field [4].

However, some of the (prospective) users are having privacy concerns and “sensitivity regarding data gathered with wearables” [17]. One could say that “personal information has never been this prone to risk given the current advancement in technologies especially in personal devices” that collect vast amounts of data, which in turn could be used to “infer sensitive personal information” [30]. Before this new technology became an integral part of many people’s lives, personal health-related information was exclusively stored in hospitals or health care provider’s systems [16]. One could argue that the information stored in a fitness tracker is even more thorough. The devices can meticulously record the number of steps we took, the geo-locations of where we did it, the calories we burned during this activity and how well we slept

© Springer International Publishing AG, part of Springer Nature 2018
G. Meiselwitz (Ed.): SCSM 2018, LNCS 10913, pp. 431–444, 2018.
https://doi.org/10.1007/978-3-319-91521-0_31

afterwards. The number of potential ways of utilizing all the data is rising with its amount and diversity.

The "problem" with privacy and data security is not new and is becoming more and more urgent with increasing digitalization. It is especially present in the context of the web and social media. One way to counteract or at least regulate the handling of personal data is an appropriate legislation [12]. Of course, in times of digitalization and globalization it is not enough to regulate data privacy solely in one's own country. Transnational corporations are active in many parts of the world and not every country can necessarily ensure an appropriate consumer protection. For example, smart watches or fitness trackers by Apple or Fitbit are very popular on the European market; however, their headquarters are located in the USA. How is the transitional data exchange regulated?

On May 25th, 2018, the General Data Protection Regulation (GDRP) will be implemented and might improve the current status of data security in Europe. It is intended to unify the data protection within the European Union and make it stronger as compared to the former data protection directive from almost 25 years ago. The regulation will be enforceable after two-year transition period, directly binding and applicable. The applying lex loci solutions ("law of the place of performance") means that even though the new regulation is applicable within the EU, it will also concern non-European companies, as long as their services or goods are being supplied on the European market. Ergo, it will also concern non-European fitness trackers' producers.

The increasing interest in data privacy can be recognized not only in the legal environment but in the scientific research as well. A search in the Scopus database (for peer-reviewed literature) for publications on data privacy and the Internet in general (Fig. 1) shows increasing number of publications on this topic, with a quite significant increase since 2013. Which aspects of data privacy in the context of fitness

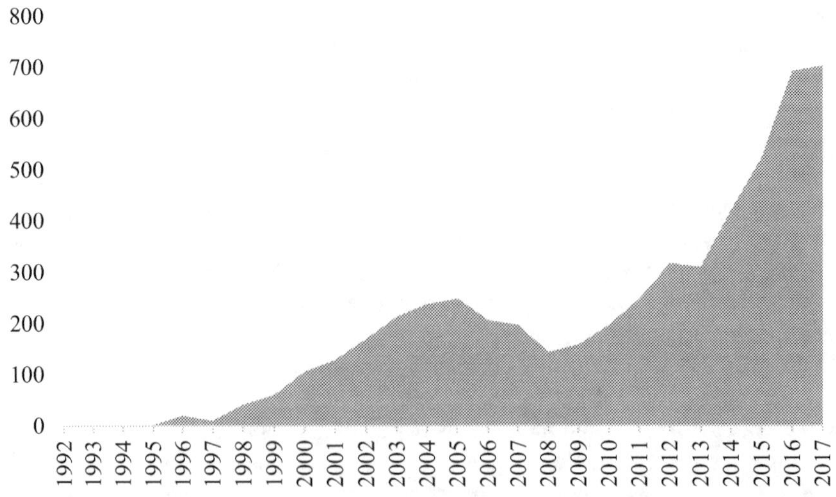

Fig. 1. Number of publications on privacy and the Internet indexed by Scopus.

trackers are the researchers interested in? What methods do they use? And, do they refer to legal sources?

This theoretical study (Fig. 2) is supposed to shed light on the status of (health-related) data privacy regulations, which also increasingly concern the manufacturers and service providers of fitness trackers. Hence, the first research question is (RQ1a): What is the legal status quo of data privacy in European Union with focus on fitness trackers? Also, since many manufacturers and services providers are located in the USA, the following question arises (RQ1b): How is the data transfer between EU and USA regulated? Finally, we want to take a look at the research trends on this particular topic and therefore formulate the final research question (RQ2): What is the state of scientific research on data privacy and fitness trackers?

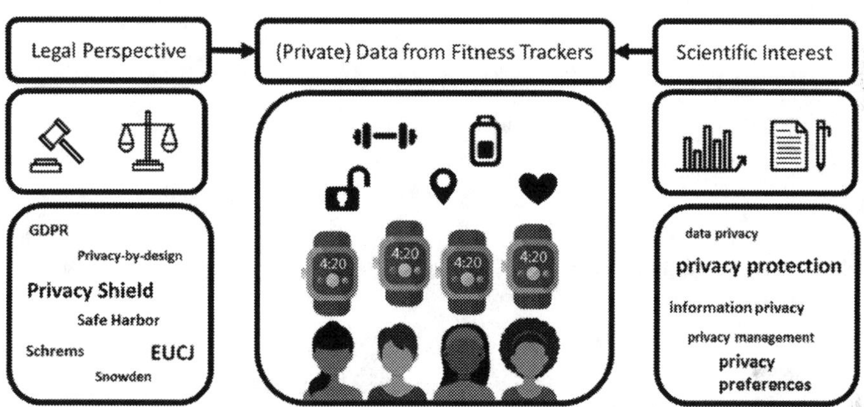

Fig. 2. Scope of our theoretical research.

2 Methods

The research procedure for the first part of this paper, the legal perspective, included literature and internet research. The basis for the following discourse is composed of US-American and EU legal regulations, reports and press releases by authorities, scientific articles (focused on law and economy), and news articles by renowned news outlets.

The second part of this work includes a review of scientific literature on data privacy and fitness trackers. To identify, analyze and synthesize relevant research in this particular field, a structured literature review was conducted. Therefore, we looked for publications focusing on both, data privacy, security or protection and fitness trackers or smart watches (Fig. 3). As shown in Fig. 3, these topics have only been combined in scientific research since about 2015. There is much more scientific interest in the intersection of social media and data privacy, as well as the new General Data Protection Regulation itself (not related to fitness trackers or similar devices). Therefore, it is no surprise that our search in the two scientific databases "Web of Science" and "Scopus" yielded a total of 23 results as of February 2018.

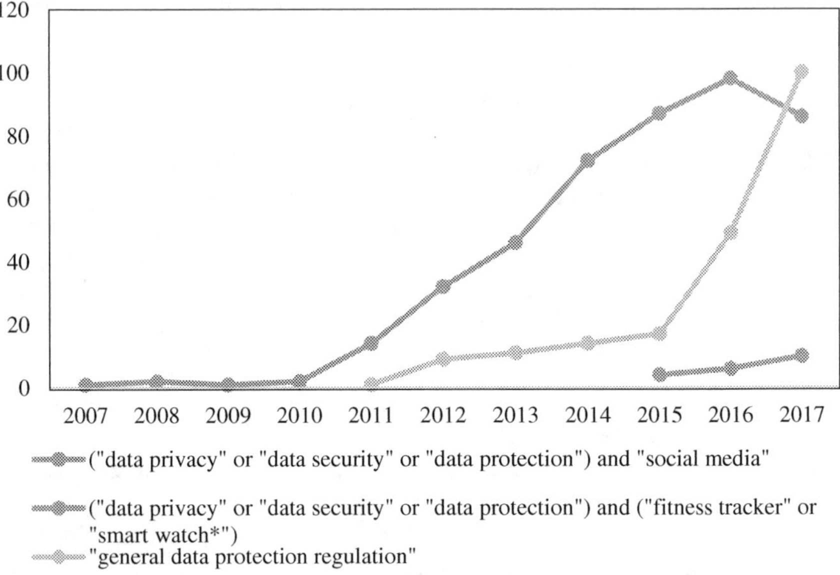

Fig. 3. Number of publications per year on privacy and social media, privacy and fitness trackers or smart watches, and on the GDPR, indexed by Scopus.

Articles included for our review had to be directly relevant to the topic and peer reviewed. We included results regardless of the age of the material, country of origin or language. We did not limit our search to theoretical, qualitative, or quantitative research as the sample was small to begin with. We excluded, however, four articles, because they were deemed irrelevant for our research question, due to either focusing on another, very specialized topic or using one of the keywords as a negative keyword, hence, expressly not talking about it. Nine articles were of a technical nature, documenting or discussing the development of a system or technical solution for data privacy in wearable technology and were excluded as well. The remaining ten articles, eight in the English, one in the German and one in the Turkish language, were analyzed regarding theories, methods and results concerning privacy and privacy protection of health data generated by wearables.

3 Results

3.1 Legal Perspective

The technological development "makes it possible for companies to collect, process and interlink data in an expanded way. They increasingly tend to use these data for various purposes, such a personalized services and marketing. As a result of technological development, along with globalization, new and increased challenges for personal data protection laws emerged" [21, 28]. The increasing privacy risks may in turn decrease people's trust in companies that collect data for their services and this "lack of trust can slow down the development of the innovative use and adoption of new technologies" [21, 28]. Especially when such sensitive data like health information is involved, the new technology brings as many possibilities as it does bring fear about one's most intimate sphere. Great advances in Big Data technology facilitate development of personal health management, health care delivery, health-related research and population health surveillance. Until now, the legal system was lagging way behind these technological and commercial developments [18], whether we look at the countries with common or with the civil law traditions. Most of the privacy protection regulations for (health-related) data "were drafted in the twentieth century for technology available at that time (…) and are outdated in the era of Big Data" (e.g. Data Protection Directive 95/46/EC; Directive 2002/58/EC; Data Protection Act 1998) [18]. However, there is still hope that the privacy and other "fundamental rights of data subjects" can be safeguarded [18, p. 38] and many voices in the literature and in the politics see the new European General Data Protection Regulation (GDPR) as the game changer.

The focus of this research paper is set on the so-called fitness trackers or similar wearable technology that enables monitoring of physical activity, sleep pattern, heart rate etc. This data does not strictly fall into "health information" collected in the medical field, but still with its increasing spectrum (including geolocation, name, IP address, email, phone number, social network, etc.) one can create a quite accurate image of one (quantified) self. Therefore, the concerns about the data privacy, personality rights and the (imminent) danger of mass surveillance might be justified.

Legal Concerns Regarding Fitness Trackers. In 2016, Norway's Consumer Council (NCC) accused Fitbit (USA), Jawbone (USA), Garmin (Switzerland) and Mio Technology (Taiwan) of braking local laws governing the handling of consumer data [2, 33]. Even though Norway is not an EU Member State, it needs to implement some of the European directives, including the Data Protection Directive from 1995. This means that the potential data privacy violations concern, at least from the legal perspective, the whole European economic zone. According to NCC, the companies gathered too many data, did not disclose how many third parties have access to it or how long it will be kept. In general, "anyone who used them [fitness trackers] gave up data on asymmetrical and obscure terms" [2]. This way the basic privacy principles are being neglected and the accumulated information can be "exploited for direct marketing and price-discrimination purposes" [2].

The complaint was based on NCC's analysis including an examination of the functionality of the trackers, the terms and conditions, privacy policies and the degree of control provided to users over the data collected [33]. Further allegations included the lacking provision of the users with proper notice about changes in terms and conditions or insufficient explanation of how data, including sensitive personal data such as heart rate, is collected and shared with third parties [33]. In general, since this type of technology is still evolving, the NCC advises incorporating consumer-protective measures in the product design as a standard in order to enhance consumers' trust [33]. This "privacy by design" will be inevitable for companies targeting European market anyway, when the General Data Protection Regulation is in force.

The concerns about privacy and personality rights relate not only to private consumers but increasingly to the corporate environment as well. The new trend for corporate wellness or corporate health management (aiming at improved employee health and lower medical insurance premiums) could be very lucrative for fitness tracking manufacturers and service providers. However, with the new regulation in sight their business model could face some obstacles. According to the EU advisory panel, employers should not be allowed to issue workers with fitness trackers or similar monitoring devices and should "be barred from accessing data from their devices their employees wear" [14]. For the authority, even a transparency regarding the usage of the data and the possibility of opting out of any data sharing are not sufficient, since "given the unequal relationship between employers and employees, (…) workers were probably never able to give legally valid consent to have their data shared" [14]. According to the new GDPR, for any kind of employee tracking, the businesses should select the most data privacy friendly solutions available [14]. Time will tell, which of the fitness tracker providers (if any) will be the chosen one.

In 2016, the German Federal Commissioner for Data Protection and Freedom of the Information also shared some concerns about the personal data while using fitness trackers [3]. Again criticized were the terms and conditions of the manufacturers and service providers for their form and vagueness, as well as the fact that, to some extent, the data is being shared with third parties (for marketing or research purposes) and its faith does not really remain in consumers' control anymore. Finally, the consumer often does not have the possibility to autonomously erase all the accumulated data linked to his or her account. The authority also sees the new GDPR as future solution for all these concerns.

Most of the popular fitness tracker manufacturers are based in non-EU countries. Therefore, another critical point in the debate on data privacy is the data transfer outside the European Union, for example in the USA (hosting headquarters for many of the big market players). When supplying the EU-market, companies need to comply with European data protection regulations. When transferring data from EU, it must be ensured that it will be equally "protected" at the new destination. In the following, a short comparison of data privacy principles in the USA and EU will be presented to point out that such transfer, given the status quo of data protection legislature, is not unproblematic.

Data Privacy Regulations in the USA and the EU. Terry [26] argues that the current developments in consumer electronics including wearable devices are "disrupting healthcare data markets by encouraging consumers to themselves collect and curate data," which in turn reveals the shortcomings of provided healthcare data protection and, especially, the flaws of domain-limited data protection that is prevalent in the USA. This is one of the biggest differences between the US and the European data protection regulations that are not limited to one specific domain. The data privacy laws can be compared regarding three aspects: the horizontal reach (public and private domains that are being regulated), vertical attributes (what data custodian behaviors they regulated), and their enforcement (investigation and penalties) [26].

When considering the European General Data Protection Regulation, it has a very broad horizontal and vertical applicability. As for the horizontal reach, it concerns all sectors of the economy (not only, e.g. the health-related domain) and all "personal data" as well as all stakeholders controlling or processing it. As for the vertical attributes, the "Fair Information Practice Principles-like protective standards [apply] throughout the lifespan of data" [26]. Terry describes two phases of possible interaction with data—the "upstream" (when the data is being collected) and the "downstream" (the subsequent data processing and/or disclosure). The GDPR aims at protecting the personal data during both phases. The data collection (upstream) needs to be limited to a legitimate purpose and as minimized as possible. The data processing (downstream) needs to be fair, lawful, transparent, and it should follow certain storage, quality, security, and integrity as well as confidentiality limitations [26].

In comparison, most of these data protection principles are absent in the US laws, starting with a quite limited horizontal protection [26] (sector-by-sector basis regulation, with different statues for the public and private sector) [24, 28]. Furthermore, it is also very limited in its vertical reach, since most of the regulations only utilize downstream protection (hence, regulate what happens with the data after it was collected), such as confidentiality security and breach notification [26]. Terry names HIPAA (Health Insurance Portability and Accountability Act of 1996) as one of the typical US data protection laws. HIPAA is domain-specific, the domain being defined by the healthcare data custodians (the health insurers and health providers) and not by the data type (e.g., healthcare data) and provides only downstream protection [26]. This leads to a quite big gap in the protection of health-related (personal) data, since the regulation does not apply to most of the healthcare data controlled or processed by entities outside the traditional healthcare environment [26].

The need to close this data protection gap can only become more urgent, when we consider the current trends in the health/lifestyle sector. As for 2016, approx. 200,000 mobile health apps were available for smartphones, of which a not insignificant part interacts with wearables [26, 27]. However, relatively few of these products are supplied by "traditional healthcare providers" so that the data will not be protected by HIPAA's privacy rules [25, 26]. That is why the data privacy regulations in the US are not as comprehensive as they are in EU. Therefore, the question arises, how is the data transfer between USA and EU regulated? And, does it provide adequate protection? Next, a short history of trans-Atlantic agreements for data transfer and some data privacy disputes, which helped shape the GDPR, will be presented.

A Quarter Century of Data Protection Faux Pas. In May 2018 the General Data Protection Regulation will come into effect and after almost 25 years replace the Data Protection Directive. In contrast to the directive from 1995, the new regulation is immediately applicable and enforceable in every EU Member State. An EU directive only sets certain requirements and goals that need to be implemented by Member States in their legislature. With the new regulation, the data controllers and processors will be "required to emphasize transparency, security and accountability, while (…) standardizing and strengthening the right of European citizens to data privacy" [19].

The European Commission made the initial proposal of the new regulation in January 2012 [8]. “A critical observer might note that the ideas behind the Regulation and the Directive go back to 2012 and that already all circumstances within which they were drafted have in the meantime changed substantially” [7]. However, during this time the European Court of Justice ruled in several cases leading to fundamental decisions within data privacy case law (e.g., right to be forgotten, extraterritoriality, international data transfer) [7] that pointed out important data security and privacy issues and helped shape the new GDPR (Fig. 4).

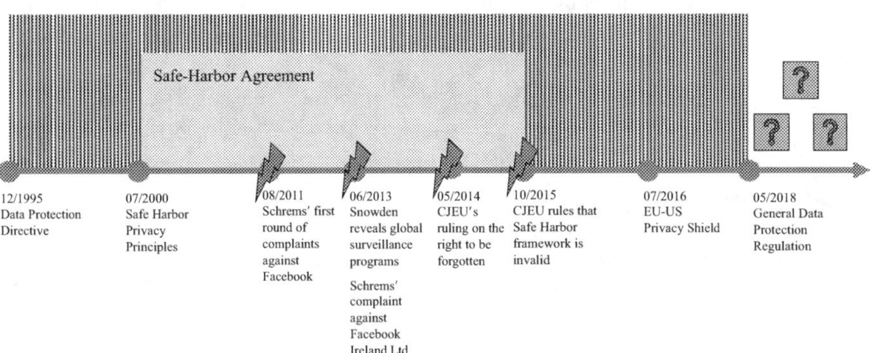

Fig. 4. Data Privacy regulations and selected disputes in the EU since 1995.

Some of the turning points were Edward Snowden’s disclosure of the large-scale espionage by NSA, also targeting European personal data, or Max Schrems’ campaign against Facebook, which lead to more questions about handling of European personal data by Apple, Skype, Microsoft and Yahoo! [15]. After the European Court of Justice ruled in Schrems’ favor, the international Safe Harbor privacy principles (agreed upon by European Commission and the US authorities) regulating data exchanges between Europe and the US, were overturned by the European Court of Justice [15]. Apparently the agreement enabled US public authorities’ interferences with the fundamental rights of persons by accessing their data [28].

The US-EU Safe Harbor program was developed in the year 2000 in order to bridge the “differences between the US and the EU data protection approaches and to provide US organizations with streamlined means to comply with” Data Protection Directive from 1995 [28]. In 2016, less than one year after the Safe Harbor agreement was overruled, the European Commission and the US Government agreed on a new framework for data exchange, the EU-US Privacy Shield. From the beginning it was challenged by civil rights organizations and privacy groups. The Privacy Shield framework includes updates of the former Safe Harbor framework to fulfill the requirements set by the CJEU’s ruling [28]. Since then, USA is within the countries recognized by the European Commission as providing “adequate” protection for personal data (limited to the Privacy Shield framework) [10, 31]. Other countries that the European Commission has so far recognized are Andorra, Argentina, Canada (commercial organizations), Faroe Islands, Guernsey, Israel, Isle of Man, Jersey, New Zealand, Switzerland, and Uruguay [10].

The GDPR requires the European Commission to regularly review its adequacy decisions. This is one of the improvements implemented due to the Schrems' case. Until Schrems' action, the Safe Harbor agreement "had never been subject to an actual review by the Commission" [31], adding up to 15 years of insufficient data transfer regulation being in force. The European Commission evaluated its adequacy decision approximately one year after the agreement was reached in an annual report and, as for October 2017, it confirmed the adequacy of the EU-US Privacy Shield [9]: "(…) the Commission concludes that the United States continues to ensure an adequate level of protection for personal data transferred under the Privacy Shield from the Union to organizations in the United States."

As for the workings of the Privacy Shield, the decision by US-based companies to join the program is entirely voluntary and leads to their public commitment to comply with the Privacy Shield Principles through (annual) self-certification (enforceable under US law) [20], which is practically the same procedure as for the Safe Harbor. Since 2016, over 2,000 companies joined the Privacy Shield program through self-certification (including, for example, the fitness tracker manufacturer Fitbit Inc.).

Still, the faith of this program remains uncertain as several actions against European Commission's decision (about the Privacy Shield) had been brought to the European Court of Justice [5, 6]. Even though the new GDPR seems to improve the data privacy situation, especially by including such upgrades as "privacy by design" or "right to be forgotten," the EU-US Privacy Shield agreement raises some questions about GDPR's adequate enforcement, e.g., when data is being transferred in the USA. The Members of European Parliament also expressed concerns about the agreement, especially after "new rules allowing the US National Security Agency (NSA) to share private data with other US agencies without court oversight [or] recent revelations about surveillance activities by a US electronic communications service provider" came to light [11]. The Parliament acknowledges "the significant improvements made compared to the former EU-US Safe Harbor, but there are clearly deficiencies that remain to be urgently resolved to provide legal certainty for the citizens and businesses that depend on this agreement" [11].

As we can see in Fig. 3, the future after the new GDPR is in force remains uncertain. Even though the regulation has the potential to significantly improve the European data protection, including health-related and personal data accumulated with wearable tracking devices, the regulation of trans-Atlantic data transfer is still raising many concerns. The question is whether the few improvements and a new name make it just a wolf in sheep's clothing, or an actual "adequate" solution. With actions against the agreement [5, 6], the concerns will be hopefully resolved by EUCJ's ruling. In the following, the outcomes of the literature review on studies concerning data privacy and fitness trackers will be summarized.

3.2 Current Research on Data Privacy and Fitness Trackers

Firstly, it should be said that only one paper [32] explicitly mentions the GDPR. They clearly state, that "[i]n most countries, laws that govern the collection, storage, analysis, processing, reuse, and sharing of data (…) fail to adequately address the privacy challenges associated with human tagging technologies" because they were "enacted decades ago" [32]. They mention the new regulation in positive light. Ghazinour et al. [13] refer to the HIPAA regulating the use of health-related data in the USA, however, as already described in our legal part of the study, this regulation only addresses medical institutions and is not applicable for wearables. Altpeter [1] mentions the E-Health Law in Germany, which regulates the data privacy in medical sector (therefore, as for its applicability, it is comparable to the HIPAA). All reviewed publications, however, are concerned about data privacy regarding the use of fitness trackers, smart watches or other wearable technology with biodata tracking functions.

Rosenbaum et al. [23] try to assess the current situation and potential future developments, benefits and risks in retail marketing. They remark that "individualized 'data mining' enables delivery of personalized product recommendations and offerings," [23] but also "may disrupt the traditional view of consumer consent" [23] to this new kind of data collection. While activity tracking surely could have many benefits for retail marketing, the authors also recognize risk, apart from health-related data breaches, identity fraud or harassment, in misinterpreting health-related information and finally endangering the costumer due to false product recommendations. They state that "consumer-oriented nutrigenomics currently does not fit neatly into existing legal categories" [23] and encourage further research into the "dark side" of these new technologies before they are utilized.

Bostanci [4] identifies malware, breach of privacy, for example when handling data in medical facilities, connection dependency, efficient data processing, and incompatibility of analysis tools and systems as ethical and technological threats and challenges for the future of wearable technologies. Meanwhile, Altpeter [1] also mentions the emotions that consumers and practitioners, who do not want their patrons to lose their trust in them, might have in the e-health sector. He emphasizes that the fear of security gaps should not hold advances of a digital health system and its advantages back.

Ghazinour et al. [13] criticize the "current binary standard" for data collection as it "leaves the user no options on selecting their privacy preferences on their data and if they do not agree to the terms, they cannot use the device" [13]. They propose a model that lets users decide about the privacy preferences for every data item.

Torre et al. [29] extend this issue by thematizing the problem of inference attacks by third parties which are granted access to health and activity data by the user. They present their idea of connecting an "Adaptive Inference Discovery Service" with personal data management functionalities to respect and take into account "the individuals' perception of privacy" [29]. In the next step of their study [30] they apply this framework in a case study with data from 49 users and predict different aspects such as weight, steps, gender and smoking with an accuracy of 50.2–99.9%. Hereby, Torre et al. show how users could be assisted in deciding which privacy settings are optimal to reduce inference risk. Of course, first of all, users need to be made aware of the risk of inference when allowing third parties to use their sensitive health and activity data.

Finally, there are also user studies which try to give insight on the opinion, perception and behavior of the users themselves. After an online survey, focus group interviews and an in-depth interview with 12 users, Yoon et al. [34] reported that power-users had fewer concerns regarding privacy ("unnecessary anxiety") than non-power-users ("vague fear") – contrary to their expectations and previous findings [34, p. 545].

Lehto and Lehto [16] asked ten participants of qualitative interviews about the sensitivity of their health data and their willingness to share data with different parties. They found that "information collected with wearable devices is not perceived as sensitive or private" while "health information stored in patient medical records is considered to be very sensitive and private" [16]. Therefore, almost all interviewees did not want to share their data with social media (9/10) but were willing to share it freely with the doctor or medical research (10/10). Eight of ten participants would share it with occupational health services and seven with the device manufacturer. Lehto and Lehto [16] conclude that handling of tracked data "needs to be described clearly and transparently to mitigate any privacy concerns from the individuals" [16] and that "[d]evice makers need to consider how and when location data is being collected as this causes many privacy concerns that can impact use and adoption of these devices" [16].

In another attempt to understand the privacy concerns of fitness tracker users, Lidynia et al. [17] conducted an online survey (n = 82). Participants preferred to keep logged data to themselves and not on external servers—sharing activity data online was not favored either. Lidynia et al. [17] admit, however, that their sample is rather small and participants were relatively young. They recommend applying their methods, the privacy paradox and the privacy calculus to a bigger and more representative sample.

4 Discussion

The legal perspective on the data privacy showed that this is an increasingly important topic, especially when such devises like fitness trackers collecting not only general personal data, but more and more health-related information, are concerned. With the new General Data Protection Regulation the European data privacy environment is changing for the better. However, is the "new" EU-US Privacy Shield agreement keeping up with this improvement? Or does it perpetuate old issues under a new name? The level and range of data privacy regulations in USA are hardly comparable to the ones in European Union. The increasing involvement of private persons in disputes about (their) personal data as well as the assistance of national data privacy authorities is somewhat reassuring that inadequate regulations violating the fundamental rights of EU citizens will be under fire. Hopefully, the decision making and emendation of these regulations will occur more quickly than it was common until now. The legislative process and formally correct execution of legal procedures take time; however, the time is running up much faster when new technologies are involved.

When compared to data privacy authorities and legislators, there is only a slight interest in data privacy and fitness trackers or similar wearables among scholars. The research on this particular topic seems to be increasing; however, it is still nascent. Very few studies address legal regulations and only one refers to the GDPR. Most of the studies are rather theoretical, defining data privacy frameworks or summarizing benefits or challenges of wearable devices. Four of the reviewed studies were more user-oriented (case study, online survey, qualitative interviews). But still, the rather small sample sizes of user-oriented investigations and quite general studies otherwise indicate that this is an early stage of research within this domain.

5 Limitations and Future Research

For the future research on the legal perspective, we would recommend a more detailed analysis of current disputes between data privacy authorities and the European Commission (regarding the EU-US Privacy Shield) or fitness tracking manufacturers/service providers (regarding violations of data privacy regulations). In this study we only focused on the trans-Atlantic data transfer and respective agreements between EU and USA. An investigation of further bilateral agreements and data transfers as well as data privacy situation in, for example, China would be an interesting aspect to investigate in the future.

Regarding the scientific research on fitness trackers and data privacy, there appear to remain many gaps that could be closed in the future. Firstly, more reference to the legal situation would be beneficial and relevant for practice. Secondly, a more extensive user-oriented research going beyond users' privacy preferences would give scholars and practitioners more relevant insights. In this respect, such aspects as users' knowledge (or lack of it) about what happens with their data (and their respective attitudes toward it), or knowledge about what (data privacy) rights are actually due to them, would be interesting.

References

1. Altpeter, B.: E-health as a component of holistic therapy optimization. (E-Health als Bestandteil ganzheitlicher Therapieoptimierung), Diabetologe **13**(1), 29–37 (2017)
2. BBC: Privacy complaint for fitness wristband makers. http://www.bbc.com. Accessed 20 Feb 2018
3. BFDI: Datenschutz bei Gesundheits-Apps und Wearables mangelhaft. https://www.bfdi.bund.de. Accessed 20 Feb 2018
4. Bostanci, E.: Medical wearable technologies: applications, problems and solutions. In: 2015 Medical Technologies National Conference, pp. 50–53 (2015)
5. Curia: Case T-670/16, Digital Rights Ireland v. European Commission. http://curia.europa.eu. Accessed 22 Feb 2018
6. Curia: Case T-738/16, La Quadrature du Net and Others v. European Commission. http://curia.europa.eu. Accessed 22 Feb 2018
7. De Hert, P., Papakonstantinou, V.: The new general data protection regulation: Still a sound system for the protection of individuals? Comput. Law Secur. Rev. **32**, 170–194 (2016)
8. EC: Commission proposes a comprehensive reform of data protection rules to increase users' control of their data and to cut costs for businesses. http://europa.eu/rapid/press-release_IP-12-46_en.htm. Accessed 21 Feb 2018

9. EC: Report from the commission to the European Parliament and the Council on the first annual review of the functioning of the EU–U.S. Privacy Shield. https://ec.europa.eu. Accessed 21 Feb 2018
10. EC: Adequacy of the protection of personal data in non-EU countries. https://ec.europa.eu. Accessed 22 Feb 2018
11. European Parliament: Data Privacy Shield: MEPs alarmed at undermining of privacy safeguards in the US. http://www.europarl.europa.eu. Accessed 22 Feb 2018
12. Fietkiewicz, K.J., Lins, E.: New media and new territories for european law: competition in the market for social networking services. In: Knautz, K., Baran, K.S. (eds.) Facets of Facebook: Use and Users, pp. 285–324. De Gruyter Saur, Berlin, Germany, Boston, MA (2016)
13. Ghazinour, K., Shirima, E., Parne, V.R., Bhoomreddy, A.: A model to protect sharing sensitive information in smart watches. Procedia Comput. Sci. **113**, 105–112 (2017)
14. Kahn, J.: Fitness tracking startups are sweating due to EU privacy regulators. privacy regulators worry companies could abuse access to data, https://www.bloomberg.com. Accessed 20 Feb 2018
15. Krystlik, J.: With GDPR, preparation is everything. Comput. Fraud Secur. **7**, 5–8 (2017)
16. Lehto, M., Lehto, M.: Health information privacy of activity trackers. In: European Conference on Information Warfare and Security, ECWS, pp. 243–251 (2017)
17. Lidynia, C., Brauner, P., Ziefle, M.: A step in the right direction – understanding privacy concerns and perceived sensitivity of fitness trackers. Adv. Intell. Syst. Comput. **608**, 42–53 (2018)
18. Mendelson, D., Mendelson, D.: Legal protections for personal health information on the age of Big Data – a proposal for regulatory framework. Ethics Med. Public Health **3**, 37–55 (2017)
19. O'Connor, Y., Rowan, W., Lynch, L., Heavin, C.: Privacy by design: informed consent and internet of things for smart health. Procedia Comput. Sci. **113**, 653–658 (2017)
20. Privacy Shield Framework. https://www.privacyshield.gov/article?id=How-to-Join-Privacy-Shield-part-1. Accessed 21 Feb 2018
21. Reding, V.: The upcoming data protection reform for the european union. Int. Data Priv. Law **1**(1), 3–5 (2010)
22. Reinerman-Jones, L., Harris, J., Watson, A.: Considerations for using fitness trackers in psychophysiology research. In: Yamamoto, S. (ed.) HIMI 2017. LNCS, vol. 10273, pp. 598–606. Springer, Cham (2017). https://doi.org/10.1007/978-3-319-58521-5_47
23. Rosenbaum, M.S., Ramírez, G.C., Edwards, K., Kim, J., Campbell, J.M., Bickle, M.C.: The digitization of health care retailing. J. Res. Interact. Mark. **11**(4), 432–446 (2017)
24. Schwartz, P.M.: The EU-U.S. privacy collision: a turn to institutions and procedures. Harvard Law Rev. **126**(7), 1966–2009 (2013)
25. Terry, N.: Mobile health: assessing the barriers. Chest **147**(5), 1429–1434 (2015)
26. Terry, N.: Existential challenges for healthcare data protection in the United States. Ethics Med. Public Health **3**, 19–27 (2017)
27. The Economist: Things are looking app: mobile health apps are becoming more capable and potentially rather useful, https://www.economist.com. Accessed 21 Feb 2018
28. Tikkinen-Piri, C., Rohunen, A., Markkula, J.: EU general data protection regulation: changes and implications for personal data collecting companies. Comput. Law Secur. Rev. **34**, 134–153 (2018)
29. Torre, I., Koceva, F., Sanchez, O. R., Adorni, G.: A framework for personal data protection in the IoT. In: 11th International Conference for Internet Technology and Secured Transactions, ICITST 2016, pp. 384–391 (2016)

30. Torre, I., Sanchez, O. R., Koceva, F., Adorni, G.: Supporting users to take informed decisions on privacy setting of personal devices. Personal and Ubiquitous Computing, pp. 1–20 (2017)
31. Van den Bulck, P.: Transfers of personal data to third countries. ERA Forum **18**, 229–247 (2017)
32. Voas, J., Kshetri, N.: Human tagging. Computer **50**(10), 78–85 (2017)
33. Xie, N.: Norway: Consumer Council addresses "transparency issues" in fitness wristbands. https://www.dataguidance.com. Accessed 20 Feb 2018
34. Yoon, H., Shin, D.H., Kim, H.: Health information tailoring and data privacy in a smart watch as a preventive health tool. In: Kurosu, Masaaki (ed.) HCI 2015. LNCS, vol. 9171, pp. 537–548. Springer, Cham (2015). https://doi.org/10.1007/978-3-319-21006-3_51

Changing Perspectives: Is It Sufficient to Detect Social Bots?

Christian Grimme(✉), Dennis Assenmacher(✉), and Lena Adam

University of Münster, 48149 Münster, Germany
{christian.grimme,dennis.assenmacher,lena.adam}@uni-muenster.de

Abstract. The identification of automated activitiy in social media, specifically the detection of social bots, has become one of the major tasks within the field of social media computation. Recently published classification algorithms and frameworks focus on the identification of single bot accounts. Within different Twitter experiments, we show that these classifiers can be bypassed by hybrid approaches, which on a first glance may motivate further research for more sophisticated techniques. However, we pose the question, whether the detection of single bot accounts is a necessary condition for identifying malicious, strategic attacks on public opinion. Or is it more productive to concentrate on detecting strategies?

Keywords: Social bots · Online propaganda · Social media analysis Social media computation

1 Introduction

Automation in social media has received enormous attention in scientific and public discussions. Scientific papers [1–5] up to newspapers [6–9] – recently also in Germany [10,11] – report on the threats posed by automated accounts as well as on the identification of automated profiles during election campaigns like the Brexit vote [12] or the last US Presidential election [13]. Specifically the term "Social Bot" stands synonym for malicious activities, which aim for manipulation of public opinion or even elections. Consequently and rather straightforward, science focuses on mechanisms to detect these automated profiles based on their individual behavior. Besides descriptive observation techniques, a plethora of automated techniques are available to identify social bots, ranging from machine learning approaches to very simple activity indicators. Basic approaches [14] merely analyze the frequency of an account's activity (a social bot is postulated, if an activity threshold is passed), sophisticated approaches try to identify behavioral patterns of automated accounts. Probably the most well-known approach of the latter class is the Botometer (formerly known as BotOrNot) service provided by the Indiana University [5,15].

All approaches, simple up to complex, follow rules that usually describe fully automated behavior of social media accounts. If a human partly or temporarily

© Springer International Publishing AG, part of Springer Nature 2018
G. Meiselwitz (Ed.): SCSM 2018, LNCS 10913, pp. 445–461, 2018.
https://doi.org/10.1007/978-3-319-91521-0_32

manages an account, the indicators as well as the trained (machine learning) models become vague and imprecise in their detection performance. Especially for machine learning approaches, another problem occurs: trained with limited (and manually gathered) ground truth, these methods specialize to detect exposed behavioral and metadata patterns for a given set of accounts within a fixed time interval. Due to high dynamics and changing usage of social media accounts, exposed patterns of these profiles may change also rapidly. This leads to varying accuracy of the trained detection mechanisms and eventually, the (at least temporary) inability to detect before-known social bot accounts.

To empirically support our argument, we first conduct two experiments to highlight the volatility of social bot detection mechanisms under changing usage patterns for social media accounts. Exemplarily, we concentrate on Botometer as the most prominent and rather advanced detection technique. In a first experiment, we construct fully automated social bots, which can be easily detected by simple indicators and Botometer alike, and successively integrate human behavior. During the bots' activity, we analyze the detection performance of Botometer over time. In a second experiment, we implement a set of 30 social bots that actively befriend to Twitter users and expose human like behavior. After a month of constant and fully automated behavior the small bot net starts massive action to promote a topic. Here, we also track the detection performance of Botometer.

Starting from these experimental insights and the discussion of current detection techniques, we pose the principal question, how detection mechanisms for social bots contribute to the prevention of manipulation or propaganda via social media. We propose a shift of perspective from detecting simple account properties towards identifying coordinated strategies, i.e., orchestrated activities of multiple (automated, semi-automated or human-steered) accounts. This shift from the micro-level of social bot detection to the macro-level of strategy detection is a by far greater challenge to research, but certainly of greater importance.

This work is structured as follows: The next section highlights some established and current developments in social bot detection and proposes a taxonomy that identifies two main overall streams of methodology: inferential and descriptive analysis. Thereafter, an experimental study on Botometer as current inferential detection mechanism is presented. Based on this, we pose the principle question, whether detecting automation patterns in single accounts is helpful after all. Based on two case studies on campaigns observed during the German general election in September 2017, we propose a change of perspective towards detecting orchestrated behavior of actors in social media.

2 Detection of Social Bots

With the "Rise of Social Bots" – this wording is also a reference to one of the most recent and influential reviews on the topic [5] – research tackled the detection of automated social media profiles. Early social bot realizations and also many current implementations are simple and merely focused on content amplification. Consequently, detection approaches for this type of bots monitor the activities of suspicious accounts and set (usually rather arbitrary) thresholds for defining accounts as social bots. Interestingly, a lot of current research is still based on these methods [13,14].

Current social bot implementations are far more sophisticated. Accounts are created to resemble human accounts and social bots mimic human behavior on the meta data level, i.e., they automatically vary their activity profile, follow a day-night-cycle or befriend and even communicate (in a simple manner) with other accounts. Although there are limits in intelligent interaction [16], with the before mentioned rudimentary techniques, social bots are not detectable anymore. Even human observers may be deluded by these obfuscation techniques. Sophisticated automatic detection mechanisms however, can analyze multiple aspects of the meta data over time and are (sometimes) able to find suspicious patterns in behavior for classifying accounts. Others analyze the behavior of many accounts over time with respect to predefined indicators. Thus, in contrast to Ferrara et al. [5], we divide the current detection techniques in only two classes.

2.1 Inferential Approaches

The first class of detection approaches is based on the analysis of data from account activities in social media and tries to infer representative patterns for social bot behavior. Sometimes, methods of machine learning are applied to automatically deduce features and rule sets. Those rule sets are then used on not yet classified accounts to get some rating. An early detection mechanism contained in this class is not based on machine learning but manually defines rules for befriending behavior of social bots [17]. Yang et al. [18] also use feature extraction techniques from representative behavioral features of human and robotic accounts in the RenRen network to identify meaningful discrepancies of both classes. Based on this, an online sybil detection system for automated accounts is implemented. Another method by Clark et al. [19] tries to identify automated activity on Twitter by focussing on language analysis. The approach identifies natural (human) language patterns to indirectly distill automated produced content. The currently most popular approach for classifying single Twitter accounts is the Botometer (formerly known as BotOrNot) web service[1] provided by the Indiana University [15,20]. Based on more than 1,000 features used in a random forest classifier, a given Twitter account is analyzed and rated in an interval of 0 (human) and 1 (social bot). This rating can be interpreted as probability for the specific account for being a social bot (or not).

Overall, inference-based methods implicitly assume underlying common characteristics of social bot behavior that need to be explored and described by fixed rule sets. To generate these rules, an annotated data set (ground truth) is needed to extract representative features for human and social bot behavior.

[1] https://botometer.iuni.iu.edu.

All approaches focus on classifying single user accounts in social networks to detect the type of actor (human or machine) behind the curtain.

2.2 Descriptive Approaches

Different from inferential approaches, the second class of descriptive approaches comprises usually manual observations of specific campaigns in social networks. Examples of such case studies are the detection of a Ukrainian bot net by Hegelich and Janetzko [21]. The authors analyze a large dataset of Twitter posts and metadata by applying frequency indicators and clustering methods. From these insights, they extract evidence for a large bot net that was active during the Ukrainian revolution in 2014. In the same way, using tools from descriptive data analysis, Eccheverria and Zhou [22] identified a large social bot network, which posted Star Wars quotations – probably just to age the Twitter accounts for later use in campaigns. An early clustering approach by Cao et al. [23] for detecting similar behavior in accounts can also be considered as descriptive method. The authors provide a so-called SynchroTrap, which detects loosely synchronized actions of accounts in the context of campaigns. The basic assumption is, that a campaign needs a central, thus synchronized activity of multiple social bot accounts.

A major advantage of the descriptive approaches is their openness towards new and yet unknown strategies. However, they demand an (usually a-posteriori) identification of campaigns. Even in current approaches, it is necessary to integrate human intelligence for the selection of indicators as well as for the interpretation of results. Once a campaign is identified, actors can be investigated and bots can be separated from human accounts.

2.3 A Comment on both Classes

The approaches of the defined classes differ in their perspective on social bot detection. The inferential perspective assumes universal patterns to be identified for social bots. The descriptive perspective works case-based and tries to identify social bots from a group of accounts that participate in an observed campaign. Although all approaches have the same goal, the descriptive approaches are inherently context-related. The initial restriction on a topic or campaign indirectly restricts the amount of accounts that has to be considered for detecting social bots. Still, the approaches of the inferential class are predominant in literature and current discussion. They work in a rather context-free manner by identifying bot characteristics for single accounts. On the one hand, this can be of advantage, as these methods are directly applicable to social media accounts. On the other hand, the missing context implies the absence of important indications that could support or falsify the detection result. In the following section, we investigate this ambivalence for the most commonly used indicator Botometer.

3 Experiments

To get first insights into the performance of current bot detection mechanisms, we conducted several Twitter-based experiments. Therefore we used fully automated and hybrid bot approaches to check, whether those mechanisms are capable to appropriately identify bots. The bot types used in the following experiments are motivated by a taxonomy published by Grimme et al. [16]. They assume three classes of bots ranging from simple automation (for broadcasting and multiplication of content) via human-like acting bots (possibly also containing a hybrid component) to intelligent acting (and content producing) bots. To show that current bot detection mechanisms struggle to appropriately identify social bots, we restrict ourselves to the first two classes. For the third class no productive realization is known yet.

3.1 Experimental Setup

The different experiments are based on a propitiatory social bot framework, that is capable of realizing the before mentioned simple and hybrid bot types of the taxonomy introduced in [16]. Figure 1 visualizes the three core components of the proposed framework: *Account, Bot, and Human.* The account component depends on the underlying social media platform. The remaining components of the framework do not explicitly focus on a specific platform and can be regarded in a more abstract way. The account can be accessed and interacted with, by either the fully automated bot component via an application programming interface (API) or by the human via a web/mobile client. It has to be emphasized that the functionality that can be realized by the automated bot component mainly depends on the provided functionalities of the platform's API. In case of the Twitter platform, the API provides full access to all functionalities that can be used within the web-frontend. Therefore all natural account interactions can be mimicked by the bot component.

The used framework can be adjusted in two different dimensions. The *hybridization* dimension specifies to what extend the bot component, and the human component should interact with the social media platform account. Figure 1 displays an equal share of bot and human interaction.

The *steering/orchestration* dimension adjusts the proportion of the individual components. The bot component may consist of different automation mechanisms. A rather simple functionality would be the repetitive multiplication of social media posts (retweeting). Hence this scenario indicates a small steering share for the bot component. For a higher bot steering factor, we could add a day-night-cycle or automated and intelligent following mechanisms. The human component can also be vertically adjusted. Within a simple scenario, human interaction could be reduced to a minimum, such as specifying which kind of content should be promoted or retweeted. In contrast, a prominent human interaction scenario would realize an automated spreading of original but predefined postings. In such a case the social bot needs a variety of tweets as input, which have to be manually created and curated by humans.

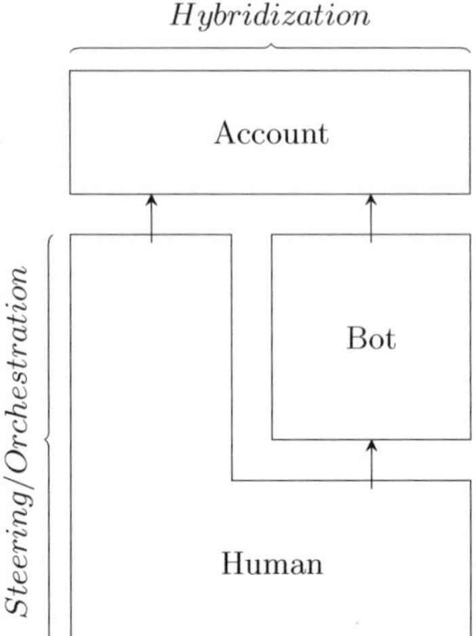

Fig. 1. Conceptual bot framework.

Within our experiments, we utilize the Botometer service to analyze the scores for different Twitter accounts [15]. Botometer is a classification system which determines the probability for a given Twitter account being a social bot. Applying supervised learning techniques such as random forests, the system learns a classifier by using 1,000 different account related features. Those features are divided into six different categories: *user*, *friends*, *network*, *content*, *timing*, and *sentiment*. For each category the learning algorithm predicts a bot likelihood. Additionally, an aggregated bot score that considers all available features is provided by the service.

3.2 Pre-experiment: Botometer

The first experiment aims for the analysis of Botometer scores of bot-accounts which expose different behavior over time. Furthermore, we want to examine whether and to which extend human interaction in terms of hybridization is able to bias the assessment of the Botometer scores. Therefore, the experiment is divided into three phases:

Phase 1: At the beginning three different bot accounts are started with new and empty profiles. Each bot account follows a simple retweeting strategy. In this case the bots retweet posts containing the hashtag *#bitcoin* without adding additional texts or comments. Therefore, candidate tweets related to the hashtag *#bitcoin* are picked via the Twitter streaming API. Each bot retweets random posts from the candidate list. Furthermore, the bots follow no day-night-cycle. Their retweet actions are strictly set to specific points in time. Additionally, we set the bot activity to 50 retweets per day. All these regular and simple settings ensure, that a clearly automated basic behavior

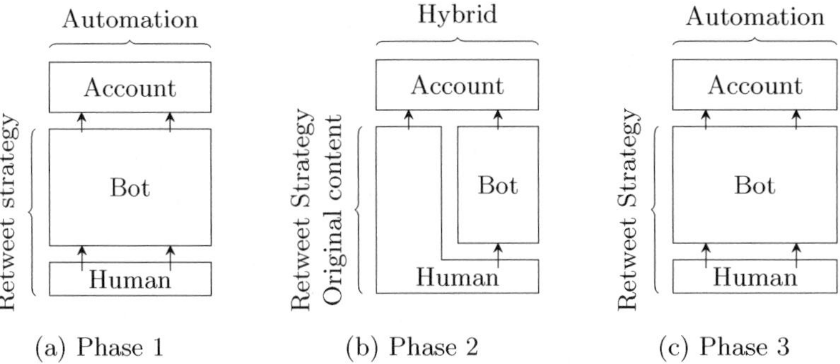

Fig. 2. Conceptual view of the Botometer experiment, divided in three phases of behavior.

is exposed by the accounts. We expect Botometer to detect these accounts as social bots with >0.5 probability. Apart from setting up the bot scripts, there is no human interaction in the first phase of the experiment. Using this fully automated configuration, which is schematically shown in Fig. 2(a), the bots ran for two weeks.

Phase 2: After the initial fully automated phase, two of the three bots are manually curated (starting February 5th). Manual interaction is, for example, tweeting, retweeting, liking of posts related to current incidents, like discussions about soccer games, the weather, or TV series. The manual human intervention follows a typical daily-life structure. An exemplary activity pattern is manual interaction in the morning, at lunch time, and in the evening. With human intervention, the bots do up to ten "human actions" per day, in addition to their basic retweet-strategy. As shown in Fig. 2(b) the two accounts are controlled in a hybrid way now. The human intervention is also part of the hybridization-axis, since the human-controlled actions are done directly through the web interface of the account. Using this configuration the bots run two additional weeks. As a baseline, the third bot still follows the simple retweet-strategy, described in Phase 1.

Phase 3: After two weeks, the human intervention is stopped, and the bot behavior changes back to the configuration of Phase 1, refer to Fig. 2(c).

At each phase of the experiment the Botometer score of the bots is calculated on a hourly basis.

Figure 3 shows the development of the Botometer scores of the three bots during the four weeks of the experiment. To display the score per day, the mean of the hourly scores is calculated. Additionally, a regression line for each bot has been computed, in order to analyze the trend of the account classification.

For all social bots, the Botometer score of the simple retweeting phase 1 converges to a score of 0.5. A score of 1.0 in this case means that the account is most certainly controlled by a bot, where a score of 0.0 means, that the account apparently only contains human-steered interactions. The authors of Botometer state, that a score around 0.5 enables to no precise statement, whether the account is steered by a human, or a bot [15]. Hence the behavior of the Botometer measurement for our simple bots is astonishing. Obviously, already the very simple and regular implementation of activity leads to the inability to classify the accounts. At the same time, we find that the start of phase 2 shows no change of the score development. With some inter-bot variance, the *overall* Botometer score converges to a range of 0.3 to 0.5 at the end of the experiment. To get more information on the effects on our hybrid interaction, we take a deeper look at the sub-scores of Botometer. Exemplary, the development for three of the five sub-scores is shown in Fig. 3.

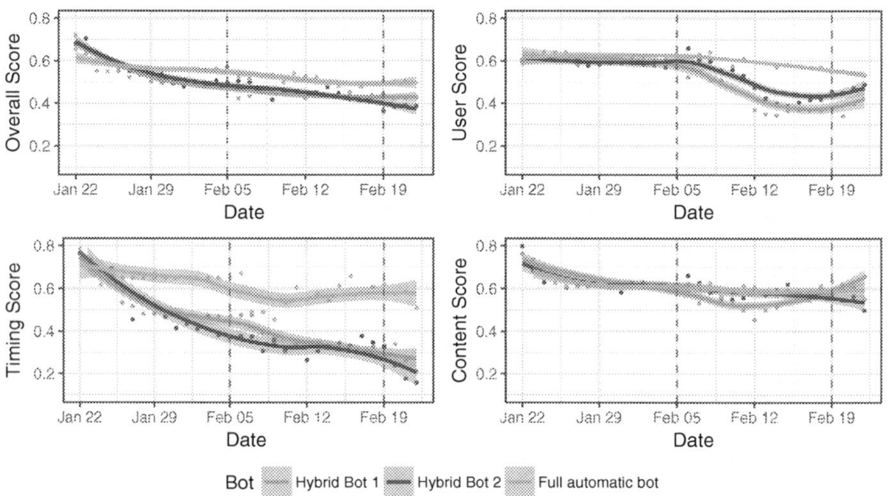

Fig. 3. Botometer scores over time, including trends. Top left: overall score; top right: user score; bottom left: timing score; bottom right: content-related score.

Considering the *user* sub-score, it is obvious, that the start of human intervention on February 5th leads to a strong decrease of the scores for the hybrid bots. The score of the fully automated account never drops below 0.5. After February 19th – the end of human intervention – the user scores of the two hybrid bots increase again. Amongst other features the *user* sub-score takes into account the features "number of tweets/retweets/mentions/replies (per hour and total)" [15]. Certainly, these features change significantly during the human intervention in Phase 2. Another feature, which may lead to a decrease of all the accounts, is the continuously changing "age of the account".

An even more obvious change of the score range in phase 2, is noticeable in the *Timing* sub-score. This score is based on calculation of time ranges between two consecutive tweets/retweets/mentions. The human intervention in phase 2 massively improves the scores of the hybrid bots. The timing sub-score dropped under 0.4, whereas the value of the fully automated account ranges about 0.5.

Within the third phase of the experiment, the scores of the hybrid bots decrease even further and reach a score range of around 0.3 to 0.2. This might be caused by the change in tweeting activity at the transition from phase 2 to phase 3.

Analyzing the sub-score *Content* indicates that the human intervention seems to have almost no impact on the features of this score. Within all phases of the experiment, the mean of the three values ranges between 0.7 and 0.4. Since the content is changed from merely retweeting bitcoin tweets to original text post, pictures, etc., this behavior is surprising. An explanation of this behavior could be the fact, that Botometer is trained on English profiles and content. The bots tweeted mainly in German, so the available detection patterns are possibly not able to properly classify the content.

The sub-scores *Friends* and *Networking* (not shown here) have no impact on the overall-score as well. This might be due to the fact that the human intervention was limited on posting activities. No network activities have been done, neither by the automated nor by the human influenced account. The sub-score *Sentiment*, is – like the sub-score content – composed of different text-based features. Furthermore, there is no observable difference between the scores of the automated and the hybrid accounts. This might again be, due to the fact that the algorithm is trained on English data.

3.3 A Social-Bot-Driven Campaign

The second experiment has been conducted between January 5 and February 5 in 2018. Within this study we investigate the impact of a coordinated strategy to push a predefined hashtag or topic, respectively. The main goal is to check, if bot accounts that are part of the attack, can be detected by the Botometer service and whether our attack is able to actually trigger a new trend on the twitter platform. In order to conduct the experiment we constructed a hashtag that should encourage users to actively join the twitter conversation. To ensure user's participation, we tried to gamify the whole setting: using the hashtag #songmoji, Twitter users are asked to post titles of different songs, only by relying on emoticons. Figure 5 shows an exemplary songmoji which was prepared in advance of our study. The complete experiment was conducted in two different phases, namely

1. building a follower network and
2. pushing the predefined hashtag by spreading tweets through the network.

Figure 4 visualizes both phases within the conceptual view of our proposed bot framework.

For the first phase, we created 30 distinct twitter accounts, each of them consisting of different meta-data such as profile image, hobbies and user location. Within a period of 28 days, all bot accounts automatically increased their reach by following twitter accounts which tweeted about different predefined topics. We focused on trending German hashtags, since the experiment was aimed to a German audience. It should be emphasized that during the first phase, the accounts only retweeted content. None of the accounts actively tweeted any original content.

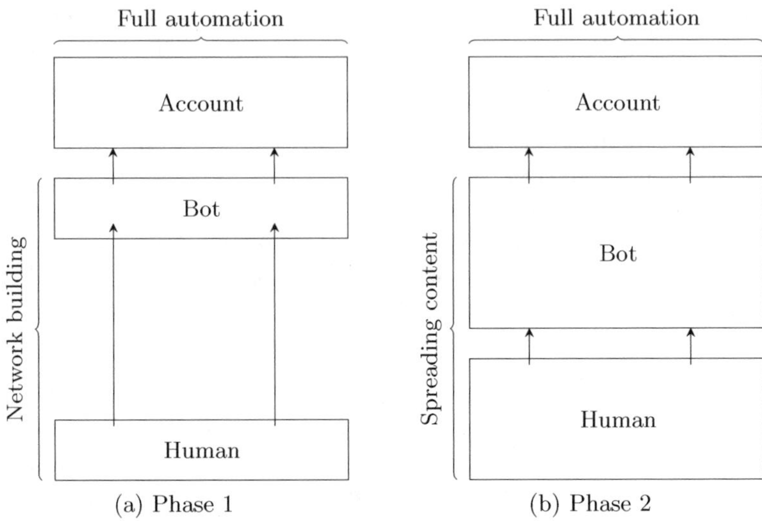

Fig. 4. Conceptual view of the campaign experiment.

In preparation to the second phase, a set of 120 unique #songmoji tweets was manually created. This pool of original tweets was used by the social bots to massively spread the hashtag through their follower network. Additionally, all bots automatically liked tweets published by users which adapted the #songmoji hashtag. Furthermore our bots retweeted #songmoji tweets, which were posted by other users. In order to avoid that our bots would be banned by Twitter, because of content spamming, we restricted the actual tweet and retweet frequency to a high but human achievable number of 75 posts per day.

Within Fig. 6 the average Botometer scores of all 30 bots over the experimental duration (until the accounts were suspended by Twitter) are visualized. For almost all scores, there is a significant drop, starting at the beginning of the second phase. Especially the average user score drops to a minimum of 0.25. This drop can be explained by the fact that within the second phase, the bots initially started to spread the original tweets that were manually created beforehand.

Fig. 5. Example of a predefined "Songmoji".

Due to the fact that Botometer's user feature measures, amongst other, the number of tweets and retweets of an account, it is not a surprising result that this score drops most. We also observed that at the beginning of the second phase, many users, which showed the willingness to participate at our emoticon game, followed our bot account. Hence, we can also explain the drop of Botometer's Friend and Network score. All in all, we see that an automated, coordinated strategy, executed by more or less simple but orchestrated bot programs cannot be detected by the Botometer service at an individual account level.

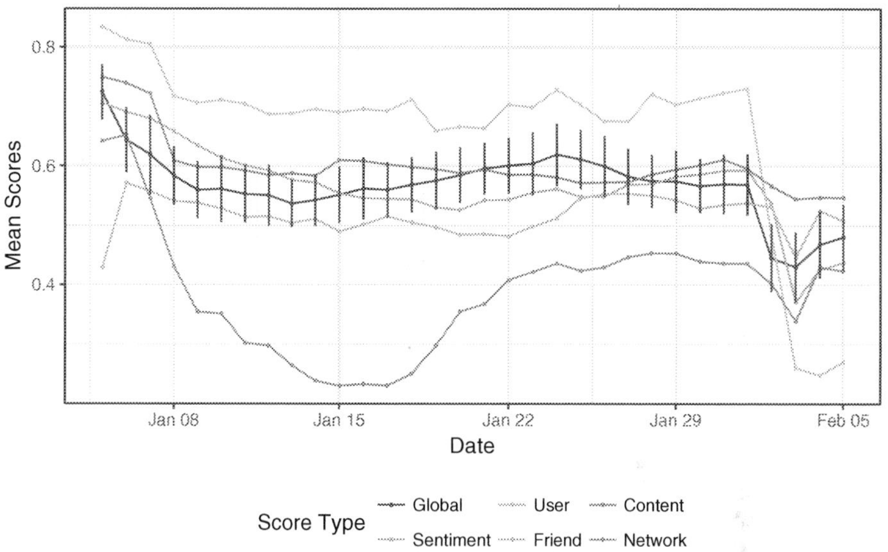

Fig. 6. Average bot scores over time including all average sub-scores.

Although the results indicate that the Botometer service was not able to individually classify our bot accounts correctly, all of them were suspended by Twitter after two days of spreading the hashtag. In our case it was not the Twitter platform itself that detected the bots, but other Twitter users. In contrast to individually analyzing each bot and its actions, the participating users noticed the aggressive behavior of the bot net, e.g., that all of their #songmoji tweets were instantly liked by several bot accounts. Some users reported the accounts to Twitter, which resulted in a ban of the accounts to temporarily prevent them from tweeting. An exemplary user reaction leading to the ban can be seen in Fig. 7.

Fig. 7. Detection of our bot army by a user (translated from German, anonymized).

4 On the Importance of Strategy Detection

The previously presented experiments on detection approaches for social bots suggest two main conclusions:

1. Although there are tools available, which base on state-of-the-art pattern recognition, their detection quality is depending on previously learned patterns. Obviously, it is easy to create social bots that bypass these patterns in a largely automated fashion. When human interaction is combined with automatic behavior, profiles cannot reliably be classified anymore by these approaches.
2. Human analytic capabilities are in principal able to detect social bot behavior, as our second experiment demonstrated. The humans, however, do not only focus on specific patterns in single account behavior (micro level). They observe macro effects of multiple automated agents as unusual behavior and sort out the actors participating in a campaign.

While the first conclusion may motivate further research to find even more sophisticated approaches for social bot detection, the second conclusion certainly challenges the current way of social bot detection. Current social bot detection is merely the identification of possible vehicles for information or disinformation in social media. Manipulation or propaganda, however, is the result of applying complex strategies or campaigns in and between social media channels as well as in the "real world". Therein multiple types of content may be used by multiple types of users and groups over long or short periods of time. Often, social media campaigns are accompanied by information and campaigns outside social media.

Considering all this, we wonder: Is it necessary to know a single social bot account, and how do we identify specific threats or strategic attacks to public opinion? And even more pointed: Does it really matter, what kind of actor – human or social bot – is part of a malicious campaign?

Here, we demonstrate our argument with two identified orchestrated campaigns during the German governmental election in September 2017. With the help of multiple indicators, their combination, and the integration of human intelligence, we identified and verified two coordinated (luckily unsuccessful) manipulative attacks. We find that it is of minor importance, whether the participating accounts are automated or not; the challenging task is to identify the orchestrated behavior of accounts.

4.1 Case 1: A Troll Attack to the TV Debate of Candidates

In this case study, we present a short summary of an analysis of Twitter usage by troll accounts during the TV debate between the German chancellor Angela Merkel and her contender Martin Schulz (social democrats), with an emphasis on detecting organized communication.

As data source we use German language tweets from the Twitter Gardenhose stream (1% sample) and from the Decahose stream (a fair 10% sample of all tweets), which contain topic-related hash tags (for details refer to [24]). For this case study, we gathered data between 6:00 pm and 11:59 am on September 3, 2017, resulting in 111,317 tweets.

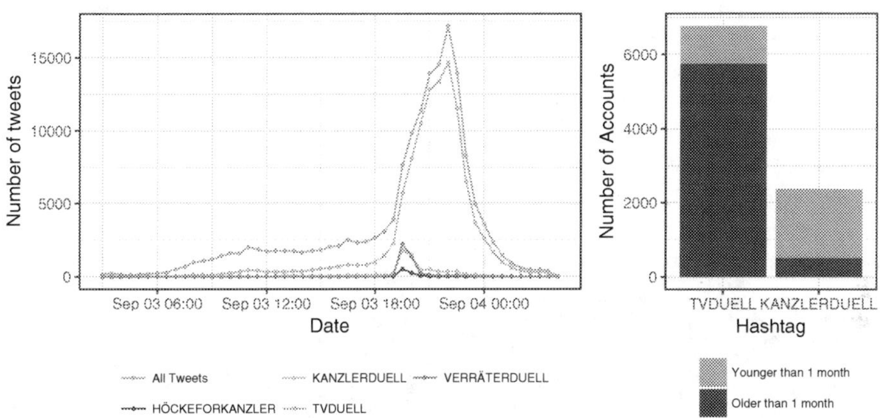

Fig. 8. Important indicators for the first case-study. The figure on the left hand side shows a time series of the activities during the TV debate. The figure on the right hand side shows the proportion of new and old accounts active for two hashtags.

In contrast to existing studies, we employ multiple indicators, some of which are the tweet/retweet relation, the age of twitter accounts, trending hashtag frequency and time series for a descriptive analysis. As a first result, we find that a very high number of new accounts simultaneously tried to push the new hashtag #verräterduell (traitor duel) by combining it with the already existing (and during the TV debate trending) hashtags #kanzlerduell (chancellor duel). The accounts are younger than one month and have mostly been used for retweeting existing content (without commenting it), to a fraction of 79%. Figure 8 (right) shows the disproportionately high amount of young accounts for the hashtag #kanzlerduell compared to the major hashtag #tvduell for the considered observation. Additionally, Fig. 8 (left) gives an impression of the development of several hashtags over time. The campaign is visible as a small activity peak at the beginning of the overall activity peak on Twitter just before the TV debate started.

We presume that what we have documented, was an attempt of an orchestrated attack by human-steered accounts on Twitter that tried to establish a pejorative hashtag hooked onto a neutral one by means of about 380 Twitter accounts, many of which have been established just for being used for this or similar purposes during the election phase. Interestingly, our findings are confirmed by an investigative BuzzFeed publication that refers to an inside report of chat groups that planned to push the mentioned hashtags [25].

4.2 Case 2: A Social Bot Campaign During the German General Election

The second analysis was also performed in the context of the German general election and focuses on the activity of social bots, which distribute advertisement for programmatic details of a (small) German party (Freie Wähler). Although the distribution of political advertisement is ethically unproblematic in principal, the respective party proclaimed not to use social bots for campaigns and demanded the flagging of automated profiles in social networks.

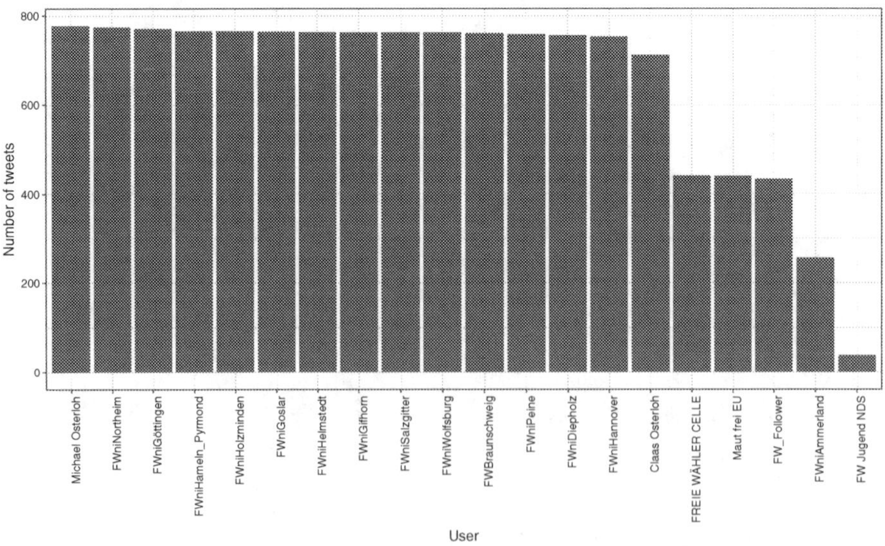

Fig. 9. Most active user accounts for #freiewaehler

German language tweets containing general-election-related hashtags were taken from the Twitter Gardenhose (1% sample) and Decahose (fair 10% sample) streams starting at September 10, 2017 until September 25, 2017 (one day after the election), resulting in about 5.5 million tweets.

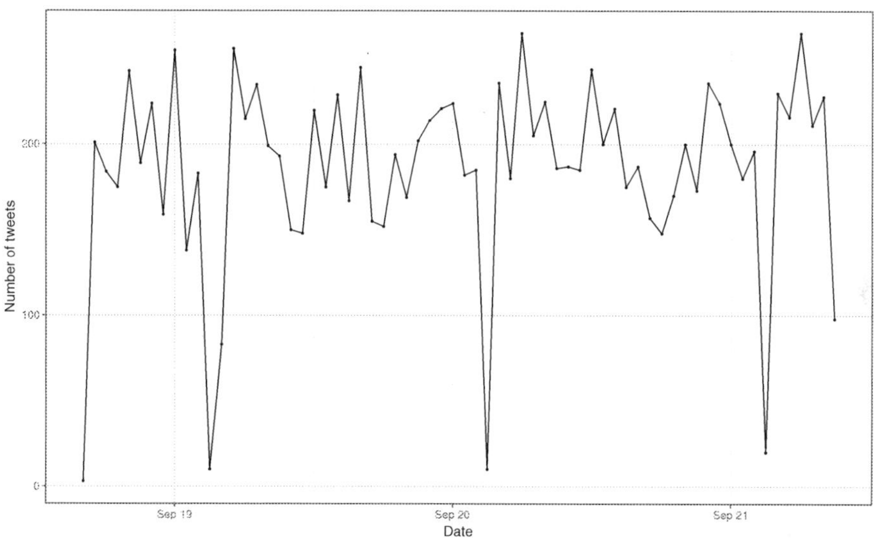

Fig. 10. Number of tweets for #freiewaehler over time.

The indicators in Figs. 9 and 10 expose clear patterns of automated behavior. The first indicator simply measures the overall activities for the 20 most active accounts. Interestingly, at least the eighteen most active accounts expose very similar activity behavior. Additional proof of automated actions is provided by the activity time series. We notice a regular drop of activity to almost zero activity at 3:00 am every night. This is caused by a standard network reset procedure at this time. Note, that the use of social bots in this context was later confirmed by the responsible candidate of the respective party – after he was confronted with our findings.

5 Discussion and Future Directions

The cases shown above highlight campaigns, which were conducted in two extreme ways. One used almost certainly only human actors (trolls). The other one applied social bots to spread content. Both campaigns, however, were centrally coordinated and followed a specific goal, namely spreading ideological content on Twitter to reach a larger audience. In that process, the vehicles for content distribution – humans or bots – are only of secondary interest. The foremost challenge is to identify the strategy as such. This would have not been possible by analyzing arbitrary user accounts using current detection techniques. With a lot of luck, we would have found some of the very simple bots applied in case 2. The first campaign – promoted by humans – would have been undiscovered. After discovering the campaigns however, we were able to perform a detailed and forensic analysis of the contributing accounts, classifying them as troll or automated accounts, and even finding the responsible actors behind the campaigns.

Therefore we strongly suggest a shift of perspective in current bot detection. As inherently included (but not strictly pursued) by the descriptive approaches and partly addressed by a very recent work of Varol et al. [26], we believe that automated strategy and campaign detection is of major importance for defending against malicious attacks of social bots and human actors alike.

The scientific challenges are to identify patterns in campaigns and attacks rather than in behavior of single actors. This certainly requires – apart from longitudinal observations (time dimension) – to consider data from multiple social media/online platforms (spatial dimension). In the end, this can provide methods, which are able to deal with human-driven, fully automated as well as hybrid campaigns and attacks in cyberspace.

Acknowledgement. This work is part of the PropStop project, which is funded by the German Federal Ministry of Education and Research (FKZ 16KIS0495K). The authors are also supported members of the ERCIS network.

References

1. Boshmaf, Y., Muslukhov, I., Beznosov, K., Ripeanu, M.: The socialbot network: when bots socialize for fame and money. In: Proceedings of the 27th Annual Computer Security Applications Conference, ACSAC 2011, pp. 93–102. ACM, New York (2011)
2. Boshmaf, Y., Muslukhov, I., Beznosov, K., Ripeanu, M.: Key challenges in defending against malicious socialbots. In: Proceedings of the 5th USENIX Conference on Large-Scale Exploits and Emergent Threats, LEET 2012, pp. 1–4. USENIX Association, Berkeley (2012)
3. Messias, J., Schmidt, L., Oliveira, R., Benevenuto, F.: You followed my bot! Transforming robots into influential users in Twitter, First Monday (2013)
4. Maréchal, N.: Automation, algorithms, and politics — when bots tweet: toward a normative framework for bots on social networking sites (feature). Int. J. Commun. **10**, 10 (2016)
5. Ferrara, E., Varol, O., Davis, C., Menczer, F., Flammini, A.: The rise of social bots, vol. 59, pp. 96–104 (2016)
6. Tynan, D.: Social Spam is taking over the Internet, Apr 2012. http://www.itworld.com/article/2832566/it-management/social-spam-is-taking-over-the-internet.html
7. Fredheim, R.: Putin's bot army - part one: a bit about bots (2013). http://quantifyingmemory.blogspot.co.uk/2013/06/putins-bots-part-one-bit-about-bots.html
8. Elliott, C.: The readers' editor on pro-Russia trolling below the line on Ukraine stories, May 2014. http://www.theguardian.com/commentisfree/2014/may/04/pro-russia-trolls-ukraine-guardian-online

9. Ohlheiser, A.: Trolls turned tay, microsoft's fun millennial ai bot, into a genocidal maniac, March 2016
10. Rosenbach, M.S.: Internet-Kommentare von Automaten: AfD will im Wahlkampf Meinungsroboter einsetzen, October 2016
11. Pfaffenzeller, M.: Bundestagswahlkampf: CDU erwägt Einsatz von Chatbots, March 2017
12. Howard, P.N., Kollanyi, B.: Bots, #StrongerIn, and #Brexit: Computational Propaganda during the UK-EU Referendum (2016)
13. Kollanyi, B., Howard, P.N., Woolley, S.C.: Bots and automation over Twitter during the US election. Technical report Data Memo 2016.4, Project on Computational Propaganda, Oxford, UK (2016). www.politicalbots.org
14. Neudert, L.M.N.: Computational propaganda in Germany: a cautionary Tale. Project on Computational Propaganda. Technical report (2017). www.politicalbots.org
15. Varol, O., Ferrara, E., Davis, C.A., Menczer, F., Flammini, A.: Online Human-Bot Interactions: Detection, Estimation, and Characterization (2017)
16. Grimme, C., Preuss, M., Adam, L., Trautmann, H.: Social bots: human-like by means of human control? Big Data **5**(4), 279–293 (2017)
17. Paradise, A., Puzis, R., Shabtai, A.: Anti-reconnaissance tools: detecting targeted socialbots. IEEE Internet Comput. **18**(5), 11–19 (2014)
18. Yang, Z., Wilson, C., Wang, X., Gao, T., Zhao, B.Y., Dai, Y.: Uncovering social network sybils in the wild. ACM Trans. Knowl. Discov. Data **8**(1), 2:1–2:29 (2014)
19. Clark, E.M., Williams, J.R., Galbraith, R.A., Jones, C.A., Danforth, C.M., Dodds, P.S.: Sifting robotic from organic text: a natural language approach for detecting automation on Twitter. J. Comput. Sci. **16**, 1–7 (2016)
20. Davis, C.A., Varol, O., Ferrara, E., Flammini, A., Menczer, F.: Botornot: A system to evaluate social bots. CoRR abs/1602.00975 (2016)
21. Hegelich, S., Janetzko, D.: Are social bots on twitter political actors? empirical evidence from a Ukrainian social botnet. In: International AAAI Conference on Web and Social Media, pp. 579–582 (2016)
22. Echeverría, J., Zhou, S.: The 'Star Wars' botnet with >350k Twitter bots. CoRR abs/1701.02405 (2017)
23. Cao, Q., Yang, X., Yu, J., Palow, C.: Uncovering large groups of active malicious accounts in online social networks. In: Proceedings of the 2014 ACM SIGSAC Conference on Computer and Communications Security, CCS 2014, pp. 477–488. ACM, New York (2014)
24. Grimme, C., Assenmacher, D., Adam, L., Preuss, M., Stockdiek, J.F.H.L.: Bundestagswahl 2017: Social-Media-Angriff auf das #kanzlerduell? Technical report 2017.1, Project PropStop, Münster, Germany (2017). www.propstop.de
25. Schmehl, K.: Diese geheimen Chats zeigen, wer hinter dem Meme-Angriff #Verräterduell aufs TV-Duell steckt, September 2017
26. Varol, O., Ferrara, E., Menczer, F., Flammini, A.: Early detection of promoted campaigns on social media. EPJ Data Sci. **6**(1), 13 (2017)

Towards the Design of a Forensic Tool for Mobile Data Visualization

Karen Kemp and Subrata Acharya(✉)

Department of Computer and Information Sciences, Towson University, Towson, MD, USA
kkempl@students.towson.edu, sacharya@towson.edu

Abstract. The growing popularity and use of mobile devices over the past decade has provided law enforcement agencies with new types of evidence to aid them in solving crimes. These devices can store a great deal of data that can be instrumental in digital investigations. Equally important is the acquisition and analysis of this data, which can be used to connect individuals or organizations to an incident or crime involving cell phone communication.

Data obtain from cell phones and SIM cards are useful and meaningful because it can help investigators connect individuals who may have been collaborating or cooperating about a criminal activity; for instance, in the scenario of drug trafficking and/or terrorism events. In situations such as these where the data is not standalone, the visualization of this information is of high importance. Law enforcement officials must be able to not only collect data from cell phones but also understand the big picture; in other words, to determine how the data is connected and correlated to better draw conclusions about crimes and other topics of forensic interest. A good data visualization tool would allow them to make connections they otherwise might not have seen. While there exist many forensic tools for mobile phone data collection, the current software available to transform this information into a meaningful presentation is limited. To this effect, this research aims at addressing the need to create a mobile forensic tool that involves semantic data analysis to provide real-time data visualization information.

Keywords: Forensic · Visualization · Mobile · Semantic · Communication

1 Introduction

The goal of this research is to create a visualization framework for analysis of mobile phone data. The aim is to create a forensic tool that accepts various types of mobile data as input and presents them in the form of linked graphs, diagrams, and other types of digital drawings that can help connect and correlate the information in real-time. Instead of focusing on quantitative data, such as the number of times suspect A contacted suspect B, the proposed tool will be used to semantically analyze various dimensions of metrics and draw correlations from the incoming dataset. In other words, the tool will allow users to infer information from text messages, images, and other correlative datasets. The tool would be able to determine discussion subjects for individuals and potentially link them to crimes occurred or ongoing investigations.

© Springer International Publishing AG, part of Springer Nature 2018
G. Meiselwitz (Ed.): SCSM 2018, LNCS 10913, pp. 462–470, 2018.
https://doi.org/10.1007/978-3-319-91521-0_33

This would provide an efficient method for drawing conclusions from a large distributed dataset, based on numerous seized cell phones or SIM cards. It would also enable inferences as to whether a crime occurred or if two suspects are linked and could enable the automated analysis in a timely manner, without having to manually search through tens of thousands of messages.

The proposed tool will be built on the software base of an existing network analyzer visualization tool but will have a modified dataset that would include mobile phone data characteristics. There will be some overlap in the type of data that can be visualized, such as images, videos, and browse logs, which can be collected from a typical network traffic analysis application. The tool will also be able to accept source/destination and/or sender/receiver data from cell phones and visualize it in various output formats.

2 Related Work

This section will review several different tools for mobile phone data acquisition, analysis and/or visualization. In order to develop a tool that analyzes cell phone data, we will need to first acquire the data. The *National Institute of Standards and Technology [NIST]* has reference test data for cell phone forensic tools that can be used to populate SIM cards. There exist many tools that can retrieve data from mobile devices and SIM cards for forensic purposes, and we have included the most relevant tools in our evaluation. These tools have capabilities of performing logical data acquisition, which deals with files and directories in the phone and SIM card, and physical data acquisition, which is a memory dump of the device.

The *Cellebrite UFED* system is a standalone device that performs both physical and logical extraction of cell phone data [7]. The device has a simple user interface that guides analysts through the process. The user needs to know the make and model of the phone and choose it from a menu on the UFED screen. It will display the type of cable needed to connect the phone, or alternatively a SIM card can be inserted into the device. The user can then select what type of data they would like to acquire from the device. This device can extract deleted information such as call history, text messages, and phone book entries, and could enable access to internal data, such as the previous SIM cards used on the phone. The device also has a "*target*" port for inserting a USB drive to collect the data. The *Cellebrite UFED* comes with a feature that allows the user to navigate the data once it is uploaded and generates access report.

The *Encase Neutrino* is a mobile forensics device that can extract and analyze data from multiple cell phones at the same time [8]. The device is placed in a bag that blocks wireless signals. This helps to determine the position of the last cell towers used. Similar to *Cellebrite UFED*, it comes with multiple cables to connect devices to a port inside the bag, and it has a phone wizard that tells the user which connector to use. The Encase Neutrino includes a SIM card reader as well. *Encase Neutrino* has the capability of accessing unallocated space on certain devices.

The *Micro Systemation XRY* mobile forensic system is another device that extracts data from phones and SIM cards and generates reports from the data [9]. The small device is connected to a computer using a USB cable, and it includes many different phone cables. The user navigates the system using the software included with the device. The interface has tabs for viewing different types of data, such as audio, pictures, and SMS. The XRY creates encrypted files from the data stored on the phone.

During our evaluation, we inferred that there is limited development on existing cell phone data visualization tools. One of the current proprietary tools is *THREADS* [1], a product that is used to analyze cell phone forensic data and connect the information based on phone contacts and communications made between devices. It can show patterns and correlations among the data, determine organizational structures of crime groups, determine who is making frequent calls to the same number, and analyze organized relationships and events. The tool provides data visualization with linkage charts, interactive time lines, diagrams and reports to present the information to the security analyst. In addition to being proprietary in nature (no access to source code), the tool also does not provide contextual analysis on the dataset provided.

There are, however, several open source tools for network data visualization. One such tool is *Graphviz*, a graph visualization application that presents input data in a variety of different graph and diagram layouts to display semantic relationships [13]. The software is able to generate the graphs from external data sources [2]. *Cytoscape*, is another open source tool that is designed to visualize large scale networks. It was originally designed to be used in biological research, such as for visualizing molecular interactions and gene expression profiles [3]. The tool works as a web service client and can import data from external databases (but is format specific).

The Time-based Network Visualizer (TNV) is an open source network traffic analysis tool that visualizes packets and links between hosts. The tool enables capture of live packets and/or open saved *pcap* data that it exports to a database [4]. The data is presented in a visualization matrix that shows port activity and linked communications between hosts, which allows the user to determine communication patterns and correlation among the data [5]. The visualization is also essentially designed around a timeline, which lets the user view trends in network activity and relationships between hosts over time [6]. The tool is not designed to work on data from mobile devices. Figure 1 shows a visualization layout of the tool [17].

Otter, is another general-purpose, data independent, network visualization tool that was developed by CAIDA (Cooperative Association for Internet Data Analysis). The tool can handle any formatted data set as long as it consists of links and nodes [11]. *Otter* can handle datasets with several hundred nodes, and it has been used in applications such as visualizing paths of routing tables in the form of graphs. However, the tool does have limitations. Since users can specify the type of data structure, *Otter* must consistently be modified in order to keep up with the different types of data that might be used as input [11].

Also, related to cell phone data visualization is an open source file format called the *Advanced Forensic Format 4*, which is used for sharing evidence and analyzing forensic data. The developers of this tool redesigned the architecture of the original Advanced Forensic Format to allow it to store multiple data types [12]. The AFF4 is essentially a library that can be implemented in tools and programs that analyze forensic data. The library can store several different types of evidence from mobile devices in a single archive and allows user to manage the information [12].

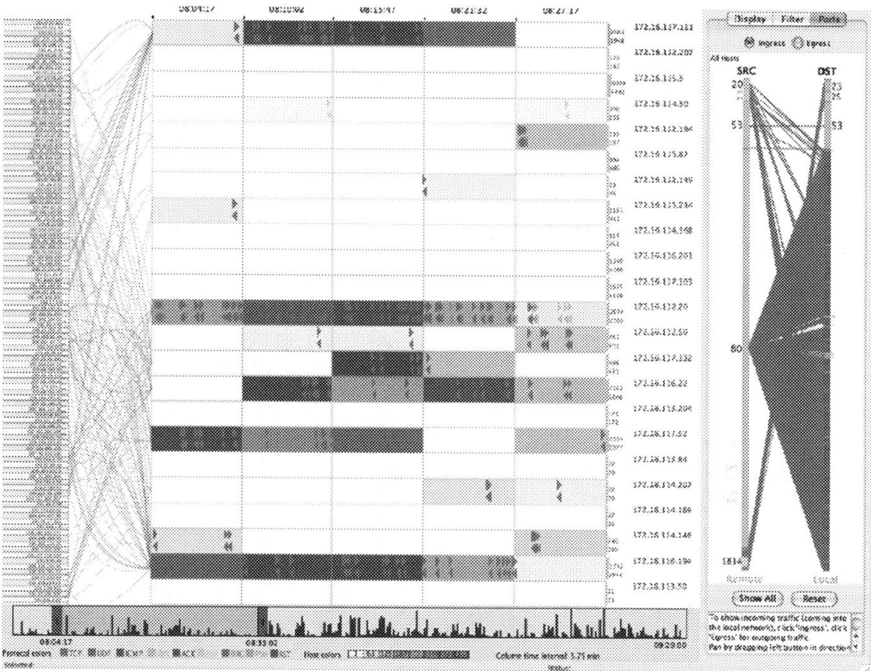

Fig. 1. Visualization layout

3 Proposed Approach

After conducting extensive evaluation of current approaches, the research plan was to build upon an existing open source data visualization tool and modify the code so that it will accept mobile phone data and address the limitations to visualize mobile data. We have studied the existing tools for data visualization and based on comparative analysis we have determined that TNV has the most intersections to a mobile data visualization based on the metrics of *time to compute* and *communication methods to visualize* the presented data. TNV analyzes *pcap* files, which, like mobile phone data, contains source, destination, and service components. Moreover, TNV's strength to analyze data over a specific range would enables accurate search results for the proposed tool.

We have modified the TNV tool to account for all the characteristics of mobile phone data. For example, collaboration is an important factor in cell phone forensics and needs to be considered in the analysis of the data. Text messages, multimedia messages, phone calls, contact lists, and call logs are all examples of data that will need to be accepted as input. Since, the code has already been written for data visualization, there is no need to change the presentation of the dataset. We can select existing graphs or charts that would work well with different aspects of the dataset; for example, a linked chart would be useful to display which devices communicated with each other regarding a given subject matter. Since TNV does not provides semantic analysis of mobile data, it will need to do more than simply link callers and messages based on frequency of communication. The actual contents of the messages will need to be analyzed to determine the significance of the communication between two or more individuals. Two examples of data that will be analyzed in this manner are text messages and image files.

The important features of text messages are the sender information, receiver information, text and/or characters sent between the communication link. We have also implemented a text parser to derive meaning from the available messages. The parser would analyze the text to determine words, phrases, and subject matters that are being discussed. It would be able to represent the meaning of the message independent of the syntax used [15]. For example, the program could search text messages to determine if two people discussed a person or a street name that is relevant to an investigation. It is important to note, due to the limitation of the maximum size of the text message, the operation would be completed within a feasible real-time computational interval.

Comparing images, however, is a time intensive process. Every image would need to be broken down into a matrix of pixels, each of which has a specific RGB (red, green, blue) ratio. The program would compare images by reviewing the RGB ratios and inferring on the differences and match ratios. Two images with a high percentage of pixels having similar match ratios are likely to be based on the same subject. The latency of this aspect of the program would also depend on the resolution of the image compared.

The graphing aspect of the tool will be used to visualize the inferences that were made from the data. Individual messages and callers will be represented as nodes and will be linked if it is determined that they have communicated. While a tool like *THREADS* shows the number of times two people talked, the proposed tool will be concerned with the semantics of the communication. We test the strength of the designed tool using mobile phone test case data from the NIST input dataset. The SIM-fill reference test data is an open source, Java based application that includes information that can be used to populate SIM cards for research on mobile forensic tools [10]. The reference test data consists of three XML files, each of which includes data to populate one SIM. The code for the SIMfill test data [14, 16] is modifiable to address evaluation and user needs [15]. The focus of the tool is to maximize accuracy and minimize false positives generated during the semantic analysis phase.

4 Evaluation

As part of the evaluation of the proposed tool, we will be assessing the accuracy of semantic analysis on the SIMfill dataset from NIST. The data is used as a standard for assessing mobile forensic tools. The dataset consists of three XML files, each containing a sample of SMS messages. Each text has a sender/receiver timestamp and body of the message. We have designed a SMS packet class, so data could be seamlessly imported to the tool. Figure 2 displays a sample SIMfill dataset.

```
- <!--
   Second part of a long outgoing message, sent, with French text and a small image
  -->
- <submit>
   <status>sent</status>
  - <sca>
   - <address>
       <ton>international</ton>
       <npi>telephone</npi>
       <number>19703769313</number>
     </address>
   </sca>
   <rd>true</rd>
   <srr>false</srr>
   <rp>false</rp>
   <mr>122</mr>
  - <da>
   - <address>
       <ton>international</ton>
       <npi>data</npi>
       <number>33123456789</number>
     </address>
   </da>
   <dcs>default</dcs>
  - <udh>
     <smlimg pos="18" file="./data/smlimg.bmp"/>
     <concat ref="21" max="2" seq="2"/>
   </udh>
   <ud>{ avec une image []</ud>
  </submit>
```


Fig. 2. NIST SIMfill dataset

The goal is to semantically analyze various types of mobile data in order to present correlation patterns, which could be used to detect or prevent criminal activity. We will evaluate the accuracy of the tool by determining the precision with which the tool is able to detect the relation between two callers or messages. In other words, if the test case data contains information showing that two people sent each other messages discussing a third individual, the tool will be able to accurately represent this information in the visualization format. Figure 3 displays an example tool display with 50,000 network packet communication.

Based on the sample NIST data, the tool can display correlation information to create accurate representations between individuals as displayed in Fig. 4. To determine false positives, the tool detects relationships among the data that did not exist in the given dataset. For example, if the tool shows that two people talked about a subject and the "*subject*" was just a word that looked similar (but had a different meaning), then that would be discarded as a false positive. The goal is to minimize all instances of such occurrences in the overall correlation set.

The tool creates a payload analysis dialogue as displayed in Fig. 5. A time interval is set for analysis of each packet. The tool parses the text and retrieves a list of words with higher context (frequency, etc.). The wordlist is appended periodically, and the tool returns the results based on the appropriate sender-receiver information. The tool

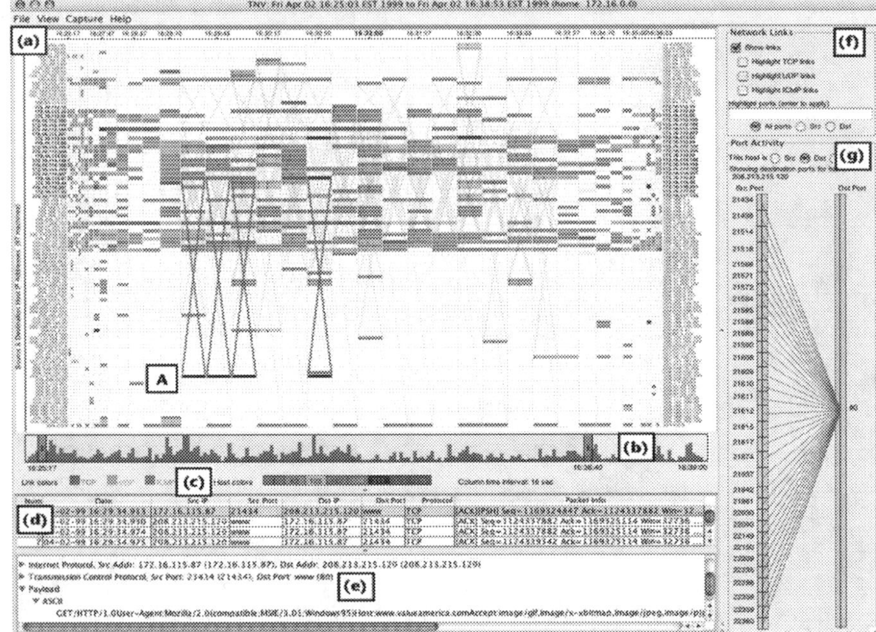

2 TNV showing 50,000 network packets. (a) The main visualization matrix, along with details of a selected host A, including network links with Web (TCP port 80) activity. Other TNV features include (b) the navigation and data overview mechanism; (c) the legend panel; (d) a table of packets for the selected host; (e) packet details for a selected row in the table; (f) the emphasis filtering panel; and (g) the selected host's port activity.

Fig. 3. Example tool display with 50K network packets

accurately displays the correlation event since the list has been predefined (uncommon words will be ignored if they are not on the list). Another aspect that has been analyzed is the efficiency of the tool. The tool can determine the latency for analyzing and drawing inferences from different types of data. Data images will have higher latency based on the type of file and what is required to correlate the data. Additionally, larger datasets will have higher latency of operation. Finally, the tool would be enhanced by including a training module to facilitate the fast population of the keyword dataset. This would enable quick review and reduce the false positives during correlation.

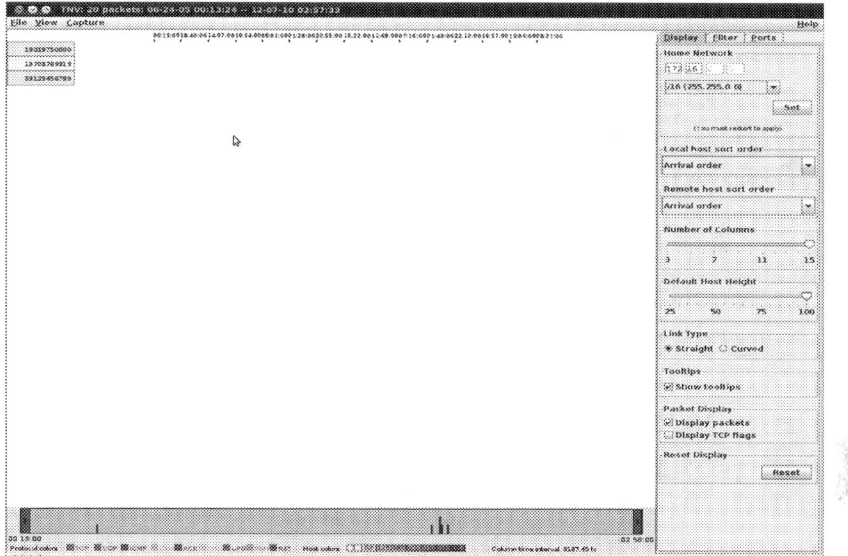

Fig. 4. Tool display with NIST SIMfill1 dataset

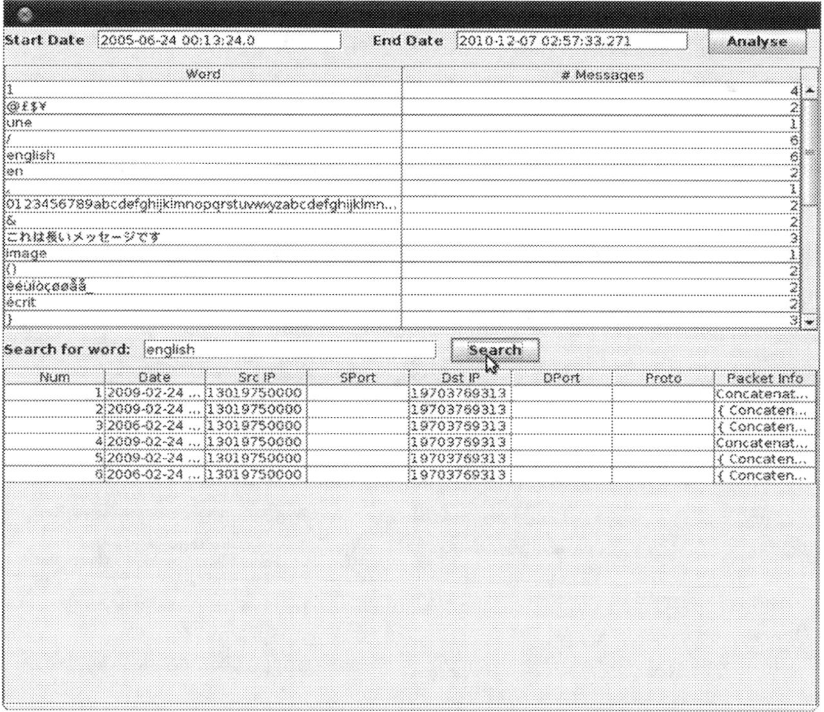

Fig. 5. Example of payload

5 Conclusion

The goal of this research is to enable law enforcement agencies conduct accurate evaluation of forensic mobile data obtained for investigative purposes. Graphs, linkage charts, and other types of visualization methods will enable correlation (or lack of correlation) amongst data to generate accurate inferences from the evidence provided. The proposed tool aims to fill the gap for semantic analysis of mobile phone data. As future work, we plan to conduct evaluation of the proposed tool on larger datasets and include diverse application inputs operating on various mobile platforms.

References

1. THREADS: http://www.directhitinc.com. Accessed 2017
2. Graphviz: http://www.graphviz.org. Accessed 2017
3. Cytoscape: http://www.cytoscape.org. Accessed 2017
4. TNV, Computer Network Traffic Visualization Tool. http://tnv.sourceforge.net/index.php. Accessed 2017
5. Goodall, J.R., Lutters, W.G.: Focusing on context in network traffic analysis. IEEE Comput. Graph. Appl. **26**(2), 72–80 (2006)
6. Goodall, J.R., Lutters, W.G.: Preserving the big picture: visual network traffic analysis with TNV. Presented at the 2005 Workshop on Visualization for Computer Security, Minneapolis, MN (2005)
7. Cellebrite: http://www.cellebrite.com. Accessed 2017
8. Encase Neutrino: https://www.guidancesoftware.com/mobile-cellphone-forensics-software-neutrino.htm. Accessed 2017
9. Micro Systemation XRY. https://www.msab.com/products/xry. Accessed 2017
10. Jansen, W., Aurelien, D.: Mobile Forensic Reference Materials: A Methodology and Reification, National Institute of Standards and Technology, Gaithersburg, MD, NISTIR 7617, October 2009
11. Huffaker, B., Nemeth, E., Claffy, K.: Otter: a general purpose network visualization tool, Cooperative Association for Internet Data Analysis (CAIDA), ISOC Inet 1999. http://www.caida.org/tools/visualization/otter/paper/
12. Cohen, M., Garfinkel, S., Schatz, B.: Extending the advanced forensic format to accommodate multiple data sources, logical evidence, arbitrary information and forensic workflow. Digit. Investig. **6**, S57–S68 (2009). http://simson.net/page/Main_Page. Accessed 2017
13. Pietriga, E.: Semantic web data visualization with graph style sheets. In: ACM Symposium on Software Visualization, Brighton, UK (2006)
14. Jansen, W., Delaitre, A.: Guide to SIMfill Use and Development, National Institute of Standards and Technology, Gaithersburg, MD, NISTIR 7658, February 2009
15. Shi, L., Mihalcea, R.: An algorithm for open text semantic parsing. In: Proceedings of the ROMAND 2004 Workshop on "Robust Methods in Analysis of Natural Language Data", Geneva, Switzerland, August 2004
16. SIMfill software downloads. http://csrc.nist.gov/groups/SNS/mobile_security/mobile_forensics_software.html. Accessed 2017
17. McRee, R.: Security visualization: what you don't see can hurt you. ISSA, June (2008)

Social Media Policies in UK Higher Education Institutions – An Overview

Rebecca Lees(✉)

Kingston University, Kingston Hill KT2 7LB, UK
b.lees@kingston.ac.uk

Abstract. Social media has brought about a new communication landscape and this has far reaching implications for higher education and academic practice. With this comes a need for a sound governance structure, and this paper aims to investigate the prevalence and content of social media policies within the UK higher education environment. Governance documents from all publicly funded universities were gathered and analyzed for accessibility through readability statistics, and a thematic analysis of the content compared to the main themes from the current body of literature pertaining to social media policies. The results suggest that a large proportion of UK institutions lack an explicit social media governance document and those that do exist score reasonably high on the Flesch-Kincaid Grade Level and SMOG readability measures. The content suggests that whilst higher education institutions are providing guidance for use of social media *at* work, little direction is provided for the use of social media *for* work. The paper then concludes by considering issues for policy implementation within the UK higher education environment.

Keywords: Policy analysis · Readability · Education governance
Flesch-Kincaid grade level · SMOG · UK higher education

1 The Rise of Social Media Use Within Education

Social media has brought about a new communication landscape [1] and the popularity of this new technology has shaped a new world of collaboration and communication [2]. The terms social media and social network are now so commonly used they are part of the everyday vernacular and for many, an integral part of daily interactions with others. Recent statistics from the UK show two thirds of all adults using the internet for social networking activities and almost half to upload user-generated content [3]. Within the 16–24 age group, typical undergraduate entry age, this activity increases to 96% engaging in social networking and 72% posting social media content [3].

From the academic perspective, a report on Teaching and Learning use of Social Media reported that over 90% of academics surveyed (n = 1943) used social media technologies, either as part of their classroom activities or in a professional capacity [4]. Even applicants and potential students have the expectation that social media is an integral part of everyday university life [5] and it is now considered a permanent and necessary feature in higher education, with use continuing to grow year on year [6].

© Springer International Publishing AG, part of Springer Nature 2018
G. Meiselwitz (Ed.): SCSM 2018, LNCS 10913, pp. 471–483, 2018.
https://doi.org/10.1007/978-3-319-91521-0_34

1.1 Social Media Use Amongst HE Stakeholders

Social media has been reported as having a significant impact on all aspects of academic practice [7, 8] and the stakeholders with whom institutions interact [9] to the point where it is now considered "no longer an option in higher education" [10, p. 1]. The main (generic) benefits focus on speed of communication [11], creating immediate and interactive dialogue [12] across a greater number of channels, with reduced participation and production cost.

Changes in educational practice that impact the institutional, staff and student perspective include teaching and learning [8, 13, 14], research and collaboration [7, 15], communication [8, 15], self-promotion [7], academic publishing and citations [16], alumni relations [10], and marketing and recruitment [17, 18]. From the student perspective, many aspects of student life are now facilitated online, enabling students to connect with their campus, their peers and university services [19] and give feedback and enter discussions with staff and peers [20] and access improved transition support [21, 22].

1.2 Risks Inherent with Social Media Use in Education

Despite its benefits, social media offers a double-edged potential [21] and there are risks associated with these perceived benefits. Of particular concern to academics are issues of privacy and the integrity of student work [6], and professional identity [23]. Institutions are wary that bad information or negative events will spread just as fast as positive messages, and given they have reduced control over what is being said, risks to the institution brand image and reputation are high [24]. However, despite the risks inherent in this form of media, institutions have been slow in constructing effective social media governance, and although this is a growing area in the literature, there are several studies already noting the lack of accessible social media policy documents [25, 26] and support within higher education [22, 27] particularly in the form of training and guidance for both academics [22] and students [28].

1.3 Social Media Governance Content

The study of social media governance mechanisms is a relatively new field in the literature, since much of the relevant documentation and artefacts were created and updated between 2008 and 2015 [29]. The field has since gained momentum since the number of individuals using social media on a daily basis meant its presence has started to infiltrate into the work environment.

Social media governance is defined as "the formal or informal frameworks which regulate the actions of the members of an organization within the social web" [30, p. 1033] and comprises a variety of instruments including policies, guidelines and regulations. It is widely acknowledged that a social media policy is a vital organizational tool to provide direction in the use of social media [29, 31] even if the company is not using social media in a formal way or at all [17] since it is likely that their employees are [32].

A study focusing on the content of social media policies [33] suggested eight essential elements which should be included: employee access, account management, acceptable use, employee conduct, content, security, legal issues and citizen conduct. Subsequent studies have expanded this list to include explicit items on organization goals, personal vs official use, and monitoring [34], social screening [17] and individuals' privacy settings [34, 35].

Studies that focus on governance documents in education have commented on a general lack of inclusion of learning and teaching specific provisions [36, 37]. These concerns include intellectual property rights, plagiarism and academic freedom [7, 8]. Although some studies acknowledge the need for stakeholders to confirm they have read a given policy [17, 34] there was an overwhelming lack of attention paid to the need to review and update such documents as a key item in a social media policy inclusion.

1.4 Accessibility of Social Media Governance Documents

To be implemented, governance documents need to be accessible, both in terms of the literacy needs of the reader in order to understand its content, and the format in which the documents are presented. Documents available online in HTML format are preferable to those that need downloading, usually as a PDF document, since HTML offers a more natural format for reading online, with easier page navigability and searching [26], whilst documents that need downloading interrupt the flow of online browsing, and depending on the file format may also need additional software to read it [38].

A recent study of social media documents in the US suggests that policy reading levels should be below 12th grade level, and preferably below 10th grade, to ensure all stakeholders can comprehend the policy content [26]. Since similar adult literacy patterns are seen in the UK as in the US, this is a sensible benchmark to place on the UK documents gathered here. Whilst it is not unreasonable to assume academics have overall higher reading abilities, many documents aimed at guiding social media usage in higher education focus on all university stakeholders and therefore encompass a wide range of students, employees and visitors who do not all possess the same literacy abilities.

Overall, the literature suggests that social media governance content has tended to have a negative focus, highlighting mitigation of risk and emphasizing sanctions for negative behavior rather than suggestions for positive behavior. Organizations need to be careful that policies don't ignore the potential opportunities and benefits that social media may bring by focused so intently on avoiding reputational risk.

2 Methodology

2.1 Data Collection

So far, much of the research undertaken has focused on the American education system. This study takes a UK-centric approach and examined social media policies from all public universities within England, Scotland, Wales and Northern Ireland.

The documentary data was gathered via an internet search. An institution population list was obtained from the Higher Education Statistics Agency (HESA) which gave the Institution ID, Region Code and Provider Name for publicly registered higher education providers based in the UK. The HESA data contained a list of 164 institutions, which included one private university along with 163 publicly funded institutions. To this list, five additional private institutions with degree awarding powers were added, with the final distribution shown in Table 1.

Table 1. Distribution of UK higher education providers

Region	Number of institutions
England	130
Wales	9
Scotland	19
Northern Ireland	5
Private	6
Total	169

The documents were identified through internet searches both for specific and broad search strings focusing on social media or social networking governance documents. Since the study was concerned with analyzing documents which were designed to direct, guide and support social media use within institution, both policies and guidelines were identified. Materials classed as policies were either whole explicit policies or sections of a larger document that was explicitly named, e.g.: Social Media Policy inside/as an appendix to an IT Policy. Guidelines were again those which were explicitly named as such and the other category comprised documents which contained sections related to the intentional use of social media but where it was subsumed into a different policy, e.g.: Acceptable Use, IT Regulations, Code of Conduct or Student Charter.

This approach of taking policy, guidelines or SM sections in larger IT documents has been used previously in the literature [26] and therefore was an appropriate stance to take here. Table 2 shows the number of universities that provided an explicit document related to the use of social media within their institution across the UK.

Table 2. Institutions with explicit Social Media Governance (SMG) documentation

Location	England	Wales	Scotland	Northern Ireland	Private	Total
With SMG documentation	79 (61%)	8 (89%)	18 (95%)	3 (60%)	2 (33%)	110 (65%)
Without SMG documentation	50 (39%)	1 (11%)	1 (5%)	2 (40%)	4 (67%)	58 (35%)
No. of HEIs	129	9	19	5	6	169

2.2 Policy Content Analysis

Analysis of the governance documents obtained was done via a thematic analysis. The aim was to identify the trends and foci apparent in the documents and thus determine how they were aligned with those found in the literature for both general and education-focused social media policy studies.

2.3 Readability Measures

In order to analyze the data collected a range of readability measures were calculated for each document. These included the quantitative scores Flesch-Kincaid Grade Level and Simple Measure of Gobbledygook (SMOG), along with Word Count and Number of Pages. Readability reflects "the ease of understanding or comprehension due to the style of writing" [39, p. 3] and such measures have been used extensively in the literature to determine how difficult it is to understand a given piece of text in written English [26, 40, 41]. Whilst they provide an objective, calculated measure to compare a range of text documents, they have been subject to criticism in that they rely on the premise that shorter sentences and words are easier to understand than longer ones, and that non-narrative elements that may ease readability such as use of white space, bullets, paragraphs, images and page layout are not included in these calculations [42, 43].

However, despite these deficiencies reading scores remain a popular analytical tool, primarily because they are easy to use and apply, and do not require any reader input in order to be calculated [39, 42, 43].

3 Results

In total 142 documents were collected from 110 institutions. The number of UK higher education institutions lacking an explicit social media governance document was one-third. For these institutions, it doesn't mean that no such document exists; indeed, many had a set of IT regulations or an Acceptable Use Policy in place that contained references to social media, but these were contextual points, for example references to bullying either in person or online, rather than a full consideration of online activity, its application, restrictions, support, and implications of use. The majority of institutions who did have a relevant document in place only had one, and where multiple documents existed, these were either policies or guidelines aimed at the same audience, or were the same document but aimed at a different audience (Fig. 1).

3.1 Document Scope

Of the documents gathered guidelines were more prevalent than policies, and nearly half of all documents were staff-focused, with the remainder equally split between those aimed at just students or all institution members. Documents aimed at either staff or students were more likely to be guidelines rather than policies, although for those aimed at all users the opposite was the case. Whilst the policy/guideline split was fairly even for staff, for students guidelines were almost three times more likely to be used (Table 3).

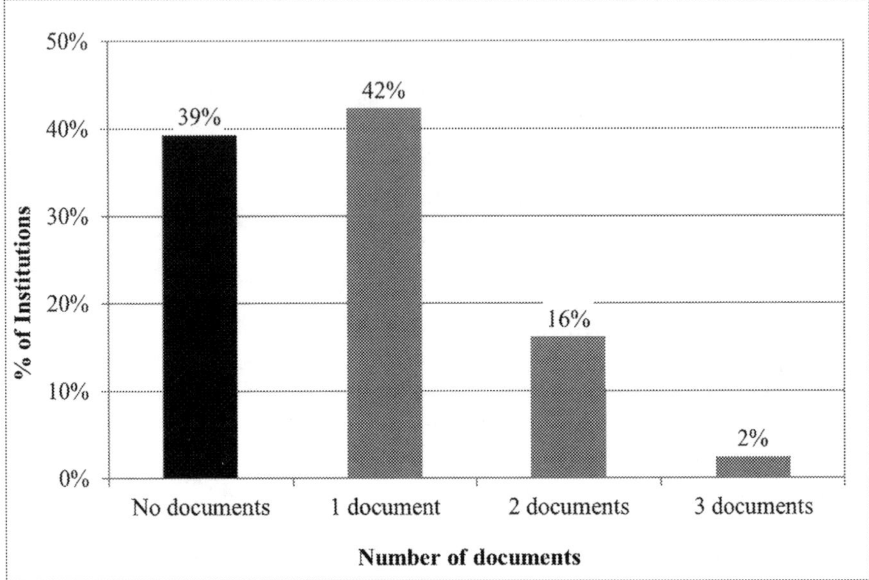

Fig. 1. Number of governance documents available per institution

Table 3. Audience and type of document

		Intended audience of document			Total
		Staff	Students	All users	
Type of document	Guidelines	26%	20%	9%	56%
	Policy	23%	8%	14%	44%
	Total	49%	28%	23%	100%

These statistics, along with the proportion of institutions without a relevant document, reflect the literature suggesting that institutions are lagging behind when it comes to providing explicit social media related governance [25, 26]. Where they are implemented, the majority is in guideline format and therefore lacks the official, enforceable behavior directives that policies provide [10].

3.2 Document Accessibility

All documents were either available to read online in a HTML format or were downloadable as a PDF document. The overwhelming majority of policies had to be downloaded rather than read online (Fig. 2). For guidelines, whilst not as large a difference, the opposite was the case. Each format has implications for accessibility and therefore likelihood of use, updating and version control.

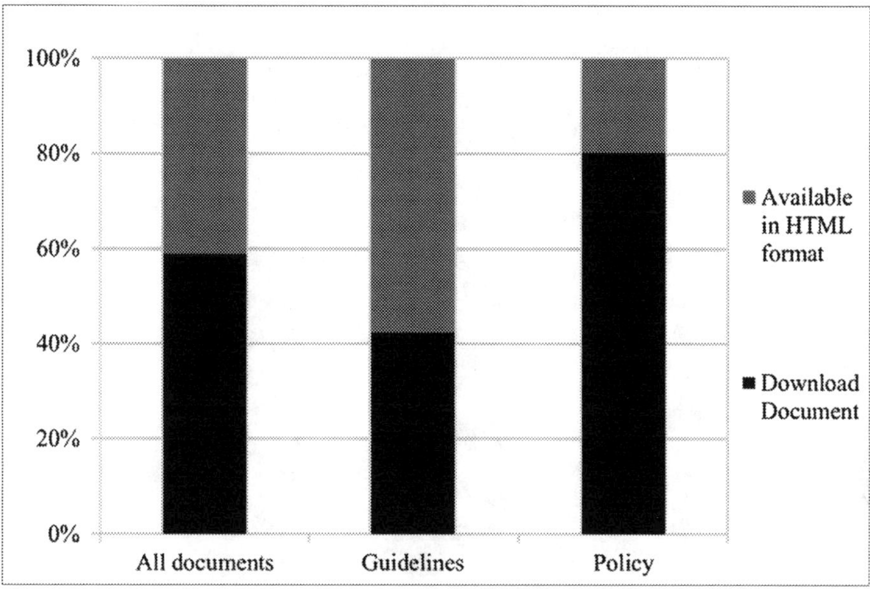

Fig. 2. Available format of document

When looking at whether these documents included some form of versioning or history, for example a date of version/update, owning department, or point of contact, nearly two thirds of all documents lacked even the basic information (Fig. 3). Guidelines overwhelmingly lacked any history or versioning, with four out of five documents having no detail, and whilst policies were more likely to have the information than not, only about 60% did.

This is a critical point, since once a document with no history is downloaded, the reader has little way of knowing if it is the latest and most relevant version, which can be problematic if it is being relied upon for formal guidance.

Taken together these results suggest that policies are more likely to be a downloadable document, but likely have some versioning history, whilst guidelines are much more likely to be available online but less likely to have any history attached. Given the speed with which social media technologies develop, updating and reviewing policies would seem to be a vital component of organizational governance which is currently being overlooked.

3.3 Document Readability

The Flesch-Kincaid grade level for the sample ranged from 6 to 18.8, with an average of 10.6 across all documents gathered, and whilst the average level was just at the recommended maximum, many of the documents were well over this level (Table 4). The SMOG values, whilst slightly higher given their mode of calculation, followed a similar pattern. Staff-targeted documents had a slightly higher grade level than those aimed at either students or all university members.

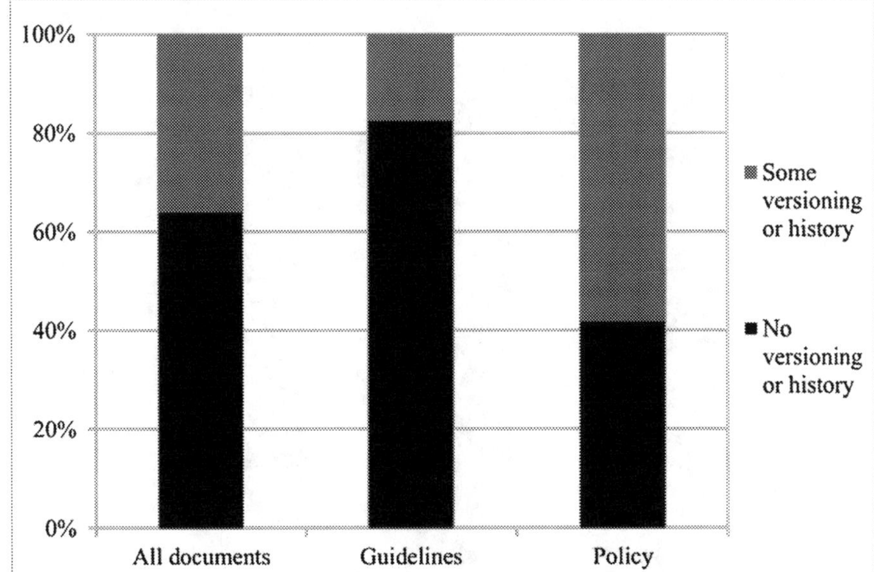

Fig. 3. Versioning history of documents

Table 4. Document readability statistics by audience

Audience	Average Flesch-Kincaid grade level	Average of SMOG index	Average word count	Average number of pages
Staff (n = 69)	10.8	13.3	1638	5.6
Students (n = 40)	10.3	12.8	1247	4.6
All users (n = 33)	10.4	12.7	2083	6.0
Total	10.6	13.0	1621	5.4

While the documents aimed at all members tended to be longer with more sentences in the document (not unsurprising given the wider audience), they had on average fewer words per sentence making them slightly easier to read than those aimed at just staff or students.

Whilst Table 5 suggests almost no difference in the reading levels or overall length for documents classed as either guidelines or policies, there was a much more marked difference between documents available online compared to those that needed downloading (Table 6). Online documents on average required a reading level two grades higher than those downloaded as a PDF, despite being shorter as measured in equivalent pages. This would seem to contradict the literature [38] which suggests that information available online is more accessible. Whilst downloading documents are

Table 5. Document readability statistics by document type

Document type	Average Flesch-Kincaid grade level	Average of SMOG index	Average word count	Average number of pages
Guidelines (n = 79)	10.6	13.1	1510	5.2
Policy (n = 63)	10.5	12.9	1757	5.7
Total	10.6	13.0	1621	5.4

Table 6. Document readability statistics by format availability

Format availability	Average Flesch-Kincaid grade level	Average of SMOG index	Average word count	Average number of pages
Online (n = 57)	12.0	14.2	1155	4.0
Download (n = 85)	9.6	12.2	1940	6.3
Total	10.6	13.0	1621	5.4

claimed to break the natural flow of working online, the evidence here suggests they are easier to read and therefore may be preferable.

In an effort to explain this difference, the proportion of documents using hyperlinks was examined, based on the potential for link addresses, Facebook page names or twitter handles which may use non-standard words impacting the overall grade level. Whilst the proportion of online documents using links was slightly higher than the downloadable documents at 65% to 56%, this difference was not felt to be a major contributor to the different statistics and therefore further research is needed.

3.4 Document Content Analysis

When analyzing the content of the documents, there was surprisingly little attention paid to the teaching and learning use of social media. The thematic analysis identified four overarching themes:

- Institutional Governance: scope, purpose, monitoring principles and compliance
- Institutional Account Management: strategic and operational management
- Online Behavior: Content, Conduct, Copyright, Confidentiality and Consent
- Risk Mitigation: resources, institutional and personal reputation and liability

Not all themes were as equally prominent in the policies as they were with the guidelines. Of prime importance to both sets of documents were content and conduct guidance, focusing on issues of privacy, in terms of not breaching copyright, confidentiality and having consent to post, and respectful behavior that avoided discrimination, harassment or bullying. Compliance was prominently featured in policy documents, but had much less emphasis in guidelines, whilst making the most of the opportunities social media offered was important in guideline documents but much less so in policies.

When considering the document audience, personal reputation and protecting yourself online were heavily featured in student targeted documents, but considerably less so in those aimed at staff or all users. Issues of transparency, disclaimers and directives for branding were prominent in documents targeted to staff, whilst account management and placing the onus of being personally responsible for online activity were featured more so in documents aimed at all users.

Most of the documents lack content that identify them as education-focused, so whilst they provide parameters which guide general social media behavior, there was little that would directly guide such behavior that involved aspects of teaching, learning, research or educational marketing activity.

4 Conclusions

Overall, this overview suggests that guidelines are more likely to be shorter documents and available online, whereas policies tended to be longer and presented as a downloadable PDF file. On average, those files available online measured as harder to read via the readability scores than the PDFs, and guidelines were generally harder to read than policies (except when aimed at students). Documents aimed at all users were on average the longest, followed by staff and then students. All documents had average reading requirements around the Flesch-Kincaid grade level of 10–11, mirroring results seen in previous studies [26]. Whilst academics would likely not find issue with reading such documents, the ability of students and general, non-academic staff to comprehend their contents may be impaired.

Clearly, one suggestion would be to work on reducing the readability levels of the policies, maybe by use of collaborative policy development groups involving a range of relevant stakeholders with different backgrounds to ensure all relevant groups can understand the content. Since the findings suggest downloadable documents are easier to read, the addition of full versioning and history information should be included as standard in all future policies and guidelines.

The content of the governance documents, whilst providing substantial guidance about how to use social media whilst *at* work, provided almost no information on how to use social media *for* work within a higher education institution. Contextualized direction for different stakeholder groups using social media as engagement and communication tools, such as academic debate or assessment, students looking for or giving feedback, or course leaders working with prospective students via social networking, was absent with nearly all guidance bordering on the generic. In order to be truly useful, institutional governance managers need to consider the local uses of social media within the education environment and modify their documentation accordingly. Such an approach combined with training on the impact of social media within education, both negative and positive, would support the understanding of using social media technology within the higher education environment.

These results contribute to the debate over reading ease and accessibility and how easily accessible they are to their intended audience, and direction on how to use social media within the educational context.

References

1. Kietzmann, J.H., Hermkens, K., McCarthy, I.P., Silvestre, B.S.: Social media? Get serious! Understanding the functional building blocks of social media. Bus. Horiz. **54**(3), 241–251 (2011)
2. Cheung, C.M., Chiu, P.Y., Lee, M.K.: Online social networks: why do students use facebook? Comput. Hum. Behav. **27**(4), 1337–1343 (2011)
3. Office of National Statistics. https://www.ons.gov.uk/peoplepopulationandcommunity/householdcharacteristics/homeinternetandsocialmediausage/datasets/internetaccesshouseholdsandindividualsreferencetables. Accessed 16 Mar 2018
4. Moran, M., Seaman, J., Tinti-Kane, H.: Teaching, Learning, and Sharing: How Today's Higher Education Faculty Use Social Media. Babson Survey Research Group, Boston (2011)
5. Ratliff, A.F.: Are they listening? Social media on campuses of higher education. J. Aust. NZ Stud. Serv. Assoc. **38**, 65–69 (2011)
6. Seaman, J., Tinti-Kane, H.: Social Media for Teaching and Learning. Pearson Learning Systems, London (2013)
7. Lupton, D.: 'Feeling better connected': academics' use of social media. News and Media Research Centre (2015)
8. Spallek, H., Turner, S.P., Donate-Bartfield, E., Chambers, D., McAndrew, M., Zarkowski, P., Karimbux, N.: Social media in the dental school environment, part A: benefits, challenges, and recommendations for use. J. Dent. Educ. **79**(10), 1140–1152 (2015)
9. Palmer, S.: Characterisation of the use of Twitter by Australian Universities. J. High. Educ. Policy Manag. **35**(4), 333–344 (2013)
10. Ahlquist, J.: Social Media Policy vs Guidelines vs Best Practice in Higher Education. http://www.josieahlquist.com/2013/09/30/policyguidebestpractice/?utm_campaign=shareaholic&utm_medium=printfriendly&utm_source=tool. Accessed 16 Mar 2018
11. Pomerantz, J., Hank, C., Sugimoto, C.R.: The state of social media policies in higher education. PLoS One **10**(5), e0127485 (2015)
12. Bertot, J.C., Jaeger, P.T., Hansen, D.: The impact of polices on government social media usage: issues, challenges, and recommendations. Gov. Inf. Q. **29**(1), 30–40 (2012)
13. Rodriguez, J.E.: Social media use in higher education: key areas to consider for educators. J. Online Learn. Teach. **7**(4), 539–550 (2011)
14. Selwyn, N.: Social media in higher education. Europa World Learn. 1–10 (2012)
15. Gruzd, A., Staves, K., Wilk, A.: Connected scholars: examining the role of social media in research practices of faculty using the UTAUT model. Comput. Hum. Behav. **28**(6), 2340–2350 (2012)
16. Priem, J., Piwowar, H.A., Hemminger, B.M.: Altmetrics in the wild: using social media to explore scholarly impact. ArXiv.org (2012). https://arxiv.org/abs/1203.4745. Accessed 16 Mar 2018
17. Wright, A.D.: Polish your social media policy. HR Mag. **58**(8), (2013)
18. Jeske, D., Shultz, K.S.: Using social media content for screening in recruitment and selection: pros and cons. Work Employ. Soc. **30**(3), 535–546 (2016)

19. Rowan-Kenyon, H.T., Martínez Alemán, A.M., Gin, K., Blakeley, B., Gismondi, A., Lewis, J., McCready, A., Zepp, D., Knight, S.: Social media in higher education. ASHE High. Educ. Rep. **42**(5), 7–128 (2016)
20. Zailskaite-Jakste, L., Kuvykaite, R.: Implementation of communication in social media by promoting studies at higher education institutions. Eng. Econ. **23**(2), 174–188 (2012)
21. Junco, R.: The need for student social media policies. Educause Rev. **46**(1), 60–61 (2011)
22. Joosten, T., Pasquini, L., Harness, L.: Guiding social media at our institutions. Plan. High. Educ. **41**(2), 1–11 (2013)
23. Chen, B., Bryer, T.: Investigating instructional strategies for using social media. Int. Rev. Res. Open Distance Learn. **13**(1), 87–104 (2012)
24. Bryer, T.A.: Designing social media strategies for effective citizen engagement: a case example and model. Natl. Civic Rev. **102**(1), 43–50 (2013)
25. Henry, R., Webb, C.: A survey of social media policies in US dental schools. J. Dent. Educ. **78**(6), 850–856 (2014)
26. Meiselwitz, G., Wang, Y.: Evaluations of policies for social networking use in higher education. In: Proceedings of EdMedia 2016–World Conference on Educational Media and Technology, pp. 424–429. Association for the Advancement of Computing in Education (AACE), Vancouver (2016)
27. Joosten, T., Pasquini, L., Harness, L.: Guiding social media at our institutions. Plan. High. Educ. **41**(2), 125 (2013)
28. Malesky, L.A., Peters, C.: Defining appropriate professional behavior for faculty and university students on social networking websites. High. Educ. **63**(1), 135–151 (2012)
29. Johnston, J.: 'Loose tweets sink fleets' and other sage advice: social media governance, policies and guidelines. J. Public Aff. **15**(2), 175–187 (2015)
30. Zerfass, A., Fink, S., Linke, A.: Social media governance: regulatory frameworks as drivers of success in online communications. In: 14th International Public Relations Research Conference on Pushing the Envelope in Public Relations Theory and Research and Advancing Practice, pp. 1026–1047. Institute for Public Relations, Gainesville (2011)
31. Fuduric, M., Mandelli, A.: Communicating social media policies: evaluation of current practices. J. Commun. Manag. **18**(2), 158–175 (2014)
32. Olmstead, K., Lampe, C., Ellison, N.B.: Social media and the workplace. Pew Research Center (2016)
33. Hrdinová, J., Helbig, N., Peters, C.S.: Designing social media policy for government: eight essential elements. Center for Technology in Government, University at Albany, Albany (2010)
34. McHale, R.: Navigating Social Media Legal Risks: Safeguarding Your Business. Que Publishing, Indianapolis (2012)
35. Dodd, M.D., Stacks, D.W.: Organizational social media policies and best practice recommendations. In: Social Media and Strategic Communications, pp. 159–179. Palgrave Macmillan, London (2013)
36. Pasquini, L.A., Evangelopoulos, N.: Organizational identity, meaning, and values: analysis of social media guideline and policy documents. In: Proceedings of the 2015 International Conference on Social Media and Society. ACM (2015)
37. Reed, A.S.: The current state of US higher education social media policies with regard to teaching and learning: a document review needs assessment. University of North Texas (2013)
38. Nielsen, J.: Top ten mistakes in web design (2011). http://www.nngroup.com/articles/top-10-mistakes-web-design. Accessed 16 Mar 2018

39. Klare, G.R.: Measurement of Readability. Iowa St. (1963). Ermakova, T., Krasnova, H., Fabian, B.: Exploring the Impact of Readability of Privacy Policies on Users' Trust, ECIS (2016)
40. Mandic, C.G., Rudd, R., Hehir, T., Acevedo-Garcia, D.: Readability of special education procedural safeguards. J. Spec. Educ. **45**(4), 195–203 (2012)
41. Paasche-Orlow, M.K., Taylor, H.A., Brancati, F.L.: Readability standards for informed-consent forms as compared with actual readability. N. Engl. J. Med. **348**(8), 721–726 (2003)
42. Jones, M.J., Shoemaker, P.A.: Accounting narratives: a review of empirical studies of content and readability. J. Account. Lit. **13**, 142 (1994)
43. Redish, J.: Readability formulas have even more limitations than Klare discusses. ACM J. Comput. Doc. **24**(3), 132–137 (2000)

Is It Really Fake? – Towards an Understanding of Fake News in Social Media Communication

Judith Meinert(✉), Milad Mirbabaie, Sebastian Dungs, and Ahmet Aker

University of Duisburg-Essen, Duisburg, Germany
{judith.meinert,milad.mirbabaie,sebastian.dungs, ahmet.aker}@uni-due.de

Abstract. This paper outlines the development of Fake News and seeks to clarify different perspectives regarding the term within Social Media communication. Current information systems, such as Social Media platforms, allow real-time communication, enabling people to produce and spread false information and rumors within a few seconds, potentially reaching a wide audience. This, in turn, could have negative impacts on politics, society, and business. To demystify Fake News and create a common understanding, we analyzed the literature on Fake News and summarized existing articles as well as strategies tested to detect Fake News. We conclude that detection methods mostly perform binary classifications based on linguistic features without providing explanations or further information to the user.

Keywords: Fake news · Fake news detection · Social media
Social media analysis · Social media analytics

1 Introduction

Today, the term "Fake News" is omnipresent and often discussed in both media and research. But to what does it refer to? What kind of information and news are included and where does it come from? The 2016 US presidential election campaign brought maximum attention to the phenomenon of intentionally using false information for political reasons. A famous example is the "Pizzagate" scandal, which was provoked by misinformation shared on Social Media about presidential candidate Hillary Clinton's connection to a child pornography ring acting in a pizzeria that ended up with a shooting [1, 2]. Furthermore, the use and distribution of Fake News discrediting presidential candidate Hillary Clinton is supposed to have influenced the actual election results [1–5].

Ever since, Fake News has had a profound impact on politics, democracy, society, and economy [6] and the ability to trigger actions, outcomes and consequences, in particular, if they are spread through Social Media [7, 8]. The issue is maximized by the fact that in the US most of the news consumption takes place through Social Media [8, 9]. Reports also show that among the consumed pieces of information, a vast amount of

© Springer International Publishing AG, part of Springer Nature 2018
G. Meiselwitz (Ed.): SCSM 2018, LNCS 10913, pp. 484–497, 2018.
https://doi.org/10.1007/978-3-319-91521-0_35

news is reported by alternative media types which are regarded as a source of misinformation and propaganda due to the loss of commonly held standards regarding mediated information and the absence of easily decipherable credibility cues [10].

In the US, for example, Twitter users reference news reported by alternative media sites as often as news produced by professional news media. The authors report, however, that this was different for EU countries such as the UK, Germany, and France. In these countries, alternative news websites were only referred to five to 12 percent of the time [11]. The emergence of Fake News is additionally fraught with risks while most Americans limit their political participation during elections to sharing memes, pictures, quotes, and statements about their favored candidate [4]. However, the phenomenon of spreading false information is not limited to the US; in Europe, fake stories were published as well, particularly about refugees, refugee policy, and politicians [12].

While researchers have proved the success of Fake News in terms of its distribution and impact on Social Media [2, 6, 7], one question remains unanswered: Why is Fake News successful and why do recipients believe in such misinformation without further fact-checking?

With the radical examples that occurred in the US election, an important issue became how to detect Fake News in Social Media [3]. Facebook founder Mark Zuckerberg emphasized (in a statement on Facebook), that it is important to find “better technical systems to detect what people will flag as false before they do it themselves.” But, what methods have already been developed and tested, and what can be improved for the future?

Once Fake News is pervasive, businesses, public institutions, and governments have to react efficiently and quickly to label or delete Fake News published in Social Media and manage the situation [13–16]. For instance, in June 2017, Germany passed a law (“Netzwerkdurchsetzungsgesetz”) [17] that requires the deletion of Fake News within 24 hours. Accordingly, Facebook started a fact-checking collaboration with the research initiative Corrective, which allows users to indicate potentially false news to get it double checked. In addition, Facebook and Google both began to work on improving the detection of Fake News, for instance by optimizing the news feed algorithm or identifying URLs of potential Fake News distributors [18].

However, at this point, we also have to investigate the potential risks of flagging something as Fake News. It is a balancing act not to behave in a manipulative or suggestive way but rather to support recipients of Social Media communication with valid credibility ratings for information posted.

While Fake News is currently an almost-universal topic, and research has already been conducted on this topic, no overview of Fake News and Fake News detection, including underlying psychological mechanisms, exists. Research articles or essays focus either on technical aspects and solutions [2, 19] or highlighting current events [7].

Therefore, we will put the lens on Fake News and seek to understand its origin and emergence by examining its definitions, use, and interpretation in current research articles. Furthermore, we summarize methods and strategies which are already applied for detecting Fake News and relate them to situational and contextual conditions. We continue by discussing the (potential) limits of Fake News detection and round up our article with a conclusion including recommendations for further research.

2 How Is Fake News Defined and Interpreted in the Literature?

There have been many discussions about the definition and meaning of the term Fake News. While the definitions present in the literature have many similarities as to the meaning of the term, the definitions still differ in some respects. One definition by Douglas and colleagues refers to Fake News as a "deliberate publication of fictitious information, hoaxes, and propaganda" [19, p. 36]. A similar understanding is stated by Klein and Wueller, who describe Fake News as an "online publication of intentionally or knowingly false statement of fact" [20, p. 6], while the publishers of Fake News do not necessarily have to believe that their asserted facts are correct, which outlines the importance of intentionality [6].

Moreover, Allcott and Gentzkow [1] describe Fake News as news articles that are intentionally and verifiably false and mislead the reader. However, they make some exceptions, for instance, accidental errors of reporting, rumors which are not related to an article, conspiracy theories and incorrect political statements. Excluding statements which do not originate from articles illustrates a differentiation between Fake News and rumors. Thus, rumors are seen as information spread through Social Media which are unverified at the time of publication [21, 22], such as in crisis situations [23–25]. An excellent review regarding rumors and related issues is provided by Zubiaga et al. [21].

However, the importance of the intention to deceive is underlined by deliberately factoring out mistakes which occur by ignoring a lack of verification of information sources [26]. To be precise, this indicates the differentiation between misinformation and disinformation. Misinformation refers to false information that is not intentionally inaccurate, but rather a result of misinterpretation or a lack of source verification. In contrast, disinformation is used to describe content that is fabricated to be misleading [27].

Some authors focus more on the financial aspect of distributing false information. According to Silverman [28], Fake News is thoroughly false information solely created for financial gain to boost attention. This form of misinformation is referred to as click-baiting, which is applied to achieve financial goals by publishing attention-grabbing, misleading headlines to increase traffic to a connected website. Klein and Wueller [20] concur with that point of view. They state that false facts are typically published on websites and spread on Social Media for profit or social influence.

Another point of discussion is satire. On the one hand, it can be regarded as a form of Fake News which is accepted as accurate by many observers [29]. On the other hand, satirical websites use humor and exaggeration to criticize social and political issues, so the primary intention is not to make the public believe their news is accurate [20]. However, most of the authors exclude satire from the Fake News category, because it is not produced to achieve financial or political benefits [5].

Besides a lack of the factual basis in news stories, Berghel [2] presented typical characteristics to indicate Fake News; these include hidden or blurred authorship or imprint and the use of account names which sound similar to recognized news portals. Moreover, McClain adds that Fake News stories try "to imitate the style and appearance of real news articles" [30, p. 1].

In sum, Fake News is described as news articles that contain false, discrediting or whitewashed information with the intention to manipulate and deceive recipients. Publication of such false information is mainly motivated by financial or political interests [1, 22, 31] and spreading is accelerated by the popularity of Social Media sites.

3 Where Does Fake News Come From?

In recent years, the usage of Social Media platforms has grown tremendously, leading to a change in peoples' work and lives and resulting in increased online human interaction. Due to the development of Social Media, users are not only able to consume information, they can produce and share content [32, 33]. Furthermore, mass media no longer functions as a gatekeeper of information [1]. This results in more opportunities for content production as well as reaching a potentially broad audience using informal and privately hosted Social Media channels.

In general, Social Media provides a communication space without gatekeepers, filtering options, or the control for quality standards, so not only private users can produce content and information, but also groups, organizations, parties, and politicians [34]. These preconditions can lead to the publication or sharing of information that is not validated and potentially untrue [2] and because of this able to negatively impact users' perceptions and opinions.

Fake News, which reflect the negative side of Social Media communication, is gaining significant popularity nowadays [1, 2, 35]. However, manipulating information is not a new phenomenon. It is currently receiving attention [2, 20] because of its accelerated means to share and distribute intentionally faked content. But, with a detailed look at the history and context of false information in the media, it can be stated that the use of false information took place for a long time. Historical examples indicate that people have always manipulated information and stories to achieve specific goals. For instance, one famous example from the past was the so-called "Great Moon Hoax", a series of articles describing the existence of human beings on the moon and published by the New York Sun in 1835 to increase the paper's circulation [1, 27, 35].

In later years, the term Fake News was used to refer to comedic programs engaging in political satire [36] or general parodies of professional news [37]. Broussard described Fake News as an outcome of combining entertainment and information in media content, thus creating a third genre called "infotainment." The researcher also stressed that Fake News could help audiences understand complex political information through the humorous way it was presented [38].

Due to the rise of Social Media platforms, the term Fake News has gone through a substantial transformation [36]; it is now commonly seen as a form of misinformation that benefits from the fast pace of information dissemination on social networks [39]. This refers to the versatile possibilities to share content on Social Media through the connected structure of the network. Furthermore, site vendors encourage the spread of information by allowing users to broadcast content to their personal networks using a single mouse click. Combined with the ubiquitous mobile accessibility of Social Media applications, the rate of information distribution through Social Media is considerable.

Consequently, a “new political and cultural climate” [30, p. 1] arose in which the prevalence of Fake News and alternate facts grew significantly. With the US elections at the end of 2016 and the frequent use of manipulated news stories as a powerful part of the campaign strategy [1–5], Fake News has achieved great public interest [35].

Apparently, false information with the intent to manipulate recipients has long been used. Nowadays, the engaging features of Social Media, namely the ease of sharing content and social connections, have become main reasons for an increased emergence of Fake News [39].

4 Why Do Recipients Believe Fake News?

The important role of Social Media is supported by the fact that many people use it as their only source of news and political information [30] without utilizing professionally edited media, such as newspapers and magazines [40]. This results in an immense impact of (false) information spread on Social Media.

Besides, not turning to traditional information sources makes recipients more vulnerable to manipulation. The logic of social networks includes high connectivity, whereby faked content can quickly go viral by receiving thousands of likes and shares [26]. This, in turn, can create a potentially misleading impression of trust in a piece of information.

The preconditions of Social Media - everyone can produce content at any time - lead to vast amounts of information. Users are unable to process everything in an elaborated way, because they are confronted with information overload and limited cognitive capacities [41, 42]. Social Media communication, in general, was found to be processed more peripherally [43]. Accordingly, recipients are possibly guided by simple heuristic rules, for instance, applying the Bandwagon heuristic, described as “If others think that something is good, then I should, too” [44, p. 83]. This implicit rule was already found to be influential for ratings and reviews in e-commerce [45] and could possibly serve as explanation for why recipients believe in Fake News if it is shared and liked a lot by others.

In line with this, Pennycook and colleagues [6] revealed that recipients’ tendency to rely on Fake News is strengthened by perceptions of familiarity due to prior exposure. Due to high connectivity, current articles are widely distributed in Social Media by sharing and liking activities of members of individuals’ personal networks, which also leads to potential repeated receptions. Through these repetitions, recipients tend to be guided by the rule “I saw this before so its probability true” [6, p. 8]. This process probably takes place unconsciously, since explicit warnings did not change this behavior. Additionally, Fake News headlines which have been presented before were rated as more credible, even combined with a warning message, compared to Fake News headlines which were viewed for the first time and not accompanied by a warning.

Moreover, if users believe that a website or news account is journalistic, they are easily persuaded and believe everything stated by this source [35]. Often, Fake News producers exploit this by employing credible-sounding names for the Fake News sources like *CNN_politics* or *The Denver Guardian* or by using an article design which is derived from journalistic sources [6]. Another example is a faked Twitter account with the name of former New York mayor Giuliani [5]. It is intricate for the recipients to identify the account as fake, so that source and message are mainly perceived as credible due to the supposed reputation of the account.

A further aspect refers to the tendency that people are striving for consistency in their attitudes, behavior, and self-perception and thus favor information which is in line with their opinions. Psychological mechanisms like cognitive dissonance theory and belief disconfirmation paradigm [46] state that persons, who are confronted with conflicting news, perceive feelings of stress, which often result in a rejection of the conflicting information to defend and justify their prior beliefs. This behavior can be transferred to Social Media consumption as users prefer to receive information that confirms already existing views [39] and mostly avoid conflicting information. This behavioral pattern is strengthened by technical features of Social Media applications. For instance, Facebook uses filtering and search algorithms that limit users' news feed content to previously consumed topics. Similar techniques are applied in other networks and other contexts. As a result, filter bubbles are created, wherein users only read and share information they already believe in. Due to this so-called confirmation bias, malformed worldviews and echo chambers can be formed even when the disinformation is disproved [4, 47].

It is a crucial finding that the identification and correction of false information do not necessarily change peoples' beliefs because they have already made up their minds. According to Berghel, since a Fake News story is posted online, "the story already had legs" [2, p. 82]. As a correction or deletion of manipulated content can even backfire and entrench users in their initial beliefs [39], it is difficult to assess how to deal with identified Fake News stories and which strategy could be used to present it to the users efficiently.

Since Fake News often appeals to emotions instead of being supported by evidence or facts, it is even easier for users to rely on this kind of information because less cognitive effort is needed to make a judgment or form an opinion on something. As outlined earlier, the 2016 US elections played an important role in the development and transformation of today's understanding of Fake News. It has been found that for the voters in the elections (especially those who voted for Donald Trump), verifiable and reliable facts get outweighed by emotional headlines and news [48]. Taken together, the emotional impact of Fake News should not be disregarded in the discussion.

5 What Strategies Can Be Used to Detect Fake News?

As mentioned before, it is imperative for governments, public institutions and businesses to detect Fake News. In our article, we highlight two existing ways for detecting Fake News. The identification of false information can either be tackled by manual efforts based on experts or crowd knowledge, or by using automated approaches to identify check-worthy claims and perform a veracity check [49]. For the first approach, Social Media users can be involved by being asked to flag all potentially Fake News articles to be checked later by journalists or research organizations such as Corrective in Germany.

Naturally, manual Fake News detection is mostly unfeasible or at least time consuming due to the vast amount of content generated on Social Media. Therefore, automated approaches are more suited for the task of detecting Fake News systematically. These approaches can be categorized by their primary features' sources, i.e., some approaches rely on linguistic cues, while others perform network analyses to detect behavioral patterns. In either method, after feature extraction, machine learning algorithms are used to tackle the problem. In fact, they indicate if something is fake or factual news based on the features. In the following examples of automated Fake News detection, concepts taken from the literature are reviewed and discussed. Note that our review is short and aims to give just a taste regarding approaches to tackle the problem automatically.

By using linguistic features to classify scientific publications into fraudulent or genuine material, Markowitz and Hancock reached an accuracy rate of 71.4% [50]. The most descriptive features in the discussed dataset were found to be adjective, amplifier, and diminisher as well as certainty term frequencies. Identified relevant features are also found to be useful in fake review detection as well as in research related to reality monitoring. Hardalov et al. [26] used a combination of linguistic, credibility and semantic features to determine real from Fake News. Linguistic features in their work include (weighted) n-grams and normalized number of unique words per article. Credibility features were adopted from the literature and included capitalization, punctuation, pronoun usage and sentiment polarity features generated from lexicons. Text semantics were analyzed using embedding vectors trained on DBPedia. All feature categories were tested independently and in combination based on self-created datasets. In two out of three cases, the best performance was achieved using all available features.

Besides using linguistic or contextual features to detect false information, argumentation and textual structure can be used for the analyses. Lendavi and Reichel [51] investigated how contradictions in rumorous sequences of micro-posts can be detected by analyzing posts at the level of text similarity only. The authors argue that vocabulary and token sequence overlap scores can be used to generate cues to veracity assessment, even for short and noisy texts. In addition, Ma et al. [52] expanded on previous work by observing changes in linguistic properties of messages over the lifetime of a rumor. Using SVM (support vector machine) based on time series features, they were able to show reasonable success in the early detection of an emerging rumor.

Another approach is presented by Conroy et al. [31] who argue that the best results in Fake News detection could be achieved by combining linguistic and network features. This is because in the literature, both feature categories are used in topic-specific studies.

To utilize the information provided by knowledge networks like DBPedia, Ciampaglia et al. [18] continued with the proposition to map the fact-checking task to the well-known task of finding the shortest path in a graph. In that case, a shorter path indicates a higher probability of a truthful statement. It should be noted that the latter approach is limited by the requirement that the knowledge graph must include the topic in question. In the case of emerging topics, that requirement will hardly ever be met in practice. However, methods used for Fake News detection are highly dependent on the specific case and related conditions. So, currency, time, duration and topic area have to be considered when selecting the method.

A further feature which could be exploited are pictures accompanying a piece of information. Accordingly, Jin et al. [49] include news articles' images in the Fake

News detection process. Based on a multimedia dataset, the authors explore various visual and statistical image features to predict respective articles' veracity. Promising results were achieved by comparing the distance of a set of event-related images to the general set containing images of all events. Moreover, within another research project, Jin et al. [53] proposed a Fake News detection method utilizing the credibility propagation network built by exploiting conflicting viewpoints extracted from tweets.

Some factors that come with news articles are not yet extensively included in strategies for Fake News identification. In this line, Shu et al. [54] state that social context features of news articles are underused in Fake News detection in Social Media. These features are categorized as user-based, referring to characteristics of the user profile like number of followers, followings, or postings, post-based, which includes postings related to the Fake News article, and network-based, which describes a cluster of user groups depending on their reaction to the article or their relationships with each other (e.g., the following structure). The authors advise researchers to consider those features appropriately when performing Fake News detection.

Table 1 summarizes the discussed articles and their approaches for detecting Fake News. It is evident that current Fake News detection approaches commonly focus on linguistic features. While these show promising results in their respective domains, other feature categories are underused in the literature. The network's structure could be used to detect spreading patterns of Fake News and include temporal information to improve prediction accuracy. In domains where multimedia content is prevalent, the analysis should be extended to include visual features accordingly. Finally, current approaches perform binary Fake News classification only. Future work could explore the possibility of probabilistic classification yielding a Fake News score on a continuous scale.

Table 1. Methods for Fake News detection used in the literature (*denotes proposition rather than actual application of method)

Author/Method	Linguistic features	Semantic features	Credibility features	Network features	Visual and statistical image features	Social context features
Markowitz and Hancock [50]	X					
Hardalov et al. [26]	X	X	X			
Lendvai and Reichel [51]	X					
Ma et al. [52]	X					
Conroy et al. [31]	X*			X*		
Ciampaglia et al. [18]				X		
Jin et al. [49]					X	
Shu et al. [54]						X*

6 What Are Ethical Borders of Fake News Detection?

After the US elections, often seen as the climax of the rise of Fake News and a post-truth age, the call for more observation and control in Social Media came up to optimize the identification of Fake News. However, defining truth and identifying the truthfulness of information is difficult.

One possible approach for defining the truth is to adopt Appleman and Sundar's definition; they refer to the "veracity of the content of communication" [55, p. 63]. In fact, this means if the included information could be proved, the message is true, and if not, it will be labeled as fake. However, it should be considered that even a concept like veracity is situated on a continuum [31] and a binary decision between true or false - as it is performed by today's automated methods - is probably not (always) sufficient. For instance, even if satire and parody do not intentionally deceive recipients, it happens nonetheless, because the content is not clearly and absolutely true [29]. As a result, the recipient of the information has to be considered as an influencing factor of how information is processed and perceived. Using a strict binary definition for Fake News detection, satire had to be labeled as fake content. However, the effects on society related to flagging or deleting satirical communication and media are unclear. Typically, satire is used for criticizing political events or actions. Removing or blocking this content from the discussion because some recipients are potentially unable to get the joke, understand the message, or double-check the information with other sources has to be considered carefully.

Furthermore, figuring out the author's real intention is probably difficult. How should anyone be able to find out if the author shared false information because he or she misinterpreted the facts or intended to manipulate the audience [27]? From a practical perspective, misinformation and disinformation are not highly selective, and the differentiation is hard to pinpoint using an objective viewpoint.

Overall, there is a balancing act between governmental supervision and freedom of speech and expression [14, 16]. Most authors are quite critical of governmental control over the media, especially when also considering historical examples of misuse. Once media regulation methods are applied, they can also be extended or encroached within a change in government [17].

Regarding potential risks, some articles highlight the importance of strengthening recipients' media competence instead of building up governmental control [17, 19] which would positively contribute to support users in identifying satire as well as false information and getting to know suitable fact-checking methods.

Moreover, Tufekci [56] draws attention to the fact that some Social Media companies like Facebook have a monopoly position concerning insights into data patterns. It would bode well if those companies were to collaborate with researchers to evaluate relationships and samples with the aim of applying measures for optimized Fake News identification. Including researchers and independent organizations would lead to more objectivity in the process. However, given the commercial nature of these networks, making user data accessible to researchers would contradict their core business model. Consequently, legislation is needed for the scientific community to gain a right to access large-scale usage data.

The importance of an appropriate societal debate about this matter is also highlighted in current events. For instance, Facebook has recently been criticized for its selection of Trending Topics (this function is not available in all countries), since conservative news articles seem to be incorporated less frequently. This is possibly caused by an opposite political attitude of the employees of Facebook, who were selecting the articles to appear in this section [3]. Besides potentially unknown biases such as individual perspectives on a controversial topic, financial aspects should be considered, especially in the case of dominant, globally positioned Social Media companies.

Generally, the limitation of expression and publication of opinions and information come along with restricting freedom of speech. Marking postings as fake especially contradicts the idea that Social Media represents a platform where everyone has an opportunity to express opinions and thoughts [3].

7 Conclusion

By summarizing the existing literature, we shed light on the phenomenon of Fake News. Publishing false information (for any purpose) is not a new phenomenon, but contextual conditions, speed of distribution, and potential message range have changed immensely over time. The rise of Social Media enables rapid distribution of information so that the impact of Fake News (information, stories, etc.) has grown. Social Media is increasingly used as the only source for gaining political information and news [40]. At the same time, Fake News has become an influential tool for elections and society, especially since the US elections in 2016 [1]. Considering the negative implications, the importance of detecting and "fighting" Fake News rises continuously.

Several automated detection approaches using standard machine learning techniques have been developed. Most commonly, these approaches perform binary classification that relies on extracting of linguistic features to determine news veracity. However, those methods have been tested in domain-specific datasets only, potentially limiting their generalizability. Furthermore, as was discussed above, a binary classification may be insufficient for real-world applications.

Besides working on technical solutions, psychological factors can be considered to optimize methods for Fake News detection. For instance, Berghel [2] emphasizes the need to present the classification process as well as the reasons for how and why a news story is indicated as fake to the user. This should help in overcoming the unconscious use of heuristic rules. Additionally, users have to learn to not overly rely on account names when judging articles' veracity, as they can easily be fabricated to resemble official news agencies.

Moreover, research in the area of recommender systems showed that people tend to accept recommendations for products more if they were accompanied by explanations [57]. These explanations encompass information including the kind of data a recommendation is based on. Transferred to Fake News detection, explaining to users why a story is faked, which facts are presented wrong, and where they can get further information could be a promising approach along with raised attention, acceptance and trust from the recipient's perspective. Additionally, explanations can support the improvement of the detection process; for example, mistakes could be found more easily.

Overall, the need for developing and applying methods to efficiently detect intentionally published Fake News stories increased with the use of Social Media and potentially unlimited and fast-running possibilities to produce and spread information. Further research is needed to improve practical used mechanisms to overcome existing difficulties like users' reactance, unclear definitions of truth, and ethical considerations around restricting or limiting the extent of user expressions.

Acknowledgements. This work is supported by the German Research Foundation (DFG) under grant No. GRK 2167, Research Training Group "User-Centred Social Media". We also thank our student assistant Annika Deubel for supporting us with the literature review.

References

1. Allcott, H., Gentzkow, M.: Social media and fake news in the 2016 election. J. Econ. Perspect. **31**(2), 211–236 (2017). https://doi.org/10.3386/w23089
2. Berghel, H.: Lies, damn lies, and fake news. Computer **50**(2), 80–85 (2017). https://doi.org/10.1109/MC.2017.56
3. Isaac, M.: Facebook, in cross hairs after election, is said to question its influence. The New York Times (2016). https://www.nytimes.com/2016/11/14/technology/facebook-is-said-to-question-its-influence-in-election.html. Accessed 31 Jan 2018
4. Ott, B.L.: The age of Twitter: Donald J. Trump and the politics of debasement. Crit. Stud. Media Commun. **34**(1), 59–68 (2017). https://doi.org/10.1080/15295036.2016.1266686
5. Rogers, K., Bromwich, J.E.: The Hoaxes, Fake News, and Misinformation We Saw on Election Day. The New York Times (2016). https://www.nytimes.com/2016/11/09/us/politics/debunk-fake-news-election-day.html. Accessed 6 Feb 2018
6. Pennycook, G., Cannon, T.D., Rand, D.G.: Prior Exposure Increases Perceived Accuracy of Fake News. Social Science Research Network (2017). https://papers.ssrn.com/sol3/papers.cfm?abstract_id=2958246. Accessed 6 Feb 2018
7. Howard, P.N., Bolsover, G., Kollanyi, B., Bradshaw, S., Neudert, L.M.: Junk news and bots during the US election: What were Michigan voters sharing over Twitter. Data Memo, January 2017. Project on Computational Propaganda, Oxford (2017). http://comprop.oii.ox.ac.uk/2017/03/26/junk-news-and-bots-during-the-uselection-what-were-michigan-voters-sharing-over-twitter. Accessed 6 Feb 2018
8. Mitchell, A., Gottfried, J., Matsa, K.E.: Millennials and political news. Pew Research Center (2015). http://www.journalism.org/2015/06/01/millennials-political-news/. Accessed 6 Feb 2018
9. Stieglitz, S., Mirbabaie, M., Ross, B., Neuberger, C.: Social media analytics-challenges in topic discovery, data collection, and data preparation. Int. J. Inf. Manag. **39**, 156–168 (2018). https://doi.org/10.1016/j.ijinfomgt.2017.12.002
10. Starbird, K.: Examining the alternative media ecosystem through the production of alternative narratives of mass shooting events on Twitter. In: ICWSM, pp. 230–239 (2017)
11. Gallacher, J.D., Kaminska, M., Kollanyi, B., Yasseri, T., Howard, P.N.: Social Media and News Sources during the 2017 UK General Election. Data Memo, June 2017. Project on Computational Propaganda, Oxford (2017). http://comprop.oii.ox.ac.uk/wp-content/uploads/sites/89/2017/06/Social-Media-and-News-Sources-during-the-2017-UK-General-Election.pdf. Accessed 6 Feb 2018

12. Connolly, K., Chrisafis, A., McPherson, P., Kirchgaessner, S., Haas, B., Phillips, D., Hunt, E., Safi, M.: Fake news: an insidious trend that's fast becoming a global problem. The Guardian (2016). https://www.theguardian.com/media/2016/dec/02/fake-news-facebook-us-election-around-the-world. Accessed 31 Jan 2018
13. Gabriel, R., Röhrs, H.-P.: Trends, Chancen und Risiken von Social-Media-Anwendungen – eine kritische Betrachtung. In: Gabriel, R., Röhrs, H.-P. (eds.) Social Media, pp. 219–243. Springer, Heidelberg (2017). https://doi.org/10.1007/978-3-662-53991-0_9
14. Goodman, E.: How has media policy responded to fake news? Media Policy Blog (2017). http://blogs.lse.ac.uk/mediapolicyproject/2017/02/07/how-has-media-policy-responded-to-fake-news/. Accessed 31 Jan 2018
15. Mirbabaie, M., Ehnis, C., Stieglitz, S., Bunker, D.: Communication roles in public events. In: Doolin, B., Lamprou, E., Mitev, N., McLeod, L. (eds.) Working Conference on Information Systems and Organizations, pp. 207–218. Springer, Heidelberg (2014). https://doi.org/10.1007/978-3-662-45708-5_13
16. Mirbabaie, M., Zapatka, E.: Sensemaking in social media crisis communication - a case study on the Brussels bombings in 2016. In: Proceedings of the Twenty-Fifth European Conference on Information Systems (ECIS) (2017). https://aisel.aisnet.org/ecis2017_rp/138/. Accessed 6 Feb 2018
17. BMJV Aktuelle Gesetzgebungsverfahren. Gesetz zur Verbesserung der Rechtsdurchsetzung in sozialen Netzwerken (Netzwerkdurchsetzungsgesetz – NetzDG) (2017). https://www.bmjv.de/SharedDocs/Gesetzgebungsverfahren/Dokumente/BGBl_NetzDG.html;jsessnid=111BF7BB5DA1C6A0A4D6F8912345D764.1_cid324?nn=6712350. Accessed 8 Feb 2018
18. Ciampaglia, G.L., Shiralkar, P., Rocha, L.M., Bollen, J., Menczer, F., Flammini, A.: Computational fact checking from knowledge networks. PLoS ONE **10**(6), e0128193 (2015). https://doi.org/10.1371/journal.pone.0128193
19. Douglas, K., Ang, C.S., Deravi, F.: Farewell to truth? Conspiracy theories and fake news on social media. Psychologist **30**, 36–42 (2017)
20. Klein, D.O., Wueller, J.R.: Fake news: a legal perspective. J. Internet Law **20**(10), 6–13 (2017)
21. Zubiaga, A., Aker, A., Bontcheva, K., Liakata, M., Procter, R.: Detection and resolution of rumours in social media: a survey. ACM Comput. Surv. (2017). https://arxiv.org/pdf/1704.00656.pdf. Accessed 6 Feb 2018
22. Starbird, K., Spiro, E., Edwards, I., Zhou, K., Maddock, J., Narasimhan, S.: Could this be true? I think so! Expressed uncertainty in online rumoring. In: CHI 2016 Proceedings of the 2016 CHI Conference on Human factors in Computing Systems, pp. 360–371. ACM (2016). https://doi.org/10.1145/2858036.2858551
23. Oh, O., Agrawal, M., Rao, H.R.: Community intelligence and social media services: a rumor theoretic analysis of tweets during social crises. MIS Q. **37**(2), 407–426 (2013)
24. Stieglitz, S., Bunker, D., Mirbabaie, M., Ehnis, C.: Sense-making in social media during extreme events. J. Conting. Crisis Manag. **26**, 1–12 (2017). https://doi.org/10.1111/1468-5973.12193
25. Stieglitz, S., Mirbabaie, M., Milde, M.: Social positions and collective sense-making in crisis communication. Int. J. Hum.-Comput. Interact. (2018). https://doi.org/10.1080/10447318.2018.1427830
26. Hardalov, M., Koychev, I., Nakov, P.: In search of credible news. In: Dichev, C., Agre, G. (eds.) International Conference on Artificial Intelligence: Methodology, Systems, and Applications, pp. 172–180. Springer, Cham (2016). https://doi.org/10.1007/978-3-319-44748-3_17

27. Jack, C.: Lexicon of Lies: Terms for Problematic Information. Data and Society Research Institute, New York (2017). https://datasociety.net/pubs/oh/DataAndSociety_LexiconofLies.pdf. Accessed 6 Feb 2018
28. Silverman, C.: This Analysis Shows How Viral Fake Election News Stories Outperformed Real News on Facebook. BuzzFeed News (2016). https://www.buzzfeed.com/craigsilverman/viral-fake-election-news-outperformed-real-news-on-facebook?utm_term=.oopAlP795#.wtEYJ9gba. Accessed 31 Jan 2018
29. Rubin, V., Conroy, N., Chen, Y., Cornwell, S.: Fake news or truth? using satirical cues to detect potentially misleading news. In: Proceedings of the Second Workshop on Computational Approaches to Deception Detection, pp. 7–17 (2016)
30. McClain, C.R.: Practices and promises of Facebook for science outreach: becoming a "Nerd of Trust". PLoS Biol. **15**(6), e2002020 (2017). https://doi.org/10.1371/journal.pbio.2002020
31. Conroy, N.J., Rubin, V.L., Chen, Y.: Automatic deception detection: methods for finding fake news. Proc. Assoc. Inf. Sci. Tech. **52**(1), 1–4 (2015). https://doi.org/10.1002/pra2.2015.145052010082
32. Gil de Zúñiga, H., Molyneux, L., Zheng, P.: Social media, political expression, and political participation: panel analysis of lagged and concurrent relationships. J. Commun. **64**(4), 612–634 (2014). https://doi.org/10.1111/jcom.12103
33. Stieglitz, S., Brockmann, T., Dang-Xuan, L.: Usage of social media for political communication. In: Proceedings of 16th Pacific Asia Conference on Information Systems, Ho Chi Minh City, Vietnam (2012)
34. Jong, W., Dückers, M.L.A.: Self-correcting mechanisms and echo-effects in social media: an analysis of the "gunman in the newsroom" crisis. Comput. Hum. Behav. **59**, 334–341 (2016). https://doi.org/10.1016/j.chb.2016.02.032
35. Vargo, C.J., Guo, L., Amazeen, M.A.: The agenda-setting power of fake news: a big data analysis of the online media landscape from 2014 to 2016. New Media Soc. **10**(6), 1–22 (2017). https://doi.org/10.1177/1461444817712086
36. Gerhart, N., Torres, R., Negahban, A.: Combatting fake news: an investigation of individuals' information verification behaviors on social networking sites. In: Twenty-Third Americas Conference on Information Systems, Boston (2017)
37. Day, A., Thompson, E.: Live from New York, it's the fake news! Saturday night live and the (Non)politics of parody. Pop. Commun. **10**(1–2), 170–182 (2012). https://doi.org/10.1080/15405702.2012.638582
38. Broussard, P.L.: Fake news, real hip: rhetorical dimensions of ironic communication in mass media. Unpublished thesis, The University of Tennessee at Chattanooga, Tennessee (2013)
39. Lazer, D., Baum, M., Grinberg, N., Friedland, L., Joseph, K., Hobbs, W., Mattsson, C.: Combating Fake News: An Agenda for Research and Action (2017). http://www.sipotra.it/wp-content/uploads/2017/06/Combating-Fake-News.pdf. Accessed 31 Jan 2018
40. Elyashar, A., Bendahan, J., Puzis, R.: Is the Online Discussion Manipulated? Quantifying the Online Discussion Authenticity within Online Social Media. arXiv preprint arXiv:1708.02763 (2017)
41. Lang, A.: The limited capacity model of mediated message processing. J. Commun. **50**(1), 46–70 (2000). https://doi.org/10.1111/j.1460-2466.2000.tb02833.x
42. Nyhan, B., Reifler, J.: Displacing misinformation about events: an experimental test of causal corrections. J. Exp. Polit. Sci. **2**(1), 81–93 (2015). https://doi.org/10.1017/XPS.2014.22
43. Metzger, M.J., Flanagin, A.J., Medders, R.B.: Social and heuristic approaches to credibility evaluation online. J. Commun. **60**(3), 413–439 (2010). https://doi.org/10.1111/j.1460-2466.2010.01488.x

44. Sundar, S.S.: The MAIN model: a heuristic approach to understanding technology effects on credibility. In: Metzger, M., Flanagin, A. (eds.) Digital Media, Youth, and Credibility, pp. 73–100. MIT Press, Cambridge (2008)
45. Sundar, S.S., Oeldorf-Hirsch, A., Xu, Q.: The bandwagon effect of collaborative filtering technology. In CHI 2008 Extended Abstracts on Human Factors in Computing Systems, pp. 3453–3458. ACM (2008). https://doi.org/10.1145/1358628.1358873
46. Festinger, L.: A Theory of Cognitive Dissonance, vol. 2. Stanford University Press, Palo Alto (1962)
47. Tan, E.E.G., Ang, B.: Clickbait: Fake News and Role of the State. RSIS Commentaries, 026-17 (2017)
48. Gross, M.: The dangers of a post-truth world. Curr. Biol. **27**(1), R1–R4 (2017). https://doi.org/10.1016/j.cub.2016.12.034
49. Jin, Z., Cao, J., Zhang, Y., Zhou, J., Tian, Q.: Novel visual and statistical image features for microblogs news verification. IEEE Trans. Multimed. **19**(3), 598–608 (2017). https://doi.org/10.1109/TMM.2016.2617078
50. Markowitz, D.M., Hancock, J.T.: Linguistic traces of a scientific fraud: the case of Diederik Stapel. PLoS ONE **9**(8), e105937 (2014). https://doi.org/10.1371/journal.pone.0105937
51. Lendvai, P., Reichel, U.D.: Contradiction Detection for Rumorous Claims. arXiv preprint arXiv:1611.02588 (2016)
52. Ma, J., Gao, W., Wei, Z., Lu, Y., Wong, K.F.: Detect rumors using time series of social context information on microblogging websites. In: Proceedings of the 24th ACM International on Conference on Information and Knowledge Management, pp. 1751–1754. ACM, New York (2015). https://doi.org/10.1145/2806416.2806607
53. Jin, Z., Cao, J., Zhang, Y., Luo, J.: News verification by exploiting conflicting social viewpoints in microblogs. In: Proceedings of the Thirtieth AAAI Conference on Artificial Intelligence (AAAI 2016) (2016). https://www.aaai.org/ocs/index.php/AAAI/AAAI16/paper/view/12128/12049. Accessed 6 Feb 2018
54. Shu, K., Sliva, A., Wang, S., Tang, J., Liu, H.: Fake news detection on social media: a data mining perspective. ACM SIGKDD Explor. Newsl. **19**(1), 22–36 (2017)
55. Appelman, A., Sundar, S.S.: Measuring message credibility: construction and validation of an exclusive scale. Journal. Mass Commun. Q. **93**(1), 59–79 (2016). https://doi.org/10.1177/1077699015606057
56. Tufekci, Z.: Mark Zuckerberg is in Denial. The New York Times (2016). https://www.nytimes.com/2016/11/15/opinion/mark-zuckerberg-is-in-denial.html. Accessed 31 Jan 2018
57. Tintarev, N., Masthoff, J.: Evaluating the effectiveness of explanations for recommender systems. User Model. User-Adap. Interact. **22**(4–5), 399–439 (2012). https://doi.org/10.1007/s11257-011-9117-5

CyberActivist: Tool for Raising Awareness on Privacy and Security of Social Media Use for Activists

Borislav Tadic(✉), Markus Rohde, and Volker Wulf

Information Systems and New Media, University of Siegen, 57068 Siegen, Germany
borislav@tadic.biz,
{markus.rohde,volker.wulf}@uni-siegen.de

Abstract. Bosnia-Herzegovina (BH) and its entity Republika Srpska (RS) are among the most fragile democratic environments in Europe. In the first phase of our long-term participatory design case study, we engaged the some of the main activists in BH/RS, providing a structured picture of their practices in recent years, concrete needs and the various constraints under which they act. Our research highlighted importance and utilization of the social media for the activism in the region, but also problems such as limited budgets and know-how of the activists, intensive outsourcing practices, and a lack of awareness regarding data privacy and cyber security. Due to the perspective of BH/RS, the rising number of threats and impact incidents, and activist experiences from other unstable regions, we propose a more structured approach to privacy and security within activist circles and non-profit organizations. As the initial step in the second phase of our study, we offered a prototype of the free web application "CyberActivist" to BH/RS activists for user tests. Based on their qualitative feedback we defined the functional and non-functional requirements on further improvement of this privacy and security awareness tool. In the next phase, we will technically address their direct feedback, as well as design recommendations from relevant research and user experience literature. We also plan to propose design method improvements, design corresponding privacy and security trainings and to further internationalize the tool.

Keywords: ICT · Tool · Security · Privacy · Anonymity · Social media Awareness · Activist · Activism · Non-profit · Political · Facebook Bosnia · Srpska

1 Introduction

Bosnia-Herzegovina (BH) and its entity Republika Srpska (RS) are among the most fragile democratic environments in Europe. The relationship between this political environment, the kinds of activism that seem to be prevalent, and how best to support them is in the focus of our research. Our research follows the methodological concept of long-term design case studies, as it was elaborated for practice-oriented design research [3–5]. Design case studies are ethnographically informed studies that are

© Springer International Publishing AG, part of Springer Nature 2018
G. Meiselwitz (Ed.): SCSM 2018, LNCS 10913, pp. 498–510, 2018.
https://doi.org/10.1007/978-3-319-91521-0_36

"describing the original social practices, the design discourse, the design options considered, the appropriation process, the effectiveness of the artifacts' functions and the emerging new social practices" [3]. They are based on a participatory and cyclic approach of analyzing social practices in a pre-study, creating and implementing design solutions and evaluating the appropriation practices of users. This paper presents essential insights from the analytical pre-study and participatory design phase of a long-term design case study that is still ongoing.

In the first phase of our design case study, we identified the main activists in RS, providing a structured picture of their practices in recent years, concrete needs and the various constraints under which they act [1]. Empirical investigations of social media use and qualitative interviews with the country's activists indicate their strong interest in information and communication technology (ICT). Especially social media in the region is even more relevant since it basically becomes the only vehicle for activism other than direct action. Benefits for the use of ICT and social media by activists include e.g. more efficient access to their target group, easier information sharing with the general population, and quicker reaction to spontaneous "offline" activities [cf. 1, 22, 25]. At the same time, research highlighted problems of the activists such as limited budgets and know-how, intensive outsourcing practices, and a significant lack of awareness regarding data security. Although our activists are digitally very active and consequently ICT-literate, they are largely self-taught, being neither ICT-professionals nor "digital natives". After we conducted problem-centric interviews with six cyber activists, we clustered their needs and our observations in the following categories:

(1) a structured approach to cyber security, data privacy and anonymity within activist circles and the NPO sector
(2) specialized trainings tailored for cyber activists, the specific region and based on available resources
(3) support for practices enhancing self-learning and knowledge transfer within the specific BH/RS setting.
(4) sustainable models within ICT outsourcing and use of external freelancers within cyber activism.

Due to the perspective of BH/RS, the rising number and impact of privacy and security incidents, and an increasing relevance of social media and activist experiences from e.g. Turkey or "Arab Spring", we believe that a more structured approach to privacy and security within BH/RS activist circles and non-profit organizations is needed. Aiming to address these developments and elements (1)–(3) listed above, in the second phase of our design case study, we decided to implement a prototype of a web application named "CyberActivist" for awareness in the areas of privacy and security. Following a participatory design approach like Caveat [6] or Come_IN [2], we made our prototype software available to the BH/RS activists for a test in a real-world practice ultimately leading to documentation of clearly articulated requirements for improvement of their communication practices and the tool itself.

In Sect. 2 of this paper, we are looking at the related state-of-the-art work regarding the social media impact within global activism and the related privacy and security considerations. Section 3 provides an overview of the functionalities of the tool and Sect. 4 follows with the summarized outcome of the BH/RS activist experiences during and after the test of the "CyberActivist" prototype. Last section provides an outlook on planned next steps and research possibilities in this context.

2 Related Work

Social media based movements and their members leave behind digital footprints that authoritarian powers can exploit for the surveillance and oppression [7], e.g. using provocateurs and bots [32, 34]. [34] looked at social media focusing on one side with insider threat prediction and prevention, connecting malevolent insiders and predisposition towards computer crime with personality trait of narcissism. At other side, regardless of national scope, an important social threat is based on user generated content exploitation and leads to political affiliation profiling. Activists are a very relevant group here, esp. within authoritarian systems and even with potential employers. According to [32], human resource departments increasingly use social media screening, which produces negative reactions of the candidates in the US. If this would be the case in BH/RS, where non-employment is high and cyber activists can be marked as the opponents of the regime, they might be having additional difficulties finding jobs, if they are not careful with the information published online. [33] argues that many "difficulties associated with the protection of digital privacy are rooted in the framing of privacy as a predominantly individual responsibility". This is very visible regarding Terms and Conditions of social media; although users of social media platforms are poorly informed about the changes in the privacy policies, it is often "setting forth the expectation that the user has been educated enough to now make decisions in their best interest".

Social media relevance in regard to the privacy and security differs over activist heritage [29], age, gender [18, 19], habits [28], and changes over time [30]. [29] conducted a comparative study on social media use with focus on privacy aspects within 5 nations. Although a majority of users stated that is "important to prevent risks that might arise from privacy related behavior", they had significantly different implementations, such as anonymizing their identity or self-disclosure. Mentioned implementations might easily be customized to address the needs of activists of other nations. Study participants reported that they had not yet experienced many privacy violations. In our case, RS activists have also their specific attitude, similar to the part of the attitudes from [29] which must be considered within the tools supporting their engagement. With effectiveness and practicality in mind, we implemented a prototype of a web application "CyberActivist" for awareness in the areas of privacy and security of social media, described in detail in the next section. It also might be used in other geographic contexts, similar to implementations of [29]. [18] has shown how different population structures have a different understanding of privacy, its enforcement and importance in the social media context. This may very well apply to our activists. [28] focused on undergraduate students' experiences with social network system privacy. Students worried about their privacy being violated by someone physically locating them still felt comfortable sharing their personal information. More media literacy leads to better awareness about risks of sharing information on social media. This supports the thesis on need for specialized training for activists identified in [1]. [30] compared

Facebook users to understand how their privacy and disclosure behavior changed between 2005–2011. Besides concluding that users exhibited volatile privacy-seeking behavior, from less disclosure in the first years to an increase towards the end of the study, they warned from the often non-transparent "silent listeners". Due to the increase in amount and scope of personal information that users revealed privately to other connected profiles, more information is available to Facebook itself, third-party apps, and indirectly advertisers. Authors of this paper assume that these findings are becoming even more relevant for numerous cyber activists, if we extend the list of "silent listeners" to state-related apparatus and highlight low privacy awareness of the activists present on Facebook (e.g. low interest in terms and conditions).

Following a participatory design approach [cf. 2, 6], our implementation was tested by the activists in their real-world practice. This led to the tuning of our tool based on direct interaction, and ultimately improved activist communication practices. We also orientated us on insights of e.g. [24, 26] and recommendations from best practices such as [9]. [26] proposed a framework including an open source implementation with semantic, hierarchical scoring structure for raising the awareness of social media users with respect to the information that is disclosed and that can be inferred by third parties with access to their data. It enables users to browse over different privacy-related aspects considering both information that is explicitly mentioned in users' shared content, as well as implicit information, that may be inferred from it. [24] also claims that ICT and social media enabled better access to personal and location information of another person, and activists may not be aware of the possibilities here. Despite having regulatory policies, it is possible to extract quite exact location information of a person over time by using volunteered or contributed geographic information available from social media sites (e.g. GeoAPI of Twitter).

Although privacy and security requirements are sometimes in conflict, we can reasonably raise both aspects using tailored approaches [20, 27] and by creating visibility over vulnerabilities of an activist or his environment [23]. It is also important to consider differentiation of the social groups in their attitude towards privacy and security when developing ICT solutions [33] and unconventional approaches to promote privacy and security such as using celebrity engagement in social media [21]. Taking the example of one group of human rights activists, [33] highlights the importance of developing a collective approach to address their digital privacy and security needs. Digital security strategies cannot remove all threats; they can only mitigate their effects and deal with numerous elements such as authentication on Facebook. We included the question about the Facebook authentication into the *Self-assessment* within our prototype (see next section of this paper). [23] introduced methods for determining the amount of information that can be ascertained using only publicly accessible data and provides a framework for determining a user's web footprint. Threat of user's attributes that may be inferred by an adversary using only public sources of information has been reconfirmed by analysis across multiple social networks. The same method can be applied by cyber activists and other individuals to assess and act upon their own exposure in the public media.

3 Web Application "CyberActivist"

The development of the initial version of our web application "CyberActivist" (in English and Serbo-Croatian language) took six months in 2016, using HTML, JavaScript and CSS. One of the paper authors has written the whole source code of the initial version of the prototype that was provided to the activists for the test.

Primary functions of the tool are: to enable self-assessment of privacy and security in the context of social media and make results transparent to the user, then dynamically point to open, external, self-learning resources esp. in areas marked as "blind spots" and volunteering opportunities.

"CyberActivist" consists of four sections, which are represented by the icons on the primary screen after the application start: *Self-assessment*, *Self-learning*, *Contribute*, and *My Profile*. In addition, there is information about the so called *cyber safe score* of the self-assessment, visible only after the performed self-assessment, and hyperlinks to two information pages: *About the application* and *How does this application work.*

Self-assessment (Fig. 2) and *My Profile* (Fig. 4) sections enable users to gain transparency about the risks within their social media environment and to see how they are positioned regarding these risks. We are using easily understandable, user-centric language, knowing the average ICT proficiency of the target group, to help them gain insight and derive appropriate action. Section *Self-assessment* contains nine groups of questions: 25 general questions, applicable to most social media platforms, then specific platform questions on Facebook (11 questions), Google/Youtube (8), Twitter (6), Whatsapp (4), Viber (4), Skype (4), Instagram (6) and one group reserved for other platforms such as Linkedin (3), which can be answered through multiple-choice text options (e.g. "yes", "no" and "I do not know"). An example of a question is "Do I know who will be accessing information I have put on social media?". The question groups are focusing on the most frequently mentioned ICT tools and social media platforms mentioned by the activists in [1] and publicly available ranking information [cf. 8]. When results are saved, they are being recorded on the activist's device using the local storage functionality of HTML and not transmitted to any remote server. The selection of questions and their formulation have been based on experience of one of the authors of this paper, as well as on similar international questionnaires and assessments such as [cf. 9–17]. The Section *My Profile* shows the data about the user available within browser he uses, e.g. whether Java is activated or what is the geographic location. It also enables the user to set the language of the application.

The main screen shows a so called "*cyber safe score*" (Fig. 1). This score is calculated based on the number of positive ("plus" point) and negative answers ("minus" point) from the self-assessment with the maximum score of 64 points being achievable. An example of the positive/negative answer is "I have/have not latest version of Twitter installed on my devices". On the main application screen, the user is also being given an instruction to perform a self-assessment before being able to use the application's full functionality and find out "how the tool actually works".

Section *Self-learning* (Fig. 3) offers a customized array of reading materials based on the *cyber safe score* and improvement areas. Most materials are articles published by the relevant social media platforms, non-profit organizations or media with direct

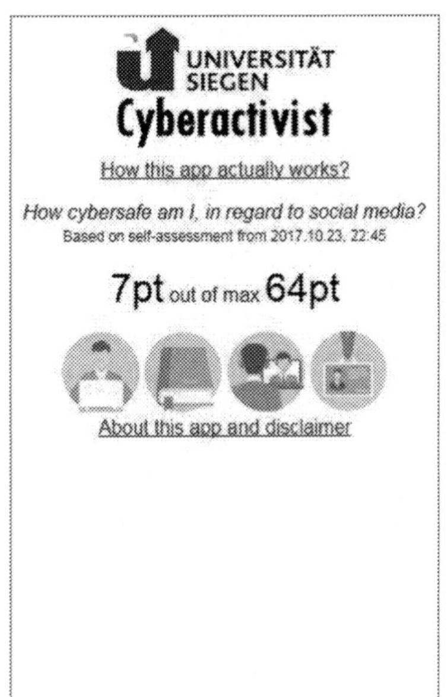

Fig. 1. Main screen showing sections and *cyber safe score*

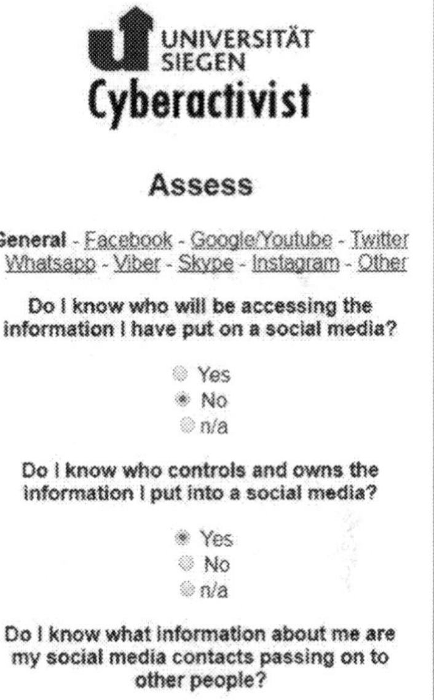

Fig. 2. *Self-assessment* section/questionnaire

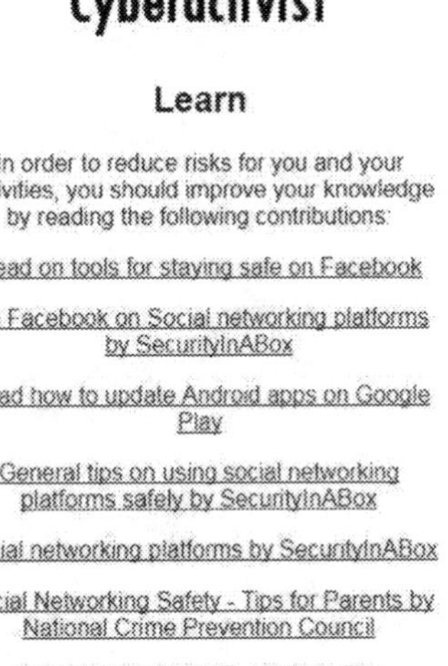

Fig. 3. *Self-learning* section/recommended reading

Fig. 4. *My Profile* section

actionable advice on improving security, privacy and anonymity. In the case of the mentioned Twitter answer example, it would be a reading material related to "software patching" or "privacy and security settings of Twitter". It supports preferred way of (self-)learning of the BH/RS cyber activists, caused by resource limitations (e.g. training budget). Every click in this section opens an additional web browser window and shows the original web page outside the "CyberActivist" application.

The *Contribute* section aims at knowledge sharing and multiplication effects, providing a non-customized list of organizations and websites providing privacy and security advice to activists, e.g. "TacticalTech" [36]. The list is based on the selection of the authors, based on the background of BH/RS activists.

"Cyberactivist" does not collect, process or send any information about the users or their online behavior to the author or any other subject. The application does not use cookies. All links included in the *Self-learning* section are to third party websites, which have separate privacy policies and the authors therefore have no responsibility or liability for their content or activities.

The format of the application - web-based, platform independent, free - is also chosen based on the activists' usage of phones and PCs as primary hardware. Making the "CyberActivist" source code open, with no modification and expansion constraints, improves its reach among activists. After completion, the authors and their academic institution plan to publish and keep the software free and open source providing a clear value adding to the activist and developer community.

4 Participatory Design: Feedback and Possible Improvements

After the development of the application, we have shared a link to the prototype for the test with the selected activists. We contacted all the activists who participated in our former research [cf. 1] and additional new activists we identified monitoring social media activities in the BH/RS.

Five activists responded to our invitation (Table 1). We asked them to test the application and did not provide them with any information besides that the web application is focused on privacy and security. They tested the application on one day, but did not invest longer than an hour of their time. Neither usage data nor self-assessment results were transmitted to the paper authors during or after the test. Activists also committed to the interview in the Serbo-Croatian language after the test, to document their impressions and feedback on possible tool improvements. The activists provided us with almost four hours of responses which were digitally audio-recorded in five separate sessions between May and September of 2017. One activist complemented his audio statement with an e-mail response. Skype with an audio recording plug-in was used as an interview tool. The key findings of our interviews were transcribed in English language and comprise approximately 50 pages.

All activists suggested that the application is simple. They all also agree that the purpose, background methodology, and the user interface of the "CyberActivist" application has to be further sharpened. There is a need to further optimize the main screen. Brad posed a question: "Is the tool meant for single use or for reuse?".

Table 1. Interviewed activists/participatory design phase

Pseudonym	Birth year	Role/Active since	Participated in our earlier research [1]
Brad	1980	Project Manager at local NPO, 2006	Yes
Ela	1984	Project Manager at the local branch of an international NPO, 2008	Yes
Adam	1981	Member of international NPO focused on the RS, 2008, located in Austria	No
Kevin	1981	Local journalist/an individual activist	No
Alena	1980	Individual activist for disabled population	No

Kevin did not even open sections *Self-learning, Contribute, and My profile* as access to these sections was not visible or intuitively displayed. With regard to navigation within the app, Alena suggested that a "Go Back" key is missing.

Adam suggested establishing separate scores for security and for privacy; as referenced in Sect. 2 of this paper, security and privacy aspects are not always correlated. The methodology to calculate the *cyber safe score* raised many questions among activists. Originally planned as the simple, high-level information of displaying general protection status, *cyber safe score* did not fulfill its purpose. The score was unclear for most activists (e.g. Ela: "I got 35 out of 65 points…" - what does it concretely mean, where are my weaknesses, what do I need to improve). The outcome from the self-assessment should be visible immediately, and not only later through links in the section *Self-learning*. The outcome should be explained in more descriptive language, rather than only by a number. Adam considers himself experienced within security and got only 2 points after the self-assessment. The other activist did not understand the logic of adding "plus" and "minus" points.

Most activists tested the tool on the laptop or desktop computer, not on the mobile device. However, Adam suggested that our application should be further customized based on the platform used (e.g. screen resolution, native user interface). The platform should also influence the offered advice in the *Self-learning* section. Differentiation between PCs and mobile devices in the answers within the *Self-assessment* section are also proposed, as usage patterns are differing.

Regarding the *Self-assessment* section, Adam commented that 25 questions in the general part of this section might be too much and proposed separation over several screens/pages. Another idea would be to show the progress of the questionnaire ("how much I still have to go?"). Almost all activists felt that there are lots of repetitions of the similar questions (e.g. same formulation "did you perform an update for… Twitter, Facebook, Whatsapp…"), however they meant that the "questions are clear". Several questions in this section contain formulation "Do I or my organization use…"; Kevin suggested to clearly separate the two, as the answer may differ. Kevin's proposal was also to add the answer option "I don't care/It's not important" to existing possible answers "yes/no/don't know" in the self-questionnaire. Kevin also suggested reconsidering which questions are suitable for the "general questions" category. For him the question "do I trust my connections" would be differently answered for different social media platforms, e.g. for Facebook and Twitter. Two or more predefined answers are offered for every question in the *Self-assessment* section based on the multiple-choice logic. Alena claimed that there is no need for any choice to be marked as default, as it is with the choice "I don't know" in our case. Activists also suggested adding or rephrasing some questions such as "how to add to the group on social media, limiting member's access" or "would your identity disclosure jeopardize your close people/relatives". They claim that is positive that a person is not asked on all tools if they do not own an account on this specific social media.

The first improvement proposal for the *Self-learning* section was that the introduction text should not be shown if the self-assessment is not done. Some of the activists such as Adam did not notice the correlation between the *Self-assessment* and *Self-learning* sections. Activists also claimed that the explanation of the results is needed, such as "…because you don't understand X, you need to read Y and Z". Therefore, a clear link needs to be established between "negative" answers from *the Self-assessment* section, "minus" points of the cyber safe score and the proposed reading materials in the *Self-learning* section. Optimally, related reading materials should be grouped. Authored privacy and security advice is welcome, according to Ela.

Looking at the *Contribution* section, Adam asked whether the listed organizations want/need help or volunteers at all. The others found this section useful as it is. As BH/RS NPOs and activists are struggling with resources [1], Brad suggested an additional feature "find/engage an expert" (e.g. specialist for IT security or video production). He also proposed to integrate some "advertisement" in the tool such as „you are an IT expert - do you want to help and engage in our activities?".

The information in the section *My profile* was found to be useful, however not always self-explanatory (e.g. web browser information as "user agent string").

Brad suggested the replacement of the term "activist" with "socially responsible person", due to "negative connotation" of the term. In general, activists asked that tool's goals, benefits and "flow" are described more clearly in the tool itself (e.g. are results of self-assessment sent somewhere for analysis, how is the score calculated). In addition, better instructions on the tool proper usage are welcome.

In addition, all activists suggested that the used text for a Serbo-Croatian version can be improved. Activists advised the use of fewer Anglicisms in the text and less synonyms esp. in technical context (e.g. "data privacy" vs "data protection"). They also made proposals on how to increase readability, through consistent use of the local alphabet (e.g. "č vs c"), adequate font size and text margins on the different platforms. The *Self-learning* section was referred by Ela as useful as it's good to point to sources and practices from other countries. Other activists were only partially satisfied with the fact that all reading materials offered by the tool as a result of the self-assessment are in English (and not in Serbo-Croatian). This feedback is a good reminder that text quality and thorough localization of the tool plays an important role for acceptance among the activists.

This very qualitative feedback from the activists gathered specific functional and non-functional software requirements and enabled multiple ways of improving the tool. Several ideas for tool improvements are coming from state-of-the art research, e.g. aligning it to models such as "privacy nudge" [20, 27], considering integration with approaches such as "FaceCloak" [31] or adding features such as "celebrity cause" [21], which might be considered in future work and tool adaptations. Research on behavioral decisions and soft paternalism to design mechanisms led to development of so-called "privacy nudge" for Facebook users [27]. This alarm reminds Facebook users to consider the content and context of the information before posting them, helping individuals avoid regrettable online disclosures. Nudges provide visual cues about the audience for a post, time delays before a post is published and gives users feedback about their posts. Adaptation of this nudging might prevent activists' unintended disclosure. [20] also argues the idea of nudging the user with "Privacy Nudge" to help people make better privacy choices and decisions on online social networks. The proposed model will nudge users while posting by calculating Privacy Score and accessing last modified privacy settings for users which will alert users to adjust their privacy settings. FaceCloak protects user privacy on a social media by shielding a user's personal information while maintaining usability of the site's services [31]. This Firefox browser extension for the Facebook provides fake information to the social media and by storing sensitive information in encrypted form on a separate server. Although oriented on one platform only, it is an interesting concept that could be a measure related to our "*cyber safe score*". Celebrities, such as movie actors, often take up an active interest in the "good causes" such as prevention of engagement of children as soldiers in Africa. Their posts on the cause in the social media help draw attention to the cause among their numerous followers. This might be an opportunity for cyber activists, also in the context of awareness for protection of their privacy and security and lobbying for e.g. less surveillance in authoritative societies [21].

The authors themselves also identified ideas on improving the tool, such as those improving user experience, building a more intuitive graphical user interface and adding relevant information sources.

5 Outlook

Especially the more detailed evaluation of users' appropriation of our prototype in the practice goes beyond the scope of this paper and will be object of future research. We base our original contribution to the HCI knowledge corpus on the long-term design case study which enabled numerous insights into practices of political activists in BH/RS, which led to a tool "CyberActivist". Our presentation includes the relevant state-of-the-art research, online and offline experiences with our prototype, unfiltered feedback of the activists, and differentiation through simple, yet unique awareness and self-learning capabilities on social media.

The tool enables activists to understand, address and mitigate the privacy and security risks related to use of social media. The authors plan first to adapt the tool based on the input from the section four of this paper, and eventually to publish it cost-free in multiple languages making it available to the global activist community. This will follow an intense exchange with other HCI researchers which have worked in multiple other geopolitical regions (e.g. Middle East) and incorporation of their thoughts on applicability and target group reach. In addition, in further publications we plan to continue our design case study by observing the development of the ICT and esp. social media use in BH/RS.

Authors and the research community can further refine the underlying research method, e.g. regarding the precision of the questions asked in the interview phase, or evaluation and consolidation of sometimes opposing improvement proposals of the activists. Industry best practices such as Scrum within agile software development [cf. 35] are a great opportunity for improvement of both, our method and quality of the tool. Continuous presence of the activists in the role of the "customers" during the development "sprints" would directly increase the quality of the tool, and potentially fully remove the need for interviews after the implementation of the new tool functionalities.

Our strong belief is that the tool's impact would be raised, if activists would receive free tailored and localized training on privacy and security aspects. In the future, authors will work on the conceptualization of such trainings and/or information campaigns. We believe that this holistic and integrated socio-technical approach will serve as an open, extendable, scientifically founded and practically easily applicable awareness instrument for activists in fragile democratic contexts worldwide.

References

1. Tadic, B., Rohde, M., Wulf, V., Randall, D.: ICT use by prominent activists in Republika Srpska. In: Proceedings of the 2016 CHI Conference on Human Factors in Computing Systems - CHI 2016, pp. 3364–3377 (2016). https://doi.org/10.1145/2858036.2858153
2. Aal, K., Yerousis, G., Schubert, K., Hornung, D., Stickel, O., Wulf, V.: Come_in@palestine: adapting a German computer club concept to a Palestinian refugee camp. In: Proceedings of the 5th ACM International Conference on Collaboration Across Boundaries: Culture, Distance & Technology, CABS 2014, pp. 111–120, NY, USA. ACM, New York (2014). https://doi.org/10.1145/2631488.2631498
3. Rohde, M., Brödner, P., Stevens, G., Betz, M., Wulf, V.: Grounded design – a praxeological is research perspective. J. Inf. Technol. **32**, 163–179 (2017)
4. Wulf, V., Rohde, M., Pipek, V., Stevens, G.: Engaging with practices: design case studies as a research framework in CSCW. In: Proceedings of ACM Conference on Computer Supported Cooperative Work (CSCW 2011), pp. 505–512. ACM-Press, New York (2011)
5. Wulf, V., Müller, C., Pipek, V., Randall, D., Rohde, M., Stevens, G.: Practice-based computing: empirically-grounded conceptualizations derived from design case studies. In: Wulf, V., Schmidt, K., Randall, D. (eds.) Designing Socially Embedded Technologies in the Real-World, pp. 111–150. Springer, London (2015)
6. McPhail, B., Costantino, T., Bruckmann, D., Barclay, R., Clement, A.: Caveat exemplar: participatory design in a non-profit volunteer organisation. Comput. Support. Coop. Work **7**(3–4), 223–241 (1998). https://doi.org/10.1023/A:3A1008631020266
7. Morozov, E.: The Net Delusion: The Dark Side of Internet freedom. Public Affairs, New York, USA (2011)
8. Kallas, P.: Top 15 most popular social networking sites and apps. https://www.dreamgrow.com/top-15-most-popular-social-networking-sites. Accessed 28 Jan 2018
9. Deutschland Sicher im Netz, Sicherheitscheck. https://www.dsin-sicherheitscheck.de. Accessed 15 June 2017
10. Internet Privacy Practices Self-assessment. https://libraryfreedomproject.org/wp-content/uploads/2016/02/privacy-assessment-tool-to-print.pdf. Accessed 17 June 2017
11. Online Privacy and Security Questionnaire. http://www.cc.gatech.edu/gvu/user_surveys/survey-1998-10/questions/privacy.html. Accessed 17 June 2017

12. USAID Privacy Office, Privacy Impact Assessment. https://www.usaid.gov/sites/default/files/SocialMediaPIA.pdf. Accessed 17 June 2017
13. Academic Frontier Project. Survey on the internet security awareness. http://www.kansai-u.ac.jp/riss/en/shareduse/data/17_E_questionnaire.pdf. Accessed 17 June 2017
14. Purdue University, Information Security Questionnaire. https://www.cerias.purdue.edu/assets/pdf/k-12/questionnaire/infosec_questionnaire.pdf. Accessed 17 June 2017
15. Warwick University, Information Security Awareness Questionnaire. http://www2.warwick.ac.uk/services/gov/informationsecurity/questionnaire. Accessed 17 June 2017
16. Federal Trade Commission, Privacy impact assessments. https://www.ftc.gov/site-information/privacy-policy/privacy-impact-assessments. Accessed 17 June 2017
17. Kumaraguru, P.: Privacy and security in online social networks, NOC. https://onlinecourses.nptel.ac.in/noc16_cs07/preview. Accessed 17 June 2017
18. Madden, M.: Privacy management on social media sites. http://www.pewinternet.org/2012/02/24/privacy-management-on-social-media-sites. Accessed 17 June 2017
19. Madden, M., Lenhart, A., Cortesi, S., Gasser, U., Duggan, M., Smith, A., Beaton, M.: Teens, social media, and privacy. pp. 2–86. Pew Research Center, 21 Jg (2013)
20. Saad, T., Khan, F.: Nudging Pakistani users towards privacy on social networks. In: 2016 SAI Computing Conference (SAI), pp. 1147–1154. IEEE (2016)
21. Tsaliki, L.: Tweeting the good causes: social networking and celebrity activism. In: Marshall, P.D., Redmond, S. (eds.) A Companion to Celebrity, pp. 235–257. Wiley, Boston (2016)
22. Lynch, E.: The new social imaginary vs. the education activist: social media as a conduit for protest and resistance. Hofstra University (2017)
23. Singh, L., Yang, G.H., Sherr, M., Hian-Cheong, A., Tian, K., Zhu, J., Zhang, S.: Public information exposure detection: helping users understand their web footprints. In: Proceedings of the 2015 IEEE/ACM International Conference on Advances in Social Networks Analysis and Mining, pp. 153–161. ACM (2015)
24. Kar, B., Ghose, R.: Is my information private? geo-privacy in the world of social media. In: GIO@ GIScience, pp. 28–31 (2014)
25. Fullam, J.: Becoming a youth activist in the internet age: a case study on social media activism and identity development. Int. J. Qual. Stud. Educ. **30**(4), 406–422 (2017). https://doi.org/10.1080/09518398.2016.1250176
26. Petkos, G., Papadopoulos, S.: PScore: a framework for enhancing privacy awareness in online social networks. In: 2015 IEEE 10th International Conference on Availability, Reliability and Security (ARES), pp. 592–600 (2015)
27. Wang, Y., Leon, P.G., Scott, K., Chen. X., Acquisti, A.: Privacy nudges for social media: an exploratory Facebook study. In: Proceedings of the 22nd International Conference on World Wide Web companion, pp. 763–770. ACM (2013)
28. Magolis, D., Briggs, A.: A phenomenological investigation of social networking site privacy awareness through a media literacy lens. J. Media Lit. Educ. **8**(2), 22–34 (2016)
29. Trepte, S., Masur, P.K.: Cultural differences in media use, privacy, and self-disclosure: research report on a multicultural survey study. University of Hohenheim, Germany (2016)
30. Stutzman, F., Gross, R., Acquisti, A.: Silent listeners: the evolution of privacy and disclosure on facebook. J. Priv. Confidentiality **4**(2), 7–41 (2014)
31. Luo, W., Xie, Q., Hengartner, U.: Facecloak: an architecture for user privacy on social networking sites. In: IEEE 2009 International Conference on Computational Science and Engineering, CSE 2009 (2009). https://doi.org/10.1109/cse.2009.387
32. Drake, J.R., Hall, D., Becton, J.B., Posey, C.: Job applicants' information privacy protection responses: using social media for candidate screening. AIS Trans. Hum. Comput. Interact. **8**(4), 160–184 (2016)

33. Kazansky, B.: FCJ-195 privacy, responsibility, and human rights activism. Fibreculture J. **26**, 189–207 (2015)
34. Gritzalis, D., Kandias, M., Stavrou, V., Mitrou, L.: History of Information: the case of privacy and security in social media. In: Proceedings of the History of Information Conference, pp. 283–310, Athens, Greece (2014)
35. Schwaber, K., Beedle, M.: Agile Software Development with Scrum. Pearson International Edition, USA (2002)
36. Tactical Technology Collective. https://tacticaltech.org. Accessed 28 Jan 2018

Workplace Sexual Harassment on Social Media

Jennifer Wohlert(✉)

Institute of Entrepreneurship and Business Development,
Fachhochschule Lübeck, Lübeck, Germany
jennifer.wohlert@fh-luebeck.de

Abstract. Social media offers both very small companies and the Fortune 500 a variety of benefits in communication within the company as well as in dealing with customers, suppliers and business partners. At the same time, the integration of social media into everyday business life presents new challenges for companies. From an interdisciplinary point of view, this article is intended to give an overview of some negative factors that can arise through the use of social media in companies. In this context, the focus is on the negative effects on sexual harassment at work and the prevention of unethical behavior in companies.

Keywords: Social media · Sexual harassment · Workplace

1 Introduction

In a network society dominated by the Internet, there is nowadays hardly any area in which Internet-based social networks play no role at all [1]. In the course of the dynamic technological progress and the associated integration of digital channels, the importance of digital exchange via social media, especially in companies, continues to increase [2, 3]. Through the use of social networking sites, microblogging services or the intranet, both micro-enterprises and the Fortune 500 benefit from process optimization, collaborative project management and exchange with business partners, customers and suppliers [4]. At the same time, the internal use of social media in the workplace not only accelerates communication and strengthens interaction, but also brings new challenges for employees and companies [5, 6].

Particularly with regard to organizational requirements and ethical principles of companies, the use of social media has a considerable influence [6, 7].

One ethical area that has been the focus of research and public attention in recent decades is sexual harassment in companies [8, 9]. As a well-known, widespread ethical problem with far-reaching consequences for organizations, companies and employees, a lot of attention has already been paid to sexual harassment in the workplace. Although many companies have already recognized the relevance of sexual harassment in the workplace [10], the prevention of sexually harassing behavior with regard to compliant use in social media is not represented in all companies [11].

In relevant literature, the effects of sexual harassment on workers and businesses have frequently been discussed [8, 12]. In addition, the research already offers some

© Springer International Publishing AG, part of Springer Nature 2018
G. Meiselwitz (Ed.): SCSM 2018, LNCS 10913, pp. 511–520, 2018.
https://doi.org/10.1007/978-3-319-91521-0_37

studies on the different forms, characteristics of sexual harassment and the occurrence sexually harassing behavior in various environments [13, 14].

The influence on the occurrence of sexual harassment in the workplace with regard to the use of social media on the other hand has neither been researched nor empirically investigated in many areas.

According to the present problem and existing literature the aim of this article is to broaden scientific knowledge about the situational factors that influence sexual harassment at work in the context of social media use. Based on existing research results from sociology, psychology and economic ethics, an interdisciplinary perspective is to be adopted in order to investigate a multitude of influencing of sexual harassment. With the help of proposals, solutions approaches for the in-house prevention of sexual harassment via social media at the workplace are to be created.

2 Theoretical Background

2.1 Sexual Harassment

Harassment of a sexual nature can be perceived and evaluated differently by those involved and neutral persons on the basis of typically subjective and objective components [15]. On the basis of subjective assessment, it is difficult to define the concept of sexual harassment in a uniform way. Fitzgerald, Swan & Magley provide a comprehensive definition of sexual harassment as an unwanted, sexually suggestive behavior that aims at gender-specific devaluation and violation of the dignity of a target or leads directly to gender-specific devaluation [8].

Sexual harassment can be divided into three distinct categories [16]. The first form of harassment based on gender, which contains unwelcome comments and comments that insult people because of their gender and gender-specific degrading comments. The second form of unwanted sexual attention, which is defined by undesirable behaviors, such as the communication of sexual desires, as well as by behavior and commentary, which desires sexual acts. The third form of sexual coercion which, in addition to the psychological pressure exerted, primarily describes undesirable physical harassment, such as physical contact [16, 17]. All forms of sexual harassment can occur both offline and online [17]. One variant of sexual harassment via online channels, which has increased considerably in recent years, is so-called textual harassment. Textual harassment refers to the writing and sending of inappropriate and unsolicited messages [18, 19].

In the psychological models that examine different variables to explain the occurrence of sexual harassment, a theoretical model emerged from a socio-psychological point of view, which is used to explain the occurrence of sexual harassment [20]. Pryor et al. provide an explanation of the emergence of sexual harassment in their model on the likelihood of the occurrence of sexual harassment. In the model was described that multiple factors influence the occurrence of sexually harassing behavior [21–23] and links the occurrence of sexual harassment with multiple factors from personal and situational components [24]. The model has been widely used over the past twenty years to explain the factors influencing sexual harassment [17, 22, 23, 25] and is also used in this article as a basis for identifying contextual factors influencing sexual harassment.

Due to the fact that the subject matter of this article is related to the workplace, special attention is paid to the sphere of influence of the organization in this article.

In the organizational context, sexual harassment at the workplace is defined by the U. S. Equal Employment Opportunity Commission as follows: "Unwanted sexual advances, requests for sexual favors and other verbal or physical conduct of a sexual nature […], where such conduct expressly or implicitly impairs a person's employment or performance inappropriately or creates an intimidating, hostile or offensive working atmosphere." [26]. In contrast to the general definition of sexual harassment, eeoc integrates the influence of sexual harassment on work performance and working atmosphere into the definition of sexual harassment. Sexual harassment in the workplace can be defined as unethical behavior, since sexual harassment is a violation of laws and social norms [27]. Violations of an organization's ethical values in the form of sexual harassment may occur among peers or quid pro quo, e.g. by superiors [28].

2.2 Social Media

Social media is the umbrella term for "Internet-based applications that build on the ideological and technological foundations of Web 2.0, and that allow the creation and exchange of user-generated content" [29, p. 61]. Social media platforms provide a mechanism for users that enables them to network, communicate and interact digitally with the aid of various features [30]. The platforms are offered in many different forms, such as blogs and microblogs, platforms for media exchange, social networks and forums [31].

The use of social media platforms in the workplace is widespread [11] thereby, the majority of employees of those companies that use social media consider social media to be useful for their work [32]. Basically, the social media platforms used in companies can be divided into two types of use. The first type is for internal communication, such as the intranet and the second type for communication with the focus on external parties, such as Facebook or Twitter [33]. The internal use of social media enables the networking of employees and cross-company interaction [34]. The added value of in-house use of social media can be quantified through enterprise-wide collaboration, cross-functional knowledge exchange, interdisciplinary innovation management, preadaptive agility enhancement and activating change management [5]. External social media platforms as Facebook or Twitter allow the use of several social media applications to communicate with customers, suppliers or the public [33], whereby the external platforms are often also used for communication within the workplace [11].

Communication via digital channels and, in particular, the use of social media at the workplace has led to a change in the workplace. The traditional workplace, which is characterized by typical physical characteristics such as classic office space, has changed over the past few decades due to the integration and increasing use of digital channels. The new concept of workplaces, which is characterized by digitization, is not only limited to physical components, but is also being extended in that the new concept of workplaces is increasingly being decoupled from the fixed physical definition of the location [35]. Integration and the increasing use of digital channels [36] will increase the flexibility of workers to work from almost anywhere and at any time [35]. Home working models, work in coworking spaces, crowdworking and cloudworking can be realized in this way and contribute to a location-independent and time-flexible way of working.

3 Propositions

The following proposals are intended to provide a framework for an overview of the specific situational factors that influence the behavior of sexual harassment in the workplace through social media.

According to a PEW Research Study 2016, many employees have already used external social media platforms, such as Facebook, for work-related topics [11]. The dilemma with the use of external social platforms such as Facebook in the workplace is above all that the user presents his or her profile as a business card. While in reality it is possible to switch between private and business personality, there is a hybrid through the Facebook profile [37, 38]. Social media accordingly supports the dissolution of the borders between work and leisure [39]. By blurring these areas, professional and private contacts are increasingly merging. Consequently, the ambivalent use of social media strengthens the blurring of the boundaries between private and professional life [40].

Proposition 1: The use of external social media platforms leads to a blurring of private and professional life.

The fusion of private and professional life through the use of social media presents companies with a multitude of new challenges. Since people have different roles depending on the location and recipient of the interaction, they behave differently in the work environment than in their private surroundings. This schizophrenic social behavior is limited by the transparency and speed of the flow of information via social media platforms and blurs these roles [38]. In addition, codes of conduct and values of social media that violate the Code of Conduct can be transferred to employees' behavior [41]. The blurring of boundaries resulting from the use of social media platforms can therefore have a negative impact on the ethical behavior of employees through the disappearance of social roles.

Proposition 2: The blurring of the boundaries between the private and professional lifes can lead to unethical behavior in the workplace.

Social media is the most commonly used online communication channel for online harassment [11]. Basically, the use of digital channels for sexual harassment is interesting for several reasons. Social media appear to offer potential victims less protection through blurred legal boundaries, lack of supervisory bodies and sanctions [17].

Another factor that leads to unethical behavior in the use of social media is restrictive legal aspects and regulations in the social media platforms. Unclear legislation and competences create a seemingly lawless area in which the perpetrators apparently have no legal restrictions to fear [42]. In addition, Kaptein assumes that the risk of unethical behavior is even higher, the greater the discretion of the employees and less the organizational frame of reference of an organization [43].

Proposition 3: The use of social media at the workplace can lead to unethical behavior due to unclear regulations.

Social media platforms create opportunities for communication and interaction. Communication via social media basically comprises the exchange and transmission of information, which, in contrast to offline communication, takes place with the help of a medium. Offline communication is typically characterized by simultaneousness, i.e. synchronicity of exchange and aspects of non-verbal conversation such as body language, eye contact, tone of voice and posture [44]. The use of social media, implies a number of peculiarities compared to offline communication due to media integration and the typical features of digital channels. Due to the media usage and the lack of physical contact between communication and interaction participants, typical non-verbal conversation characteristics are decoupled from communication. Accordingly, communication via social media has a higher potential to cause misunderstandings and different intentions between the interlocutors [45]. In addition, the lack of non-verbal conversation features in communication can lead to misinterpretation between two communication partners, this misinterpretation was integrated into the research on sexual harassment behavior by Stockdale in 1993. Misinterpretations in communication in combination with other factors, such as the aggressive sexual value system in the person, can lead to sexual harassment [21]. Accordingly, communication via some social media in combination with personal disposition can lead to unethical behavior.

Proposition 4: Communication via social media at the workplace has negative effects on the ethical behavior of employees.

The perceived organizational tolerance is an important internal factor for the occurrence of sexual harassment at work [20]. If a company's tolerance of sexual harassment by employees is considered relatively high, this increases the likelihood of sexual harassment [20]. Organizational tolerance can be defined more precisely by internal corporate values and ethical cultures of the organization or company.

Proposition 5: High organizational tolerance has a negative impact on the ethical behavior of employees.

In the ethical organizational context, the ethical culture and the ethical climate have a particular impact on the ethical behavior of the employees [43].

The ethical organizational approach perceived by employees can be subdivided into the constructs of ethical culture and ethical climate in business ethics literature. Ethical culture primarily defines the aspects of ethical behavior [43]. Previous research has focused primarily on the significance of an organization's ethical culture [46]. A company's ethical culture has a significant influence on the reporting of employee misconduct [47].

Ethical-cultural guidelines of an organization or a company create the basis for the behavior, actions and interactions of the employees among themselves and beyond that the basis for actions of the management. The ethical climate of an organization summarizes those aspects that influence the ethical behavior of an organization [48]. It describes the fundamental view of an organization as to which behavior is considered ethically correct and what consequences result from ethical problems [49]. In cases where unethical behavior, such as sexual harassment, appears to be tolerated by management and superiors, the likelihood of individuals with personal disposition being more likely to have sexual harassment tendencies is higher [20].

Proposition 6: Ethical culture and the ethical climate of an organization have an influence on the ethical behavior of the people involved.

At the individual level, behavioral patterns of ethical or unethical conformity can be described by the moral identity of a person. Moral Identity refers to the self-regulatory mechanism of a person who is regulated by environmental influences and motivates moral actions [50]. If a person's moral identity is high, this person can strengthen his or her ethical behavior and suppress unethical behavior [51]. Factors influencing moral identity are primarily social references. In the corporate context, superiors in particular influence a person's moral identity [52]. Typically, employees develop their moral identity based on the ethical or unethical behavior of their supervisor. Accordingly, unethical behavior of superiors can lead to a reduction in the self-regulation of employees in the course of the moral identity of employees [51].

Proposition 7: The handling of sexual harassment and social media by superiors influences the ethical behavior of employees

4 Prevention

The prevention of sexual harassment at work can be divided into three main dimensions: Changes in organizational-social culture in companies, education and training of employees and legislation and law [9, 53].

Within the company, the implementation of clear guidelines for the use of social media, the identification of clear consequences in case of violations and the communication of ethical guidelines by managers are important factors for the ethical, value-oriented use of social media in the company. The communication of ethical corporate cultures and the creation of an appropriate ethical working climate are intended to provide employees with orientation and a clear zero tolerance of the company towards sexual harassment. The hybrid use of social media in private and professional life implies the need to establish clear guidelines for the use of digital channels in order to comply with corporate ethics. Since liberal rules and standards in companies promote unethical behavior such as sexual harassment [54], it is necessary to define rules in a code of conduct and concrete sanctions for violations. The use of company-internal social media platforms supports the concrete assignment to work-relevant topics and the avoidance of hybrid use, which is restricted by the blurring of boundaries. At the same time, the use of social media with company-compliant guidelines implies that the standards of an external social media platform are not transferred to the behavior of employees. The official greeting of the report on non-compliance with company guidelines and the support of employees who report violations is intended to promote communication about non-compliance with corporate ethical principles and thus give the responsible authorities in the company the opportunity to intervene at an early stage in the event of unethical behavior [43].

In addition, supervisors who exert a tremendous influence on the moral identity of employees [52] should serve as role models for ethical conduct in the workplace. Accordingly, it is important to focus on training superiors for responsible, committed leadership that advocates a firm policy against unethical behavior. Workshops and seminars make it possible to sharpen the awareness of employees and especially their superiors in their function as role models for the fulfilment of internal company regulations [55].

Due to the subjective perception of sexual harassment, it is often difficult to apply the law [17]. However, it is still necessary to introduce legal restrictions as a preventive measure, since clear legislation and legal restrictions on sexually harassing behavior via social media counteract this. By pointing out boundaries and sanctions, it is possible to give guidance to employees and to show which interpersonal behavior with regard to morals and values should be lived in companies [17].

5 Conclusion

The aim of this article was to examine the negative effects on sexual harassment caused by the integration of social media in the workplace. To this end, the global areas of workplace and social media platforms were examined for some of the situational factors that influence sexual harassment. Due to the complexity of the areas of influence and factors influencing sexual harassment, only a small part of the topic was examined and a framework has been created to provide an incentive for further research.

In summary, sexual harassment in the workplace via social media platforms results from a multitude of factors resulting from personal disposition as well as from the typical characteristics of social media platforms and the company.

Even if companies consistently take action against unethical behavior, it would be utopian to claim that misconduct in the workplace, such as sexual harassment, can be completely eliminated. Even with the best preventive measures, it will most likely not be possible to completely eliminate sexual harassment in the workplace [56]. However, it is possible to take holistic preventive measures in all areas of influence to counteract sexual harassment in the best possible way. The holistic approach and the comprehensive prevention of unethical behavior enables companies and employees to use the added value of social media in the workplace efficiently instead of having to counteract negative effects.

References

1. Bommes, M., Tacke V.: Das Allgemeine und das Besondere des Netzwerkes. In: Hollstein B, Strauß, F. (Hrsg) Qualitative Netzwerkanalyse: Konzepte, Methoden, Anwendungen. Verlag für Sozialwissenschaften, Wiesbaden, pp. 37–62 (2006)
2. Icks, A., Bijedić, T., Große, J.: Mittelstand und Prävention 4.0. In: Cernavin, O., Schröter, W., Stowasser, S. (eds.) Prävention 4.0. pp. 335–354. Springer, Wiesbaden (2018)
3. El-Darwiche, B., Friedrich, R., Koster, A., Sabbagh, K., Singh, M.: Digitization for economic growth and job creation: regional and industry perspectives. In: The Global Information Technology Report, pp. 35–42 (2013)
4. Culnan, M., McHugh, J., Patrick, I., Zubillaga, J.: How large U.S. companies can use Twitter and other social media to gain business value. MIS Q. Executive **9**(4), 243–259 (2010)
5. Berger, R.: Die Studie: Wer teilt, gewinnt - zehn Thesen, wie Digitalisierung und Social Media unsere Unternehmen verändern Roland Berger, p. 20 (n.d.). https://www.tecchannel.de/a/roland-berger-10-thesen-zu-social-media,2067038. Accessed 15 Jan 2018
6. Vaast, E., Kaganer, E.: Social media affordances and governance in the workplace: an examination of organizational policies. J. Comput. Mediated Commun. **19**(1), 78–101 (2013)
7. McAfee, A.P.: Enterprise 2.0: the dawn of emergent collaboration. Eng. Manage. Rev. **47**(3), 38 (2007)
8. Fitzgerald, L.F., Drasgow, F., Hulin, C.L., Gelfand, M.J., Magley, V.J.: Antecedents and consequences of sexual harassment in organizations: a test of an integrated model. J. Appl. Psychol. **82**(4), 578–589 (1997)
9. Paludi, M., Paludi, C.: Academic and workplace sexual harassment: a handbook of social science, legal, cultural, and management perspectives. Praeger, Westport (2003)
10. Aquino, K., Thau, S.: Workplace victimization: aggression from the target's perspective. Ann. Rev. Psychol. **60**, 717–741 (2009)
11. PEW Research: Social Media and the Workplace (2016). http://www.pewinternet.org/2016/06/22/social-media-and-the-workplace/. Accessed 12 June 2016
12. Bargh, J.A., Raymond, P., Pryor, J.B., Strack, F.: Attractiveness of the underling: An automatic power sex association and its consequences for sexual harassment and aggression. J. Pers. Soc. Psychol. **68**(5), 768–781 (1995)
13. Uggen, C., Blackstone, A.: Sexual harassment as a gendered expression of power. Am. Sociol. Rev. **69**(1), 64–92 (2004)
14. Morral, A.R., Gore Kristie, L.G., Schell, T.L.: Sexual assault and sexual Harassment in the U.S. Military. RAND Corporation, Santa Monica, California (2006)
15. Felson, R.B., Tedeschi, J.T.: Violence, aggression, and coercive actions. American Psychological Association, Washington (1994)
16. Fitzgerald, L.F., Gelfand, M.J., Drasgow, F.: Measuring sexual harassment: theoretical and psychometric advances. Basic Appl. Soc. Psychol. **17**(4), 425–445 (1995)
17. Barak, A.L.: Sexual harassment on the Internet. Soc. Sci. Comput. Rev. **23**(1), 77–92 (2005). also published In: O'Toole L L, Schiffman J R, Kiter Edwards M L (Eds). (2007) Gender violence: Interdisciplinary perspectives. vol 2 pp. 181-193. New York: New York University Press
18. Baldas, T.: Textual harassment on the rise. Nat. Law J. (2009) https://www.law.com/nationallawjournal/almID/1202432261824. Accessed 12 Jan 2018
19. Parker-Pope, T.: Digital flirting: easy to do and to get caught. New York Times (2011). http://www.nytimes.com/2011/06/14/science/14well.html. Accessed 12 Jan 2018

20. Pryor, J.B., LaVite, C.M., Stoller, L.M.: A social psychological analysis of sexual harassment: the person/situation interaction. J. Vocat. Behav. **42**(1), 68–83 (1993)
21. Stockdale, M.S.: The Role of sexual misperceptions of women's friendliness in an emerging theory of sexual harassment. J. Vocat. Behav. **42**(1), 84–101 (1993)
22. Pryor, J.B., Giedd, J.L., Williams, K.B.: A social psychological model for predicting sexual harassment. J. Soc. Issues **51**(1), 69–84 (1995)
23. Aquino, K., Bradfield, M.: Perceived victimization in the workplace: the role of situational factors and victim characteristics. Organ. Sci. Arch. **11**(5), 525–537 (2000). INFORMS Institute for Operations Research and the Management Sciences (INFORMS), Linthicum, Maryland, USA
24. Pryor, J.B., Whalen, N.J.: A typology of sexual harassment: characteristics of harassers and the social circumstances under which sexual harassment occurs. In: O'Donohue, W. (ed.), Sexual Harassment: Theory, Research, and Treatment, pp. 129–151. Allyn & Bacon, Needham Heights (1997)
25. Zapf, D.: Organisational, work group related and personal causes of mobbing/bullying at work. Int. J/ Manpower **20**(1/2), 70–85 (1999)
26. U.S. Equal Employment Opportunity Commission: Facts About Sexual Harassment (n.d.). https://www.eeoc.gov/eeoc/publications/fs-sex.cfm. Accessed 15 Jan 2018
27. O'Leary-Kelly, A.M., Bowes-Sperry, L.: Sexual harassment as unethical behavior: the role of moral intensity. Hum. Resour. Manage. Rev. **11**(1–2), 73–92 (2001)
28. MacKinnon: Sexual Harassment of Working Women: A Case of Sex Discrimination, p 32. Yale University Press, New Haven (1979)
29. Kaplan, A.M., Haenlein, M.: Users of the world, unite! The challenges and opportunities of Social Media. Bus. Horiz. **53**, 59–68 (2010)
30. Correa, T., Hinsley, A.W., Gil de Zúñiga, H.: Who interacts on the Web?: The intersection of users' personality and social media use. Comput. Hum. Behav. **26**(2), 247–253 (2010)
31. Piskorki, M.J., Tommy, McCall: Mapping the social Internet. Harvard Bus. Rev. **88**(7/8), 32–34 (2010)
32. Onyechi, G.C., Abeysinghe, G.: Adoption of web based collaboration tools in the enterprise: challenges and opportunities. In: International Conference on the Current Trends in Information Technology (2009)
33. Leonardi, P.M., Huysman, M., Steinfield, C.: Enterprise social media: definition, history, and prospects for the study of social technologies in organizations. J. Comput. Mediated Commun. **19**(1) (2013)
34. Petry, T.: State-of-the-Art und Herausforderungen von Enterprise 2.0 in Unternehmen. In: Digitale Medien in Unternehmen. Hochschule RheinMain, Wiesbaden, Germany (2012)
35. Wallace, P.: The Internet in the Workplace: How New Technology is Transforming Work, Cambridge, p. 4 (2004)
36. Anandarajan, M., Paravastu, N., Caiib, N.C., Simmers, C.A.: Perceptions of personal Web usage in the workplace: AQ-methodology approach. Cyber Psychol. Behav. Issue **9**, 325–335 (2006)
37. Hull, G., Lipford, H.R., Latulipe, C.: Contextual gaps: privacy issues on Facebook. Ethics Inf. Technol. **13**(4), 289–302 (2011)
38. Qualman, E.: Socialnomics: How Social Media Transforms the Way We Live and Do Business. Wiley, Hoboken (2010)
39. Hanisch, R.: Das Ende des Projektmanagements: Wie die Digital Natives die Führung übernehmen und Unternehmen verändern. Linde Verlag GmbH, p 7 (2013)
40. Sánchez Abril, P.: Blurred boundaries: social media privacy and the twenty-first-century employee. Am. Bus. Law J. **49**(1), 63–124 (2012)

41. Henry, S.E.: Social networking for the equipment finance industry: divine or a distraction? J. Equip. Lease Financing **29**(1), 1–7 (2011)
42. Cohen, R., Hiller, J.: Internet Law and Policy, 1st edn. Prentice Hall, Upper Saddle River (2002)
43. Kaptein, M.J.: From inaction to external whistleblowing: the influence of the ethical culture of organizations on employee responses to observed wrongdoing. Bus Ethics. **98**(3), 513–530 (2011)
44. Byron, K.: Carrying too heavy a load? The communication and miscommunication of emotion by email. Acad. Manag. Rev. **33**(2), 309–327 (2008)
45. Kruger, J., Epley, N., Parker, J.: Egocentrism over e-mail: can we communicate as well as we think? J. Pers. Soc. Psychol. **89**, 925–936 (2005)
46. Victor, B., Cullen, J.B.: The organizational bases of ethical work climates. Adm. Sci. Q. **33** (1), 101–125 (1988)
47. Paine, L.S.: Managing for organizational integrity. Harvard Bus. Rev. **72**(2), 106–117 (1994)
48. Berry, B.: Organizational culture: a framework and strategies for facilitating employee whistleblowing. Empl. Responsib. Rights J. **16**(1), 1–11 (2004)
49. Treviño, L.K., Weaver, G.R.: Man aging Ethics in Business Organizations: Social Scientific Perspectives. Stanford University Press, Stanford (2003)
50. Victor, B., Cullen, J.B.: A theory and measure of ethical climate in organizations. In: Research in Corporate Social Performance and Policy, vol. 9, pp. 51–71 (1987)
51. Aquino, K., Reed II, A.: The self-importance of moral identity. J. Pers. Soc. Psychol. **83**, 1423–1440 (2002)
52. Yi Liao, X., Ho Kwong, K., Qi-tao, T.: Effects of sexual harassment on employees' family undermining: social cognitive and behavioral plasticity perspectives. Asia Pac J Manage. **33**, 959–979 (2016)
53. Jennings, P.L., Mitchell, M.S., Hannah, S.T.: The moral self: a review and integration of the literature. J. Organ. Behav. **36**(1), 104–168 (2015)
54. Sbraga, T.O.P., O'Donohue, W.: Sexual harassment. Ann. Rev. Sex Res. **11**, 258–285 (2000)
55. Fitzgerald, L.F., Hulin, C.L., Drasgow, F.: The antecedents and consequences of sexual harrassment in organizations: an integrated model. In: Keita, G.P., Hurrell Jr., J.J. (eds.) Job Stress in a Changing Workforce: Investigating Gender, Diversity, and Family Issues, pp. 55–73 (1994)
56. Paludi, M.A., Barickman, R.B.: Sexual Harassment, Work, and Education: A Resource Manual for Prevention. State University of New York Press, Albany (1998)
57. Treviño, L.K.,. Nelson, K.A,: Managing business ethics: straight talk about how to do it right (1999)

Author Index

Printed in the United States
By Bookmasters